Resilient Forest Management

Resilient Forest Management

Resilient Forest Management

Philip J. Burton

University of Northern British Columbia, Canada

Great Clarendon Street, Oxford, OX2 6DP,
United Kingdom

Oxford University Press is a department of the University of Oxford.
It furthers the University's objective of excellence in research, scholarship,
and education by publishing worldwide. Oxford is a registered trade mark of
Oxford University Press in the UK and in certain other countries

Published in the United States of America by Oxford University Press
198 Madison Avenue, New York, NY 10016, United States of America

British Library Cataloguing in Publication Data
Data available

Library of Congress Control Number: 2024939917

ISBN 9780198832997
ISBN 9780198833000 (pbk.)

DOI: 10.1093/oso/9780198832997.001.0001

Printed and bound by
CPI Group (UK) Ltd, Croydon, CR0 4YY

The manufacturer's authorised representative in the EU for product safety is Oxford
University Press España S.A. of El Parque Empresarial San Fernando de Henares, Avenida
de Castilla, 2 – 28830 Madrid (www.oup.es/en or product.safety@oup.com). OUP España
S.A. also acts as importer into Spain of products made by the manufacturer.

Dedicated to my father,
George P. Burton,
who invited us all to make the world a better place.

Preface

The need for this book emerged around the turn of the millennium. After years of research and outreach to promote more effective forest regeneration and the restoration of degraded lands, it became apparent to me that a more ecological approach was needed for Canadian forest management to be truly sustainable. That concern was shared by numerous forest scientists and practitioners, and was further explored under the auspices of Canada's Sustainable Forest Management Network of Excellence. It resulted in publication of the book *Towards Sustainable Management of the Boreal Forest* (Burton et al. 2003), for which I had the honor of organizing, editing, and contributing. A recurrent theme explored in that book was how the emulation of natural disturbances provided a better model for protecting forest biodiversity and processes than the prevailing agricultural model of silviculture. The idea that timber harvesting should match the size, frequency, and residual structure of the natural disturbance regime has since become widely accepted in much of the world.

About the same time as the 2003 book was in production, an unprecedented series of bark beetle outbreaks and wildfires was unfolding across western North America, indicating that some of the former climatic constraints were no longer limiting forest disturbances. But the disappointing forest management response to those leaps in widespread tree mortality was to harvest as much of the damaged wood as possible before it lost commercial value, typically using clearcut logging methods that took green trees as well and accelerated overall rates of habitat degradation. Similar policies were being followed on public, multipurpose forests in Australia, the USA, and Europe as well as in Canada, which prompted writing of the coauthored book *Salvage Logging and its Ecological Consequences* (Lindenmayer et al. 2008).

What aspects of forest management need to be "more ecological," and how might such changes help forestry achieve its stated mandate of sustaining all forest values? The answer is not so much about more engineered interventions and greater control, but about letting the natural complexity of forest ecosystems map their own way forward under changing conditions. This concept was developed in the book *Managing Forests as Complex Adaptive Systems* (Messier et al. 2013), to which I made a modest contribution. Yet the insights provided by that book and the preceding workshop on which it was based did not fully register with me for some years. It was the cumulative impacts of a changing climate and other socio-ecological stressors on forests, wildlife, and the land in general that made the challenge of perpetual change apparent. I took part in one regional analysis of those effects and our current understanding of their underlying causes, published as *Drivers of Landscape Change in the Northwest Boreal Region* (Sesser et al. 2019). But forest management solutions to the stresses of a rapidly changing world still remained elusive.

That journey, supported by my own research, numerous field trips and site inspections, much reading, and vigorous discussion with colleagues, has culminated in this book. It eventually became clear to me that sustainability was not enough, or rather that it was an aspirational but insufficient goal for forest managers in times of change. Instead, it is time for resilient forest management, a flexible, cautious, and humble approach to forest stewardship, one that facilitates and empowers the natural adaptive responses of the world's forests and human ingenuity to the many challenges they now face. I hope that this book provides tree huggers and tree cutters alike with some ideas on how we can better aid in the persistence of the vibrant forests, woodlands, and forest-dependent communities that mean so much to all of us.

This book is laid out in three parts. Part I outlines some fundamentals of forest management, the challenge of sustaining multiple forest values, and how resilience thinking might support that endeavor. Forest zoning is introduced as a framework for prioritizing management options. Part II explores options for promoting resilience in forest management, both in a general sense and in addressing the specific threats of climate change and forest disturbances. Part III provides a synthesis with an emphasis on restoring past damage, reviewing some promising applications through case studies, with resilient forest stewardship placed in the context of adaptive management and some basic ecological principles.

Terminology used throughout the book should be familiar to most ecologists and foresters; terms follow their standard use and definitions, which can be checked using online resources in lieu of a glossary.

Some colloquial terms and popular expressions are bracketed by quotation marks (as in "fire-stick farming" or "scientific forestry"), but again follow common usage.

Phil Burton
August 19, 2024
Terrace, British Columbia, Canada

References

Burton, P. J., Messier, C., Smith, D. W., and Adamowicz, W. L. (eds). (2003). *Towards Sustainable Management of the Boreal Forest*. Ottawa: NRC Research Press. 1039 pp.

Lindenmayer, D. B., Burton, P. J., and Franklin, J. F. (2008). *Salvage Logging and its Ecological Consequences*. Washington: Island Press. 227 pp.

Messier, C., Puettmann, K J., and Coates, K. D. (eds). (2013). *Managing Forests as Complex Adaptive Systems: Building Resilience to the Challenge of Global Change*. New York: Routledge. 353 pp.

Sesser, A. L., Rockhill, A. P., Magness, D. R., Reid, D., DeLapp, J., Burton, P. J., Schroff, E., Barber, V., and Markon, C. (eds). (2019). *Drivers of Landscape Change in the Northwest Boreal Region*. Fairbanks: University of Alaska Press. 326 pp.

Acknowledgments

Work on the manuscript for this book started during a 2017/18 sabbatical, funded by the University of Northern British Columbia (UNBC). During that time, I was hosted by David Norton at the University of Canterbury, and by Peter Bellingham at Manaaki Whenua—Landcare Research in New Zealand; I thank them and their institutions for their hospitality and support. Early thinking on the topic was enhanced by feedback received on presentations delivered at those institutions, the University of Northern British Columbia, the Bulkley Valley Research Centre, and to BC Timber Sales. Similarly, the challenges of conveying these concepts to my FSTY 408 class at UNBC helped hone my thinking.

This book reflects the work of generations of researchers and practitioners around the world. The work of colleagues with which I am familiar is perhaps highlighted, but relevant work has accelerated so much in recent years that there are undoubtedly thousands of other relevant books, papers, and technical reports that provide additional depth and nuance to the topics explored here. So I apologize in advance to friends and strangers alike for not citing all of the important research and insights out there.

Earlier versions of individual chapters were reviewed by the following individuals:

Chapter 1: Hugues Massicotte and Rodney Keenan

Chapter 2: Christian Messier and Vic Adamowicz

Chapter 3: Kathy Lewis and Art Fredeen

Chapter 4: Mark Dale and Klaus Puettmann

Chapter 5: Suzanne Simard and Sybille Haeussler

Chapter 6: Peter Jackson, Dave Spittlehouse, and Sylvie Gauthier

Chapter 7: Dan Kneeshaw and Tony D'Amato

Chapter 8: Rupert Seidl, Yan Boulanger, and Brian Buma

Chapter 9: Jack Putz, John Stanturf, and Lars Walker

Chapter 10: David Lindenmayer and Chris Johnson

Chapters 11 and 12 (originally one chapter): Dave Coates, Per Angelstam, and Evelyn Hamilton.

I have tried to incorporate the feedback provided by those reviews, and my thinking probably reflects innumerable discussions with those same colleagues over the years. I thank them for their input, and also appreciate the suggestions made by Craig Allen, Jürgen Bauhus, Anders Ekstrand, Rachel Holt, Reed Noss, Brad Pinno, Hans Pretzch, Claudia Romero, and Landon Shepherd, among others.

Unless otherwise attributed, photographs and illustrations are by the author. I am grateful to several friends and colleagues who willingly made additional photographs and graphic materials freely available for use here, for which Nicole Burton filled some gaps. Carla Burton provided feedback and quality control as she reviewed, proofread, and reference-checked all chapters. Furthermore, Carla provided me with the necessary moral support to undertake and complete this multi-year project, for which I am immensely grateful. All errors and omissions or any unfounded assertions are strictly my own, and should not be construed to represent the thinking of my many associates.

Finally, I would like to thank the editorial team at Oxford University Press, Ian Sherman, Charles Bath, Katie Lakina, Hilary Walford, James Oates and Kanimozhi Ramamurthy. This work would not have seen the light of day without their guidance, feedback, and patience.

Contents

Part III   Synthesis and Direction   231

9   Restoring Forests for Resilience   233

10   Resilience through Adaptive Management   261

11   What Resilient Forest Management Looks Like   288

12   Forest Stewardship Based on Socio-Ecological Principles   317

Abbreviations

AAC	allowable annual cut
AHM	annual heat:moisture index
ANF	Apalachicola National Forest, Florida
AR1	First Assessment Report of the IPCC
AR6	Sixth Assessment Report of the IPCC
ASCC	Adaptive Silviculture for Climate Change
BC	British Columbia, Canada
BMP	best management practice
Btk	*Bacillus thuringiensis* var. *kurstaki*
CAI	current annual increment
CFUG	community forest user group
CMD	climate moisture deficit
CMI	climate moisture index
CSA	Canadian Standards Association
DBH	diameter at breast height, 1.3m
DD<0	degree-days less than 0 °C
DEM	digital elevation model
EDI	equity, diversity, and inclusion
EMT	extreme minimum temperature
ENGO	environmental non-governmental organization
EV	ecosystem value
FFP	uninterrupted frost-free period
FLR	forest and landscape restoration
FSC	Forest Stewardship Council
FWI	fire weather index
GCM	global circulation model
GHG	greenhouse gas
GIS	geographic information system
HRV	historical range of variability
IPCC	Intergovernmental Panel on Climate Change
JNP	Jasper National Park, Alberta
LiDAR	light detection and ranging
MAI	mean annual increment
MAP	mean annual precipitation
MAT	mean annual temperature
MBTI	Meyers–Briggs Type Indicators
MCMT	mean cold month temperature
MPB	mountain pine beetle (*Dendroctonus ponderosae*)
NCP	nature's contributions to people
NFFD	number of frost-free days
NTFP	non-timber forest product
NRV	natural range of variability
OSB	oriented strandboard
PEFC	Programme for the Endorsement of Forest Certification
POD	potential [wildfire] operations delineation
PPF	production possibility frontier
RCM	regional circulation model
RCP	representative concentration pathway scenario for IPCC assessment
REDD+	Reducing Emissions from Deforestation and Forest Degradation
RIL	reduced-impact logging
SCS	suitable climate space
SDG	sustainable development goal of the UN
SFI	Sustainable Forestry Initiative
SFM	sustainable forest management
SHM	summer heat:moisture index
sp., spp.	species
SSP	shared socio-economic pathway
SRES	Special Report on Emissions Scenarios by IPCC
TEK	traditional ecological knowledge
TFL	tree farm license (in BC)
TSA	timber supply area (in BC)
UK	United Kingdom
UN	United Nations
US	United States of America
VAM	vescicular-arbuscular mycorrhizae
VPD	vapor pressure deficit
WUI	wildland–urban interface
Y2Y	Yellowstsone to Yukon conservation initiative

PART I

Fundamentals

The history of forest management around the world has been marked by many successes and challenges in its pursuit of sustainability. Perhaps most daunting have been the many expectations we place on forests to deliver and yet conserve a wide array of ecosystem goods and services, both material and non-material. Timber harvesting remains the driving economic force behind active forest management, but with impacts on biodiversity, carbon stocks, watershed function, and many other forest values and uses. The bundling of compatible forest uses and the zoning of those that conflict with each other is one way forward. Forest zones, uses, and compatible combinations of forest values can be loosely categorized as protected forests, those intensively managed for timber production, and an intermediate category of multipurpose forests. Yet the full extension of sustained-yield timber management to sustainable forest management that actively promotes all forest values remains elusive.

Sustainability is a challenging goal at the best of times but can be seriously compromised under conditions of rapid global and socio-ecological change. With anthropogenic climate change and its impacts now recognized, the time is ripe for a paradigm shift away from popular expectations of sustainable development. Lessons can be garnered from how forest ecosystems and socio-economic institutions have undergone cycles of growth, maturity, collapse, and reorganization in the past. An understanding of complex adaptive systems provides an underlying set of principles and requirements for a more resilient approach to forest management. Socio-ecological sustainability remains an aspirational goal but must reject expectations of constancy to embrace change and adaptability.

CHAPTER 1

The Evolving Scope of Forest Management

The future ain't what it used to be.
popularized by Yogi Berra, 1974, paraphrasing Riding and Graves (1937)

1.1 Introduction

Forest management encompasses the policies and actions associated with purposeful human interventions in the world's tree-dominated ecosystems. Those ecosystems include, but are not limited to, the planet's tropical moist forests, the temperate deciduous forests, and the needle-leaved forests of boreal and montane regions that consist of more or less continuous tree cover. Although vegetation ecologists, biogeographers, and various national jurisdictions use different definitions for delineating forests (Ghazoul 2015), international reporting conventions refer to "forest and other wooded land" where woody species capable of attaining heights greater than 5 meters occupy more than 10 percent cover in patches at least 0.5 hectares in area (FAO 2015). Without being restricted to any one definition, the principles and issues of forest management extend to open woodlands, parklands and savannas, swamps, mangroves, bamboo groves, successional stages in which trees may not be obvious, and the many human undertakings to establish trees, sometimes as part of agricultural and urban environments.

This book explores options for improving the prospects for forest persistence, health, productivity, and integrity under the many pressures that forests are experiencing today. This persistence under pressure and following various forms of stress, disruption, or disturbance is referred to as resilience. As such, forest resilience and its promotion comprise a goal and a recurrent theme of this book. Described as an attribute of ecological systems in the 1970s, resilience and related concepts (see Chapter 4) have widespread application in engineering, business, and community planning as well as natural resource management (Gunderson 2000; Walker and Salt 2006). It can be argued that trees and forest ecosystems are inherently resilient by virtue of evolutionarily selected attributes contributing to survival, dispersal, regeneration, genetic recombination, phenotypic plasticity, and diversity. Forests have persisted, recovered, or reconfigured themselves despite meteor impacts, shifting continents, and repeated glaciations. But we now have billions of people creating an unprecedented draw upon forests for the many roles they fill in our lives, some directly and many indirectly.

Many of the world's trees impress us with their stature, and we value Earth's forests for the many products and ecological services they provide for humankind (see Chapter 2). Trees and forests have variously provided us with cover, food, fuel, and construction materials, while we have alternately nurtured and exploited selected tree species, making our relationship a complex one. The modern science of forestry focuses on growing and harvesting trees for wood and fiber production. But the discussion that follows also recognizes the importance—and long-standing traditions—of manipulating forests to promote non-wood goods and services, as well as non-utilitarian forest values that are being increasingly appreciated in the postmodern era.

Resilient Forest Management. Philip J. Burton, Oxford University Press. © Philip J. Burton (2025). DOI: 10.1093/oso/9780198832997.003.0001

Individual trees often have life spans greater than that of any human being, and forests throughout the world thus seem to provide a steadfast backdrop to our lives. Consideration over longer time frames reveals that forests are in fact quite dynamic, undergoing birth, growth, and death over multiple spatial scales. Subject to the influence of year-to-year, decade-to-decade, and century-to-century differences in weather, glaciation, sea level, natural disturbance (fires, windstorms, insect outbreaks, landslides, and so on), and human exploitation, forests have variously expanded and contracted throughout history and prehistory (Hooghiemstra 2002). Many of those drivers of change have been accelerating in recent decades and are expected to become more accentuated in the decades to come. Those drivers include a general warming of Earth's atmosphere, a growing human population, and increased intercontinental transport of products and pests (Steffen et al. 2005; Mery et al. 2010). Earth's forests are not threatened with obliteration at this time, but it is certain that they will not be the same in the future as they have been in the past. At the same time, evolving concepts of social and environmental justice suggest that forest management must likewise change with the changing world (Farcy et al. 2019).

This first chapter sets the stage for further discussion by outlining the history of forest use and management, and highlights some recent trends in the evolution of forestry and conservation. The status of the world's forests is reviewed, and the scope of forest management is further developed in the context of international conservation and development goals. An expansive definition of forest management is applied to three broad categories of forest land in Chapter 2, thereby providing the basis for many of the examples and case studies presented.

Whereas we once could simply harvest or extract forest benefits provided by the bounty of nature, most nations now subscribe to the need to steward their forest resource responsibly, and to sustain the ongoing generation of forest goods and services. But this commitment to sustainability needs to be taken to the next level, as elaborated upon in Chapter 3, if forests and our use of them are to adapt to a rapidly changing milieu. Strategies and options for dealing with climate change, invasive organisms, altered disturbance regimes, and unknown levels of socio-economic risk and uncertainty are here developed under the general umbrella of managing forests for resilience. Note that the focus here is on forests, not on the wood-products sector or the science of forestry. Nonetheless, it is axiomatic that resilient forests are a necessary, if not sufficient, requirement for resilience in the forest products sector. Some of the connections between resilient forests, resilient forest management, and resilient approaches to conservation and protected area management are explored in this volume as well.

1.2 Some history

The history of human interactions with Earth's forests is a fascinating and surprisingly interdependent one. Our use of forests has gone through several cycles of exploitation and conservation at different times in different regions of the planet. Some convergence in approaches to forest management and governance have occurred, but management philosophies continue to evolve. Recent summaries by Vogt et al. (2007), Sands (2013), Ghazoul (2015), and Innes (2017) provide well-documented chronicles, which were drawn upon for much of the material summarized below.

Human history is rife with examples of deforestation and soil degradation associated with the cutting of trees for timber and fuelwood, and with the conversion of forest and woodland to support cultivated crops and pastures instead (Williams 2000). Such transformations took place in Neolithic times in China, India, the Middle East, and Europe, but they continue today, particularly in the tropics. Recent decades have seen the expansion of agriculture (employing both animal husbandry and crop cultivation, at both subsistence and commercial scales) in tropical Africa and South America. Mangrove swamps continue to be converted to shrimp farms and diverse lowland rainforests are being replaced with homogenous oil-palm plantations in southeast Asia (Stibig et al. 2014). Much tropical deforestation in the twentieth century was driven by the demand for sugar, bananas, beef, and coffee in wealthy northern economies.

Unregulated exploitation of forest resources and the associated loss of forest cover have been a recurrent theme in human history. It seems to be human nature to exploit the natural environment to levels of devastation and extinction; sometimes we have been able to draw ourselves back from the brink of destruction, but sometimes not (Diamond 2005). There is evidence that some forests have been able to recover where and when human pressures are reduced, while others have recovered only partially (for example, in structure, but not composition), and some formerly forested lands remain in a degraded, primarily un-treed state. To the trained eye, our ecological footprint is evident in the many deletions and additions to regional floras and faunas, such that few of the world's forest ecosystems today have the same composition as they did before the arrival of human beings.

The impacts of forest loss on fuelwood security, access to construction timbers, soil erosion, and water supplies were recorded by the ancient Babylonians and Greeks (Ghazoul 2015; Innes 2017). When wood shortages arose and forest cover contracted, many cultures independently designated forest reserves for wildlife, for naval use, and as sacred groves where tree-cutting was curtailed (Colding and Folke 2001; Vogt et al. 2007). Forest protection and management for hunted wildlife ("game" for the enjoyment of the ruling class) often predated forest management for timber or fuelwood supplies and extended to non-wooded habitats as well. Coppicing, thinning, and the planting of trees were practiced in prehistory, while the Romans documented systems of pruning and tree breeding (Innes 2017). Forest management practices developed in medieval Europe included the refinement of coppicing and pollarding techniques to assure sustainable supplies of fuelwood and the desired quality of basketry materials (Vogt et al. 2007).

Many peoples around the world practiced swidden (slash and burn) agriculture, and the propagation of fruit trees and other desirable plants used for food, fuel, fiber, and animal fodder. These management practices were sometimes so pervasive and persistent that the preindustrial forest can be described as having co-evolved with human culture (for example, in the Mediterranean Basin (Nocentini and Coll 2013)) or as being one large garden (for example, in the Amazon Basin and in Central America (Erickson 2008; Ford and Nigh 2016)). The purposeful use of fire on the land in support of hunting, warfare, and land-use conversion (for example, for mining or agriculture) has often served a destructive role with respect to forests. But it has also served maintenance or constructive roles to keep some forests, parklands, and savannas with a desired balance of openness and shade to promote palatable forage for hunted wildlife, to encourage selected food plants, and to keep trails and village perimeters safer from attack (Boyd 1999). The Indigenous peoples of Australia were especially adept at this form of "fire-stick farming," in which they used fire to nurture the food sources and woodland structures most amenable to reliable hunting, gathering, travel, and safety (Bird et al. 2008).

Despite scattered traditions of forest conservation and renewal, the juggernaut of agricultural expansion and the need for timber prevailed in most parts of the world, variously to support shipbuilding, mining and smelting, and the manufacture of ceramics and glass, and to meet the housing, cooking, and heating needs of a growing human population. As a result, widespread deforestation and real timber shortages characterized many parts of the populated world at different periods of history. From ancient times through to the modern day, powerful states and prosperous economies have typically made up for timber shortages by exploiting the resources available to them through colonization, conquest, or trade, essentially "exporting" deforestation, even while protecting their own forests. Where timber imports were unfeasible, greater emphasis was placed on conservation: official decrees by smaller, landlocked city states in western and central Europe as early as the 1300s restricted levels of woodcutting to amounts that could be regrown (Innes 2017), establishing concepts of sustainability that we embrace today. Over a period of several centuries, we saw the development in central Europe of the set of principles and practices that we today recognize as "sustained-yield management" and "scientific forestry." With its roots in the monasteries of medieval Europe

(Vogt et al. 2007), and widely practiced and formalized by the eighteenth and nineteenth centuries, silvicultural systems of forest-stand manipulation (through felling, tree propagation, thinning, and pruning) were implemented to assure a steady supply of various wood products from a designated area of forest land. The different attributes, growth, and regeneration needs of different tree species were incorporated into these plans and practices, as was recognition that forest practices were constrained by climate and the limits of some soils.

The conceptual breakthrough marking the development of sustainable wood production should not be underestimated. Born of the same idealistic fervor in the eighteenth century that we recognize as the Age of Enlightenment and that spawned the French Revolution (Hütte 2000), the idea of curtailing and managing what had been exploitative practices so as to promote a wood supply in perpetuity was indeed a revolutionary concept. Nonetheless, many preliterate societies had developed similar conservation and resource-nurturing practices, often in the form of taboos and informal cultural norms, in their stewardship of fisheries, wildlife, and selected (often "sacred") habitats (Colding and Folke 2001). Translating such traditions of intergenerational thinking into quantitative forest planning and management for sustainable wood production represents the birth of modern sustainability policies, partly owing to its incorporation in written regulations, formal texts, management plans, and the training of professionals. One might also detect a degree of romanticism in these early schools of forestry, in which the stated goal was not just timber for a king's navy, but also the restoration of forests that had once been prevalent but had become rare or degraded (Powers 1999; Batho and Garcia 2006)—perhaps reflecting a nostalgia for the bounty of pre-exploitation forests and more natural landscapes (Kennedy 1985).

The singular success of growing predictable amounts of wood over predictable periods of time was quickly exported around the world as part of the European colonial enterprise. Unfortunately, colonial forestry initiatives typically perpetuated the European feudal practices of excluding people from using their local forests, and did not recognize local ecological knowledge or the legitimacy of long-standing access rights (Ghazoul 2015; Innes 2017). Nonetheless, some of the best examples of multiple crops of timber being grown according to rigorously followed silvicultural management plans are found not only in Europe (for example, oak forests and pine plantations in France (Clément et al. 2012)), but also in southern Asia and the East Indies (for example, teak plantations in Indonesia (Brockerhoff et al. 2008)). Similar examples of regulated forest harvest and renewal have precedents in eighteenth- and nineteenth-century Japan, but were absent or less widespread in Korea and Imperial China during the same period (Totman 1989; Saito 2009). Sustained-yield forestry was taught in the forestry schools of the United States, Canada, and Australia from the late nineteenth century, but had little bearing on forest policies and practices so long as broad expanses of wild timber were there to be logged. The logging of primary and old-growth forests, even in locations with poor prospects for forest renewal, continues in some jurisdictions of those countries and elsewhere around the world to this day.

1.3 Managing for more than trees

Popular support for sustained-yield forestry started to wane in much of the world during the 1960s and 1970s, associated with a growing environmental awareness (Kennedy 1985). Where large tracts of public land were concerned, it was felt that managed wood production must also accommodate and support public values of watershed protection, wildlife habitat, livestock grazing, and recreational activities such as hunting, fishing, hiking, and camping (Hall 1963; Fedkiw 1998). In practice, accommodation of non-timber concerns was accomplished primarily by imposing constraints and restrictions on timber development planning, because the harvesting of wood remained the principal management activity and the only one with a significant revenue stream. Where democratic traditions and transparent decision-making prevail, each stage or level of forest planning provides opportunities to consider multiple forest values, public input, and alternative management options. A hierarchy of policy (by government), strategic ("big picture") planning, tactical (spatially explicit) planning, and operational (annual, site-specific) planning is common to the administration of public forests in most developed countries, with public input nominally welcome at each level (Tittler et al. 2001).

Although environmental protests became widespread in the 1970s, it was not until the 1980s and 1990s that modern forestry operations were acknowledged to have negative impacts on some endangered species, fisheries, recreational opportunities, and traditional livelihoods around the world (Marchak 1995; Myers 1995). These concerns were especially prevalent where wild forests were still being logged for the first time, where tropical forests were being lost to agriculture, and where centuries-old trees were being replaced by uniform plantations destined for harvest in a few decades. Different aspects of industrial forestry became levers for protest and reform in different parts of the world: protecting the endangered northern spotted owl (*Strix occidentalis caurina*) and salmon-spawning rivers in the Pacific Northwest of the US (Craig 1987; Thomas et al. 1988), outrage over forest-regeneration failures in Canada (Swift 1983), objections to the use of herbicides and pesticides in Europe (Kardell 1980), while the Australian environmental movement mobilized against the clear-felling of native rainforest and its conversion to pine plantations (Hutton and Connors 1999).

In many cases, protests focused on the short-term and aesthetic impacts of clear-cut logging rather than against forest management per se (Spurr 1981; Devall 1993; Keenan and Kimmins 1993), with the public remaining largely ignorant of the overall forest regulation agenda and its many corollaries. For example, if a given volume of wood is to come from a series of partial cuts rather than clear-cuts, a larger total area of forest is impacted and requires a larger road system to be built and maintained. But the debate between even-aged and uneven-aged silviculture and whether it is better to assure sustainable levels of growing stock within each forest stand or across a forest estate as a whole has raged amongst foresters (not just between foresters and environmentalists) since the 1800s (Sands 2013).

Public concerns, coupled with a growing understanding of the functioning of forest ecosystems, spawned a broad series of forest reforms and advocacy for different approaches to forest management (see Box 1.1) that blossomed primarily in the 1990s. Some of those approaches—for example, "new forestry" and "variable retention"—were motivated by opposition to clear-cut logging on the basis of both aesthetic and ecological impacts. Other approaches, such as "ecosystem-based management" (also known as "ecosystem management" in the US), "wholistic forest management," and "systemic silviculture," sought to take a broader view of the composition, structure, and function of forest ecosystems rather than just considering trees and the direct factors affecting their regeneration and growth. "Sustainable forest management" (SFM) became widely adopted internationally as one implementation of the principles of sustainable development as articulated by the World Commission on Environment and Development (WCED 1987), and was included in Agenda 21 of the 1992 Earth Summit held in Rio de Janeiro (Parson et al. 1992; Grubb et al. 1993). The "Rio Forest Principles" then became the basis for a widespread set of SFM criteria and indicators, which have been adopted in global forest certification programs (see Chapter 10, particularly Section 10.2), government policies, and national reporting activities (MPWG 2015; Siry et al. 2018). Sustainable forest management has been described as an extension of the principles of sustained yield to the full range of forest goods and services (Adamowicz and Burton 2003; Higman et al. 2005). It is widely accepted that any of these alternative paradigms for forest management require structured programs of evaluation and continuous improvement, popularly known as "adaptive management" (see Chapter 10).

Foresters have been among the first to understand that human manipulations of a forest ecosystem must be compatible with the underlying evolutionary history and adaptations of its components (Toumey 1928). Nonetheless, land-use conversion and exploitative logging operations have continued to destroy many important ecosystems, and habitat loss continues to threaten the viability of many rare plant and animal species. The Nature Conservancy in the US called for the protection of representative plant communities in the 1980s as a "coarse-filter"

Box 1.1 Some forest management paradigms

Many different schools of practice have emerged over the years to promote forest sustainability. Each has its own priorities, information needs, strengths and weaknesses, as outlined in Table 1.1.

Table 1.1 A comparison of some alternative and complementary forest management paradigms

Paradigm	Priority values	Foundation	Strengths	Weaknesses	References
Sustained-yield forestry	Wood production	Forest inventory, growth, and yield; "scientific forestry"	Assured wood supply; invokes a basic understanding of forest-stand development	Homogenization of wild forests to regulated ones, other values often compromised; requires rigorous forest regeneration and protection	Mason (1927); Orchard (1953); Stoddard (1959); US Congress (1960); Leslie (1966)
Multiple-use forest management	Wood, wildlife, water, livestock, recreation	Sustained yield, assured rights of various stakeholders to use public lands	Several utilitarian values accommodated	Resource conflicts, especially when applied to small holdings	US Congress (1960); Hall (1963); Brunig (1970); Koch and Kennedy (1991)
Sustainable forest management	Economic, social, ecological	Sustainable development, sustained yield of all forest values	Multiple utilitarian, biocentric, and social values are actively supported	Planning and stakeholder consensus gets complicated; many foresters poorly prepared	Adamowicz and Burton (2003); Lindenmayer and Franklin (2003); von Gadow et al. (2012); MPWG (2015); Innes and Tikina (2017)
New forestry, variable retention forestry	Wood, wildlife, biodiversity	Avoid clear-cutting; old and mature forest habitat continuity	Old-forest attributes retained in young forest	Compromised timber harvesting operations and silvicultural performance	Swanson and Franklin (1992); Gustafsson et al. (2012)
Plantation forestry	Wood production, forest cover	Sustained yield, the agronomic model	The basis for afforestation and for assured reforestation; optimized wood production	Other values often compromised, intensive management inputs, low stand diversity, vulnerable to disturbance, pests, disease, soil exhaustion	Boyle et al. (1999); Evans and Turnbull (2004); Bauhus et al. (2010)
Continuous-cover forestry, close-to-nature forestry	Wood, wildlife, water, recreation, biodiversity	Avoid clear-cutting; maintain aesthetics and watershed protection of mature tree canopies	Mature forest cover maintained, benefitting local hydrology and forest-dependent species	Reduced presence or productivity of shade-intolerant tree species; no place for open-habitat organisms; dense road network must be maintained	Garfitt (1995); Pommerening and Murphy (2004); Pukkala and von Gadow (2012); Bürgi (2015)

Ecosystem management, ecosystem-based management, holistic forest management, systemic silviculture	Biodiversity, ecological integrity, wildlife, social values	Biodiversity a priority; human use restricted to maintaining ecosystem integrity first	Hierarchical consideration of ecological and social opportunities and constraints	Knowledge limited, so typically falls back on coarse-filter conservation approaches; ongoing debate regarding priority to ecocentric and anthropocentric needs	Hammond (1991); Christensen et al. (1996); Grumbine (1997); Kohm and Franklin (1997); Slocombe (1998); Gauthier et al. (2009); Ciancio and Nocentini (2011)
Emulation of natural disturbances	Biodiversity, ecological integrity, aesthetics	Coarse-filter conservation biology	Maintains structural diversity and keeps habitats within range of natural variability; may help logging look more natural	Usually overlaid on natural disturbances, does not replace them; can become overly prescriptive with limited reference information	Hunter (1993); Bergeron and Harvey (1997); Angelstam (1998); Landres et al. (1999); Perera et al. (2004); North and Keeton (2008); Long (2009)
Restoration forestry	Historic forest composition and structure	Preindustrial templates for natural, healthy forests	Rehabilitation of degraded lands, reverse effects of human exploitation	Tends to be backward looking; full ecological restoration is difficult to attain	Pilarski (1994); Fanta (1997); Parrotta (2002); Sarr et al. (2004); Chazdon (2008); Stanturf et al. (2014); Stanturf (2016)
Community forestry	Local governance, watershed protection, wildlife habitat, non-timber forest products	Traditional forest uses; governance, and ecological knowledge; presumption of greater commitment to long-term sustainability	Better protection of non-timber values and resources; local support for forest protection and management; local retention of economic gains	Governance and interaction with government oversight can be cumbersome; areas often too small to be viable for generation of revenue from timber on an annual basis	Pagdee et al. (2006); Charnley and Poe (2007); de Jong et al. (2010); Gilmour (2016)

approach to conserving the many species (especially invertebrates, fungi, non-vascular plants, and microbes) for which there was no time and limited expertise with which to inventory and develop individual species protection plans (Noss 1987). With a growing emphasis on protecting biological diversity in recent decades, the emulation of natural disturbances has been suggested as a logical forest management extension of this approach, and an all-purpose template for the maintenance of indigenous biodiversity in forests subject to logging (Hunter 1993; Attiwill 1994; Swanson et al. 1994). Such an approach encourages the survival of all forest species and processes within the natural range of variability experienced during their evolution. Ecological approaches to forest management that further harness the inherent genetic, population, community, and ecosystem processes of forests have now been collated in comprehensive reference books (e.g. Larocque 2016; Franklin et al. 2018). Those sources and the primary research on which they are based are drawn upon throughout this book, with some guiding principles summarized in Chapter 12.

Kimmins (1991) grouped management for several of the non-utilitarian and biocentric forest values as "social forestry," the pinnacle of forest management evolution in his opinion. It can be debated whether "values" exist outside the human perspective (as explored briefly in Chapter 2). But perhaps the most social of the forest management paradigms has not yet been discussed—namely, community-based forestry (see Box 1.1). Although more properly thought of as a form of forest tenure or forest governance, it has also become a social movement in recent decades, with several important implications to the sustainability and resilience of forests as well as communities (de Jong et al. 2010). Community forests have provided a means of resolving some of the long-standing tensions between centralized governmental or industrial control of public forest resources and the strong desire of local community members to maintain traditional access to those resources and to employment opportunities in their commercial management. Furthermore, the ability of small towns and villages (situated in or adjacent to the forest) to participate in forest planning, operations, and monitoring takes advantage of considerable local knowledge, keeps more "eyes on" the forest, and helps reduce unauthorized and destructive forms of forest use (Gilmour 2016). It is also assumed (though not always justifiably) that local communities have a greater commitment to long-term sustainability than private companies that can direct investment elsewhere (see Box 2.4), or governments that are subject to pressures from urban populations, corporate support, corruption, or political expediency.

With recognition of a warming planet, exploding human demands, and the widespread damage of invasive species, there has been a call to manage forests as complex adaptive systems (Messier et al. 2013; Filotas et al. 2014). As elaborated upon in Chapters 4 and 5, this novel perspective is based on the protection and promotion of multi-scaled diversity (not just of component species, but of processes and species interactions as well) as a requirement for forest sustainability and persistence. There is recognition that optimization of any one forest component or forest value will not only compromise other forest components and values (e.g. Cumming et al. 1994), but can threaten forest integrity and adaptive resilience too (Wagner et al. 2014; Franklin 2016). Weighing those trade-offs, assessing their implications, and dealing equitably with stakeholders associated with any given decision means that managing a forest as a complex adaptive system implies working with a complex socio-ecological system. As the history of forest use, exploitation, management, and mismanagement clearly shows us—and as the simple definition of "management" makes unavoidable—"forest management" ultimately consists of people management (see Chapter 12).

An overview of trends in forest management is incomplete without noting the many efforts to protect forests and improve forest management in previous eras. Sometimes motivated by aesthetics and "biophilia" (Wilson 1984), or by a utilitarianism that was more far-sighted than was popular at the time, there have always been visionaries who recognized the need for more responsible forest stewardship. There is a legacy of advocacy and experimentation to support forest integrity and persistence, often working outside mainstream forest policy and orthodox forestry education. Examples include writings, actions, or successful demonstration projects undertaken by popular romantics (e.g., Henry David Thoreau in the northeastern US (Botkin 2001)), utilitarian conservationists (e.g., George Perkins Marsh (Lowenthal 2009)), philosophical biologists (e.g., Aldo Leopold (Leopold 1949)), environmental activists (e.g., Chico Mendes in Brazil (Revkin 2004)), progressive estate-owners (e.g., John Wardle in New Zealand (Wardle 2016)), ostracized professional foresters (e.g., Herb Hammond in British Columbia (Hammond 1991)), and scientists and academics (e.g., Chris Maser in the US Pacific Northwest (Maser 1990)). There are many examples, often small and unheralded, of well-managed forests that would seem to be successful examples of sound conservation and sustainable forestry, some of which are highlighted in text boxes and case studies in the following chapters. The diversity of creative approaches that have organically developed to protect and promote resilient forests provides evidence that managing forests for sustainability and resilience is possible, and that there are many ways of doing so (see Chapter 11).

1.4 Trends in forest loss and use

Forest cover today is estimated to be about two-thirds of what it was several thousand years ago (see Figure 1.1). This level of global forest area perhaps contradicts the popular perception of a modern deforestation crisis (Spilsbury 2010), especially in the tropics (Laurance 1999). As described above, deforestation is not just a modern phenomenon, however, and forest loss is not an irreversible process, as much of today's forest land was once treeless and cultivated. Examples include the near-complete encroachment of tropical rainforest following the collapse of the Mayan civilization in Central America over a thousand years ago (Mueller et al. 2010), and documented expansions of forest cover in Europe after the collapse of the Roman Empire and again after the Black Death of the late Middle Ages (Williams 2000). A similar resurgence of Amazonian rainforest followed disease-induced depopulation in the sixteenth century (Piperno et al. 2015), and widespread recovery of temperate forests in the eastern US occurred in the twentieth century when agricultural production moved further west (MacCleery 1992; Foster 1995). The great migration of human beings from rural areas into the world's cities that has been observed over recent decades also seems to be associated with growth in forest cover (e.g. Parés-Ramos

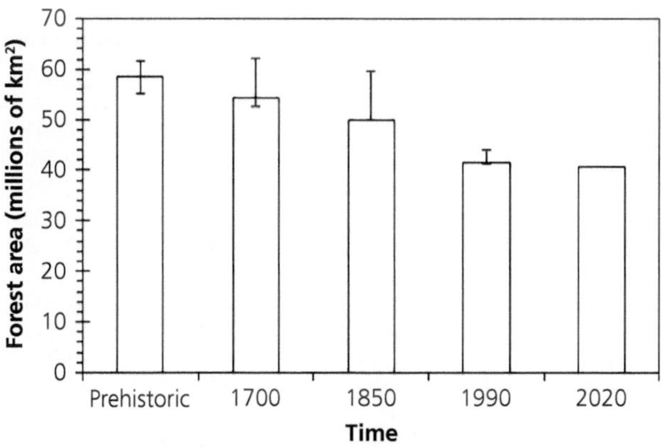

Figure 1.1 Estimates of global forest cover (natural and planted) over approximately the last 8,000 years. Columns show best estimates as recorded in the History Database of the Global Environment (HYDE v2.0, Goldewijk 2001), supplemented with 2020 values reported to the Food and Agriculture Organization of the United Nations (FAO 2020). Error bars (fiducial lines) denote the range in other published sources (as summarized by Goldewijk 2001).

et al. 2008). Forest loss has been further offset by significant reforestation efforts in Europe, China, and elsewhere in the modern era, culminating in the growing worldwide importance of forest plantations in recent decades (see Table 1.2). Trends of deforestation associated with agricultural expansion and industrialization, followed by forest recovery and reforestation associated with urbanization, have repeated themselves around the world. Coupled with greater use of other materials for fuel and construction, and more efficient agriculture (Sands 2013), this recurrent pattern in history has been described as the "forest transition" (Mather 2007; Rudel et al. 2010). Over longer periods of time, many of the world's boreal, temperate, and montane forests were displaced by repeated glaciations during the Pleistocene, but recovered much of their former area during interglacial periods (Pielou 1992). In short, forest area in the big picture is dynamic—growing and contracting in response to natural (especially climatic) and human factors—and forests have shown remarkable resilience in their ability to recover and expand in the face of change.

The general ability of trees to dominate the vegetation where climate is suitable does not mean that forest area and integrity have not been significantly compromised in recent history. For example, forest harvesting and conversion in recent decades is typically accompanied by widespread dissection (by roads) and fragmentation, leading to extensive effects that compromise the stability and habitat value of much of the land that remains under nominal forest cover (Murcia 1995). Taking such processes of ecological degradation into account, Bryant et al. (1997) estimated that the true area of large and relatively intact fully functioning native forest ecosystems is only 22 percent of the original forest extent. Forest edges, secondary forest, and planted forests around the world typically are exposed to the human pressures of hunting and food or fuelwood gathering, the human-induced introduction of exotic weeds and pests, and the greater likelihood of further damage from wind, fire, and logging. Though not reflected in the general forest area statistics, these factors pose serious threats to many aspects of forest biodiversity, sustainability, and resilience, especially in the tropics (Laurance et al. 2011; Haddad et al. 2015).

Forest ownership around the world and its implications are also worth considering. It is estimated

Table 1.2 Reported areas of naturally regenerating and planted forest

Subregion	Naturally regenerating forest (thousands of hectares (Kha))		Planted forest (thousands of hectares (Kha))		Increase in proportion of planted forest (%)
	1990	2020	1990	2020	
South America	966,621	823,941	7,046	20,245	231
Russian Federation	796,299	796,431	12,651	18,880	48
North America	698,721	676,632	22,596	45,785	102
Western and central Africa	355,885	303,441	956	2,269	177
South and southeast Asia	313,562	264,578	12,949	31,469	168
Eastern and southern Africa	339,874	288,639	6,161	7,139	36
East Asia	152,423	173,264	57,483	98,139	32
Oceania	181,705	179,949	2,784	4,812	73
Europe (excluding Russia)	116,352	118,819	41,743	55,004	20
Western and central Asia	44,965	49,288	3,757	5,621	33
Northern Africa	38,542	33,168	1,383	1,983	63
Central America	27,928	22,014	74	391	560
Caribbean	5,451	7,008	479	851	34
Total	4,038,328	3,737,172	170,062	292,588	80

Note and sources: Data and sub-region groupings from FAO (2020). All values include rounding, with several countries contributing missing or estimated values; 1990 values are updated from those previously reported to FAO and as summarized by Keenan et al. (2015).

that 22 percent of the world's forests are privately owned, which includes land owned by individuals and families, business entities and institutions, and local, tribal, and Indigenous communities (FAO 2020). This proportion is expanding, as governments privatize forest lands or transfer them to Indigenous and community groups (Whiteman et al. 2015). Privately held lands can experience a wide range of forest management approaches, ranging from timber liquidation and land-use conversion to biodiversity conservation and full ecosystem protection. Public forest ownership prevails in most parts of the world, except non-Russian Europe and Central America, where public ownership is 46 percent and 37 percent, respectively (FAO 2020). Public control of forest lands is generally considered desirable in order to prevent liquidation of the forest resource and to protect non-timber values, but the effectiveness of this policy depends on the governance structure, integrity, and capacity of governments. Over 2 billion ha or 54 percent of global forest land is covered by management plans, which implies some level of policy commitment to maintaining forest cover and its sustainable management. Noteworthy is that less than 25 percent of the forest area in Africa and South America is covered by forest management plans, the same regions subject to the highest levels of deforestation (FAO 2020). At the turn of the millennium, forest land under some form of license, contract, or concession to extract timber (but not necessarily to regenerate trees or undertake any further management) ranged from 6 percent of the forest area in Venezuela to 79 percent in the Republic of Congo (White and Martin 2002). By comparison, industrial rights to timber lands have been extended to 57 percent of the forest area in Canada and 60 percent of the forest in Indonesia.

National reports to the Food and Agriculture Organization of the United Nations (FAO) claim that 18.8 percent of global forest area is legally protected in parks and reserves (see Figure 1.2). When this is combined with forest lands managed primarily for biodiversity values, it is estimated that 29.9 percent of the world's forests have strong levels of nature protection (FAO 2020). Despite general worldwide increases in protected areas over the last several decades (see Section 1.5), it is noteworthy that there has been a steady decline in protected forests and biodiversity-emphasis forest in Central America since 1990, and in western and central Africa since 2010 (FAO 2020). Twenty-nine percent of the world's forest area is primarily managed for wood production, while 19 percent is described as having a multiple-use emphasis (see Figure 1.2); this contrasts with 26 percent reported by the FAO in 2015 with only 17 percent of tropical forest area assigned to multiple uses at that time (Sloan and Sayer 2015). Other non-timber objectives include soil and water conservation, biodiversity, and various social and other priorities, which collectively dominate 32 percent of the world's forest area, although with wide variation among countries and regions (see Figure 1.2). Incomplete and inadequate forest inventories, poorly developed human and institutional capacity, continued population growth, agricultural expansion, and widespread low-level corruption in many tropical and low-income countries continue to threaten the integrity of these designated management priorities and the sustainability of all forests (Sloan and Sayer 2015; van Hensbergen 2016).

Logs harvested from the world's forests today are used for posts and poles, sawn into solid lumber for building construction, and are variously peeled, chipped, or pulverized for reconstitution as laminated or composite paneling. Manufacturers of doors, cabinets, furniture, musical instruments, and other crafts are steady consumers of specialty woods. Wood pulp is manufactured into absorbent sanitary products, paper, and packaging. Solid wood continues to be used for heating and cooking in much of the world, while sawdust compressed into wood pellets is increasingly being burned in both residential stoves and municipal heating and electricity-generating plants. Wood cellulose and lignin can be further broken down and turned into methanol and a myriad of other biochemical products. Wood from any one forest is not usually streamed into all of those uses, and the importance of each product stream varies with tree species and in different parts of the world.

Worldwide consumption of logs for use as construction materials currently exceeds their use for pulp and paper products, which is greater than their

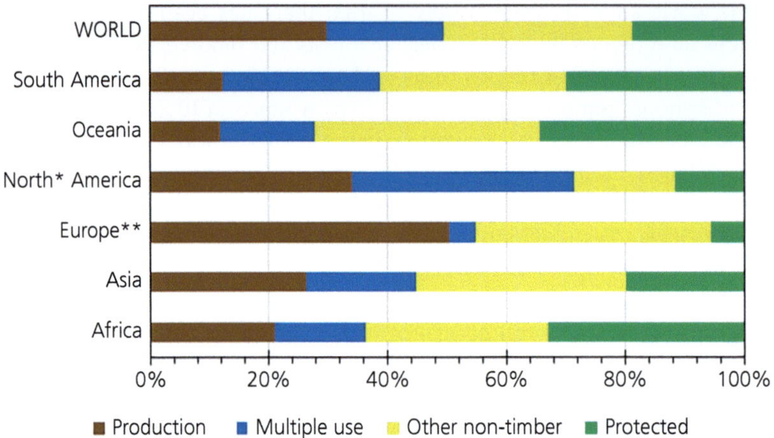

Figure 1.2 Proportion of forest land area classified by primary forest management objective or emphasis, summarized by region. "Other non-timber" objectives include soil and water protection, biodiversity conservation, and social services (such as recreation, tourism, education, research, and cultural sites), among others. Data from FAO (2020).

*North America includes Central America and Caribbean Islands; **Europe includes all of the Russian Federation*

use for fuel (bioenergy; see Figure 1.3). In much of the developed world, wood fiber is allocated to its most efficient use in vertically integrated forest products companies or through long-term contractual arrangements among different manufacturers (Carlsson and Rönqvist 2005). For example, the largest and best-quality logs ("peelers") are often allocated to the highest value stream of veneers and plywood manufacture, with others destined for sawing into dimension lumber. Chips from the log-squaring process and log ends (and sometimes branches and entire small trees) constitute the feedstock for the pulp and paper sector. Waste products are increasingly utilized as well: bark, log ends, offcuts, and sawdust are often used to fire boilers in pulp mills and in bioenergy plants, while sawdust has new value in the form of wood pellets. Global demand for newsprint has declined as electronic news dissemination continues to grow, but the need for cardboard and other packaging materials is growing steadily as intercontinental trade increases and more people shop online and have products delivered to their homes (Jonsson 2011). Many pulp mills that primarily used to serve the newspaper market are now multipurpose bioenergy and biorefining plants, generating electricity, biofuels, and a range of specialized organic chemicals (Stuart and El-Halwagi 2013). Statistics

indicate that fuelwood makes up 29 percent of wood use worldwide (see Figure 1.3), but very little of this is traded in formal markets and it is probably underestimated (Broadhead and Killmann 2008). While household use of fuelwood in some developing economies is declining with the adoption of modern fossil fuel and electrical alternatives, many northern nations have recently adopted policies promoting the use of renewable fuels and the development of bioenergy, so global fuelwood use continues to rise. Collectively, the international export of wood products generated US$244 billion in 2020 (FAO 2021), in addition to making significant contributions to domestic economic activity and employment.

Any summary of how forest products are produced, used, and documented points to a number of considerations potentially important to forest managers. In much of the world, forest management, forest products, and the so-called forestry sector continue to be dominated by the harvesting and processing of logs, with large corporate interests invested in fiber resources. There are ongoing non-market demands for forest goods and services that are being increasingly recognized, but these are much more difficult to document and evaluate. To the degree that those uses are not compatible with the harvesting of trees, there are often competing

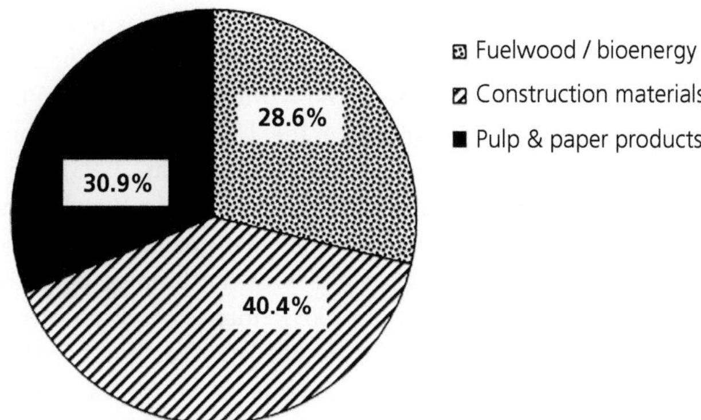

Figure 1.3 Global breakdown of wood and fiber products derived from almost seven billion m^3 of trees processed in 2020 (FAOSTAT-Forestry Database (FAO 2021), with conversion factors following Anon. (2010)).

interests for the governance and use of forest land (as discussed further in Chapter 2). Finally, it is clear there are great disparities in progress toward forest sustainability, particularly between developed and developing nations. Technological innovation, disposable household income, and shifting value systems (that may or may not be reflected in government policies) can influence the use, type, and amount of forest found today in different locations around the planet.

1.5 Protected forests

Another important dimension of forest conservation is to guarantee that specific tracts of land will remain free of logging, settlement, or conversion to agriculture. The stewardship of such conservation lands is often omitted in discussions of forest management, as they are treated as "unmanaged" from the perspective of timber production. Protecting forests and forest lands from conversion to agricultural, urban, or industrial uses is increasingly critical as the world undergoes a joint extinction and climate crisis. Contiguous tracts of indigenous tree cover are important as habitat to many known and unknown species. We must also store as much carbon as possible in wood and soils to avert climate disaster, and continuous forest cover is our best guarantee for protecting watersheds and the increasingly valuable resource of clean, reliable, fresh water.

Forest reserves have existed for a long time (as mentioned in Section 1.2), often primarily as habitat for the hunting of deer, swine, and upland game birds by the aristocracy. With frequent wars and growth in populations and economic activity, many European jurisdictions protected standing timber for naval uses or designated industrial (for example, mining and smelting) activities (Powers 1999; Vogt et al. 2007). The second kind of "forest reserve" designation has a more recent history, with the objective of reserving land (not just its current timber) for forests, forestry, or forest-dwelling wildlife in perpetuity, rather than allowing it to be settled and converted to agriculture once the timber has been removed. Looking beyond the life of trees making up the current forest cover, this kind of forest reserve has had a much more profound impact on the long-term fate of forests in many countries around the world, though primarily those that were colonized and developed in the last few centuries. In particular, it was during the late nineteenth century that large areas of land were excluded from pre-emption and privatization as homesteads during European colonization, and instead were designated as publically held timber lands in Australia, Canada, New Zealand, and the US. Many of these lands were kept under government control to protect water catchments or public grazing opportunities as well as timber production, while others were set aside strictly for military training. State forests were declared in Australia in the 1870s (Frost 1997), provincial forests in Canada in the 1880s (Hodgins et al. 1982), and national forests in the US in 1891 (Steen 2004). In eastern Africa, complex networks of forest reserves and wildlife reserves were implemented by British and German authorities, more to assure colonial access to timber and game resources than to forestall threats of settlement and agriculture, thereby contributing to considerable displacement, privation, and discontent among the traditional users of those lands (Neumann 2002).

The development of national parks, initiated by the United States with its designation of Yellowstone National Park in Wyoming in 1872, marks another major step in the protection and management of forests (Gissibl et al. 2012). Quickly emulated by other British settler societies in Australia (1879), Canada (1885), and New Zealand (1887), the concept of government protection for areas of scenic beauty and public recreation was gradually taken up by other nations around the world in the twentieth century. With trees and forests recognized as contributing to those aesthetics, the prohibition of felling and the suppression of wildfires were high on the list of management imperatives for all such parks and similar land reserves. The protection and viewing of wildlife were also important, providing another incentive to safeguard forest habitats.

The further protection of representative ecological diversity—not just scenic beauty and recreational opportunities—became a goal of parks and protected area networks only in recent decades. As part of its call for sustainable development, the Brundtland Commission (WCED 1987) concluded that protected areas around the globe would have to triple in area (estimated to be 3.5 to 4.0 percent of the land area at that time) in order to protect the world's biodiversity. Plans for implementation of the United Nations Convention on Biological Diversity (Anon. 1993) stemming from the 1992 Earth Summit in Rio de Janeiro eventually established the goal of 10 percent protection for all of the world's biomes by 2010 (Jenkins and Joppa 2009), then 17 percent for all terrestrial biomes by 2020 (Aichi Biodiversity Target 11; Montesino Pouzols et al. 2014), and eventually 30 percent protection by 2030 (Kunming–Montreal Global Biodiversity Framework Target 3; Geldmann 2023). The International Union for Conservation of Nature (IUCN) recognizes seven categories of protected areas, of which five strongly restrict the felling of trees and the removal of forest cover (see Table 1.3). With these concrete goals and concerted international initiatives and support from environmental non-governmental organizations (ENGOs), the growth in all categories of protected areas over the last few decades has been remarkable (see Figure 1.4). The IUCN estimates that all protected areas now account for more than 16 percent of Earth's land area (UNEP–WCMC and IUCN 2021). Of that area, the average representation for forested biomes is approximately 13 percent (Dinerstein et al. 2017), a notable discrepancy from the 19 percent forest protection reported by the FAO (2020).

The governmental designation of forest reserves and protected areas by no means assures the conservation of forest resources or a full defense of Earth's forest biodiversity. The current distribution of parks and protected areas does a poor job of representing

Table 1.3 Categories of protected areas recognized by the International Union for Conservation of Nature (IUCN), and interpretations regarding allowable forest modification and management

Category	Description	Direction regarding forest modification
Ia	Strict nature reserves	Prohibited
Ib	Wilderness areas	Minimized, no road construction, retain naturalness
II	National parks	Minimized, retain ecological processes
III	Natural monuments or features	Prohibited or minimized (depending on feature)
IV	Habitat/species management areas	Allowed to address species/habitat requirements
V	Protected landscapes/seascapes	Permitted/encouraged to safeguard traditional practices
VI	Protected areas with sustainable use of natural resources	Permitted if non-industrial, practiced on <25–50% of area

Source: Dudley (2008).

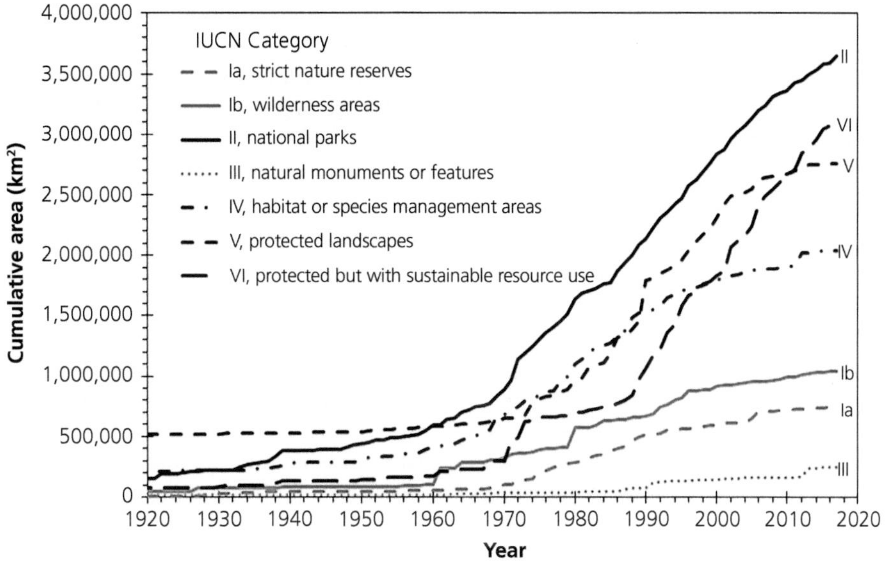

Figure 1.4 The global trend in the designation of terrestrial protected areas, by IUCN category. Total protected area is almost double the amount shown here, as another 12,775,838 km² were considered protected by national governments in 2017, but without an IUCN category being reported (data from the World Database on Protected Areas (UNEP-WCMC and IUCN 2018)).

biodiversity hotspots and regions of the planet characterized by a low human footprint (Leroux et al. 2010). Park management practices in the past (many of which continue to this day) have included predator control, fire suppression, and the prohibition of Indigenous and traditional land uses, leading to the deterioration of ecosystem integrity and the very features that so-called protection was expected to safeguard (Marsh and Hodgins 1998). Other ecosystem management practices, including the use of prescribed fire, animal culls, control of invasive species, and the channeling of human activities are often needed today in order to protect or restore park values. It is widely known that official park boundaries are subject to edge effects and the influence of resource management practices in adjacent landscapes, and those boundaries do not prevent incursions (especially in developing countries) of hunters and loggers (Brandon et al. 1998). Many parks and protected areas are not large enough to maintain their natural disturbance regimes, nor to safeguard migratory wildlife and wide-ranging predators. Consequently, the "Nature Needs Half" initiative questions the effectiveness of politically negotiated goals of 10 percent, 17 percent, or even

30 percent protection, and calls for 50 percent protection of each of the world's ecoregions if we are to save most species from extinction and maintain fully intact ecosystems on this planet (Wilson 2016; Dinerstein et al. 2017).

1.6 Ongoing global changes

Many trends described above can be expected to continue in the coming decades, with some more certain than others. It remains to be seen whether some of those trends will level off or whether even more extreme changes should be expected. The demand for different forest-generated commodities, expressed at both global and local levels, typically determines whether a particular patch of forest is logged or not. Additional socio-economic factors, including subsistence use of forest resources, recreational demand, public support for the protection of rare and endangered species, and demand for agricultural and residential land, vary from nation to nation, from location to location, and over time, collectively determining the amount of protected, harvested, and deforested land found on Earth. Several research teams have taken some of these factors into consideration, along with population growth, economic development, energy demand, arable land needs, and recent initiatives, such as the Bonn Challenge (to restore 350 million hectares of deforested and degraded landscapes by 2030 (https://www.bonnchallenge.org/)) to project future wood demand and the future extent of the world's forests. Deforestation is likely to slow down, but natural forests will remain most at risk in the tropics. In general, d'Annunzio et al. (2015) project an expected increase in total forest area by 2030 in all regions except South America and Oceania. Korhonen et al. (2021) predict an overall increase in the current planted forest area of 7–9 percent to 14 percent by 2055. Wood production from planted forests is expected to increase by 43 percent, but those plantations are subject to regionally variable risks from climate change and pests (Payn et al. 2015).

Human alteration of much of Earth's biosphere is so pervasive and on such a geological scale that the modern epoch has been termed "the Anthropocene" (Corlett 2015). It is the human impact on the planet that will be the dominant force of coming changes, with our commandeering of land for agriculture and settlement, our manipulation of nutrient cycling and water resources, and our burning of fossil fuels and its resulting effects on the global climate (Ellis 2015). On the other hand, there is also a growing appreciation of the need for conservation and sustainability at household, national, and international levels. That appreciation is reflected in the growth of a citizen constituency that is demanding more careful stewardship of the natural world, and is starting to counteract individual aspirations for more affluent lifestyles and business sectors that encourage resource consumption and short-term exploitation.

Humankind and the forests of the world are certain to be living on a warmer planet in the coming decades (as elaborated upon in Chapter 6), with impacts on ecosystems, resources, economies, and lifestyles to vary considerably among regions. Global air temperatures will average somewhere between 1.4 and 4.4°C higher than preindustrial levels within the next several decades, while droughts, storms, floods, wildfires, and insect outbreaks will

probably be more frequent and severe as well (IPCC 2023). Biotic consequences of those meteorological shifts will be expressed both directly and indirectly: not only will trees and many other plant species experience drought and heat stress outside their physiological tolerances, but they will have to cope with enhanced virility of pathogens and insect pests at the same time (Dukes et al. 2009; Sturrock et al. 2011). Increased global trade and travel mean that invasive exotic species will continue to threaten the integrity of indigenous ecosystems and managed crops of all sorts. Forest floors, peatlands, fine fuels, deadwood, and tree foliage can all be expected to have lower moisture contents throughout much of the summer, at the same time as lightning frequency may increase, resulting in a greater incidence of forest fires and more frequent unstoppable "mega-fires" (Adams 2013; Stephens et al. 2014). Forest streams can be expected to become warmer and more intermittent, to the detriment of many fish, amphibian, and invertebrate species (Durance and Ormerod 2007; Brooks 2009). Plant and animal species that had responded to past climate shifts by migrating to more favorable habitats are now unable to do so, as many of the world's natural ecosystems are fragmented and isolated from each other by heavily modified human-managed ecosystems that cannot be crossed safely (Opdam and Wascher 2004).

Human responses to climate change and other global trends reflect a complex set of individual, household, community, regional, national, and international decisions and trends. People have already been demonstrating the standard vertebrate response to habitat degradation by migration, with climate refugees evacuating homelands suffering from drought, rising sea levels, or other stresses (Wennersten and Robins 2017). Overall human population is expected to reach eleven billion by 2100 (twice the level of 1993) and will continue to rise, with stable populations expected only in high-income countries starting about 2050 (United Nations 2019). The global impact of human domination of Earth will depend not only on whether population levels off at ten billion or twenty billion people, but also on per capita levels of resource consumption. Those consumption patterns, in combination with population levels, the degree to which resources and lifestyle opportunities are equitably distributed, but also the effectiveness of environmental protection (Vasseur et al. 2017), will determine the degree to which the world will be able to achieve the UN's Sustainable Development Goals (see Box 1.2). Increasing automation in manufacturing and resource-processing industries will challenge the ability of displaced people to find meaningful employment, and there will be strong regional differences in terms of the availability of labor. Socially and environmentally mature communities will be characterized by circular economies, with high ratios of renovation and restoration compared to new construction. High levels of resource reuse and recycling are expected to become the norm, rather than most manufacturing processes depending on raw materials (Geissdoerfer et al. 2017).

Box 1.2 The United Nations sustainable development goals

A set of globally transformative goals to be achieved by 2030 were set forth in a resolution of the United Nations General Assembly on September 25, 2015. These goals are a successor to the eight Millennium Development Goals for the 2000 to 2015 period, which emphasized the modernization of developing countries, with a call to "ensure environmental sustainability" as their seventh goal. The Sustainable Development Goals (also known as Agenda 2030) are also the product of a long history of international commissions, treaties, summits, and conferences that have been convened over several decades in order to coordinate international efforts to make the world a better place. Those goals can be summarized as follows (United Nations n.d.):

Goal 1: End poverty in all its forms everywhere.

Goal 2: End hunger, achieve food security and improve nutrition, and promote sustainable agriculture.

Goal 3: Ensure healthy lives and promote well-being for all at all ages.

Goal 4: Ensure inclusive and equitable quality education and promote lifelong learning opportunities for all.

Goal 5: Achieve gender equality and empower all women and girls.

Goal 6: Ensure availability and sustainable management of water and sanitation for all.

Goal 7: Ensure access to affordable, reliable, sustainable, and modern energy for all.

Goal 8: Promote sustained, inclusive, and sustainable economic growth, full and productive employment, and decent work for all.

Goal 9: Build resilient infrastructure, promote inclusive and sustainable industrialization, and foster innovation.

Goal 10: Reduce inequality within and among countries.

Goal 11: Make cities and human settlements inclusive, safe, resilient, and sustainable.

Goal 12: Ensure sustainable consumption and production patterns.

Goal 13: Take urgent action to combat climate change and its impacts.

Goal 14: Conserve and sustainably use the oceans, seas, and marine resources for sustainable development.

Goal 15: Protect, restore, and promote sustainable use of terrestrial ecosystems, sustainably manage forests, combat desertification, halt and reverse land degradation, and halt biodiversity loss.

Goal 16: Promote peaceful and inclusive societies for sustainable development, provide access to justice for all, and build effective, accountable, and inclusive institutions at all levels.

Goal 17: Strengthen the means of implementation and revitalize the global partnership for sustainable development.

Ensuring the persistence of healthy forests is integral only to Goal 15, but forests can and should play a role in achieving several other Sustainable Development Goals as well (Katila et al. 2020). The use of forests for watershed protection and streamflow regulation is conducted widely around the world, in support of Goal 6. Forests are increasingly being harnessed as a source of bioenergy (Goal 7). The most sustainable production systems for construction materials include bamboo and timber (Goal 12). By protecting the carbon stocks of old-growth

The conversion of wild forests to other land uses can be expected to slow and eventually cease, in response to both environmental concerns and the economies of silvicultural and agricultural production. We can expect increasing specialization of land uses in both the public and private sectors, with a continuation of trends that partition public lands into conservation and production forests. Demand for wood will continue to grow, especially in the areas of packaging, bioenergy (for heating, electricity generation, and possibly for transportation), and as the feedstock for refineries generating various organic compounds and biochemical products. As we shift from an economy based on fossil fuels to one based on renewable sources of energy and biochemicals, the forest products sector will face competition from agricultural sectors that produce cellulose cheaply and quickly (Ragauskas et al. 2006). At some stage in the next century, we may see equal areas of planted forests—under various intensities and forms of management—and wild, natural forests around the planet as a whole. Despite continued growth in protected areas and non-consumptive use of forests, the biodiversity crisis will persist, with accelerated extinction rates in many organism groups owing to climate change, habitat destruction, resource exploitation, and invasive species (Tilman et al. 2017). In order to address the widest possible scope of future forest management and stewardship, the following chapters therefore give consideration to the management of protected forests, multipurpose or multiple-use forests, as well as production forests.

The Millennium Ecosystem Assessment (Reid et al. 2005) offers four different scenarios for the future, describing storylines differing broadly in levels of human population, equality, resource availability, and ecological integrity for the planet (see Figure 1.5). Those storylines in which ideas and resources are shared globally can be described as a "TechnoGarden" scenario if environmental management is proactive, or require a "Global Ochestration" scenario where environmental management reacts to individual problems as they arise. If nations and regions are more isolated in their response to global change, we might see the proactive development of a place-based "Adapting Mosaic" scenario, or a more disparate and dystopic future described as an "Order from Strength" scenario if environmental management is reactive as well as locally or nationally isolated in its perspectives. These storylines have many assumptions, corollaries, and consequences that bracket some extremes of future possibilities, each of which posits a manner in which the human species and Planet Earth may respond to the environmental and social challenges of the day. There are many intermediate futures possible as well, and it is likely that examples will arise that mix and match some of the components and assumptions of these scenarios. There are initiatives and tendencies around the world and in different sectors of society today that show affinity for each of those scenarios and their many intermediates, but there is no overall trajectory or agenda to make one of those futures more likely at this time. Consequently, forest managers must fall back on the aphorism that the future is unknowable, and it behooves us to prepare for a wide array of future threats and opportunities. Nonetheless, we can anticipate an ongoing need for forests and the ecosystem services they provide, so the Millennium Ecosystem Assessment scenarios can serve as useful reference points as the following chapters explore some options for promoting resilience in the world's forests under multiple futures.

The twentieth century saw widespread adoption of a simplifying and largely agricultural model of forest management under the guise of "scientific forestry" (Puettmann et al. 2009). Along with deforestation—conversion of forests to non-forest land uses—contemporary forest management is associated with widespread biodiversity loss, compromised ecosystem services, and vulnerability to global change. In response, this book joins the growing call for forest management reform, to treat forests as complex adaptive systems (Messier et al. 2013), to undertake a more holistic and ecological approach to forestry (Seymour and Hunter 1999;

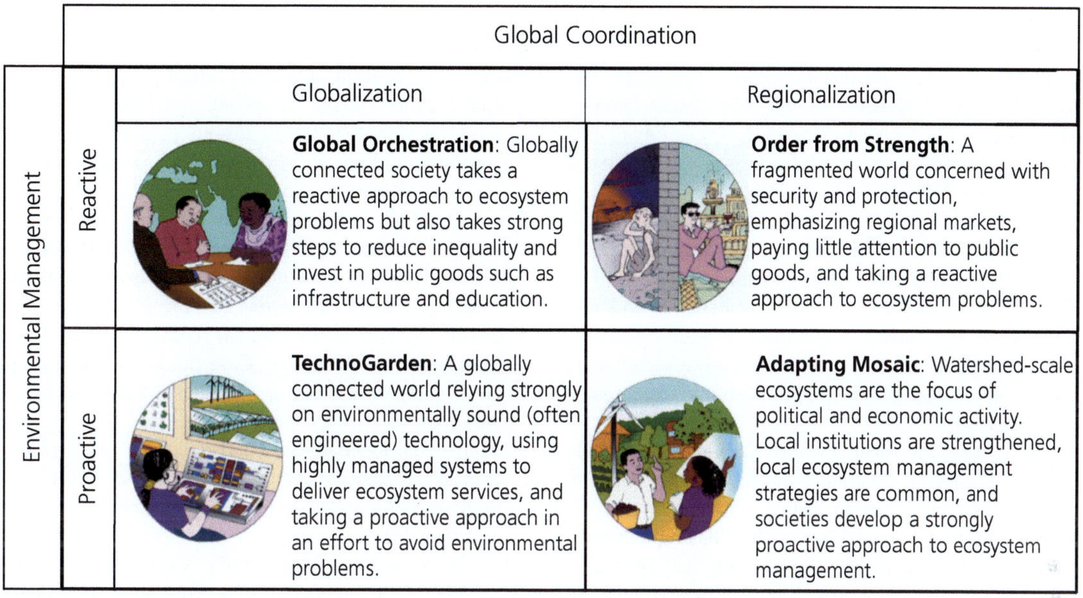

Global Coordination		
	Globalization	Regionalization
Reactive	**Global Orchestration**: Globally connected society takes a reactive approach to ecosystem problems but also takes strong steps to reduce inequality and invest in public goods such as infrastructure and education.	**Order from Strength**: A fragmented world concerned with security and protection, emphasizing regional markets, paying little attention to public goods, and taking a reactive approach to ecosystem problems.
Proactive	**TechnoGarden**: A globally connected world relying strongly on environmentally sound (often engineered) technology, using highly managed systems to deliver ecosystem services, and taking a proactive approach in an effort to avoid environmental problems.	**Adapting Mosaic**: Watershed-scale ecosystems are the focus of political and economic activity. Local institutions are strengthened, local ecosystem management strategies are common, and societies develop a strongly proactive approach to ecosystem management.

(Left axis label: Environmental Management)

Figure 1.5 Millennium Ecosystem Assessment scenario storylines (based on Reid et al. (2005); images courtesy of Millennium Ecosystem Assessment).

Franklin et al. 2018), and to adopt resilience-based ecosystem stewardship in an uncertain and rapidly changing world (Chapin et al. 2009; Triviño et al. 2023).

for forest products and forest ecosystem services will continue to rise. At the same time, forests in the Anthropocene have to be managed in the context of a biodiversity and climate-change crisis, one characterized by invasive species, rising disturbance levels, shifts in public expectations, and broad uncertainty.

Forestry isn't rocket science; it's much more complicated.

Fred Bunnell (c.1993)

Box 1.3 Key points

- The extent and composition of forests have fluxed and waned around the world over history, in response to shifts in climate and to human population, exploitation, and interventions. Natural forest continues to be lost in many parts of the world today, while the proportion of planted forest is increasing.
- Forests provide many goods and services with ecological, economic, and social benefits that support sustainable development. The harvesting and manufacture of timber and fiber products remain the primary economic activity associated with forests, but there is growing appreciation of their role in harboring biodiversity, protecting soil and water, storing carbon, and filling recreational and spiritual needs.
- There is ongoing tension between advocates of efficient timber or fiber production and those who support the protection and promotion of other forest values.
- The growing global human population and our patterns of resource consumption mean that the demand

References cited

Adamowicz, W. L., and Burton, P. J. (2003). "Sustainability and Sustainable Forest Management," in P. J. Burton, C. Messier, D. W. Smith, and W. L. Adamowicz (eds), *Towards Sustainable Management of the Boreal Forest*. Ottawa: NRC Research Press, 41–64.

Adams, M. A. (2013). "Mega-Fires, Tipping Points and Ecosystem Services: Managing Forests and Woodlands in an Uncertain Future," *Forest Ecology and Management*, 294: 250–61.

Angelstam, P. K. (1998). "Maintaining and Restoring Biodiversity in European Boreal Forests by Developing Natural Disturbances Regimes," *Journal of Vegetation Science*, 9: 593–602.

Anon. (1993). *Convention on Biological Diversity*. Montreal: United Nations Environment. Available online at https://www.cbd.int/convention/text/ (accessed January 27, 2024).

Anon. (2010). *Forest Product Conversion Factors for the UNECE Region*. Geneva Timber and Forest Discussion Paper 49. Geneva: United Nations. Available online at https://unece.org/fileadmin/DAM/timber/publications/DP-49.pdf (accessed May 13, 2023).

Attiwill, P. M. (1994). "The Disturbance of Forest Ecosystems: The Ecological Basis for Conservation Management," *Forest Ecology and Management*, 63: 247–300.

Batho, A., and Garcia, O. (2006). "De Perthuis and the Origins of Site Index: A Historical Note," *FBMIS* 1: 1–10. Available online at https://www.researchgate.net/profile/Oscar-Garcia-71/publications (accessed May 13, 2023).

Bauhus, J., van der Meer, P., and Kanninen, M. (eds). (2010). *Ecosystem Goods and Services from Plantation Forests*. London: Earthscan. 254 pp.

Bergeron, Y., and Harvey, B. (1997). "Basing Silviculture on Natural Ecosystem Dynamics: An Approach Applied to the Southern Boreal Mixedwoods of Québec," *Forest Ecology and Management*, 92: 235–42.

Bird, R. B., Bird, D. W., Codding, B. F., Parker, C. H., and Jones, J. H. (2008). "The 'Fire Stick Farming' Hypothesis: Australian Aboriginal Foraging Strategies, Biodiversity, and Anthropogenic Fire Mosaics," *Proceedings of the National Academy of Sciences*, 105/39: 14796–801.

Botkin, D. B. (2001). *No Man's Garden: Thoreau and a New Vision for Civilization and Nature*. Washington: Island Press. 288 pp.

Boyd, R.T. (ed.). (1999). *Indians, Fire, and the Land in the Pacific Northwest*. Corvallis, OR: Oregon State University Press. 313 pp.

Boyle, J. R., Winjum, J. K., Kavanagh, K., and Jensen, E. C. (eds). (1999). *Planted Forests: Contributions to the Quest for Sustainable Societies*. Dordrecht: Kluwer Academic Publishers. 472 pp.

Brandon, K., Redford, K. H., and Sanderson, S. (eds). (1998). *Parks in Peril: People, Politics, and Protected Areas*. Washington: Island Press. 532 pp.

Broadhead, J., and Killmann, W. (2008). "Forests and Energy: Key Issues." FAO Forestry Paper, No. 154. Rome: Food and Agriculture Organization of the United Nations. 56 pp. Available online at https://www.fao.org/3/i0139e/i0139e00.htm (accessed May 13, 2023).

Brockerhoff, E. G., Jactel, H., Parrotta, J. A., Quine, C. P., and Sayer, J. (2008). "Plantation Forests and Biodiversity: Oxymoron or Opportunity?" *Biodiversity and Conservation*, 17/5: 925–51.

Brooks, R. T. (2009). "Potential Impacts of Global Climate Change on the Hydrology and Ecology of Ephemeral Freshwater Systems of the Forests of the Northeastern United States," *Climatic Change*, 95/3–4: 469–83.

Brunig, E. F. (1970). "Multiple-Use Management in Germany's Forests," *Journal of Forestry*, 68/11: 718–22.

Bryant, D., Nielsen, D., and Tangley, L. (1997). *The Last Frontier Forests: Ecosystems and Economies on the Edge*. Washington: World Resources Institute. 42 pp. Available online at https://www.wri.org/research/last-frontier-forests (accessed January 26, 2024).

Bürgi, M. (2015). "Close-to-Nature Forestry," in K. J. Kirby and C. Watkins (eds), *Europe's Changing Woods and Forests: From Wildwood to Managed Landscapes*. Wallingford, UK: CAB International, 107–15.

Carlsson, D., and Rönnqvist, M. (2005). "Supply Chain Management in Forestry: Case Studies at Södra Cell AB," *European Journal of Operational Research*, 163/3: 589–616.

Chapin, F. S., Kofinas, G.P., and Folke, C. (eds). (2009). *Principles of Ecosystem Stewardship: Resilience-Based Natural Resource Management in a Changing World*. New York: Springer. 401 pp.

Charnley, S., and Poe, M. R. (2007). "Community Forestry in Theory and Practice: Where Are We Now?" *Annual Review of Anthropology*, 36: 301–36.

Chazdon, R. L. (2008). "Beyond Deforestation: Restoring Forests and Ecosystem Services on Degraded Lands," *Science*, 320/5882: 1458–60.

Christensen, N. L., Bartuska, A. M., Brown, J. H., et al. (1996). "The Report of the Ecological Society of America Committee on the Scientific Basis for Ecosystem Management," *Ecological Applications*, 6/3: 665–91.

Ciancio, O., and Nocentini, S. (2011). "Biodiversity Conservation and Systemic Silviculture: Concepts and Applications," *Plant Biosystems*, 145/2: 411–18.

Clément, V., Arnould, P., and Simon, L. (2012). "Historical Background for Sustainable Development Doctrine in France," in S.-P. Wicherek and S. Balazy (eds), *Agricultural Diversity and Sustainable Development in Europe: Examples from France and Poland*. Warsaw: SGGW Press, 17–27.

Colding, J., and Folke, C. (2001). "Social Taboos: 'Invisible' Systems of Local Resource Management and Biological Conservation," *Ecological Applications*, 11/2: 584–600.

Corlett, R. T. (2015). "The Anthropocene Concept in Ecology and Conservation," *Trends in Ecology & Evolution*, 30/1: 36–41.

Craig, B. (1987). "National Forest Planning and Anadromous Fish Protection: A Trilogy of NEPA Cases," *Journal of Environmental Law and Litigation*, 2: 255–81.

Cumming, S. G., Burton, P. J., Prahacs, S., and Garland, M. R. (1994). "Potential Conflicts between Timber Supply and Habitat Protection in the Boreal Mixedwood of Alberta, Canada: A Simulation Study," *Forest Ecology and Management*, 68/2–3: 281–302.

d'Annunzio, R., Sandker, M., Finegold, Y., and Min, Z. (2015). "Projecting Global Forest Area towards 2030," *Forest Ecology and Management*, 352: 124–33.

de Jong, W., Cornejo, C., Pacheco, P., et al. (2010). "Opportunities and Challenges for Community Forestry: Lessons from Tropical America," in G. Mery, P. Katila, G. Galloway, et al. (eds), *Forests and Society: Responding to Global Drivers of Change*. IUFRO World Series 25. Vienna: International Union of Forest Research Organizations, 299–314.

Devall, B. (ed.). (1993). *Clearcut: The Tragedy of Industrial Forestry*. San Francisco: Sierra Club Books. 291 pp.

Diamond, J. (2005). *Collapse: How Societies Choose to Fail or Succeed*. New York: Viking Press. 592 pp.

Dinerstein, E., Olson, D., Joshi, A., et al. (2017). "An Ecoregion-Based Approach to Protecting Half the Terrestrial Realm," *BioScience*, 67/6: 534–45.

Dudley, N. (ed.). (2008). *Guidelines for Applying Protected Area Management Categories*. Best Practice Protected Area Guidelines Series, No. 21. Gland, Switzerland: International Union for Conservation of Nature. 86 pp. Available online at https://portals.iucn.org/library/sites/library/files/documents/PAG-021.pdf (accessed February 19, 2024).

Dukes, J. S., Pontius, J., Orwig, D., et al. (2009). "Responses of Insect Pests, Pathogens, and Invasive Plant Species to Climate Change in the Forests of Northeastern North America: What Can We Predict?" *Canadian Journal of Forest Research*, 39/2: 231–48.

Durance, I., and Ormerod, S. J. (2007). "Climate Change Effects on Upland Stream Macroinvertebrates over a 25-Year Period," *Global Change Biology*, 13/5: 942–57.

Ellis, E. C. (2015). "Ecology in an Anthropogenic Biosphere," *Ecological Monographs*, 85/3: 287–331.

Erickson, C. L. (2008). "Amazonia: The Historical Ecology of a Domesticated Landscape," in H. Silverman and W. H. Ishell (eds), *The Handbook of South American Archaeology*. New York: Springer, 157–83.

Evans, J., and Turnbull, J. W. (2004). *Plantation Forestry in the Tropics: The Role, Silviculture, and Use of Planted Forests for Industrial, Environmental, and Agroforestry Purposes*. 3rd edn. Oxford: Oxford University Press. 480 pp.

Fanta, J. (1997). "Rehabilitating Degraded Forests in Central Europe into Self-Sustaining Forest Ecosystems," *Ecological Engineering*, 8/4: 289–97.

FAO (2015). *FRA 2015 Terms and Definitions*. Forest Resources Assessment Working Paper 180. Rome: Food and Agriculture Organization of the United Nations. 31 pp. Available online at https://www.fao.org/3/ap862e/ap862e00.pdf (accessed May 15, 2023).

FAO (2020). *Global Forest Resources Assessment 2020: Main Report*. Rome: Food and Agriculture Organization of the United Nations. https://doi.org/10.4060/ca9825en (accessed May 15, 2023).

FAO (2021). Global production and trade in forest products in 2020. FAOSTAT-Forestry Database. Food and Agriculture Organization of the United Nations, Rome. Available online at https://www.fao.org/forestry/statistics/80938/en/ (accessed May 9, 2023).

Farcy, C., Martinez de Arano, I., and Rojas-Briales, E. (2019). "Conclusions," in C. Farcy, I. Martinez de Arano, and E. Rojas-Briales (eds), *Forestry in the Midst of Global Changes*. Boca Raton, FL: CRC Press, 403–11.

Fedkiw, J. (1998). *Managing Multiple Uses on National Forests, 1905–1995*. Publication FS-628. Washington: USDA Forest Service. 284 pp.

Filotas, E., Parrott, L., Burton, P. J., et al. (2014). "Viewing Forests through the Lens of Complex Systems Science," *Ecosphere*, 5/1: 1–23.

Ford, A., and Nigh, R. (2016). *The Maya Forest Garden: Eight Millennia of Sustainable Cultivation of the Tropical Woodlands*. London: Routledge. 259 pp.

Foster, D. R. (1995). "Land-Use History and Four Hundred Years of Vegetation Change in New England," in B. L. Turner, A. G. Sal, F. G. Bernáldez, and F. di Castri (eds), *Global Land Use Change: A Perspective from the Columbia Encounter*. Madrid: Consejo Superior de Investigaciones Científicas, 253–319.

Franklin, J. F. (2016). *Understanding and Managing Forests as Ecosystems: A Reflection on 60 Years of Change, and a View to the Anthropocene*. The 2016 Pinchot Distinguished Lecture, 18 February 2016. Washington: Pinchot Institute for Conservation. Available online at http://www.pinchot.org/doc/612 (accessed January 28, 2018).

Franklin, J. F., Johnson, K. N., and Johnson, D. L. (2018). *Ecological Forest Management*. Long Grove, IL: Waveland Press. 646 pp.

Frost, W. (1997). "Farmers, Government, and the Environment: The Settlement of Australia's 'Wet Frontier', 1870–1920," *Australian Economic History Review*, 37/1: 19–38.

Garfitt, J. E. (1995). *Natural Management of Woods: Continuous Cover Forestry*. Taunton, UK: Research Studies Press. 152 pp.

Gauthier, S., Vaillancourt, M.-A., Leduc, A., et al. (eds). (2009). *Ecosystem Management in the Boreal Forest*. Quebec, Canada: Presses de l'Université du Québec. 392 pp.

Geissdoerfer, M., Savaget, P., Bocken, N. M., and Hultink, E. J. (2017). "The Circular Economy: A New Sustainability Paradigm?" *Journal of Cleaner Production*, 143: 757–68.

Geldmann, J. (2023). "Safeguarding Biodiversity Requires Understanding how to Manage Protected Areas Cost Effectively," *One Earth*, 6/2: 73–6.

Ghazoul, J. (2015). *Forests: A Very Short Introduction.* Oxford: Oxford University Press. 150 pp.

Gilmour, D. (2016). *Forty Years of Community-Based Forestry: A Review of its Extent and Effectiveness.* FAO Forestry Paper 176. Rome: Food and Agriculture Organization of the United Nations. 146 pp. Available online at https://www.fao.org/3/i5415e/i5415e.pdf (accessed May 15, 2023).

Gissibl, B., Höhler, S., and Kupper, P. (2012). "Introduction: Towards a Global History Of National Parks," in B. Gissibl, S. Höhler, and P. Kupper (eds), *Civilizing Nature: National Parks in Global Historical Perspective.* New York: Berghahn Books, 1–27.

Goldewijk, K. K. (2001). "Estimating Global Land Use Change over the Past 300 Years: The HYDE Database," *Global Biogeochemical Cycles*, 15/2: 417–33.

Grubb, M., Koch, M., Thomson, K., Sullivan, F., and Munson, A. (1993). *The "Earth Summit" Agreements: A Guide and Assessment—An Analysis of the Rio '92 UN Conference on Environment and Development.* London: Routledge. 202 pp.

Grumbine, R. E. (1997). "Reflections on 'What Is Ecosystem Management?'" *Conservation Biology*, 11/1: 41–7.

Gunderson, L. H. (2000). "Ecological Resilience—in Theory and Application," *Annual Review of Ecology and Systematics*, 31/1: 425–39.

Gustafsson, L., Baker, S. C., Bauhus, J., et al. (2012). "Retention Forestry to Maintain Multifunctional Forests: A World Perspective," *BioScience*, 62/7: 633–45.

Haddad, N. M., Brudvig, L. A., Clobert, J., et al. (2015). "Habitat Fragmentation and its Lasting Impact on Earth's Ecosystems," *Science Advances*, 1/2: e1500052.

Hall, G. R. (1963). "The Myth and Reality of Multiple Use Forestry," *Natural Resources Journal*, 3/2: 276–90.

Hammond, H. (1991). *Seeing the Forest among the Trees: The Case for Wholistic Forest Use.* Vancouver: Polestar Press. 309 pp.

Higman, S., Mayers, J., Bass, S., Judd, N., and Nussbaum, R. (2005). *The Sustainable Forestry Handbook: A Practical Guide for Tropical Forest Managers on Implementing New Standards.* 2nd edn. London: Earthscan. 332 pp.

Hodgins, B. W., Benikickson, J., and Gillis, P. (1982). "The Ontario and Quebec Experiments in Forest Reserves 1883–1930," *Journal of Forest History*, 26/1: 20–33.

Hooghiemstra, H. (2002). "The Dynamic Rainforest Ecosystem on Geological, Quaternary and Human Time Scales," in P. Verweij (ed.), *Understanding and Capturing the Multiple Values of Tropical Forests.* Wageningen: Tropenbos International, 7–19. Available online at https://www.tropenbos.org/file.php/252/tbi_proceedings_3verweij.pdf (accessed August 28, 2023).

Hunter, M. L. (1993). "Natural Fire Regimes as Spatial Models for Managing Boreal Forests," *Biological Conservation*, 65: 115–20.

Hütte, G. (2000). "Perceived Images of Various Actors Engaged in Sustainability Discussions," in K. von Gadow, T. Pukkala, and M. Tomé. (eds), *Sustainable Forest Management.* Dordrecht: Kluwer Academic Publishers, 193–216.

Hutton, D., and Connors, L. (1999). *History of the Australian Environment Movement.* Melbourne: Cambridge University Press. 325 pp.

Innes, J. L. (2017). "Sustainable Forest Management: From Concept to Practice," in J. L. Innes and A. V. Tikina (eds), *Sustainable Forest Management: From Concept to Practice.* London: Routledge, 1–32.

Innes, J. L., and Tikina, A. V. (eds). (2017). *Sustainable Forest Management: From Concept to Practice.* London: Routledge. 395 pp.

IPCC (2023). *Synthesis Report of the IPCC Sixth Assessment Report (AR6): Longer Report.* Geneva: Intergovernmental Panel on Climate Change. 85 pp. Available online at https://www.ipcc.ch/report/ar6/syr/ (accessed May 12, 2023).

Jenkins, C. N., and Joppa, L. (2009). "Expansion of the Global Terrestrial Protected Area System," *Biological Conservation*, 142/10: 2166–74.

Jonsson, R. (2011). "Trends and Possible Future Developments in Global Forest-Product Markets: Implications for the Swedish Forest Sector," *Forests*, 2/1: 147–67.

Kardell, L. (1980). "Forest Berries and Mushrooms: An Endangered Resource?" *Ambio*, 9/5: 241–7.

Katila, P., Pierce Colfer, C. J., de Jong, W., Galloway, G., Pacheco, P., and Winkel, G. (eds). (2020). *Sustainable Development Goals: Their Impacts on Forests and People.* New York: Cambridge University Press. 617 pp.

Keenan, R. J., and Kimmins, J. P. (1993). "The Ecological Effects of Clear-Cutting," *Environmental Reviews*, 1/2: 121–44.

Keenan, R. J., Reams, G. A., Achard, F., de Freitas, J. V., Grainger, A., and Lindquist, E. (2015). "Dynamics of Global Forest Area: Results from the FAO Global Forest Resources Assessment 2015," *Forest Ecology and Management*, 352: 9–20.

Kennedy, J. J. (1985). "Conceiving Forest Management as Providing for Current and Future Social Value," *Forest Ecology and Management*, 13: 121–32.

Kimmins, J. P. (1991). "The Future of the Forested Landscapes of Canada," *Forestry Chronicle*, 67/1: 14–18.

Koch, N. E., and Kennedy, J. J. (1991). "Multiple-Use Forestry for Social Values," *Ambio*, 20/7: 330–3.

Kohm, K. A., and Franklin, J. F. (eds). (1997). *Creating a Forestry for the 21st Century: The Science of Ecosystem Management.* Washington: Island Press. 475 pp.

Korhonen, J., Nepal, P., Prestemon, J. P., and Cubbage, F. W. (2021). "Projecting Global and Regional Outlooks for Planted Forests under the Shared Socio-Economic Pathways," *New Forests,* 52: 197–216.

Landres, P. B., Morgan, P., and Swanson, F. J. (1999). "Overview of the Use of Natural Variability Concepts in Managing Ecological Systems," *Ecological Applications,* 9: 1179–88.

Larocque, G. R. (ed.). (2016). *Ecological Forest Management Handbook.* Boca Raton, FL: CRC Press. 604 pp.

Laurance, W. F. (1999). "Reflections on the Tropical Deforestation Crisis," *Biological Conservation,* 91/2–3: 109–17.

Laurance, W. F., Camargo, J. L., Luizão, R. C., et al. (2011). "The Fate of Amazonian Forest Fragments: A 32-Year Investigation," *Biological Conservation,* 144/1: 56–67.

Leopold, A. (1949). *A Sand County Almanac, and Sketches Here and There.* New York: Oxford University Press. 240 pp.

Leroux, S. J., Krawchuk, M. A., Schmiegelow, F., et al. (2010). "Global Protected Areas and IUCN Designations: Do the Categories Match the Conditions?" *Biological Conservation,* 143/3: 609–16.

Leslie, A. J. (1966). "A Review of the Concept of the Normal Forest," *Australian Forestry,* 30/2: 139–47.

Lindenmayer, D. B., and Franklin, J. F. (eds). (2003). *Towards Forest Sustainability.* Melbourne: CSIRO Publishing. 231 pp.

Long, J. N. (2009). "Emulating Natural Disturbance Regimes as a Basis for Forest Management: A North American View," *Forest Ecology and Management,* 257/9: 1868–73.

Lowenthal, D. (2009). *George Perkins Marsh: Prophet of Conservation.* Seattle: University of Washington Press. 632 pp.

MacCleery, D. W. (1992). *American Forests: A History of Resiliency and Recovery.* Publication FS-540. Washington: USDA Forest Service. 59 pp. Available online at https://www.fs.usda.gov/research/treesearch/64317 (accessed May 15, 2023).

MacDicken, K. G., Sola, P., Hall, J. E., Sabogal, C., Tadoum, M., and de Wasseige, C. (2015). "Global Progress toward Sustainable Forest Management," *Forest Ecology and Management,* 352: 47–56.

Marchak, M. P. (1995). *Logging the Globe.* Montreal: McGill-Queen's University Press. 404 pp.

Marsh, J. S., and Hodgins, B. W. (eds). (1998). *Changing Parks: The History, Future and Cultural Context of Parks and Heritage Landscapes.* Toronto: Natural Heritage/Natural History Inc. 310 pp.

Maser, C. (1990). *The Redesigned Forest.* Toronto: Stoddart Publishing Co. 224 pp.

Mason, D. T. (1927). "Sustained Yield and American Forest Problems," *Journal of Forestry,* 25/6: 625–58.

Mather, A. (2007). "Recent Asian Forest Transitions in Relation to Forest Transition Theory," *International Forestry Review,* 9: 491–502.

Mery, G., Katila, P., Galloway, G., et al. (eds). (2010). *Forests and Society: Responding to Global Drivers of Change.* IUFRO World Series, Volume 25. Vienna: International Union of Forest Research Organizations. 509 pp.

Messier, C., Puettmann, K. J., and Coates, K. D. (eds). (2013). *Managing Forests as Complex Adaptive Systems: Building Resilience to the Challenge of Global Change.* New York: Routledge. 353 pp.

Montesino Pouzols, F. M., Toivonen, T., Di Minin, E., et al. (2014). "Global Protected Area Expansion Is Compromised by Projected Land-Use and Parochialism," *Nature,* 516/7531: 383–390.

MPWG (2015). *Criteria and Indicators for the Conservation and Sustainable Management of Temperate and Boreal Forests.* 5th edn. 30 pp. Montreal Process Working Group. Available online at https://montreal-process.org/The_Montreal_Process/Criteria_and_Indicators/index.shtml (accessed May 13, 2023).

Mueller, A. D., Islebe, G. A., Anselmetti, F. S., et al. (2010). "Recovery of the Forest Ecosystem in the Tropical Lowlands of Northern Guatemala after Disintegration of Classic Maya Polities," *Geology,* 38/6: 523–6.

Murcia, C. (1995). "Edge Effects in Fragmented Forests: Implications for Conservation," *Trends in Ecology & Evolution,* 10/2: 58–62.

Myers, N. (1995). "The World's Forests: Need for a Policy Appraisal," *Science,* 268/5212: 823–5.

Neumann, R. P. (2002). *Imposing Wilderness: Struggles over Livelihood and Nature Preservation in Africa.* Berkeley and Los Angeles: University of California Press. 257 pp.

Nocentini, S., and Coll, L. (2013). "Mediterranean Forests: Human Use and Complex Adaptive Systems," in C. Messier, K. J. Puettmann, and K. D. Coates (eds), *Managing Forests as Complex Adaptive Systems: Building Resilience to the Challenge of Global Change.* New York: Routledge, 214–43.

North, M. P., and Keeton, W. S. (2008). "Emulating Natural Disturbance Regimes: An Emerging Approach for Sustainable Forest Management," in, R. Lafortezza, J. Chen, G. Sanesi, and T. R. Crow (eds), *Patterns and Processes in Forest Landscapes: Multiple Use and Sustainable Management.* Dordrecht: Springer, 341–72.

Noss, R. F. (1987). "From Plant Communities to Landscapes in Conservation Inventories: A Look at the Nature Conservancy (USA)," *Biological Conservation*, 41/1: 11–37.

Opdam, P., and Wascher, D. (2004). "Climate Change Meets Habitat Fragmentation: Linking Landscape and Biogeographical Scale Levels in Research and Conservation," *Biological Conservation*, 117/3: 285–97.

Orchard, C. D. (1953). "Sustained Yield Forest Management in British Columbia," *Forestry Chronicle*, 29/1: 45–54.

Pagdee, A., Kim, Y. S., and Daugherty, P. J. (2006). "What Makes Community Forest Management Successful: A Meta-Study from Community Forests throughout the World," *Society and Natural Resources*, 19/1: 33–52.

Parés-Ramos, I., Gould, W., and Aide, T. (2008). "Agricultural Abandonment, Suburban Growth, and Forest Expansion in Puerto Rico between 1991 and 2000," *Ecology and Society*, 13/2: 1–19. Available online at http://www.ecologyandsociety.org/vol13/iss2/art1/ (accessed May 15, 2023).

Parrotta, J. A. (2002). "Restoration and Management of Degraded Tropical Forest Landscapes," in R. S. Ambasht and N. K. Ambasht (eds), *Modern Trends in Applied Terrestrial Ecology*. New York: Springer, 135–48.

Parson, E. A., Haas, P. M., and Levy, M. A. (1992). "A Summary of the Major Documents Signed at the Earth Summit and the Global Forum," *Environment: Science and Policy for Sustainable Development*, 34/8: 12–36.

Payn, T., Carnus, J. M., Freer-Smith, P., et al. (2015). "Changes in Planted Forests and Future Global Implications," *Forest Ecology and Management*, 352: 57–67.

Perera, A. H., Buse, L. J., and Weber, M. G. (eds). (2004). *Emulating Natural Forest Landscape Disturbances*. New York: Columbia University Press. 330 pp.

Pielou, E. C. (1992). *After the Ice Age: The Return of Life to Glaciated North America*. Chicago: University of Chicago Press. 376 pp.

Pilarski, M. (ed.). (1994). *Restoration Forestry: An International Guide to Sustainable Forestry Practices*. Durango, CO: Kivaki Press. 525 pp.

Piperno, D. R., McMichael, C., and Bush, M. B. (2015). "Amazonia and the Anthropocene: What was the Spatial Extent and Intensity of Human Landscape Modification in the Amazon Basin at the End of Prehistory?" *The Holocene*, 25/10: 1588–97.

Pommerening, A., and Murphy, S. T. (2004). "A Review of the History, Definitions and Methods of Continuous Cover Forestry with Special Attention to Afforestation and Restocking," *Forestry*, 77/1: 27–44.

Powers, R. F. (1999). "On the Sustainable Productivity of Planted Forests," *New Forests*, 17/1–3: 263–306.

Puettmann, K. J., Coates, K. D., and Messier, C. (2009). *A Critique of Silviculture: Managing for Complexity*. Washington: Island Press. 189 pp.

Pukkala, T., and von Gadow, K. (eds). (2012). *Continuous Cover Forestry*. 2nd edn. Dordrecht: Springer. 296 pp.

Ragauskas, A. J., Williams, C. K., Davison, B. H., et al. (2006). "The Path Forward for Biofuels and Biomaterials," *Science*, 311/5760: 484–9.

Reid, W. V., Mooney, H. A., Cropper, A., et al. (2005). *Millennium Ecosystem Assessment, 2005: Ecosystems and Human Well-Being: Synthesis*. Washington: Island Press. 137 pp. Available online at https://www.millenniumassessment.org/documents/document.356.aspx.pdf (accessed May 15, 2023).

Revkin, A. (2004). *The Burning Season: The Murder of Chico Mendes and the Fight for the Amazon Rain Forest*. Washington: Island Press. 321 pp.

Riding, R., and Graves, R. (1937). "From a Private Correspondence on Reality," *Epilogue*, 3: 107–30.

Rudel, T. K., Schneider, L., and Uriarte, M. (2010). "Forest Transitions: An Introduction," *Land Use Policy*, 27: 95–7.

Saito, O. (2009). "Forest History and the Great Divergence: China, Japan, and the West Compared," *Journal of Global History*, 4/3: 379–404.

Sands, R. (2013). "A History of Human Interaction with Forest," in R. Sands (ed.), *Forestry in a Global Context*. 2nd edn. Wallingford, UK: CAB International, 1–36.

Sarr, D., Puettmann, K., Pabst, R., Cornett, M., and Arguello, L. (2004). "Restoration Ecology: New Perspectives and Opportunities for Forestry," *Journal of Forestry*, 102/5: 20–4.

Seymour, R., and Hunter, M. (1999). "Principles of Ecological Forestry," in M. L. Hunter (ed.), *Maintaining Biodiversity in Forest Ecosystems*. Cambridge: Cambridge University Press, 22–61.

Siry, J. P., Cubbage, F. W., Potter, K. M., and McGinley, K. (2018). "Current Perspectives on Sustainable Forest Management: North America," *Current Forestry Reports*, 4: 138–49.

Sloan, S., and Sayer, J. A. (2015). "Forest Resources Assessment of 2015 Shows Positive Global Trends but Forest Loss and Degradation Persist in Poor Tropical Countries," *Forest Ecology and Management*, 352: 134–45.

Slocombe, D. S. (1998). "Defining Goals and Criteria for Ecosystem-Based Management," *Environmental Management*, 22/4: 483–93.

Spilsbury, R. (2010). *Deforestation Crisis: Can the Earth Survive?* New York: Rosen Publishing Group. 48 pp.

Spurr, S. H. (1981). "Clearcutting on National Forests," *Natural Resources Journal*, 21/2: 223–43.

Stanturf, J. A. (ed.). (2016). *Restoration of Boreal and Temperate Forests*. 2nd edn. Boca Raton, FL: CRC Press. 561 pp.

Stanturf, J. A., Palik, B. J., Williams, M. I., Dumroese, R. K., and Madsen, P. (2014). "Forest Restoration Paradigms," *Journal of Sustainable Forestry*, 33/S1: S161–S194.

Steen, H. K. (2004). *The US Forest Service: A History, Centennial Edition*. Seattle: Forest History Society and University of Washington Press. 355 pp.

Steffen, W., Sanderson, R. A., Tyson, P. D., et al. (2005). *Global Change and the Earth System: A Planet under Pressure*. Berlin: Springer. 336 pp.

Stephens, S. L., Burrows, N., Buyantuyev, A., et al. (2014). "Temperate and Boreal Forest Mega-Fires: Characteristics and Challenges," *Frontiers in Ecology and the Environment*, 12/2: 115–22.

Stibig, H. J., Achard, F., Carboni, S., Rasi, R., and Miettinen, J. (2014). "Change in Tropical Forest Cover of Southeast Asia from 1990 to 2010," *Biogeosciences*, 11/2: 247–58.

Stoddard, C. H. (1959). *Essentials of Forestry Practice*. New York: Ronald Press. 258 pp.

Stuart, P. R., and El-Halwagi, M. M. (eds). (2013). *Integrated Biorefineries: Design, Analysis and Optimization*. Boca Raton, FL: CRC Press. 852 pp.

Sturrock, R. N., Frankel, S. J., Brown, A. V., et al. (2011). "Climate Change and Forest Diseases," *Plant Pathology*, 60/1: 133–49.

Swanson, F. J., and Franklin, J. F. (1992). "New Forestry Principles from Ecosystem Analysis of Pacific Northwest Forests," *Ecological Applications*, 2/3: 262–74.

Swanson, F. J., Jones, J. A., Wallin, D. O., and Cissel, J. H. (1994). "Natural Variability—Implications for Ecosystem Management," in M. E. Jensen and P. S. Bourgeron (eds), *Eastside Forest Ecosystem Health Assessment*, ii. *Ecosystem Management: Principles and Applications*. General Technical Report PNW-GTR-318. Portland, OR: USDA Forest Service, 80–94. Available online at https://www.fs.usda.gov/pnw/pubs/pnw_gtr318.pdf#page=88 (accessed May 15, 2023).

Swift, J. (1983). *Cut and Run*. Toronto: Between the Lines. 283 pp.

Thomas, J. W., Ruggiero, L. F., Mannan, R. W., Schoen, J. W., and Lancia, R. A. (1988). "Management and Conservation of Old-Growth Forests in the United States," *Wildlife Society Bulletin*, 16/3: 252–62.

Tilman, D., Clark, M., Williams, D. R., Kimmel, K., Polasky, S., and Packer, C. (2017). "Future Threats to Biodiversity and Pathways to their Prevention," *Nature*, 546/7656: 73.

Tittler, R., Messier, C., and Burton, P. J. (2001). "Hierarchical Forest Management Planning and Sustainable Forest Management in the Boreal Forest," *Forestry Chronicle*, 77/6: 998–1005.

Totman, C. D. (1989). *The Green Archipelago: Forestry in Preindustrial Japan*. Berkeley and Los Angeles: University of California Press. 297 pp.

Toumey, J. W. (1928). *Foundations of Silviculture upon an Ecological Basis*. New York: J. Wiley and Sons. 438 pp.

Triviño, M., Potterf, M., Tijerín, J., et al. (2023). "Enhancing Resilience of Boreal Forests through Management under Global Change: A Review," *Current Landscape Ecology Reports*, 1–16.

UNEP–WCMC and IUCN (2018). *The World Database on Protected Areas (WDPA)*. Cambridge: United Nations Environment Programme, World Conservation Monitoring Centre, and International Union for Conservation of Nature, Available online at https://www.protectedplanet.net/en/thematic-areas/wdpa?tab=WDPA (accessed May 13, 2023).

UNEP–WCMC and IUCN (2021). *Protected Planet Report 2020: Tracking Progress towards Global Targets for Protected and Conserved Areas*. Cambridge, UK: United Nations Environment Programme, World Conservation Monitoring Centre, and International Union for Conservation of Nature. Available online at https://livereport.protectedplanet.net/ (accessed May 6, 2023).

United Nations (2019). *The World Population Prospects 2019: Highlights*. New York: Population Division, Department of Economic and Social Affairs, United Nations. Available online at https://www.un.org/development/desa/en/news/population/world-population-prospects-2019.html (accessed May 13, 2023).

United Nations (n.d.). *Sustainable Development Goals: 17 Goals to Transform Our World*. New York: United Nations. Available online at https://www.un.org/sustainabledevelopment/ (accessed January 27, 2024).

US Congress (1960). "An Act to Authorize and Direct that the National Forests be Managed under Principles of Multiple Use and to Produce a Sustained Yield of Products and Services, and for Other Purposes," *Public Law 86-517*, 86th Congress of the United States of America, Washington. Available online at https://www.govinfo.gov/content/pkg/STATUTE-74/pdf/STATUTE-74-Pg215.pdf (accessed January 26, 2024).

van Hensbergen, B. (2016). "Forest Concessions—Past Present and Future?" Forestry Policy and Institutions Working Paper 36. Rome: Food and Agriculture Organization of the United Nations. 67 pp. Available online at https://www.fao.org/forestry/45024-0c63724580ace381a8f8104cf24a3cff3.pdf (accessed May 15, 2023).

Vasseur, L., Horning, D., Thornbush, M., et al. (2017). "Complex Problems and Unchallenged Solutions: Bringing Ecosystem Governance to the Forefront of the UN Sustainable Development Goals," *Ambio*, 46/7: 731–42.

Vogt, K. A., Gara, R. I., Honea, J. M., et al. (2007). "Historical Perceptions and Uses of Forests," in K. A. Vogt, D. J. Vogt, R. L. Edmonds, J. Honea, T. Patel-Weynand, R. Sigurdardottir, and M. G. Andreu (eds), *Forests and Society: Sustainability and Life Cycles of Forests in Human Landscapes*. Wallingford, UK: CAB International, 1–28.

von Gadow, K., Pukkala, T., and Tomé, M. (eds). (2012). *Sustainable Forest Management*. Dordrecht: Kluwer Academic Publishers. 355 pp.

Wagner, S., Nocentini, S., Huth, F., and Hoogstra-Klein, M. (2014). "Forest Management Approaches for Coping with the Uncertainty of Climate Change: Trade-Offs in Service Provisioning and Adaptability," *Ecology and Society*, 19/1: 32.

Walker, B., and Salt, D. (2006). *Resilience Thinking: Sustaining Ecosystems and People in a Changing World*. Washington: Island Press. 174 pp.

Wardle, J. (2016). *Woodside: A Small Forest Managed on Multiple Use Principles*. Wellington: New Zealand Farm Forestry Association. 132 pp.

WCED (World Commission on Environment and Development). (1987). *Our Common Future*. Oxford: Oxford University Press. 383 pp.

Wennersten, J. R., and Robbins, D. (2017). *Rising Tides: Climate Refugees in the Twenty-First Century*. Bloomington, IN: Indiana University Press. 259 pp.

White, A., and Martin, A. (2002). *Who Owns the World's Forests? Forest Tenure and Public Forests in Transition*. Washington: Forest Trends and Center for International Environmental Law. 30 pp. Available online at https://www.cifor.org/publications/pdf_files/reports/tenurereport_whoowns.pdf (accessed May 15, 2023).

Whiteman, A., Wickramasinghe, A., and Piña, L. (2015). "Global Trends in Forest Ownership, Public Income and Expenditure on Forestry and Forestry Employment," *Forest Ecology and Management*, 352: 99–108.

Williams, M. (2000). "Dark Ages and Dark Areas: Global Deforestation in the Deep Past," *Journal of Historical Geography*, 26/1: 28–46.

Wilson, E. O. (1984). *Biophilia: The Human Bond with Other Species*. Cambridge, MA: Harvard University Press. 176 pp.

Wilson, E. O. (2016). *Half-Earth: Our Planet's Fight for Life*. New York: W.W. Norton & Company. 259 pp.

CHAPTER 2

Forest Values and Zoning

You can please some of the people all of the time, you can please all of the people some of the time, but you can't please all of the people all of the time.
John Lydgate (fifteenth-century monk and poet), popularized by Abraham Lincoln

2.1 Forest values, goods, and services

Humans use and appreciate forests for a long list of reasons. These reasons have sometimes been divided into material (serving humankind's physical needs) and non-material categories. For example, the fuelwood harvested to heat our homes and the berries picked or deer hunted for food clearly represent utilitarian human uses of a forest. In contrast, a forest's value in regulating a clean supply of the water we drink or providing habitat for the fish we eat may be less obvious. Some members of society use forests for recreation, education, or spiritual inspiration, while others do not. Some people appreciate the importance of forests in storing carbon and providing habitat for plants and animals that they may never use or even see, although other individuals view such considerations as being of lesser importance. There are analogies with Maslow's (1943) hierarchy of human needs, with basic physiological requirements taking precedence over psychological needs, recreational pursuits, and aspirations for self-actualization. Such a ranking of priorities often emerges when landowners, forest managers, and community stakeholders discuss alternatives for forest management. Many people and policies prioritize those attributes and activities contributing to immediate livelihoods and economic activity—particularly the harvesting and processing of wood products—over those with less concrete value.

Scholars, researchers, and public-policy experts have been grappling with the complex issues of terminology, categorization, and quantification of forest values in recent years. For example, an alternative to the material/non-material distinction recognizes market and non-market values, use and non-use values, and consumptive and non-consumptive resource uses (Kornatowska and Sienkiewicz 2018). But markets evolve, including the development of carbon markets and biodiversity offset investments that did not exist a few decades ago (Koh et al. 2019; Cadman and Hales 2022). Personal, community, and societal values can and do change over time. The International Society for Ecological Economics (https://www.isecoeco.org/) and the journal *Ecological Economics* (https://www.sciencedirect.com/journal/ecological-economics/) have provided platforms for the valuation of nature and ecological processes since 1989. The Intergovernmental Science-Policy Platform on Biodiversity and Ecosystem Services (IPBES; see https://www.ipbes.net/) also hosts task forces and publications addressing these issues (Pascual et al. 2022; Bridgewater and Schmeller 2023). Discussion and debate continue regarding appropriate ways to characterize and measure the many complex and overlapping ways in which we value forests, biodiversity, and nature in general (Freeman et al. 2014; Champ et al. 2017; Dasgupta 2021). Nonetheless, most public forest lands are nominally managed for a combination of multiple material and non-material values. In addition, non-industrial private forest landowners in Europe (Bieling 2004; Hugosson and Ingermarson 2004; Häyrinen et al. 2014; Sotirov et al. 2019; Bashir

Resilient Forest Management. Philip J. Burton, Oxford University Press. © Philip J. Burton (2025). DOI: 10.1093/oso/9780198832997.003.0002

et al. 2020) and North America (Fischer et al. 2010; Bengston et al. 2011; Côté et al. 2015) are usually motivated by a wide variety of environmental, recreational, and other amenity or experiential values that can be more important than revenue from timber sales (see Figure 2.1).

Distinctions between material and non-material forest values or between tangible and intangible natural assets cannot be considered rigidly and may not always be the most important consideration in forest stewardship. For example, some people may hunt forest deer for food, perhaps out of necessity as an important source of household protein; but deer hunting is also done for sport, which is still a human-centered (anthropocentric) use of the forest habitat. Social sciences research supports the notion that recreational activities and taking time to appreciate the aesthetics of one's environment are important to human well-being. Some forest values have strong cultural affinities that can be difficult to explain or to generalize. The Japanese practice of *shinrin-yoku* (forest bathing or taking in the forest atmosphere) and the appreciation of *komorebi* (sunshine filtering through the trees) have garnered international attention only recently (Miyazaki 2018). Forest-based experiences, whether alone or in groups, and whether practiced actively or passively, have demonstrable benefits

to physical and mental health (e.g. Doimo et al. 2020; see Figure 2.2). Similarly, ecological research is teaching us that most species in an ecosystem are connected in grand webs of interacting relationships (Helfield and Naiman 2001; Simard et al. 2013; Gounand et al. 2018) or provide some form of functional redundancy, such that all aspects of biological diversity (biodiversity) appear to be important to the overall smooth functioning and integrity of ecosystems (Naeem 1998; Petchey and Gaston 2002; Filotas et al. 2014). In other words, non-consumptive forest use and non-material forest values are increasingly being acknowledged as important.

We can recognize considerable variation in prioritizing forest values according to demographic, geographic, and cultural differences that end up being strongly associated with political leanings (Ameztegui et al. 2018; Birch 2020). But perhaps the greater distinction is between anthropocentric (human-centered) and biocentric (life-centered) or ecocentric (nature-centered) value systems. Looking beyond the benefits to our own species, both now and in the future, biocentrists recognize the moral worth and rights of non-human organisms (animals in particular), while ecocentrists emphasize the inherent rights of species and ecosystems to persist (Attfield 2018). A recent review for IPBES

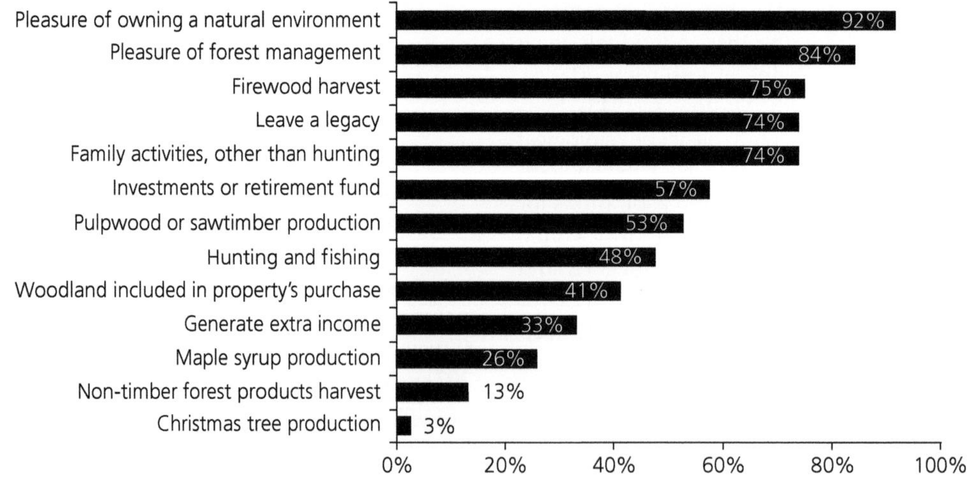

Figure 2.1 Motivations of forest owners in Quebec, Canada (Results are based on 2215 telephone interviews (from Côté et al. 2015), reproduced with permission of Elsevier).

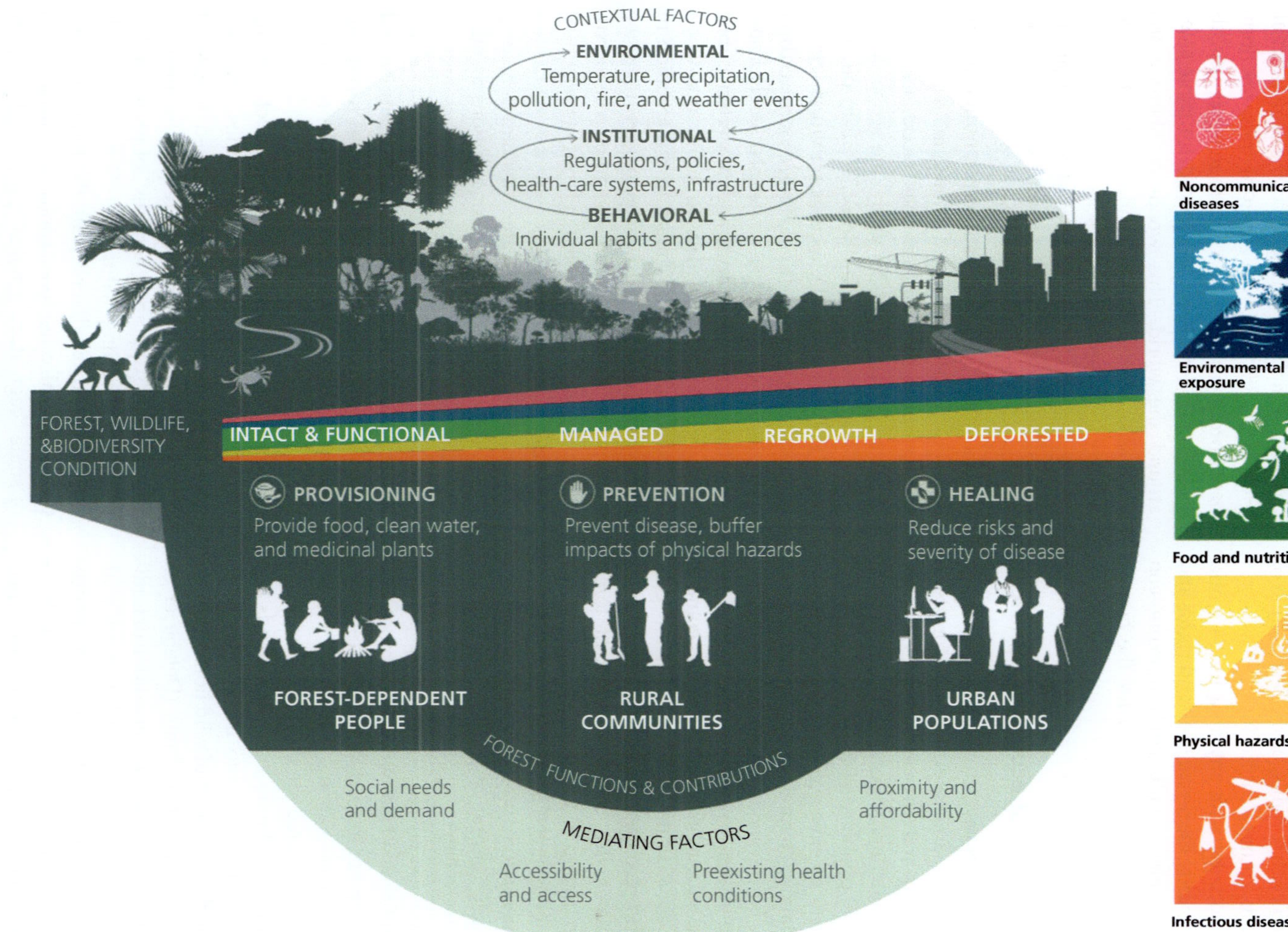

Figure 2.2 Forest and human health interactions and the factors that shape them, with an emphasis on provisioning, protection, and disease prevention services along the wildland-to-urban gradient (from Beatty et al. (2022), reproduced with permission of World Wildlife Federation US).

recognizes intrinsic (bio/ecocentric), instrumental (useful), and relational (cultural) values in natural entities of our environment (Himes et al. 2024). One might argue, however, that the mere assertion of biocentric or ecocentric rights, or the recognition of intrinsic value, is a human choice. It can thus be said that values associated with anything in the world around us, including the desire to extend rights and value to other species, are ultimately an expression of their importance or desirability for the human species.

2.2 Identifying forest ecosystem services

In recent years, efforts have increasingly been made to quantify the role and importance of forests in terms of the goods and services they provide. The foundational Millennium Ecosystem Assessment (Reid et al. 2005) popularized this work by categorizing ecosystem services according to their provisioning, regulating, cultural, and supporting roles (see Box 2.1). Forest goods (or "provisioning services") should be easy to enumerate—timber, fuelwood, berries, mushrooms, and bushmeat come to mind. These are material products that are generated by forest ecosystems and harvested from forests for human use. Even so, a list of forest goods is often limited to those with market value, those that are bought and sold on a local, regional, national, or international scale. For example, Huber (1992) lists cacti, decorative shrubs, mushrooms, herbs, wild seeds (such as pinenuts), Christmas trees, driftwood and novelty wood, and dry cones in addition to firewood, softwood sawtimber, and hardwood sawtimber in a valuation of forest products harvested from national forests of the US Southwest. Often ignored are the many traditional and subsistence uses of forests and woodlands that may not make their way into the market economy, including the gathering of wild honey (Hegde and Enters 2000), wild tree fruits (DeJene et al. 2020), and materials used in the preparation of traditional medicines (Furukawa et al. 2016). Lovrić et al. (2020) estimate that 26 percent of European households collect and utilize non-timber forest products (NTFPs), with a market value equivalent to 71 percent of the annual wood harvest, but with only 14 percent making its way into the market system.

Despite the many non-material contributions and benefits of the world's forests (see Table 2.1), their role in providing marketable wood products tends to dominate discussion of forest use and management. Emphasis is placed on the regeneration, growth, health, and harvesting of trees for fiber-based products at the core of management plans for most forests (those not designated for protection) around the world. This wood- or fiber-focused emphasis is reflected in standard definitions of forestry (for example, "a. the science of developing, caring for, or cultivating forests; b. the management of growing timber" (https://www.merriam-webster.com/dictionary/forestry)). Many professionals, researchers, and administrators addressing issues of forest science and management concern themselves solely with trees, logs, wood, and their value as commodities. Similarly, the scope of the *Forest Products Journal* (https://forestprod.org/page/FPJ), for example, focuses on wood science and technology and the marketing, economics, and business of wood-derived materials. Mechanisms for local community input and concerted public pressure are often needed to bring non-wood forest goods, uses, and values to the attention of forest policymakers and land managers. It can be challenging for non-wood issues to be considered, accommodated, or promoted in forest management policies, plans, and practices.

We can expect to see greater consideration being given to non-material forest contributions as markets are developed and economic values become associated with other forest benefits such as carbon and biodiversity (Koh et al. 2019, Cadman and Hales 2022). Nonetheless, non-material forest values can be difficult to identify and balance against the strong economic value of wood products. An ecosystem "service" is a role, often indirect, in doing something useful for the human species, while we are increasingly recognizing the need to respect diverse cultural contexts and respect for intrinsic values. The "supporting services" of photosynthesis, nutrient cycling, natural selection, and providing habitat for biodiversity are the foundation for provision of all other forest contributions. Examples of some "regulatory benefits" provided by forests include their role in carbon sequestration, oxygen generation, erosion control, and hydrological

Box 2.1 Forest ecosystem services, contributions, and benefits

It can be challenging to itemize and communicate the many socio-ecological roles that forests provide in our world (see Figure 2.3). A categorization into provisioning, regulating, cultural, and supportive services has become widely adopted and would seem to accommodate many of the contributions of forests to human well-being. However, many of these "services" are actually ecological processes rather than the final services or benefits ultimately desired by people. As such they require additional inputs of labor or capital, or are not compatible with established systems of welfare accounting (Boyd and Banzhaf 2007). All perceptions of human well-being and its contributing factors are strongly conditioned by cultural, socioeconomic, and spatiotemporal context. Consequently, the 2017 IPBES conceptual framework recognized eighteen categories of nature's contributions to people (NCPs) in three rather than four partially overlapping groups, all of which should be viewed through a cultural lens that incorporates Indigenous and local knowledge (Díaz et al. 2018). The complex 2022 IPBES methodological assessment took a more sociological approach, reclassifying NCP groups as instrumental (nature as a resource, useful to people), relational (providing meaning to human–nature interactions), and intrinsic values (free of people as valuers). These are further embedded within broad guiding principles and life goals (related to livelihood and prosperity, health and belonging, stewardship responsibility, and harmony with nature) that might further reflect different worldviews and knowledge systems (Pascual et al. 2022). Table 2.1 uses a combination of these frameworks to provide some examples of forest NCPs. Despite a broad concensus on the importance of the ecosystem services concept, its use in resource planning and environmental assessment continues to remain limited (Daily et al. 2009; Pascual et al. 2022; Bridgewater and Schmeller 2023).

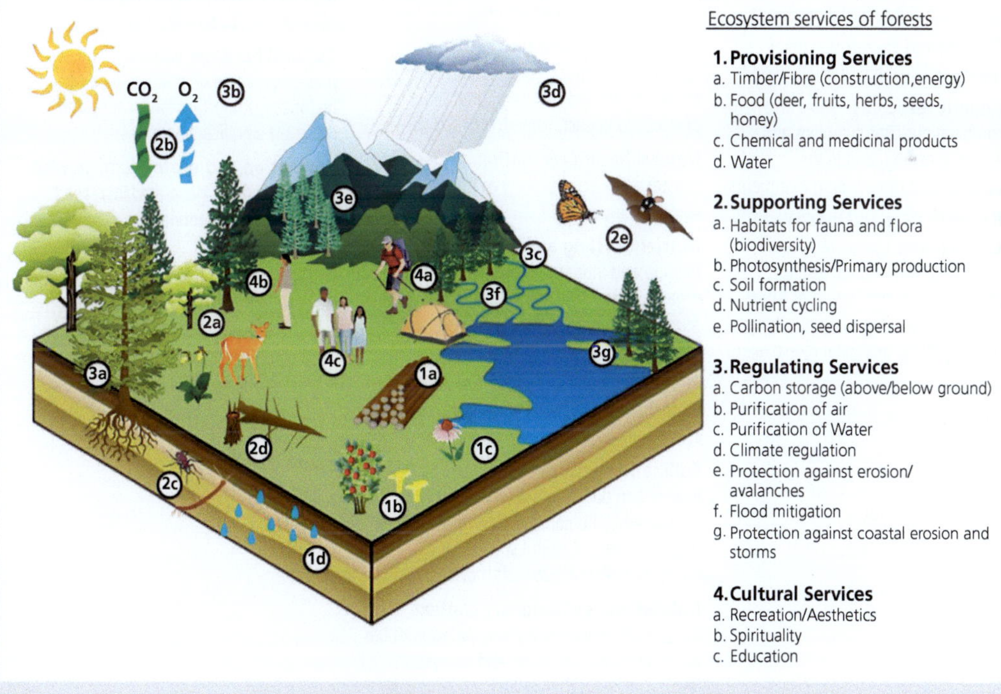

Figure 2.3 Forest ecosystem services, following the Millennium Ecosystem Assessment framework (graphic from Holzwarth et al. (2020), reproduced under the CC BY license, http://creativecommons.org/licenses/by/4.0/).

Box 2.1 *Continued*

Table 2.1 Examples of some forest ecosystem services, benefits, and values provided by forests

Provisioning, material, or instrumental benefits	Regulating or supporting benefits and contributions	Non-material, relational, and cultural contributions
Food[*]: fruits, nuts, berries, mushrooms, honey, wild forest birds and mammals hunted or trapped for meat	**Photosynthesis and primary production:** capture of solar energy into the vegetation that makes up the framework of forest ecosystems	**Cultural diversity and social relations:** many distinctive cultural beliefs and practices are intimately associated with the forests in which communities are embedded
Fiber: timber harvested for solid and engineered wood products, pulp, and paper	**Air-quality regulation:** absorption or amelioration of atmospheric O_3, NO_x, SO_2, CH_4, NH_3, and particulates	**Spiritual and religious values:** large trees, sacred groves, and quiet forests are revered for contemplation and meditation
Fuel: roundwood and fiber wastewood burned directly or pelletized or converted to charcoal for cooking, heating, or electricity generation	**Climate regulation:** CO_2 sequestration and albedo effects on air temperature, evapotranspiration effects on atmospheric moisture and precipitation	**Knowledge systems and educational values:** traditional, formal, and novel knowledge is associated with different forest types, products, and processes
Biochemicals and pharmaceuticals[*]: organic compounds (e.g. methanol, rayon) derived from cellulose and lignin; resins, bark, roots, and other plant parts used in traditional medicines	**Water cycling and purification:** interception and retention of precipitation, evapotranspiration, retention of snow and fog, gradual percolation and filtration of water through organic and mineral soil horizons, moderated runoff, aquifer recharge, and reduced flooding	**Inspiration and aesthetic values:** stimulation of art, folklore, architecture, symbols and symbolism, scenic beauty
Ornamental resources: floral greens, wildflowers, specialty woods	**Soil formation and protection:** weathering of rocks, contributions of organic matter above and below ground, reduction of raindrop impact, retention of sediments	**Sense of place:** distinctive aspects of local forests (species, structure, disturbance regime) contribute to identity and commitment to stewardship
Genetic resources: genes and gene combinations useful in plant and animal breeding, and in the development of new drug treatments	**Natural hazard regulation:** moderation of coastal wave action, snow avalanche, and landslide impacts	**Cultural heritage values:** many societies place high value on maintenance of historically important landscapes and culturally significant species
Remunerative and meaningful employment[*]: jobs and income associated with the harvesting, processing, or distribution of forest materials, and in guiding, teaching, or interpreting relational forest values	**Nutrient cycling and retention:** nitrogen fixation and inorganic mineral uptake, organic matter decomposition, and nutrient release; incorporation of nutrients in biomass before they are washed away	**Recreation and ecotourism:** growing urbanization has generated increased demand for recreational use of forests and natural ecosystems
	Pest and disease regulation: habitat for birds that help control insect pests; staggered stand compositions can minimize contagion	**Recognition of intrinsic value** of other species and the right of natural ecosystems to exist is important to many people and is central to many cultures
	Pollination and seed dispersal: forest insects and bats pollinate many fruits consumed by humans; birds and mammals disperse seeds of useful species, and facilitate forest recovery after disturbance	
	Habitat and evolutionary context[*] for biodiversity, the diversity and variation of life at genetic, species, guild, and ecosystem levels, providing the variability essential for ecological resilience and evolutionary potential	

[*] these benefits have strong relational and cultural identity roles too.
Note: Many forest ecosystem services support some combination of provisioning, regulating, or relational roles.
Source: Modified from Reid et al. (2005) and Díaz et al. (2018).

regulation. But what is often most important to people are the less tangible "relational values" or "cultural services" provided by forests as the setting for education, recreation, and inspiration (see Table 2.1). These cultural services have sometimes been referred to as "amenity values," with some forest reserves designated as amenity forests (Edlin 1963). Another cultural service that is difficult to document (or even to articulate) is the role a forest—with its full complement of plant, animal, and fungal species, its structural diversity, and its characteristic heterogeneity with site—in contributing to a "sense of place" for people who have lived and worked in and around that forest (Williams and Stewart 1998: Stedman 2003). Special features and locations embedded in a forest landscape, or even the intact forest itself, frequently rank as very important to the history and cultural identity of a community, and hence warrant recognition, protection, and maintenance (Boyd and Banzhaf 2007; Bulkan 2017).

Ecosystem services can have quite different values depending on the spatiotemporal scale of assessment: provisioning services tend to be ranked higher at local and short-term scales, while the benefits of regulatory and supporting services are primarily experienced over the long term at regional or global scales (Rodriguez et al. 2006; Zhang et al. 2007). Objective assessments of ecosystem services, and particularly the trade-offs entailed in promoting some over others, must therefore consider short- and long-term impacts, over multiple spatial scales, with attention to who benefits and who suffers from the provision or withdrawal of forest ecosystem services.

The concept of forest goods and services is not without its weaknesses. Most fundamentally, it risks equating a forest to a factory or a consultancy that functions as a sector of the economy, generating something of marketable value. That leads to another concern—namely, that it is a logical next step to monetize those goods and services, often based on their equivalence or substitution in the marketplace (e.g. Adger et al. 1995; Krieger 2001). Some environmental advocates and ecological economists have argued that we need to demonstrate the economic value provided by ecosystem services in order to raise public awareness and to

highlight trade-offs in land management options in a world where politics and business are driven by financial costs and benefits (Costanza et al. 2014; Mavsar and Varela 2014). But such efforts are subject to various errors of double-counting, sampling and measurement error, and overgeneralization (Boyd and Banzhaf 2007; Schmidt et al. 2016; Díaz et al. 2018). The entire exercise of monetizing the value of beings and ecosystems having intrinsic value is anathema to the culture and world views of many people (Lele et al. 2013). The estimation of monetary value often involves the weighing of market substitutions for ecosystem components and engineered alternatives to ecosystem components and processes that an ecocentric worldview would consider irreplaceable. For example, one could substitute the retail price of farmed beef for the equivalent weight of harvested wild moose meat to derive a dollar value for the forest-generated moose meat (Natcher et al. 2021). All such substitutions are imperfect, whether assessed nutritionally, ecologically, or experientially. Beef-equivalence is an incomplete portrayal of the value of a moose, which should include its food web roles as consumer and prey in the boreal forest ecosystem, and also its cultural or recreational value to a hunter. Monetary value is but one kind of value, needs to be viewed in context, and should not be the ultimate basis for decisions.

There are many different ways in which forest values can be inferred and measured (Freeman et al. 2014; Champ et al. 2017; Pascual et al. 2022). These can be divided into survey-based stated preference methods (consisting of "choice experiments" or "contingent valuation" based on willingness to pay), revealed preference methods (including travel cost assessments, hedonic pricing, and other cost-based methods), and benefit transfer methods (Kornatowska and Sienkiewicz 2018). Stated preference methods facilitate evaluation of biophysical and cultural forest attributes, or alternative forest management options, but are not founded on observed behavior and thus may be incorrect or biased (Barrio and Loureiro 2010; Kornatowska and Sienkiewicz 2018). Revealed travel costs to access forest-based recreational opportunities and real-estate values or happiness surveys (hedonic pricing) are well-established approaches for inferring the value of natural environments. Sometimes the value of a service provided by a forest is simply a matter of cost effectiveness, as in the case of watershed protection (Hasler et al. 2007). For example, agreements to maintain forest cover have permitted US cities such as New York, Boston, Seattle, and Portland to avoid huge costs in constructing water filtration and treatment plants (Postel and Thompson 2005). Hence another concern emerges: while such substitutions may be expensive and undesirable, the process of monetization puts engineered alternatives "on the table" as an option. Some investors and industries with a strong foundation in engineering (for example, in high-stakes sectors such as mining and petroleum development) may be willing to underwrite those substitution costs so as to exploit the resources associated with forest lands.

Substitutability, contingent valuation, and travel costs are only a few of the tools available for economic valuation, a field with many creative alternatives but little uptake in the orthodox realms of business and government (Thorsen et al. 2014; Dasgupta 2021). It is important to recognize that forest goods are generated and the functions that provide forest services are conducted with or without human utilization or appreciation, and indeed have done so before the appearance of the human species. Calling those ecosystem components and processes "goods and services," or even "forest values," is an inherently anthropocentric position, one at odds with an ecocentric perspective on the importance and value of forests. Combined with many other weaknesses, such as the widespread neglect of negative and marginal benefits and the maintenance of option value (Lele et al. 2013), any effort at the quantitative tabulation of forest benefits is bound to be incomplete. The arbitrary discounting of future value—despite the growing rarity of primary forests, wilderness, and many of the world's species—is especially at odds with the concept of intergenerational equity (Padilla 2002). An interesting alternative is to grant explicit legal (personhood) rights to natural features and ecological entities, as done for the Whanganui River in New Zealand in 2014 (Chapron et al. 2019), an argument that had already been advanced for trees and forests (Stone 2010). Nonetheless, the enumeration of ecosystem goods and services is well entrenched in the recent forest management literature, so use of such terms continues in the text that follows.

2.3 Some psychosocial considerations

In his seminal call for environmental responsibility, Aldo Leopold (1949) identified an important difference between those people who are driven by anthropocentric utilitarian values (Group A) and those who hold more ecocentric values (Group B). But perhaps just as important as this A–B cleavage or schism is what we might call a Y–Z schism between those people primarily concerned with immediate benefits (Group Y, whether focusing on resource use, profit, or gratification) and those people subscribing to the big picture of multigenerational equity, sustainability, and long-term benefits (Group Z). This view is supported by an analysis of non-industrial forest landowner motivations in Finland as being most strongly aligned according to current or future timber values and current or future experiential values (Häyrinen et al. 2014). While such categorical distinctions may not be justified, considered together, the A–B and Y–Z gradients can be used to describe different degrees of resource conservation and sustainability (see Figure 2.4). We are all "A" people in that we need the (material) basics of survival, and we are all "Y" people in that measures promoting both "A" and "B" values must be feasibly undertaken in the here and now, not just the future. We have the ability to function as "Z" people by projecting and planning for future conditions beyond that of a human lifetime, yet rarely do so. We might plan for well-being of our children or grandchildren, yet most of us do not think in terms of seven generations, as advocated in some eastern North American Indigenous traditions (Jojola 2013). As further developed throughout this book, the key in bridging to that long-term future is about keeping options open and managing for risk and resilience, not just optimizing selected values or managing primarily for those ecosystem attributes currently useful to us.

Many other institutional, demographic, cultural, and personality differences can also be expected to influence how managers and stakeholders perceive and value forests and their preferences for alternative forest management options. Decisions made in

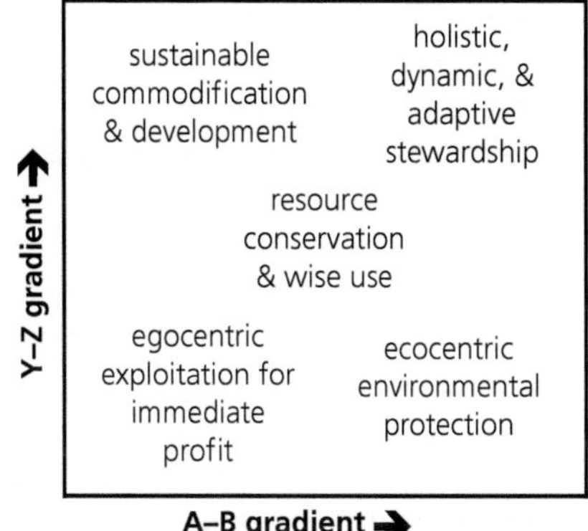

Figure 2.4 Implications of Leopold's (1949) A–B schism interpreted in conjunction with a proposed Y–Z schism with regard to forest values and their management. The A–B gradient from pure utilitarianism to ecocentrism is probably correlated with altruism and collectivism too, while the Y–Z gradient of short-term to multigenerational thinking is probably correlated with more holistic than linear thinking, and with greater appreciation of uncertainty and change rather than constancy.

large hierarchical organizations, whether corporations or government agencies, tend to be more conservative, rigid, and prescriptive in their approach to forestry (Poffenberger 1990; Kumar et al. 2007). Consequently, such organizations are less responsive to novel challenges and changing norms than smaller institutions such as community forests and small-scale consultancies. People around the world may variously value, revere, or even fear forests for a wide array of different reasons that reflect cultural norms and expectations. For example, stakeholders with roots in central Europe may have considered proper twentieth-century forestry to consist of uniform stands of spruce that are regularly thinned and pruned, from which dead wood is salvaged for firewood, and from which the primary non-timber value may be opportunities for hunting and for gathering edible mushrooms. In contrast, if one's perception of forest management is based on private woodlots in North America's New England or Quebec, the expected norm may be to promote mixtures of broadleaf trees and conifers to produce maple syrup, provide habitat for deer and grouse, and support the occasional harvest of individual high-value hardwood trees. Bring people with those backgrounds together in a new forest management context (for example, Brazil, Australia, or the US Pacific Northwest), and one can expect quite different suggestions on how a forest estate or public forest land should be appraised and managed.

Demographic and personality traits are also known to influence preferences for nature protection or resource development. Carey (1997) detected differences among engineers, district rangers, biologists, and executives with respect to detailed sequential thinking versus feeling-based integrative thinking, with implications to practicing forestry as a primarily analytical problem or an exercise in social negotiation and compromise. Such differences may reflect underlying personality differences, educational history, or their interaction in career selection and training. Similarly, conservative mindsets, or a fear of losing authority and prestige, can explain the resistance of some forest administrators to engage in community-based forest management (Kumar et al. 2007). Demographic factors (such as age, gender, education, and income) along with Meyers–Briggs Type Indicators (MBTI) of personality explained leisure travel preferences, including the value placed on outdoor recreation options, in Australia (McGuiggan and Foo 2002). Significant differences in values among motorized and non-motorized off-highway recreationists led to the practical suggestion that those activities should be kept separate on the land base (Kil et al. 2012). Simply on the basis of aesthetics, some people prefer to see forest harvesting cutblocks containing patches of uncut trees, curved boundaries, and scalloped edges emulating natural disturbances (Perera et al. 2004; Long 2009) or following principles of landscape architecture (Bell and Apostol 2007; Penn State Extension 2008). Other people consider such patterns "messy" and prefer to see uniform stands of trees with straight boundaries. Such differences are widespread, and reflect cultural norms and expectations as generated through personal history, perhaps entrained according to whether the land was experienced primarily through work (for example, on farms or in resource development) or through recreation (for example, hiking and hunting).

There are many more dimensions to personality, life stories, and cultural norms, in combination with levels of education and income, that may be just as important to ecosystem valuation and forest management preferences as the A–B or Y–Z gradients described above. It is tempting to summarize such value preferences according to the political labels of being more conservative or liberal, more right or left. But humans are much more complex than that, with different commitments to group identity, levels of risk tolerance, and degrees of rational and emotional decision-making found across the political spectrum. The research apparently does not yet exist to demonstrate how these many factors affect decision-making regarding nature conservation and the management of natural resources. However, there is a growing interest in risk perception among land managers (for example, farmers (Caveness and Kurtz 1993)), particularly with respect to climate change adaptation (e.g. Duinen et al. 2015; Eitzinger et al. 2018). One could envision a battery of socioeconomic questions and psychological tests when filling a forest manager position (as sometimes promoted in human resource management in general (e.g. Carless 2009)), or in the selection of a well-balanced representation of stakeholder values on a public advisory committee or a community forest board (see Section 2.7).

2.4 Value conflicts and trade-offs

It is physically impossible for all forest values to be maximized on every hectare of forest land. Such constraints reflect not only the inherent limitations of climate, terrain, and soils, but also the multiple competing expectations of society with respect to

priorities in forest ecosystem services and their protection or production. While landowners and managers may wish to allocate land according to its "highest and best use," this policy can easily become a market-driven or political exercise that does not prioritize long-term ecological or public well-being (Mundy and Kinnard 1998; Franklin et al. 2018). The problem of selecting and promoting a single resource use is accentuated when the same piece of land is optimal for different but largely incompatible forest values. This dilemma often applies to land with capacity for a high level of biological productivity, making it desirable for agriculture, for short-rotation forestry, and for biodiversity conservation. Furthermore, much private forest land has greater immediate value to investors for real-estate development than for its long-term ecosystem services, often resulting in the liquidation or degradation of ecological value and its withdrawal from regional timber supplies (Drummond and Loveland 2010). Although some authors have emphasized the opportunity for the "simultaneous production" of multiple forest values at the site or stand level (Dana and Fairfax 1980), implementation of multiple-use policies has been largely achieved in the form of "segregated dominant uses" in response to current economic drivers or single-interest stakeholders. In the national forests of the western US, for example, this has meant domination by cattle and timber interests (Blumm 1994), just as large portions of the productive and accessible public forest lands in Canada have been allocated for production of pulp and sawlogs (Burton et al. 2003).

The relative compatibility of different land uses or simultaneous resource uses on the same area of land (as characterized by its biophysical characteristics) can be illustrated using compatibility matrices (McHarg 1969). Table 2.2 provides such an illustration, based on a general worldwide assessment developed for the Food and Agricultural Organization of the United Nations. With so many "potentially compatible" combinations of forest use identified in the original (Kengen 1997) assessment, two of those forest-based activities have been disaggregated in Table 2.2. Separate assessments for motorized or non-motorized recreation, and intensive (industrial) or extensive (ecological) forestry,

Table 2.2 A summarization of relative compatibility of different forest land uses (forest values or forest-based activities), with timber and commercial logging disaggregated into intensive and extensive forestry, and recreation disaggregated into motorized and non-motorized activities

	Biological conservation	Hunting	Soil and water protection	Non-motorized recreation	Motorized recreation	Non-wood forest products	Subsistence and small-scale fuelwood harvesting	Extensive, ecological forest management
Intensive, industrial fiber management	X	✓	≈	X	≈	X	✓	✓
Extensive, ecological forest management	≈	≈	≈	≈	≈	✓	✓	
Subsistence and small-scale fuelwood harvesting	X	≈	≈	≈	≈	✓		
Non-wood forest products	X	≈	≈	≈	✓			
Motorized recreation	X	≈	≈	X				
Non-motorized recreation	✓	X	✓					
Soil and watershed protection	✓	✓						
Hunting	X							

✓ generally compatible or complementary; **X** generally incompatible; ≈ potentially compatible.
Source: Expanded from Kengen (1997).

Box 2.2 An evaluation of ecosystem service trade-offs based on forest stand composition

Trade-offs in the ability of forests to deliver desired ecosystem services can extend to the relative balance of different silvicultural systems in a forest estate, or even the relative balance of different tree species in a stand. Himes et al. (2020) evaluated the ecosystem services associated with different combinations of red alder (*Alnus rubra*), Douglas-fir (*Pseudotsuga menziesii*), and western hemlock (*Tsuga heterophylla*) in 35–39-year-old industrially managed plantations in northwest Oregon and southwest Washington, US. Indicators of key ecosystem services (see Figure 2.3) were selected, for which:

- (1) merchantable wood volume and (2) timber revenue denoted production services;
- (3) carbon stocks and (4) pollinator-supporting understory composition and abundance represented regulator services;
- (5) scenic beauty represented cultural services;
- (6) climate change resistance and (7) fire-resprouting understory vegetation denoted supporting services;
- (8) non-timber forest products and (9) forage for herbivores indicated some combination of provisioning, cultural, and supporting services.

Their evaluation generated unique response surfaces for each of the nine indicators for all combinations (in 10% increments of maximum values) of the three tree species. Although different combinations of tree species affected optimal delivery of each ecosystem service differently, those patterns could be generally grouped into those that favor overstory tree growth or rich understory development, particularly along the gradient of high representation by western hemlock or by red alder (see Figure 2.5).

Evaluating all nine response surfaces then permitted construction of an ecosystem services compatibility matrix (analogous to Table 2.2), based on a quantitative and repeatable approach (see Table 2.3). Himes et al. (2020) concluded that monocultures of one tree species or the other could optimize most individual ecosystem services. Scenic beauty was an exception, for which surveys and subsequent analysis revealed a preference for 47% red alder, 31% western hemlock, and 22% Douglas-fir. Mixtures of all three species do the best job of sustaining delivery of all nine ecosystem services. Scenic beauty, edible/medicinal/decorative plants, and herbivore forage support local conservation values, while carbon stocks, climate-change resistant understories, and fire-resprouting understories facilitate forest resilience, persistence, and future options.

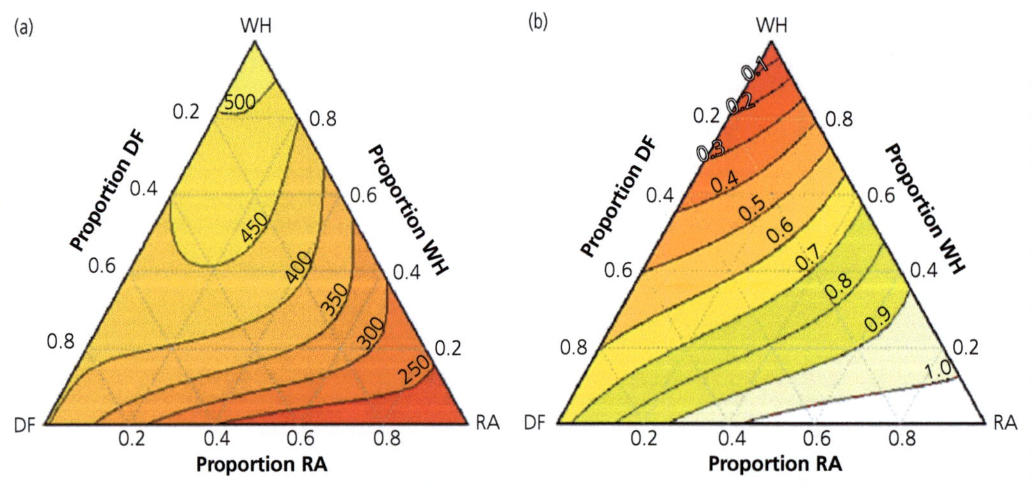

Figure 2.5 Response surfaces for two representative indicators of ecosystem services evaluated for gradients of managed second-growth forest composition in the US Pacific Northwest: (a) Merchantable timber volume production, in m^3/ha; (b) edible, medicinal, and decorative plant cover, in m^2/m^2. Axes denote the relative abundance of Douglas-fir (DF), western hemlock (WH), and red alder (RA) (from Himes et al. (2020), reproduced under the CC BY license, http://creativecommons.org/licenses/by/4.0/).

Box 2.2 *Continued*

Table 2.3 A synthesis of interactions between individual ecosystem service proxies evaluated for managed second-growth forest in the US Pacific Northwest

	Pollinator-supporting understory	Fire-resprouting understory	Climate-change resistant understory	Carbon stock	Timber revenue	Volume of merchantable wood	Herbivore forage	Edible, medicinal, and decorative plants
Scenic beauty	≈	≈	≈	≈	≈	≈	≈	≈
Edible, medicinal, and decorative plants	≈	✓	≈	≈	X	X	✓	
Herbivore forage	≈	✓	≈	≈	X	X		
Volume of merchantable wood	≈	X	≈	✓	✓			
Timber revenue	≈	X	≈	✓				
Carbon stock	X	≈	X					
Climate-change resistant understory	✓	≈						
Fire-resprouting understory	≈							

✓ positive interactions;
≈ inconsistent interactions;
X negative interactions.
Source: Himes et al. (2020).

reveal some stronger trends in compatibility or incompatibility. Typically based on expert opinion or largely biophysical considerations (see Box 2.2 for a case study), the list of values to be considered and their degree of compatibility are increasingly determined through public engagement and stakeholder workshops. For example, the step-by-step framework of structured decision-making guided by an independent facilitator can promote collaborative discussion and consensus (Johansson et al. 2018). In general, the objective of such exercises is to identify those values or ecosystem services that can or cannot be provided with acceptable levels of conflict or compromise. Compatible uses can then be bundled in one or more multipurpose zones, while leaving the option for some singularly incompatible activities to be zoned separately. In this manner, zoning—a basic instrument of urban and regional planning (Cullingworth 2017)—can be applied in a hierarchical or nested manner to help resolve the human uses and expectations of a forested land base.

2.5 Zoning as a solution

Conceptually, one could have separate land-use zones for the protection or promotion of each and every forest value. However, this would exclude large areas of land from being able to support production, conservation, and aesthetic values that could be largely compatible with each other.

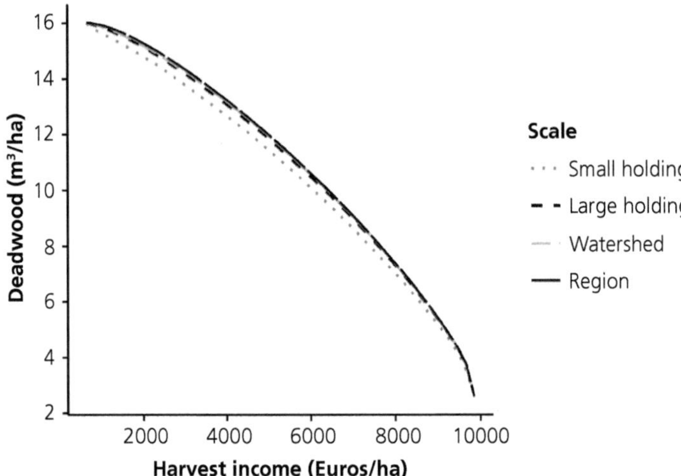

Figure 2.6 A robust (largely scale-independent) production possibility frontier (PPF) documenting the trade-off between the provisioning of deadwood (a key biodiversity resource and indicator) and income from timber sales (from Pohjanmies et al. (2019), reproduced under the CC BY license, http://creativecommons.org/licenses/by/4.0/).

Therefore, it is reasonable to bundle forest values wherever the trade-offs are negligible (or, at least, acceptable). Sophisticated analytical methods, such as monetary valuation, optimization programming (including multi-objective optimization, and empirical determination of the trade-offs associated with production possibility frontiers (see Figure 2.6)) can assist in deciding on some of those combinations, where sufficiently robust data exist to do so in a quantitative or semi-quantitative manner (Chen et al. 2016; Kaya et al. 2016). Operations research can identify theoretically optimal solutions, but it is more usual for the allocation of land uses and acceptable land-use mixtures to be based on subjective qualitative preferences and negotiated compromises among multiple interests or stakeholders. Such approaches are increasingly supported by a geographic information system (GIS) and land transformation visualization support (Sheppard and Meitner 2005; Zhang et al. 2013). However, negotiated compromises are vulnerable to the "shifting baseline syndrome," in which the extent of existing ecological degradation tends to be underestimated (Soga and Gaston 2018; Jones et al. 2020), thereby biasing negotiations in favor of further development.

Some "resource emphasis zones" designated within various public forest lands around the world have included, whether through regulation or *de facto* practice, the following categories:

- timber production zones (e.g. Jemali et al. 2015);
- areas seeded or designated for fodder production in support of livestock (Mitchell et al. 2005);
- campgrounds, roads, footpaths, and related "front-country" amenities in support of recreation and tourism (see Section 11.2);
- motorized or non-motorized recreation areas (Miller et al. 2017);
- general wildlife habitat areas for a wide array of vertebrate species (e.g. Corace et al. 2012);
- species-specific wildlife habitat sanctuaries, wintering, or denning areas (e.g. for Bengal tigers, *Panthera tigris tigris*, Khan et al. 2008; for monarch butterflies, *Danaus plexippus*, Tucker 2004; also see Section 11.3, a case study giving high priority to habitat for red-cockaded woodpecker, *Leuconotopicus borealis*);
- riparian reserve zones designed to protect aquatic values, including fish production (e.g. Young 2000);
- watershed reserves managed for reliable production of clean water for municipal or domestic use

(Postel and Thompson 2005);
- fire protection zones around forest communities (Schoennagel et al. 2009);
- scenic areas and viewpoints (e.g. Lee and Park 2012);
- old-growth forest reserves (e.g. Nagel et al. 2012);
- broadly defined nature reserves (e.g. Howard et al. 2000; Ogurtsov et al. 2018).

Most national parks and large state or provincial parks are zoned to separate high tourist traffic recreation areas ("front country") from wilderness ("back country") areas and strict nature reserves (see Section 11.2). A widespread recent trend allows Indigenous peoples to practice traditional (non-industrial) resource use and land management in these protected areas (Primack 2014; Deur and James 2020). Many state and municipal parks are often mapped as protected areas and "green space" but vary widely in the degree to which they are characterized by natural ecosystems or have been heavily modified to facilitate recreational uses in the form of sports fields, golf courses, and ski lifts. Private landholdings too can vary in the degree to which they are managed to promote commodity production, nature protection, and multiple-use philosophies.

As illustrated in Table 2.2, the forest land-use categories that tend to be most incompatible with other uses are biological conservation and industrial timber production. Other empirical examples, encompassing a broader scope of land uses, have identified activities such as mining or quarrying and residential or cottage development as having the narrowest set of compatible or potentially compatible activities (McHarg 1969; Harvey et al. 2003). Those land uses having low compatibility with other land uses are strong candidates for exclusive zoning. Despite the wide variety of special-use zones that can be recognized in forests, it is useful to recognize those that prohibit or greatly limit the removal of trees in a general "protection" category. This category typically gives priority to uncompromised primary (often old-growth) forest, watershed, biodiversity, and wildlife values, which may also accommodate non-motorized recreation. Conversely, we might recognize the timber lands designated for efficient, even-aged fiber production as "intensively managed" or "timber" lands, where

biodiversity and other ecological concerns need not unduly compromise the growing and harvesting of wood. But most other resource emphasis zones and land-use objectives can be accommodated with some level of active forest management coupled with varying degrees of environmental protection. This broad middle ground in which forests are managed according to ecological principles can be described as "multipurpose," "extensively managed," or "ecosystem-based" management zones.

Those three broad categories—of lands designated for protection, for timber production, and for ecosystem-based multiple uses—can be described as the "triad template" for forest land-use zoning (Seymour and Hunter 1992). This model (see Box 2.3) can be considered a solution to the "land sharing versus land sparing" (wildlife-friendly management versus zoned separation of conservation and production lands) debate that has largely focused on food production and agricultural management (Matson and Vitousek 2006, Fischer et al. 2014, Loconto et al. 2020). However, the triad proposal avoids the "either/or" dichotomy of integration versus zoned intensified production coupled with nature reserves by including integrated resource management in addition to the dedicated zones. Furthermore, by incorporating ecosystem-based management in the land between protected areas, the triad approach reflects the mantra that "the matrix matters" for biodiversity conservation (Ricketts 2001; Lindenmayer and Franklin 2002) by minimizing abrupt edge effects and long-lasting fragmentation in the landscape. That broad matrix characterized by ecosystem-based integrated resource management also retains a flexibility for greater direction of land to protection or production uses in the future, if warranted (Burton 1995). In this manner, a triad landscape is also compatible with a UNESCO Biosphere Reserve model, in which core protected areas are spatially buffered from most human activities and land-use changes by transitional zones of less disruptive use and management (Robertson Vernhes 1989; Ishwaran et al. 2008). As a general classification of forest land-use priorities, examples of forest management presented in some of the following chapters are variously applicable to intensively managed, integrated multipurpose, or protected forests.

Box 2.3 The triad model of forest land-use zoning and spatially identified resource management priorities

Figure 2.7 illustrates a hypothetical example for allocating a triad of forest land uses across a large landscape or region. In this scenario, the multiple use (or integrated resource management, extensive forestry, or ecosystem-based management) zone is further divided into special management areas and commodity emphasis areas in addition to the prevailing designation for standard practices. Note that protected areas have high connectivity and are largely buffered from intensive use areas by special management or other multiple-use zones. Photographic examples illustrate (a) (upper right), a protected area (in this case dominated by old-growth forest), (b) (center right), a forest landscape managed by principles of ecosystem management, and (c) (bottom right), an intensively managed, short-rotation hybrid poplar plantation. General considerations that typically guide triad zone management are summarized in Table 2.4.

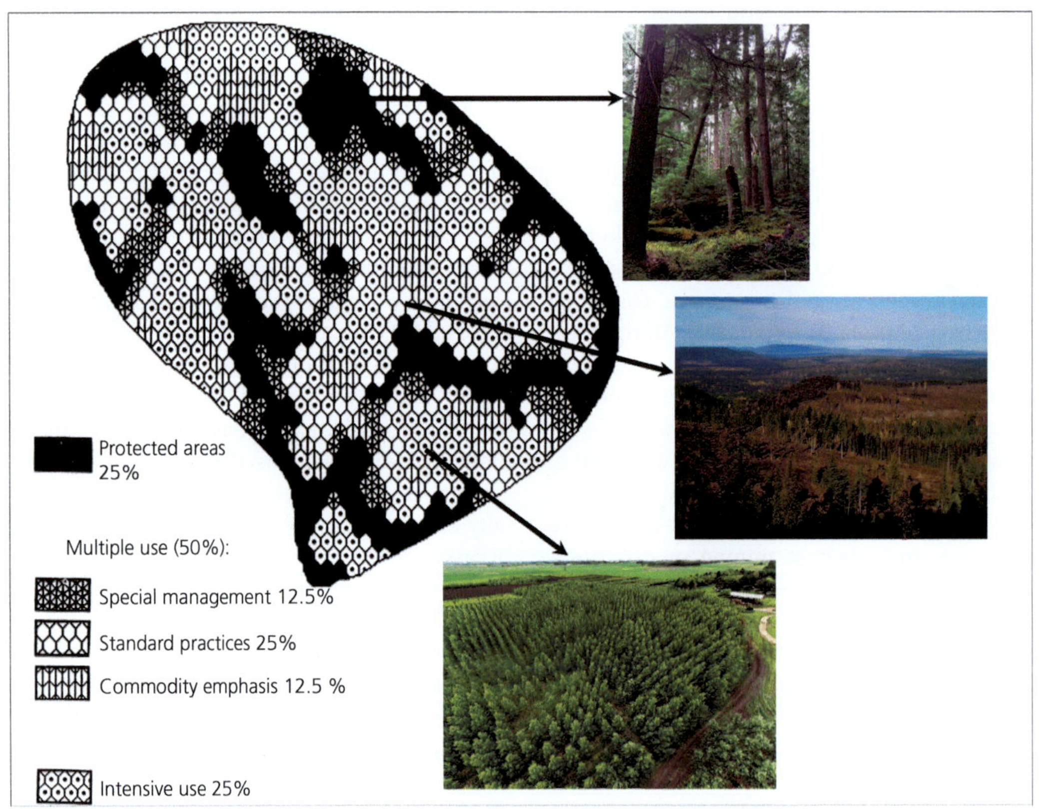

Figure 2.7 A schematic representation of triad zoning in a hypothetical landscape (from Burton (1995)). Lower-right photograph of 6-year-old plantation of hybrid poplar by Canadian Forest Service, Natural Resources Canada, is reproduced with the permission of the Department of Natural Resources, 2023.

Box 2.3 *Continued*

Table 2.4 Characteristics of the three broad triad categories of forest management emphasis

Attribute	Production	Integrated multipurpose	Conservation
Emphasis	Harvestable timber or fiber	Accommodating as many resource demands as possible	Protection of ecological values
Natural disturbances (e.g. fire)	Vigorously controlled	Variably controlled, especially near infrastructure	Usually not controlled, except near infrastructure
Fiber production	Priority	Included	Not permitted
Fuelwood gathering	Excluded, accommodated (at harvest), or controlled*	Included	Largely excluded, or limited amounts by staff***
Non-timber forest products	Accommodated*	Included	Usually excluded
Watershed protection	Provided**	Included	Priority or provided
Hunting	Accommodated*	Included	Usually excluded
Biodiversity	Accommodated*	Included	Priority
Recreation (hiking, camping)	Accommodated*	Included	Included***
Scenic beauty	Ignored	Included	Priority
Livestock grazing	Excluded	Usually included	Excluded
Quarrying, mining, and energy infrastructure	Accommodated*	Usually included	Excluded

* Provided that fiber production is not significantly compromised.
** Most of the rotation, though often not immediately after harvest.
*** Provided that conservation/biodiversity values are not significantly compromised.

The triad model of forest land-use zoning has received considerable attention and evaluation in recent decades (Betts et al. 2021; Himes et al. 2022). If greater fiber production can be coaxed from a dedicated land base through tree farming, this should relieve pressure on wild forests that can instead be protected from consumptive uses (Gladstone and Ledig 1990; Vincent and Binkley 1993; Sedjo and Botkin 1997; Paquette and Messier 2010; Tittler et al. 2012). Many forest land-use plans around the world can be seen as reflecting the triad approach (see Figure 2.8 for two examples), whether by intent, by accident, or through a post-facto classification based on widespread realities (Nitschke and Innes 2005). Many jurisdictions, in the tropics and southern hemisphere in particular but in much of Europe too, have directed timber-production efforts into intensively managed plantations, often consisting of non-native (exotic) tree species. Many such areas now have triad-like landscapes, although the "multipurpose" zones may focus on agriculture more than extensive forestry, as found in New Zealand and many other largely domesticated regions (Norton and Reid 2013). The triad model is now being implemented as an integral component of ecological forestry and is a matter of policy in some jurisdictions, such as on public land in Nova Scotia, Canada (Government of Nova Scotia n.d.).

The allocation of land under a triad model typically follows a combination of approaches. On the one hand, ecosystem protection goals may take priority, applying various international targets at the landscape or regional level, such as:

- 12 percent of the land base (as implied by the Brundtland Commission (WCED 1987));
- 17 percent to accommodate further climate-change range shifts and organismal migrations

(a)

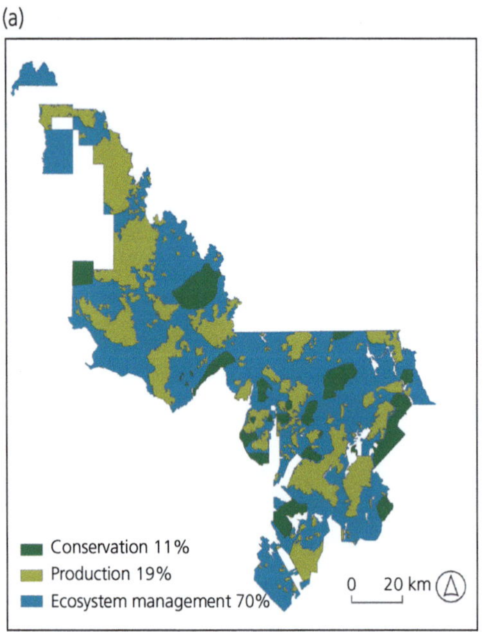

(b)

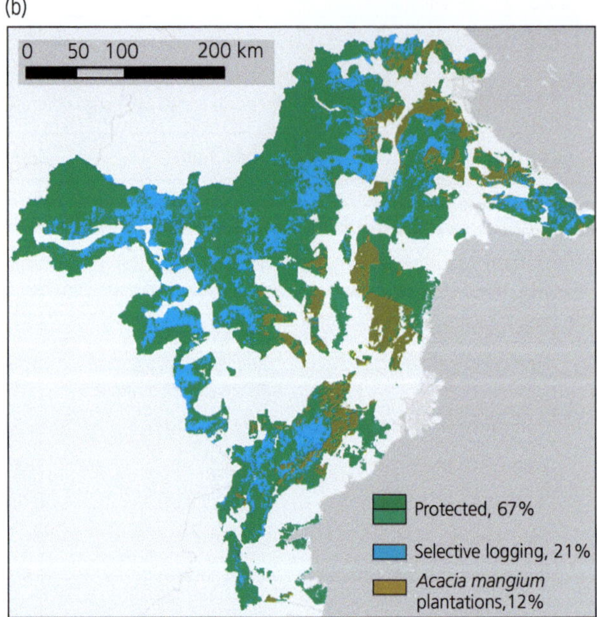

Figure 2.8 Examples of triad forest zoning: (a) The Mauricie TRIAD project in Quebec, Canada (courtesy of Christian Messier); (b) optimal distribution of conservation, selective logging, and plantation lands simulated for East Kalimantan, Indonesia (from Runting et al. (2019), courtesy of Rebecca Runting, reproduced with the permission of Springer-Nature).

(Nitschke and Innes 2008) or the Aichi Target 11 goal for implementation of the UN Convention on Biological Diversity (Watson et al. 2016);

- 30 percent targeted by the High Ambition Coalition for People and Nature (Schmidt-Traub et al. 2020; https://www.hacfornatureandpeople.org/); or
- 50 percent as argued by the "Nature Needs Half" movement (Dinerstein et al. 2017).

Recently there seems to have been some convergence on the goal of 30 percent protection by the year 2030 ("30 by 30") among national leaders, non-governmental organizations, and endorsement by the 15th Conference of the Parties (COP15), and Target 3 of the Kunming–Montreal Global Biodiversity Framework to the UN Convention on Biological Diversity (Corson and Campbell 2023; Saunders et al. 2023).

An alternative starting point may be an estimation of how much representative land can be protected (if not otherwise designated for special features) on the basis of how much of a default or target timber supply can be provided through intensively managed forests. Those estimates of boosted production and unmanaged land area equivalence can vary considerably with forest type and with management practices or assumptions. This approach is illustrated in Tables 2.5 and 2.6 for various published examples, revealing that more intensive management can multiply land productivity by an average of 9 in the southern hemisphere or 2.8 in the northern hemisphere. This means that, for a targeted level of forest land protection in the north, the desired area to be protected can be divided by 2.8 (or by more specific location- and tree-crop-specific values, as per Table 2.6 or local data) to estimate the area of intensively managed forest that would provide an equivalent timber supply to make up for land to be taken out of extractive use. To meet the 30 by 30 goal for countries in the southern hemisphere, the default timber supply could be maintained with about 5.6 percent of the land base allocated to intensive production (see Table 2.5). Additional examples with typical rotation lengths for the world's most important timber plantation species are provided in Table 7.2.

Table 2.5 Representative yield multipliers and proportional areas of intensively managed forest plantations in the southern hemisphere required to offset forgone timber harvests from a protected area target of 30% within the same country

Region/forest type	Average[*] yields ($m^3 \cdot ha^{-1} \cdot yr^{-1}$)	Plantation yields ($m^3 \cdot ha^{-1} \cdot yr^{-1}$)	Yield multiplier	Reference	Area in plantations to offset 30% protected
Argentina, compared to well-managed native forest				Cubbage et al. (2007)	
Pinus taeda	2	30	15.0		2.0%
Eucalyptus grandis	2	40	20.0		1.5%
Araucaria angustifolia	2	15	7.5		4.0%
Chile, compared to unmanaged native forest				White et al. (2021)	
Pinus radiata	1.4	29.1	20.8		1.4%
Eucalyptus globulus	1.4	23.7	16.9		1.8%
India, *Tectona grandis*	0.48	2.5	5.2	Pandey and Brown (2000)	5.8%
Indonesia, primary dipterocarps	0.63	2.38	3.8	Ruslandi et al. (2017)	7.9%
New Zealand, *Pinus radiata* compared to managed native forest					
vs managed *Agathis australis*	7	24	3.4	Richards (2005); Moore and	8.8%
vs native forest	2	24	12.0	Clinton (2015)	2.5%
New Zealand, *Pseudotsuga menziesii* compared to managed native forest					
vs managed *Agathis australis*	7	17	2.4	Belton (2005); Richards (2005);	12.4%
vs native forest	2	17	8.5		3.5%
Uruguay, compared to well-managed native forest				Cubbage et al. (2007)	
Pinus taeda	4	20	5.0		6.0%
Eucalyptus grandis	4	30	7.5		4.0%
Eucalyptus globulus	4	18	4.5		6.7%
Vietnam, *Acacia* hybrids	11	21	1.9	Frey et al. (2018)	15.7%
MEAN VALUES	3.4	20.9	9.0	15 comparisons	5.6%

[*] Average denotes natural forests or lightly managed exotic species; midpoints are given where ranges or multiple values were provided.

Such calculations of how plantation wood supply might facilitate protection of wild forests imply a substitutability of fiber from wild-grown (typically older) trees with that harvested from short-rotation plantations. Such substitutability is often not the case when it comes to lumber quality (MacPeak et al. 1990; Deresse et al. 2002; Burdon and Moore 2018). This is particularly a problem when comparing fast-growing pine and eucalypt plantations with wild tropical hardwoods. Some valuable tropical hardwoods such as the neotropical mahoganies (*Swietenia* spp.) and the African mahoganies (*Khaya* spp.) do not lend themselves to plantation production within their native ranges owing to pervasive species-adapted insect pests (Rodan et al. 1992; Opuni-Frimpong et al. 2008). Plantations are nonetheless successfully established in other locations (for example, northern Australia, Fiji),

begging the question of whether "land-sparing" calculations of subsidized protection of natural forests should extend beyond the same landscape, region, or country. It has been suggested that the super-productive plantations of tropical, subtropical, and Mediterranean climates could supply most of the world's timber and fiber needs (Sedjo 1999). Should southern countries establish enough plantations so that northern forests do not need to be logged? Without examining such global trade-offs, the discussion here and in the rest of the book assumes that it is desirable to have both natural forests and timber production in most individual landscapes or regions.

Messier et al. (2003) estimate that intensified silviculture on 15 percent of the land can provide 45 percent of current wood needs in much of Canada. A spatially explicit analysis of a forest management

Table 2.6 Representative yield multipliers and proportional areas of intensively managed forest plantations in the northern hemisphere required to offset forgone timber harvests from a protected area target of 30% within the same region

Region/forest type	Average[*] yields ($m^3 \cdot ha^{-1} \cdot yr^{-1}$)	Plantation yields ($m^3 \cdot ha^{-1} \cdot yr^{-1}$)	Yield multiplier	Reference	Area in plantations to offset 30% protected
Canada, Alberta, compared to natural *Populus tremuloides* forest					
Planted hybrid *Populus*	1.53	4.83	3.2	Anderson et al. 2012	9.5%
Planted *Populus tremuloides*	1.53	2.47	1.6		18.6%
Canada, British Columbia					
Coastal	3.85	10	2.6	Binkley 1997	11.6%
Coastal	30	40	1.3	Messier et al. 2003	22.5%
Interior	2.55	5	2.0	Binkley 1997	15.3%
Central	2.15	5.85	2.7	Nitschke and Innes 2005	11.0%
Canada, Quebec, southern	14	17	1.2	Messier et al. 2003	24.7%
Canada, Ontario, southeastern	13.55	16	1.2	Messier et al. 2003	25.4%
Canada, New Brunswick	3.2	7	2.2	Ward & Erdle 2015	13.7%
Portugal, average of 4 regions and 3 site types, compared to native *Pinus pinaster*					
Eucalyptus globulus	8.5	12	1.4	Pra et al. 2019	21.3%
Scandinavia, intensive, native spp.	2	4.5	2.25	Paquette & Messier 2010	13.3%
Hybrid *Populus*	2	23	11.5		2.6%
Exotic *Larix*	2	6.5	3.25		9.2%
US South *Pinus* taeda plantations	7.4	12	1.6	Cubbage et al. 2007	18.5%
vs natural *Pinus palustris*	4	12	3.0		10.0%
vs natural hardwoods	4	12	3.0		10.0%
US South southern pines	4	12	3.0	Gladstone & Ledig 1990	10.0%
US Pacific Northwest					
Pseudotsuga menziesii	9	21.1	2.3	Hermann & Lavender 1999	12.8%
MEAN VALUES	7.0	13.5	2.8	For 18 comparisons	14.5%

[*] Average denotes natural forests or lightly managed exotic species; midpoints are given where ranges or multiple values were provided.

unit in interior British Columbia similarly determined that 46 percent of the land base, if managed for intensive timber production, can provide 100 percent of the timber supply estimated for the same land base managed under the default integrated resource management regime (Sahajananthan et al. 1998). Other authors have made the much more conservative assumption of a 1:1 allocation of intensive forest management lands to offset protected areas. For example, Messier and Kneeshaw (1999) suggested that 15 percent of the land base under intensive management could justify putting another 15 percent under strict nature protection. Burton (1995) argued for a 25 percent allocation to both intensive production and protected areas, with 50 percent for multipurpose extensive management

to maintain adaptive capacity and flexibility based on a metaphor of genetic diversity and principles of equity. Ward and Erdle (2015) evaluated equal allocations of intensive and protected land proportions ranging from 10 percent to 35 percent. Indicators such as old forest area and large snag density generally increase in proportion to the area protected, while total timber yields, silvicultural costs, and net present value of timber vary in proportion to the area under intensive management (Ward and Erdle 2015). Nonetheless, the annual area disturbed by logging is proportional to the area under extensive management (Krcmar et al. 2003; Côté et al. 2010; Ward and Erdle 2015), so there are still trade-offs involved and no one solution that is optimal for all values.

Adoption or recognition of a triad model does not mean that there are only three land-use or resource-management emphasis zones in a landscape. For example, Burton (1995) suggests that the dominant multipurpose or extensive forest management zone could be further divided into areas subject to commodity emphasis, standard practices, and special management (see Box 2.1). In that multiple-purpose context, "commodity emphasis" might endorse use of mechanical site preparation and the planting of native tree species, while still accommodating biodiversity conservation, livestock grazing, recreational use, and so on; "special management" might focus on the needs of specific wildlife species, fuel reduction around communities, riparian protection, and so on. We also see the widespread adoption of triad zoning within the world's national parks, considered protected areas (IUCN Category II) under the more general zoning of regional land uses (Dudley et al. 2010). Most national parks have a "front country" consisting of roads, campgrounds, tourist accommodations, and other intensive uses, with the remainder of the park considered "back country" available for regulated non-motorized recreation, within which may be embedded some strict nature reserves open only to scientific and educational use (Benkhard 2004; see Section 11.2 for an example). Intensively managed timber-production zones may experience different levels of investment and modification as well, sometimes with little distinction from "commodity-emphasis" zones on multipurpose lands (as illustrated in case studies described in Sections 11.3 and 11.4). Depending on the standard management practices in the extensively managed zone, intensive management may include planting genetically improved native tree species after site preparation. Practices used in intensive management zones may promote enhanced growth of native trees with repeated brushing and spacing, or even "superintensive" plantations of fast-growing hybridized exotic trees that are fertilized so that they can be harvested on very short rotations (Messier et al. 2003; Nitschke and Innes 2005).

2.6 Resistance to management

Even forests that appear "natural," and native forests nominally "protected" from human abuse, can benefit from human stewardship to assure continued health and functioning (Franklin 2003). Altered disturbance regimes, fragmented landscapes, damaging exotic organisms, and a warming climate all bear the fingerprints of human activity on this planet, and frequently compromise the inherent resilience of ecosystems. It may seem obvious to foresters that the delivery of desired forest ecosystem services can be improved with management, but such a belief is not universal. Two particular constituencies that often express the contrary view come to mind: communities that utilize forests for subsistence or traditional uses; and environmental advocates who believe that nature should be left to take its course, particularly in parks, reserves, and other protected areas.

In reaching out to those constituencies with respect and understanding, it is worth discussing the full scope of forest management and its implications. Legislation to designate an area free from hunting, for example, is an explicit or implicit act of land management. Decisions to promote tourism and suppress fires in forested landscapes—common policies in the world's many national and state parks—do indeed constitute forest management activities. Conversely, a laissez-faire policy to allow unregulated fuelwood collection or food hunting is, in fact, also a forest management decision with predictable impacts on biodiversity. Once such policies are recognized as management decisions, the window is opened to discuss how those policies might be finessed to meet stakeholder objectives better (see Section 2.7) while enhancing the long-term ability of the forest to provide desired goods and services. Resistance to "forest management" may reflect a distrust of foresters, centralized authorities, or forest management that historically has emphasized profit-driven logging and timber production. Particularly among Indigenous communities, "scientific forestry"—whether in support of sustained-yield timber production, sustainable forest management, or resilience planning—can be considered a colonial imposition, and is distrusted as a foreign and potentially exploitative European concept (Stevenson and Webb 2003). In both Indigenous and environmentalist communities, there can be a distrust of any form of "managerialism" that might supplant the wisdom and jurisdiction of traditional practices or the sense that "nature knows best." Yet nature has always changed in the past, and has a vast potential for variation in ecosystem composition and structure, forcing us to select which face of nature we wish to sustain (Botkin 1990). In addition, ecosystems and values change over time, so approved forest policies and practices granted decades or centuries ago may no longer have broad support among contemporary stakeholders. Moving the discussion forward depends on identifying contemporary threats, clearly articulating current policy goals, clarifying forest management objectives, and collaboratively selecting stewardship activities that support, rather than conflict with, the underlying values of skeptics.

Let us briefly explore how such a discussion might unfold in each of the two situations introduced above—namely, in serving traditional communities and in introducing active management in protected areas. There are national parks in Australia, Canada, US, Africa, and elsewhere in which various levels of customary or Indigenous use of forest resources are not only practiced but often guaranteed within the context of modern nation states (e.g. Timko and Satterfield 2008; Deur and James 2020). Typical resources harvested from the forest include foodstuffs (for example, fruits, honey, mushrooms, as well as fish, birds, and mammals), deadwood gathered for heating and cooking, and medicinal materials (for example, bark from particular tree species). Such access and use may have been practiced for decades or centuries, and may be subject either to no regulatory constraints, or to various formal and informal forms of restriction (e.g. Jimoh et al. 2012). Though community members may take part in the modern cash economy, forest use and access may be important for the day-to-day survival of some families, as an expression of cultural continuity, or may be practiced to assert traditional or Indigenous rights (Stevenson and Webb 2003).

Where cultural practices, local tradition, or deference to local authority (whether elected, hereditary, or religious) is already in place, it is important to acknowledge this deference as a form of resource management (Goldtooth 2008; Jimoh et al. 2012), and to identify correctly where decision-making lies and the degree to which such jurisdictions are respected. With growing populations

and continuing commercial development on the land, people in these traditional communities are often finding it more difficult to access the desired forest resources. Discussions on the cost or rarity of certain forest products, or on the travel distances now needed to harvest them, will then lead logically to the need for conservation, monitoring, and restoration. Where commitment to place and pride in those traditions are strong, there will usually be support for measures to provide similar opportunities to future generations. The challenge is to find commonalities with the traditions of stewardship and prudence found in most cultures, concepts that are often more strongly held amongst elders and traditional leaders than among those charged with developing community support programs and the economic potential of Indigenous lands. Some programs to implement forest sustainability or to enhance forest resilience may require investments of labor or capital, yet others can be implemented by working with the appropriate representatives to effect relatively minor changes in resource harvesting practices. Public education and training opportunities may provide a means of implementing other forest conservation practices, especially if developed in collaboration with the community rather than being imposed from outside (Jacobson et al. 2015).

Working with advocates for the protection of parks and wilderness areas requires quite a different approach. In this case, an enumeration of the ways in which humankind has already altered the entire planet can be an appropriate avenue to pursue, with emphasis on how these impacts impinge on the particular area under discussion. What follows may be a discussion framed as the need for corrective action or ecological restoration. Sometimes the issue is the prevalence of exotic plants or animals that must be removed (or at least controlled) if other native species and ecological communities are to persist. As much as we all may like to avoid the deployment of synthetic compounds (herbicides, animal poisons), they are often the only practical way of protecting ground-nesting birds from rats or rare plant populations from aggressive weeds (Macdonald et al. 2003; Parks 2009). In other cases, the issue may be one of trying to compensate for the injudicious use of fire in the past, or conversely the absence of fire in fire-maintained ecosystems. Such issues can be viewed under the umbrella of ecological restoration (see Chapter 9), generally regarded as improving the naturalness of an area and "a good thing to do" in support of nature protection (Burton and Macdonald 2011), but which (somewhat paradoxically!) also requires a wide range of active interventions. Discussion with the defenders of parks and reserves often includes consideration of restoration targets: which of several possible configurations of ecosystem composition should be targeted for a forest or landscape (for example, often following a precolonial or preindustrial template; Higgs et al. 2014). Difficult conversations in which conflicting values must be prioritized often follow, including a consideration of global change (see Chapter 6) and other challenges to restoration. Comparative treatments in the form of active adaptive management (see Chapter 10), including a "no-intervention" option, can sometimes be the way to proceed.

Private landowners constitute another group that may not object to management activities per se, but that may be difficult to convince of the benefits of management reform or management coordination. The owners of production forests have made long-term investments with the expectation of predictable financial returns. Forest lands owned by individuals or families often belong to that category, but also include lands retained for the protection and enjoyment (whether consumptive or passive) of wildlife and nature (see Figure 2.1). Much like the Indigenous and subsistence users of forests described above, private landowners frequently object to outside interference with their property rights and their ability to manage the land as they wish (Rickenbach and Reed 2002; Appelstrand 2012). This behavior becomes especially problematic when there is an attempt to coordinate landscape- and regional-scale conservation plans with a diverse array of landowners and government agencies that differ in their resource management mandates and priorities. Once again, finding common ground on the need for both resilience in the short to medium term and sustainability in the long term can offer a way forward, with a diversity of educational, experimental, incentive,

and regulatory approaches often needed (Doremus 2003).

Finally, it is worth considering that word choice can be important. "Forest management" may have too much baggage as a term, implying not only a focus on timber but a top-down control-focused or authoritarian approach that has often been imposed by European colonial powers. In many cases, manipulations of forest composition or structure and constraints on human use are more acceptable if presented as "forest stewardship," with guidance rather than control implied (Franklin 2003; Chapin et al. 2009; Messier et al. 2015). This concept is further explored in Chapter 12.

2.7 The governance framework

The identification and potential promotion of different forest values, whether across a forest estate as a whole or in resource-emphasis zones, presumes a level of legal, social, and cultural legitimacy to proceed with forest management. Classical forest management education and practice make much of the issue of legal ownership (Davis et al. 2001) or tenure (Duerr et al. 1982), and meeting the objectives of the owner, as setting the direction for forest management activities. Legal access for the harvesting of forest resources is also key to most systems of forest product certification (e.g. Principle #1 of the Forest Stewardship Council (FSC 2015)). But forests are important, place-based elements of the cultural landscape for many people, a role increasingly recognized as part of complex socio-ecological systems (Fischer 2018). This means that ownership and legitimate tenure agreements are a necessary but rarely sufficient condition for successfully implementing forest management actions, especially on public or communal lands.

The language of traditional forest management, with reference to "the owner" and "the forest estate" implies a private ownership model, one in which top-down direction and decision-making are absolute. Corporate ownership, tenure, and authority can be especially problematic with respect to the long-term, place-based, sustainable stewardship of forest ecosystem services (see Box 2.4). Forest ownership or legal tenure rarely confers exclusive use or absolute decision-making ability, however. Many overlapping tenures or access rights typically exist, such as mineral rights to belowground resources, registered traplines for harvesting fur-bearing animals, and watershed (catchment) management authorities. Zoning restrictions to infrastructure development, conservation easements, and forest practices regulations may also apply. Indigenous rights for traditional uses, or universal rights for recreational activities such as hiking, tenting, berry picking, and mushroom foraging (for example, *Allemansrätten*, as guaranteed in Sweden's Constitution) also must be respected. Forests on public land may receive their management direction from a legitimate government, but elected governments will change many times before disturbed forests recover and tree crops mature. As a result, forest policies that were well supported in the past may no longer experience widespread public approval. Furthermore, the legitimacy and endorsement of any given forest management direction on public lands is confounded with public support for many other issues making up the policy platform of the governing party during the most recent election. In conclusion, most forests, even those on private land, typically have diverse communities of users, potential users, and concerned citizens who can be considered stakeholders in their sustainable management.

Box 2.4 Is sustainable stewardship of forests by for-profit corporations possible?

Let us explore the premise that some level of corporate control (whether through ownership or through government influence) is fundamentally in conflict with socio-ecological sustainability in general (Klein 2015), and the promotion of resilient forests in particular. There can be demonstrable benefits to industrial ownership of timberlands (Li and Zhang 2014). Yet there are some fundamental incompatibilities between profit-driven corporate objectives and ecosystem-based sustainability and resilience objectives. Three arguments present themselves.

Argument 1: *Corporate Premise 1A*: The most consistently tangible provisioning service provided by forests is fiber, whether for solid wood products, pulp and paper, or as a source of bioenergy or biochemicals. Thus most so-called forest companies are in the wood-fiber processing business, by which they convert the commodity of logs into profits with the greatest efficiency possible. Issues of hydrological or biodiversity protection, recreational services, or the promotion of any other non-timber value is therefore treated as a constraint to the revenue-generating potential of log production, harvesting, and processing.

Stewardship Premise 1B: In order not to compromise the health, vitality, and integrity of forests, whether natural or planted, timber should be harvested only to the extent that all other ecosystem-based values are supported (Grumbine 1994).

Conclusion 1: So long as forest interventions are driven by the processors of wood products, other values will be managed for and supported only to the degree that they support fiber production or are required by shareholders, the market, or by law.

Argument 2: *Corporate Premise 2A*: Although rarely ensconced in legislation or incorporation documents (Stout 2015), it is a fundamental goal and responsibility of corporations to maximize shareholder value, typically interpreted in terms of profits and dividends (per economist Milton Friedman 1970, *fide* Kolstad 2007). Management decisions in pursuit of that goal tend to emphasize quarterly returns and stock values, with "strategic" planning limited to windows of two to five years (Millon 2002; O'Regan and Ghobadian 2007).

Stewardship Premise 2B: It is a fundamental goal of sustainable forest management and resilient forest stewardship to assure forest productivity, health, biodiversity, and provision of ecosystem services into the indefinite future, with key decisions made at the frequency of timber crop rotations, which typically range from 25 to 120 years.

Conclusion 2: Corporate decisions, and those by government or other landowners in support of corporate priorities, are often compelled to sacrifice long-term sustainability and resilience in support of short-term benefits. Corporate priorities therefore are fundamentally in conflict with responsible forest stewardship.

Argument 3: *Corporate Premise 3A*: Under neo-liberal conditions of globalization, free trade, and investment, it is expected that capital can and should move around the globe to where it can garner the greatest returns legally permissible.

Stewardship Premise 3B: Forest sustainability, persistence, and resilience are place-based attributes, making any protection or enhancement of forest productivity, health, biodiversity, and ecosystem services being provided elsewhere ineffective as alternatives.

Conclusion 3: Different political jurisdictions (countries, states, provinces) differ in their regulatory requirements and taxation policies as well as in the biophysical conditions suitable for growing timber. Consequently, investment in forest management activities and in fiber-processing facilities is fluid, with much capital moving around the world to where it is most profitable. This frequently results in the exploitative depletion of local forest resources and little fidelity to multi-decadal or full-rotation forest management plans and commitments.

There may be exceptions to the above conclusions, but they are likely to apply where very short tree crop rotations are possible or depend on specific corporate visions and shareholder initiatives to eschew one or more of the standard corporate premises. If unbridled corporate management follows the capitalist credo of Ayn Rand that "greed is good" (Rollert 2014), then it is at odds both with resilient forest stewardship and with Mahatma Gandhi's timely admonition that "Earth provides enough to satisfy every man's need but not for every man's greed."

Even if timber production is not the sole or primary management objective, reference to the landscape-level unit of sustained yield management as "the forest" implies that the area under management is somehow isolated and discrete from its environs. Yet many forests have streams and rivers carrying water through them, as well as fish, birds, and ungulates migrating in and out of them on a seasonal basis (Gounand et al. 2018). Many people—some local, some from afar—may be pursuing subsistence and recreational activities in a forest throughout the year. Even people viewing the forest from a distance—some tax-paying citizens, some spendthrift tourists—will have opinions on its aesthetics. Consequently, it behooves any forest manager to ascertain the connections of the area under management to the surrounding landscape and region (Diaz and Apostol 1992; Bell and Apostol 2007), and the many socioeconomic and cultural connections and value expressions on the part of communities, stakeholders, and constituencies both local and distant.

The appropriate balance of managerial or professional decision-making relative to the input and authority of other government and community bodies can vary considerably around the world. Within the constraints of laws and regulations, and even if not required, it is beneficial for a broad representation of the public to be involved in identifying, defending, and promoting the many values associated with a forest (Hammersley Chambers and Beckley 2003). There is a wide range of options associated with public-involvement policies and practices, which can be generally arranged along axes describing the symmetry of information flow and the level of authority exercised by the public (see Figure 2.9). These instruments of public engagement and empowerment could also be arranged along a gradient of confrontational to good-faith collaboration, not always correlated with the one-way to two-way exchange of information. A large number of initiatives, whether on the part of industry, government, or non-governmental organizations, tend to be very one-sided. This includes many instruments and actions that are rather indirect, including the perceived will of the public (in democratic nations) as expressed in legislation and law enforcement, and as tested or verified in the courts. Good-faith consultations with the public, through open houses and feedback forms or advisory committees and negotiations, can provide opportunities for the expression of public values and concerns (including place-based local knowledge), and their subsequent incorporation in forest management plans and practices. But those discussions often occur in response to plans and policies or policy proposals already drafted, whereas it is preferable for them to be developed in collaboration with all stakeholders at the outset (Hummel and Freet 1999; Franklin et al. 2018).

Some sort of public or stakeholder participation in setting local standards is often a required component of forest certification programs (Clark and Kozar 2011), but that participation may not extend to playing a long-term role. Public advisory committees are often convened to provide a source of ongoing input and feedback. They vary widely in terms of their make-up (for example, with volunteer or elected members, representing designated sectors or the public at large), in terms of the policies for decision-making (for example, seeking consensus or not), and in the degree to which decision-makers are obligated to follow their direction (Harshaw

Figure 2.9 Various approaches to incorporating public input into forest planning and management, loosely arrayed according to levels of two-way communication and control over decision-making. Co-ownership or community ownership does not necessarily incorporate greater two-way information flow or authority unless they also embrace appropriate mechanisms for public input and participation.

and El-Lakany 2017). Many advisory committees have issues of poor representation of women, minority, and Indigenous populations, and thereby miss important and widely held perspectives (Lindgren et al. 2019). It is also frustrating for volunteers to spend many days of private time, often over several years, to come up with thoughtful recommendations on forest land use and management, only for companies or governments to "check off the box" for public consultation, without implementing many of their requests and recommendations. Frustration with the effectiveness of any of these public input instruments (for example, cynically recognized as a "talk and log" strategy) is widespread, resulting in boycotts, protests, blockades, vandalism, and lawsuits. Often such confrontational expressions of public opinion have been required to get the attention of administrators (Wilson 1998). Negotiations, mediation, and arbitration can be effective at resolving forest management conflicts, especially if moderated by professional facilitators (Johansson et al. 2018), but participants may still maintain confrontational mindsets. Even where legal rights and responsibilities are clear and are followed to the letter, there is a growing need for corporations and government agencies to have a "social license" to operate (Gunningham et al. 2004; Wang 2019). Furthermore, many laws, regulations, and policies have yet to catch up with widespread expectations (and even commitments) to respect and implement the 2007 UN Declaration on the Rights of Indigenous Peoples (Gilbert and Lennox 2019).

A fully enshrined system of respectful and collaborative shared decision-making and co-management may be the preferred instrument for reflecting public values in forest governance. The case can be made that all forestry is place-based, and so forest management decisions and governance should also be place-based (see Section 12.5). But even the most healthy and progressive system of governance still requires repeated re-scoping, consultation with constituencies, and regular renewal. Likewise, co-ownership of a forest estate or a community forest tenure does not necessarily result in the most representative and egalitarian governance, as they still depend on other forms of public engagement. Extensive public meetings, polling responses, and large commitments of volunteer time typically exclude the poor, ethnic or racial minorities, and other disadvantaged members of a community (see, e.g., Box 12.2). Like any human relationship, that between "the forest manager" and "the public" requires ongoing work and nurturing, based on mutual trust and respect. Ultimately, the effective expression of public values in forest planning and management depends on sincere and shared intentions to come up collaboratively with the most equitable and sustainable way forward despite differences in values and priorities. When fully implemented, an effective program of public empowerment in forest management not only facilitates the promotion of sustainable and resilient forests, but also supports progress to a broad number of the UN Sustainable Development Goals (see Box 1.2).

Box 2.5 Key points

- Forests have many material and non-material values, which are weighted differently by people, although the economic value of wood tends to prevail.
- Even the recognition of forest "values" and "services" reveals a human-centered perception of forests rather than an ecocentric respect for nature.
- Efforts to assign economic and legal value to non-market forest services have been diverse, but, with the exception of new markets being developed for carbon, there has been little uptake by authorities and forest decision-makers.
- Conserving or promoting all values is not possible in the same forest stand, though some values and associated uses are more compatible than others; market and non-market values will always be difficult to compare and balance.
- The spatial separation of forest uses into those that are more or less compatible can be one solution to accommodate multiple uses and expectations of a forest landscape.
- Forest zones on public land can generally be grouped into three broad categories: protected, integrated resource management or multipurpose, and intensively managed timber-production areas, recognized as the triad model of forest land use.
- Zoning and targeted management may also be required to protect or enhance identified values in parks, on Indigenous lands, and on private property.
- The expression and acceptance of values in forest land-use policies and practices depend on the legitimacy of the governance framework, with various instruments that vary in the degree of two-way information flow and the authority exercised by the public.

You can't have everything. Where would you put it?
Stand-up comedian Steven Wright (c.1985)

References cited

Adger, W. N., Brown, K., Cervigni, R., and Moran, D. (1995). "Total Economic Value of Forests in Mexico," *Ambio*, 24; 286–96.

Ameztegui, A., Solarik, K. A., Parkins, J. R., Houle, D., Messier, C., and Gravel, D. (2018). "Perceptions of Climate Change across the Canadian Forest Sector: The Key Factors of Institutional and Geographical Environment," *PLoS One*, 13/6: e0197689.

Anderson, J. A., Armstrong, G. W., Luckert, M. K., and Adamowicz, W. L. (2012). "Optimal Zoning of Forested Land Considering the Contribution of Exotic Plantations," *Mathematical and Computational Forestry & Natural Resource Sciences*, 4/2: 92–104.

Appelstrand, M. (2012). "Developments in Swedish Forest Policy and Administration: From a 'Policy of Restriction' toward a 'Policy of Cooperation'," *Scandinavian Journal of Forest Research*, 27/2: 186–99.

Attfield, R. (2018). *Environmental Ethics: A Very Short Introduction.* Oxford: Oxford University Press. 141 pp.

Barrio, M., and Loureiro, M. L. (2010). "A Meta-Analysis of Contingent Valuation Forest Studies," *Ecological Economics*, 69/5: 1023–30.

Bashir, A., Sjølie, H. K., and Solberg, B. (2020). "Determinants of Nonindustrial Private Forest Owners' Willingness to Harvest Timber in Norway," *Forests*, 11/1: 60.

Beatty, C. R., Stevenson, M., Pacheco, P., Terrana, A., Folse, M., and Cody, A. (2022). *The Vitality of Forests: Illustrating the Evidence Connecting Forests and Human Health.* Washington: World Wildlife Fund. 54 pp. Available online at https://files.worldwildlife.org/wwfcmsprod/files/Publication/file/3peoo4s5i3_VoF.8.14.22.pdf (accessed February 1, 2023).

Bell, S., and Apostol, D. (2007). *Designing Sustainable Forest Landscapes.* London: Taylor & Francis. 368 pp.

Belton, M. (2005). "Silviculture of Douglas-Fir," in M. Colley (ed), *Forestry Handbook.* Tauranga, New Zealand: New Zealand Institute of Forestry, 125–7.

Bengston, D. N., Asah, S. T., and Butler, B. J. (2011). "The Diverse Values and Motivations of Family Forest Owners in the United States: An Analysis of an Open-Ended Question in the National Woodland Owner Survey," *Small-Scale Forestry*, 10/3: 339–55.

Benkhard, B. (2004). "Relations between IUCN-Zoning and Tourism in Hungarian National Parks," in T. Sievänen, J. Erkkonen, J. Jokimäki, J. Saarinen, S. Tuulentie, and E. Virtanen (eds), *Policies, Methods and Tools for Visitor Management.* Helsinki: Finnish Forest Research Institute, 377–81.

Betts, M. G., Phalan, B. T., Wolf, C., et al. (2021). "Producing Wood at Least Cost to Biodiversity: Integrating

Triad and Sharing–Sparing Approaches to Inform Forest Landscape Management," *Biological Reviews*, 96/4: 1301–17.

Bieling, C. (2004). "Non-Industrial Private-Forest Owners: Possibilities for Increasing Adoption of Close-to-Nature Forest Management," *European Journal of Forest Research*, 123/4: 293–303.

Binkley, C. S. (1997). "Preserving Nature through Intensive Plantation Forestry: The Case for Forestland Allocation with Illustrations from British Columbia," *Forestry Chronicle*, 73/5: 553–9.

Birch, S. (2020). "Political Polarization and Environmental Attitudes: A Cross-National Analysis," *Environmental Politics*, 29/4: 697–718.

Blumm, M. C. (1994). "Public Choice Theory and the Public Lands: Why Multiple Use Failed," *Harvard Environmental Law Review*, 18: 405–32.

Botkin, D. B. (1990.) *Discordant Harmonies: A New Ecology for the Twenty-First Century*. New York: Oxford University Press. 241 pp.

Boyd, J., and Banzhaf, S. (2007). "What Are Ecosystem Services? The Need for Standardized Environmental Accounting Units," *Ecological Economics*, 63/2–3: 616–26.

Bridgewater, P., and Schmeller, D. S. (2023). "The Ninth Plenary of the Intergovernmental Platform on Biodiversity and Ecosystem Services (IPBES-9): Sustainable Use, Values, and Business (as Usual)," *Biodiversity and Conservation*, 32/1: 1–6.

Bulkan, J. (2017). "Social, Cultural and Spiritual (SCS) Needs and Values," in J. L. Innes and A. V. Tikina (eds), *Sustainable Forest Management: From Concept to Practice*. London: Routledge, 241–56.

Burdon, R. D., and Moore, J. R. (2018). "Adverse Genetic Correlations and Impacts of Silviculture Involving Wood Properties: Analysis of Issues for Radiata Pine," *Forests*, 9/6: 308.

Burton, P. J. (1995). "The Mendelian Compromise: A Vision for Equitable Land Use Allocation," *Land Use Policy*, 12/1: 63–8.

Burton, P. J., and Macdonald, S. E. (2011). "The Restorative Imperative: Assessing Objectives, Approaches and Challenges to Restoring Naturalness in Forests," *Silva Fennica*, 45/5: 843–63.

Burton, P. J., Messier, C., Weetman, G. F., Prepas, E. E., Adamowicz, W. L., and Tittler, R. (2003). "The Current State of Boreal Forestry and the Drive for Change," in P. J. Burton, C. Messier, D. W. Smith, and W. L. Adamowicz (eds), *Towards Sustainable Management of the Boreal Forest*. Ottawa: NRC Research Press, 1–40.

Cadman, T., and Hales, R. (2022). "COP26 and a Framework for Future Global Agreements on Carbon Market Integrity," *International Journal of Social Quality*, 12/1: 76–99.

Carey, A. B. (1997). *Cognitive Styles of Forest Service Scientists and Managers in the Pacific Northwest*. General Technical Report, PNW-GTR-414. Portland, OR: USDA Forest Service. 25 pp. Available online at https://www.fs.usda.gov/pnw/pubs/gtr414.pdf (accessed January 30, 2024).

Carless, S. A. (2009). "Psychological Testing for Selection Purposes: A Guide to Evidence-Based Practice for Human Resource Professionals," *International Journal of Human Resource Management*, 20/12: 2517–32.

Caveness, F. A., and Kurtz, W. B. (1993). "Agroforestry Adoption and Risk Perception by Farmers in Senegal," *Agroforestry Systems*, 21/1: 11–25.

Champ, P., Boyle, K., and Brown, T. (eds).(2017). *A Primer on Nonmarket Valuation*. 2nd edn. Dordrecht: Springer. 504 pp.

Chapin, F. S., Kofinas, G. P., and Folke, C. (eds). (2009). *Principles of Ecosystem Stewardship: Resilience-Based Natural Resource Management in a Changing World*. New York: Springer. 401 pp.

Chapron, G., Epstein, Y., and López-Bao, J. V. (2019). "A Rights Revolution for Nature," *Science*, 363/6434: 1392–3.

Chen, S., Shahi, C., and Chen, H. Y. (2016). "Economic and Ecological Trade-off Analysis of Forest Ecosystems: Options for Boreal Forests," *Environmental Reviews*, 24/3: 348–61.

Clark, M. R., and Kozar, J. S. (2011). "Comparing Sustainable Forest Management Certifications Standards: A Meta-Analysis," *Ecology and Society*, 16/1: 3.

Corace, R. G., Shartell, L. M., Schulte, L. A., Brininger, W. L., McDowell, M. K., and Kashian, D. M. (2012). "An Ecoregional Context for Forest Management on National Wildlife Refuges of the Upper Midwest, USA," *Environmental Management*, 49/2: 359–71.

Corson, C., and Campbell, L. M. (2023). "Conservation at a Crossroads: Governing by Global Targets, Innovative Financing, and Techno-Optimism or Radical Reform?" *Ecology and Society*, 28/2: 3.

Costanza, R., de Groot, R., Sutton, P., et al. (2014). "Changes in the Global Value of Ecosystem Services," *Global Environmental Change*, 26: 152–8.

Côté, M. A., Gilbert, D., and Nadeau, S. (2015). "Characterizing the Profiles, Motivations and Behaviour of Quebec's Forest Owners," *Forest Policy and Economics*, 59: 83–90.

Côté, P., Tittler, R., Messier, C., Kneeshaw, D. D., Fall, A., and Fortin, M.-J. (2010). "Comparing Different Forest Zoning Options for Landscape-Scale Management of the Boreal Forest: Possible Benefits of the TRIAD," *Forest Ecology and Management*, 259/3: 418–27.

Cubbage, F., Mac Donagh, P., Sawinski Júnior, J., et al. (2007). "Timber Investment Returns for Selected Plantations and Native Forests in South America and the Southern United States," *New Forests*, 33: 237–55.

Cullingworth, J. B. (2017). *Urban and Regional Planning in Canada*. New York: Routledge. 399 pp.

Daily, G. C., Polasky, S., Goldstein, J., et al. (2009). "Ecosystem Services in Decision Making: Time to Deliver," *Frontiers in Ecology and the Environment*, 7/1: 21–8.

Dana, S. T., and Fairfax, S. K. (1980). *Forest and Range Policy: Its Development in the United States*. 2nd edn. New York: McGraw-Hill. 458 pp.

Dasgupta, P. (2021). *The Economics of Biodiversity: The Dasgupta Review (Abridged Version)*. London: HM Treasury, Government of the United Kingdom. 99 pp. Available online via https://seea.un.org/content/economics-biodiversity-dasgupta-review-abridged-version (accessed May 20, 2023).

Davis, L. S., Johnson, K. N., Bettinger, P., and Howard, T. E. (2001). *Forest Management: To Sustain Ecological, Economic, and Social Values*. 4th edn. Long Grove, IL: Waveland Press. 804 pp.

Dejene, T., Agamy, M. S., Agúndez, D., and Martin-Pinto, P. (2020). "Ethnobotanical Survey of Wild Edible Fruit Tree Species in Lowland Areas of Ethiopia," *Forests*, 11/2: 177.

Deresse, T., Shepard, R. K., and Rice, R. W. (2002). "Longitudinal Shrinkage, Kiln-Drying Defects, and Lumber Grade Recovery of Red Pine (*Pinus resinosa* Ait.) from a 125-Year-Old Natural Stand and a 57-Year-Old Plantation," *Forest Products Journal*, 52/5: 88–93.

Deur, D., and James, J. E. (2020). "Cultivating the Imagined Wilderness: Contested Native American Plant-Gathering Traditions in America's National Parks," in N. J. Turner (ed.), *Plants, People and Places: The Roles of Ethnobotany and Ethnoecology in Indigenous Peoples' Land Rights in Canada and Beyond*. Montreal: McGill-Queens University Press, 220–37.

Diaz, N., and Apostol, D. (1992). *Forest Landscape Analysis and Design: A Process for Developing and Implementing Land Management Objectives for Landscape Patterns*. R6 ECO-TP-043-92. Portland, OR: USDA Forest Service. 118 pp. Available online at https://www.fs.usda.gov/research/treesearch/6268 (accessed January 30, 2024).

Díaz, S., Pascual, U., Stenseke, M., et al. (2018). "Assessing Nature's Contributions to People," *Science*, 359/6373: 270–2.

Dinerstein, E., Olson, D., Joshi, A., et al. (2017). "An Ecoregion-Based Approach to Protecting Half the Terrestrial Realm," *BioScience*, 67/6: 534–45.

Doimo, I., Masiero, M., and Gatto, P. (2020). "Forest and Wellbeing: Bridging Medical and Forest Research for Effective Forest-Based Initiatives," *Forests*, 11/8: 791.

Doremus, H. (2003). "A Policy Portfolio Approach to Biodiversity Protection on Private Lands," *Environmental Science & Policy*, 6/3: 217–32.

Drummond, M. A., and Loveland, T. R. (2010). "Land-Use Pressure and a Transition to Forest-Cover Loss in the Eastern United States," *BioScience*, 60/4: 286–98.

Dudley, N., Parrish, J. D., Redford, K. H., and Stolton, S. (2010). "The Revised IUCN Protected Area Management Categories: The Debate and Ways Forward," *Oryx*, 44/4: 485–90.

Duerr, W. A., Teeguarden, D. E., Christiansen, N. B., and Guttenberg, S. (eds).(1982). *Forest Resource Management: Decision-Making Principles and Cases*. Corvallis, OR: OSU Bookstores. 612 pp.

Duinen, R. V., Filatova, T., Geurts, P., and Veen, A. V. D. (2015). "Empirical Analysis of Farmers' Drought Risk Perception: Objective Factors, Personal Circumstances, and Social Influence," *Risk Analysis*, 35/4: 741–55.

Edlin, H. L. (1963). "Amenity Values in British Forestry," *Forestry*, 36/1: 65–89.

Eitzinger, A., Binder, C. R., and Meyer, M. A. (2018). "Risk Perception and Decision-Making: Do Farmers Consider Risks from Climate Change?" *Climatic Change*, 151/3–4: 507–24.

Filotas, E., Parrott, L., Burton, P. J., et al. (2014). "Viewing Forests through the Lens of Complex Systems Science," *Ecosphere*, 5/1: 1–23.

Fischer, A. P. (2018). "Forest Landscapes as Social-Ecological Systems and Implications for Management," *Landscape and Urban Planning*, 177: 138–47.

Fischer, A. P., Bliss, J., Ingemarson, F., Lidestav, G., and Lönnstedt, L. (2010). "From the Small Woodland Problem to Ecosocial Systems: The Evolution of Social Research on Small-Scale Forestry in Sweden and the USA," *Scandinavian Journal of Forest Research*, 25/4: 390–8.

Fischer, J., Abson, D. J., Butsic, V., et al. (2014). "Land Sparing versus Land Sharing: Moving Forward," *Conservation Letters*, 7/3: 149–57.

Franklin, J. F. (2003). "Challenges to Temperate Forest Stewardship: Focusing on the Future," in D. B. Lindenmayer and J. F. Franklin (eds), *Towards Forest Sustainability*. Melbourne: CSIRO Publishing, 1–13.

Franklin, J. F., Johnson, K. N., and Johnson, D. L. (2018). *Ecological Forest Management*. Long Grove, IL: Waveland Press. 646 pp.

Freeman, A. M., Herriges, J. A., and Kling, C. L. (2014). *The Measurement of Environmental and Resource Values: Theory and Methods*. Baltimore, MD: Resources for the Future Press. 478 pp.

Frey, G. E., Cubbage, F. W., Ha, T. T. T., et al. (2018). "Financial Analysis and Comparison of Smallholder

Forest and State Forest Enterprise Plantations in Central Vietnam," *International Forestry Review*, 20/2: 181–98.

FSC (2015). *FSC Principles and Criteria for Forest Stewardship*. FSC-STD-01-001 V5-2 EN. Bonn, Germany: Forest Stewardship Council. 32 pp. Available online at https://fsc.org/en/document-centre/documents/resource/392 (accessed May 18, 2021).

Furukawa, T., Kiboi, S. K., Mutiso, P. B. C., and Fujiwara, K. (2016). "Multiple Use Patterns of Medicinal Trees in an Urban Forest in Nairobi, Kenya," *Urban Forestry & Urban Greening*, 18: 34–40.

Gilbert, J., and Lennox, C. (2019). "Towards New Development Paradigms: The United Nations Declaration on the Rights of Indigenous Peoples as a Tool to Support Self-Determined Development," *International Journal of Human Rights*, 23/1–2: 104–24.

Gladstone, W. T., and Ledig, F. T. (1990). "Reducing Pressure on Natural Forests through High-Yield Forestry," *Forest Ecology and Management*, 35/1–2: 69–78.

Goldtooth, T. (2008). "Protecting the Web of Life: Indigenous Knowledge and Biojustice," in M. K. Nelson (ed.), *Original Instructions: Indigenous Teachings for a Sustainable Future*. Rochester, VT: Bear & Company, 220–8.

Gounand, I., Little, C. J., Harvey, E., and Altermatt, F. (2018). "Cross-Ecosystem Carbon Flows Connecting Ecosystems Worldwide," *Nature Communications*, 9/1: 4825.

Government of Nova Scotia (n.d.). *Ecological Forestry*. Halifax, NS: Nova Scotia Department of Natural Resources and Renewables. Available online at https://novascotia.ca/ecological-forestry/ (accessed May 25, 2023).

Grumbine, R. E. (1994). "What Is Ecosystem Management?" *Conservation Biology*, 8/1: 27–38.

Gunningham, N., Kagan, R. A., and Thornton, D. (2004). "Social License and Environmental Protection: Why Businesses Go beyond Compliance," *Law & Social Inquiry*, 29/2: 307–41.

Holzwarth, S., Thonfeld, F., Abdullahi, S., et al. (2020). "Earth Observation Based Monitoring of Forests in Germany: A Review," *Remote Sensing*, 12/21: 3570.

Hammersley Chambers, F., and Beckley, T. (2003). "Public Involvement in Sustainable Boreal Forest Management," in P. J. Burton, C. Messier, D. W. Smith, and W. L. Adamowicz (eds), *Towards Sustainable Management of the Boreal Forest*. Ottawa: NRC Research Press, 113–54.

Harshaw, H., and El-Lakany, H. (2017). "Public Participation in Forest Land-Use Decision-Making," in J. L. Innes and A. V. Tikina (eds), *Sustainable Forest Management: From Concept to Practice*. London: Routledge, 257–71.

Harvey, B. D., Nguyen-Xuan, T., Bergeron, Y., Gauthier, S., and Leduc, A. (2003). "Forest Management Planning Based on Natural Disturbances and Forest Dynamics," in P. J. Burton, C. Messier, D. W. Smith, and W. L. Adamowicz (eds), *Towards Sustainable Management of the Boreal Forest*. Ottawa: NRC Research Press, 395–432.

Hasler, B., Lundhede, T., and Martinsen, L. (2007). "Protection versus Purification: Assessing the Benefits of Drinking Water Quality," *Nordic Hydrology*, 38: 373–86.

Häyrinen, L., Mattila, O., Berghäll, S., and Toppinen, A. (2014). "Changing Objectives of Non-Industrial Private Forest Ownership: A Confirmatory Approach to Measurement Model Testing," *Canadian Journal of Forest Research*, 44/4: 290–300.

Hegde, R., and Enters, T. (2000). "Forest Products and Household Economy: A Case Study from Mudumalai Wildlife Sanctuary, Southern India," *Environmental Conservation*, 27/3: 250–9.

Helfield, J. M., and Naiman, R. J. (2001). "Effects of Salmon-Derived Nitrogen on Riparian Forest Growth and Implications for Stream Productivity," *Ecology*, 82/9: 2403–9.

Hermann, R. K., and Lavender, D. P. (1999). "Douglas-Fir Planted Forests," *New Forests*, 17: 53–70.

Higgs, E., Falk, D. A., Guerrini, A., et al. (2014). "The Changing Role of History in Restoration Ecology," *Frontiers in Ecology and the Environment*, 12: 499–506.

Himes, A., Betts, M., Messier, C., and Seymour, R. (2022). "Perspectives: Thirty Years of Triad Forestry, a Critical Clarification of Theory and Recommendations for Implementation and Testing," *Forest Ecology and Management*, 510: 120103.

Himes, A., Muraca, B., Anderson, C. B., et al. (2024). "Why Nature Matters: A Systematic Review of Intrinsic, Instrumental, and Relational Values," *BioScience*, 74/1: 25–43.

Himes, A., Puettmann, K., and Muraca, B. (2020). "Trade-offs between Ecosystem Services along Gradients of Tree Species Diversity and Values," *Ecosystem Services*, 44: 101133.

Howard, P. C., Davenport, T. R. B., Kigenyi, F. W., et al. (2000). "Protected Area Planning in the Tropics: Uganda's National System of Forest Nature Reserves," *Conservation Biology*, 14/3: 858–75.

Huber, D. W. (1992). "Utilization of Hardwoods, Fuelwood, and Special Forest Products in California, Arizona, and New Mexico," in *Ecology and Management of Oak and Associated Woodlands: Perspectives in the Southwestern United States and Northern Mexico: April 27–30, 1992, Sierra Vista, Arizona*. General Technical Report RM-GTR-218. Fort Collins, CO: USDA Forest Service, 103–8.

Hummel, M., and Freet, B. (1999). "Collaborative Processes for Improving Land Stewardship and Sustainability," in W. T. Sexton, A. J. Malk, R. C. Szaro, and N.

C. Johnson (eds), *Ecological Stewardship: A Common Reference for Ecosystem Management*, iii. New York: Elsevier, 97–129.

Hugosson, M., and Ingemarson, F. (2004). "Objectives and Motivations of Small-Scale Forest Owners: Theoretical Modelling and Qualitative Assessment," *Silva Fennica*, 38/2: 217–31.

Ishwaran, N., Persic, A., and Tri, N. H. (2008). "Concept and Practice: The Case of UNESCO Biosphere Reserves," *International Journal of Environment and Sustainable Development*, 7/2: 118–31.

Jacobson, S. K., McDuff, M. D., and Monroe, M. C. (2015). *Conservation Education and Outreach Techniques*. 2nd edn. Oxford: Oxford University Press. 427 pp.

Jemali, N. J. N. B., Shiba, M., and Zawawi, A. A. (2015). "Strategic Forest Management Options for Small-Scale Timber Harvesting on Okinawa Island, Japan," *Small-Scale Forestry*, 14/3: 351–62.

Jimoh, S. O., Ikyaagba, E. T., Alarape, A. A., Obioha, E. E., and Adeyemi, A. A. (2012). "The Role of Traditional Laws and Taboos in Wildlife Conservation in the Oban Hill Sector of Cross River National Park (CRNP), Nigeria," *Journal of Human Ecology*, 39/3: 209–19.

Johansson, J., Sandström, C., and Lundmark, T. (2018). "Inspired by Structured Decision Making: A Collaborative Approach to the Governance of Multiple Forest Values," *Ecology and Society*, 23/4: 16.

Jojola, T. (2013). "Indigenous Planning: Towards a Seven Generations Model," in R. Walker, T. Jojola, and D. Natcher (eds), *Reclaiming Indigenous Planning*. Montreal: McGill-Queen's University Press, 457–72.

Jones, L. P., Turvey, S. T., Massimino, D., and Papworth, S. K. (2020). "Investigating the Implications of Shifting Baseline Syndrome on Conservation," *People and Nature*, 2/4: 1131–44.

Kolstad, I. (2007). "Why Firms Should not Always Maximize Profits," *Journal of Business Ethics*, 76: 137–45.

Kaya, A., Bettinger, P., Boston, K., et al. (2016). "Optimisation in Forest Management," *Current Forestry Reports*, 2/1: 1–17.

Kengen, S. (1997). *Forest Valuation for Decision-Making: Lessons of Experience and Proposals for Improvement*. Rome: Food and Agriculture Organization of the United Nations. 151 pp. Available online at http://www.fao.org/3/w3641e/w3641e.pdf (accessed May 7, 2021).

Khan, J. A., Kumar, S. I. A. R. A., Khan, A., et al. (2008). "An Ecological Study in the Buffer Zone of the Corbett Tiger Reserve: Prey Abundance and Habitat Conditions," *International Journal of Ecology and Environmental Sciences*, 34/2: 121–31.

Kil, N., Holland, S. M., and Stein, T. V. (2012). "Identifying Differences between Off-Highway Vehicle (OHV) and Non-OHV User Groups for Recreation Resource Planning," *Environmental Management*, 50/3: 365–80.

Klein, N. (2015). *This Changes Everything: Capitalism vs the Climate*. Toronto: Penguin Random House. 564 pp.

Koh, N. S., Hahn, T., and Boonstra, W. J. (2019). "How Much of a Market Is Involved in a Biodiversity Offset? A Typology of Biodiversity Offset Policies," *Journal of Environmental Management*, 232: 679–91.

Kornatowska, B., and Sienkiewicz, J. (2018). "Forest Ecosystem Services–Assessment Methods," *Folia Forestalia Polonica*, 60/4: 248–60.

Krcmar, E., Vertinsky, I., and Van Kooten, G. C. (2003). "Modelling Alternative Zoning Strategies in Forest Management," *International Transactions in Operational Research*, 10/5: 483–98.

Krieger, D. J. (2001). *Economic Value of Forest Ecosystem Services: A Review*. Washington: Wilderness Society. 30 pp. Available online at http://www.truevaluemetrics.org/DBpdfs/EcoSystem/The-Wilderness-Society-Ecosystem-Services-Value.pdf (accessed February 6, 2021).

Kumar, S., Kant, S., and Amburgey, T. L. (2007). "Public Agencies and Collaborative Management Approaches: Examining Resistance among Administrative Professionals," *Administration & Society*, 39/5: 569–610.

Lee, G. G., and Park, C. W. (2012). "A Zoning Method for Forest Landscape Management by Visual Quality Assessment," *Journal of Korean Society of Forest Science*, 101/1: 148–57.

Lele, S., Springate-Baginski, O., Lakerveld, R., Deb, D., and Dash, P. (2013). "Ecosystem Services: Origins, Contributions, Pitfalls, and Alternatives," *Conservation and Society*, 11/4: 343–58.

Leopold, A. (1949). *A Sand County Almanac and Sketches Here and There*. New York: Oxford University Press. 240 pp.

Li, Y., and Zhang, D. (2014). "Industrial Timberland Ownership and Financial Performance of US Forest Products Companies," *Forest Science*, 60/3: 569–78.

Lindenmayer, D. B., and Franklin, J. F. (2002). *Conserving Forest Biodiversity: A Comprehensive Multiscaled Approach*. Washington: Island Press. 351 pp.

Lindgren, A., Robson, J. P., Reed, M. G., et al. (2019). *Engaging the Public in Sustainable Forest Management in Canada: Results from a National Survey of Advisory Committees*. Information Report LAU-X-142E. Quebec: Canadian Forest Service. 79 pp.

Loconto, A., Desquilbet, M., Moreau, T., Couvet, D., and Dorin, B. (2020). "The Land Sparing–Land Sharing Controversy: Tracing the Politics of Knowledge," *Land Use Policy*, 96: 103610.

Long, J. N. (2009). "Emulating Natural Disturbance Regimes as a Basis for Forest Management: A North

American View," *Forest Ecology and Management*, 257/9: 1868–73.

Lovrić, M., Da Re, R., Vidale, E., et al. (2020). "Non-Wood Forest Products in Europe: A Quantitative Overview," *Forest Policy and Economics*, 116: 102175.

Macdonald, I. A. W., Reaser, J. K., Bright, C., et al. (eds).(2003). *Invasive Alien Species in Southern Africa: National Reports & Directory of Resources*. Cape Town: Global Invasive Species Programme. 89 pp. Available online at https://www.doi.gov/sites/doi.gov/files/uploads/forging_cooperation_in_southern_africa.pdf (accessed February 1, 2024).

McGuiggan, R., and Foo, J. A. (2002). "Sun and Surf or Adventure: Who Plays what Tourist Roles?" *Asia–Pacific Advances in Consumer Research*, 5: 414–21.

McHarg, I. L. (1969). *Design with Nature*. Garden City, NY: Natural History Press. 198 pp.

MacPeak, M. D., Burkart, L. F., and Weldon, D. (1990). "Comparison of Grade, Yield, and Mechanical Properties of Lumber Produced from Young Fast-Grown and Older Slow-Grown Planted Slash Pine," *Forest Products Journal*, 40/1: 11–14.

Maslow, A. H. (1943). "A Theory of Human Motivation," *Psychological Review*, 50/4: 370–96.

Matson, P. A., and Vitousek, P. M. (2006). "Agricultural Intensification: Will Land Spared from Farming be Land Spared for Nature?" *Conservation Biology*, 20/3: 709–10.

Mavsar, R., and Varela, E. (2014). "Why Should We Estimate the Value of Ecosystem Services?" in B. J. Thorsen, R. Mavsar, L. Tyrväinen, I. Prokofieva, and A. Stenger (eds), *The Provision of Forest Ecosystem Services*, i. *Quantifying and Valuing Non-Marketed Ecosystem Services*. Joensuu, Finland: European Forest Institute, 41–6. Available online at https://efi.int/sites/default/files/files/publication-bank/2018/efi_wsctu5_vol1_2014.pdf (accessed June 3, 2023).

Messier, C., Bigué, B., and Bernier, L. (2003). "Using Fast-Growing Plantations to Promote Forest Ecosystem Protection in Canada," *Unasylva*, 54/3: 59–63.

Messier, C., and Kneeshaw, D. D. (1999). "Thinking and Acting Differently for Sustainable Management of the Boreal Forest," *Forestry Chronicle*, 75/6: 929–38.

Messier, C., Puettmann, K., Chazdon, R., et al. (2015). "From Management to Stewardship: Viewing Forests as Complex Adaptive Systems in an Uncertain World," *Conservation Letters*, 8/5: 368–77.

Miller, A. D., Vaske, J. J., Squires, J. R., Olson, L. E., and Roberts, E. K. (2017). "Does Zoning Winter Recreationists Reduce Recreation Conflict?" *Environmental Management*, 59/1: 50–67.

Millon, D. (2002). "Why Is Corporate Management Obsessed with Quarterly Earnings and what Should Be Done about it?" *George Washington Law Review*, 70: 890–920.

Mitchell, J. E., Ffolliott, P. F., and Patton-Mallory, M. (2005). "Back to the Future: Forest Service Rangeland Research and Management," *Rangelands*, 27/3: 19–28.

Miyazaki, Y. (2018). *Shinrin-yoku: The Japanese Way of Forest Bathing for Health and Relaxation*. London: Octopus Publishing Group. 192 pp.

Moore, J., and Clinton, P. (2015). "Enhancing the Productivity of Radiata Pine Forestry within Environmental Limits," *New Zealand Journal of Forestry*, 60/3: 35–41.

Mundy, B., and Kinnard, W. N. (1998). "The New Noneconomics: Public Interest Value, Market Value, and Economic Use," *Appraisal Journal*, 66/2: 207.

Naeem, S. (1998). "Species Redundancy and Ecosystem Reliability," *Conservation Biology*, 12/1: 39–45.

Nagel, T. A., Diaci, J., Rozenbergar, D., Rugani, T., and Firm, D. (2012). "Old-Growth Forest Reserves in Slovenia: The Past, Present, and Future," *Schweizerische Zeitschrift für Forstwesen*, 163/6: 240–6.

Natcher, D., Ingram, S., Bogdan, A. M., and Rice, A. (2021). "Conservation and Indigenous Subsistence Hunting in the Peace River Region of Canada," *Human Ecology*, 49: 109–20.

Nitschke, C. R., and Innes, J. L. (2005). "The Application of Forest Zoning as an Alternative to Multiple Use Forestry," in J. L. Innes, G. M. Hickey, and H. E. Hoen (eds), *Forestry and Environmental Change: Socioeconomic and Political Dimensions*. Wallingford, UK: CAB International, 97–124.

Nitschke, C. R., and Innes, J. L. (2008). "Integrating Climate Change into Forest Management in South-Central British Columbia: An Assessment of Landscape Vulnerability and Development of a Climate-Smart Framework," *Forest Ecology and Management*, 256/3: 313–27.

Norton, D., and Reid, N. (2013). *Nature and Farming: Sustaining Native Biodiversity in Agricultural Landscapes*. Melbourne: CSIRO Publishing. 304 pp.

Ogurtsov, S. S., Zheltukhin, A. S., and Kotlov, I. P. (2018). "Daily Activity Patterns of Large and Medium-Sized Mammals Based on Camera Traps Data in the Central Forest Nature Reserve, Valdai Upland, Russia," *Nature Conservation Research*, Заповедная наука, 3/2: 68–88.

Opuni-Frimpong, E., Karnosky, D. F., Storer, A. J., and Cobbinah, J. R. (2008). "Silvicultural Systems for Plantation Mahogany in Africa: Influences of Canopy Shade on Tree Growth and Pest Damage," *Forest Ecology and Management*, 255/2: 328–33.

O'Regan, N., and Ghobadian, A. (2007). "Formal Strategic Planning: Annual Raindance or Wheel of Success?" *Strategic Change*, 16/1–2: 11–22.

Padilla, E. (2002). "Intergenerational Equity and Sustainability," *Ecological Economics*, 41/1: 69–83.

Pandey, D., and Brown, C. (2000). "Teak: A Global Overview," *Unasylva*, 51: 3–13.

Paquette, A., and Messier, C. (2010). "The Role of Plantations in Managing the World's Forests in the Anthropocene," *Frontiers in Ecology and the Environment*, 8/1: 27–34.

Parks, J. (2009). *The War on Pests: Dealing to Key Pest Plants and Animals that Threaten Native Species: A Landowners Guide for Banks Peninsula and Kaitorete Spit.* Christchurch, New Zealand: Banks Peninsula Conservation Trust and Environment Canterbury. 74 pp. Available online at https://www.ecan.govt.nz/document/download/?uri=1172438 (accessed February 1, 2024).

Pascual, U., Balvanera, P., Christie, M., et al. (eds).(2022). *Summary for Policymakers of the Methodological Assessment Report on the Diverse Values and Valuation of Nature: Summary for Policymakers.* Intergovernmental Science–Policy Platform on Biodiversity and Ecosystem Services. Bonn, Germany: IPBES Secretariat. Available online at https://zenodo.org/record/7410287 (accessed June 3, 2023).

Penn State Extension (2008). *Planning for Beauty and Enjoyment.* Forest Stewardship Bulletin Number 8. Harrisburg, PA: Pennsylvania State University. 7 pp. Available online at https://extension.psu.edu/forest-stewardship-planning-for-beauty-and-enjoyment (accessed 3 May 23, 2023).

Perera, A. H., Buse, L. J., Weber, M. G., and Crow, T. R. (2004). "Emulating Natural Forest Landscape Disturbances: A Synthesis," in A. H. Perera, L. J. Buse, and M. G. Weber (eds), *Emulating Natural Forest Landscape Disturbances.* New York: Columbia University Press, 265–74.

Petchey, O. L., and Gaston, K. J. (2002). "Functional Diversity (FD), Species Richness and Community Composition," *Ecology Letters*, 5/3: 402–11.

Poffenberger, M. (1990). "Facilitating Change in Forestry Bureaucracies," in M. Poffenberger (ed.), *Keepers of the Forest: Land Management Alternatives in Southeast Asia.* West Hartford, CT: Kumarian Press, 101–18.

Pohjanmies, T., Eyvindson, K., and Mönkkönen, M. (2019). "Forest Management Optimization across Spatial Scales to Reconcile Economic and Conservation Objectives," *PloS One*, 14/6: e0218213.

Postel, S. L., and Thompson, B. H. (2005). "Watershed Protection: Capturing the Benefits of Nature's Water Supply Services," *Natural Resources Forum*, 29/2: 98–108.

Pra, A., Masiero, M., Barreiro, S., et al. (2019). "Forest Plantations in Southwestern Europe: A Comparative Trend Analysis on Investment Returns, Markets and Policies," *Forest Policy and Economics*, 109: 102000.

Primack, R. B. (2014). *Essentials of Conservation Biology.* 6th edn. Sunderland, MA: Sinauer Associates. 603 pp.

Reid W. V., Mooney, H. A., Cropper, A., et al. (2005). *Ecosystems and Human Well-Being: Synthesis.* A Report of the Millennium Ecosystem Assessment. Washington: Island Press. 137 pp. Available online at https://www.millenniumassessment.org/documents/document.356.aspx.pdf (accessed February 1, 2024).

Richards, C. R. (2005). "Management of Indigenous Forest," in M. Colley (ed.), *Forestry Handbook.* Tauranga, New Zealand: New Zealand Institute of Forestry, 29–35.

Rickenbach, M. G., and Reed, A. S. (2002). "Cross-Boundary Cooperation in a Watershed Context: The Sentiments of Private Forest Landowners," *Environmental Management*, 30/4: 584–94.

Ricketts, T. H. (2001). "The Matrix Matters: Effective Isolation in Fragmented Landscapes," *American Naturalist*, 158/1: 87–99.

Robertson Vernhes, J. (1989). "Biosphere Reserves: The Beginnings, the Present and the Future Challenges," in W. P. Gregg., S. L. Krugman, and J. D. Wood (eds), *Proceedings of the Symposium on Biosphere Reserves, Fourth World Wilderness Congress, September 14–17, 1987, Estes Park, Colorado, USA.* Atlanta, GA: National Park Service, US Department of the Interior, 7–20.

Rodan, B. D., Newton, A. C., and Verissimo, A. (1992). "Mahogany Conservation: Status and Policy Initiatives," *Environmental Conservation*, 19/4: 331–42.

Rodríguez, J. P., Beard, T. D., Bennett, E. M., et al. (2006). "Trade-offs across Space, Time, and Ecosystem Services," *Ecology and Society*, 11/1: 28.

Rollert, J. P. (2014). "Greed is Good: A 300-Year History of a Dangerous Idea," *The Atlantic*, April 7. Available online at https://www.theatlantic.com/business/archive/2014/04/greed-is-good-a-300-year-history-of-a-dangerous-idea/360265/ (accessed May 1, 2023).

Runting, R. K., Ruslandi, Griscom, B. W., et al. (2019). "Larger Gains from Improved Management over Sparing–Sharing for Tropical Forests," *Nature Sustainability*, 2/1: 53–61.

Ruslandi, Romero, C., and Putz, F. E. (2017). "Financial Viability and Carbon Payment Potential of Large-Scale Silvicultural Intensification in Logged Dipterocarp Forests in Indonesia," *Forest Policy and Economics*, 85: 95–102.

Sahajananthan, S., Haley, D., and Nelson, J. (1998). "Planning for Sustainable Forests in British Columbia through Land Use Zoning." *Canadian Public Policy*, 24: S73–S81.

Saunders, S. P., Grand, J., Bateman, B. L., et al. (2023). "Integrating Climate-Change Refugia into 30 by 30 Conservation Planning in North America," *Frontiers in Ecology and the Environment*, 21/2: 77–84.

Schmidt, S., Manceur, A. M., and Seppelt, R. (2016). "Uncertainty of Monetary Valued Ecosystem Services:

Value Transfer Functions for Global Mapping," *PloS One*, 11/3: e0148524.

Schmidt-Traub, G., Locke, H., Gao, J., et al. (2020). "Integrating Climate, Biodiversity, and Sustainable Land-Use Strategies: Innovations from China," *National Science Review*, 8/7: nwaa139.

Schoennagel, T., Nelson, C. R., Theobald, D. M., Carnwath, G. C., and Chapman, T. B. (2009). "Implementation of National Fire Plan Treatments near the Wildland–Urban Interface in the Western United States," *Proceedings of the National Academy of Sciences*, 106/26: 10706–11.

Sedjo, R. A. (1999). "The Potential of High-Yield Plantation Forestry for Meeting Timber Needs," *New Forests*, 17: 339–59.

Sedjo, R. A., and Botkin, D. (1997). "Using Forest Plantations to Spare Natural Forests," *Environment: Science and Policy for Sustainable Development*, 39/10: 14–30.

Seymour, R. S., and Hunter, M. L. (1992). *New Forestry in Eastern Spruce-Fir Forests: Principles and Applications to Maine*. Miscellaneous Publication 716. Orono, ME: Maine Agricultural Experiment Station, University of Maine. 36 pp.

Sheppard, S. R., and Meitner, M. (2005). "Using Multi-Criteria Analysis and Visualisation for Sustainable Forest Management Planning with Stakeholder Groups," *Forest Ecology and Management*, 207/1–2: 171–87.

Simard, S., Martin, K., Vyse, A., and Larson, B. (2013). "Meta-Networks of Fungi, Flauna and Flora as Agents of Complex Adaptive Systems," in C. Messier, K. J. Puettmann, and K. D. Coates (eds), *Managing Forests as Complex Adaptive Systems: Building Resilience to the Challenge of Global Change*. New York: Routledge, 133–64.

Soga, M., and Gaston, K. J. (2018). "Shifting Baseline Syndrome: Causes, Consequences, and Implications," *Frontiers in Ecology and the Environment*, 16/4: 222–30.

Sotirov, M., Sallnäs, O., and Eriksson, L. O. (2019). "Forest Owner Behavioral Models, Policy Changes, and Forest Management: An Agent-Based Framework for Studying the Provision of Forest Ecosystem Goods and Services at the Landscape Level," *Forest Policy and Economics*, 103: 79–89.

Stedman, R. C. (2003). "Sense of Place and Forest Science: Toward a Program of Quantitative Research," *Forest Science*, 49/6: 822–9.

Stevenson, M. G., and Webb, J. (2003). "Just Another Stakeholder? First Nations and Sustainable Forest Management in Canada's Boreal Forest," in P. J. Burton, C. Messier, D. W. Smith, and W. L. Adamowicz (eds), *Towards Sustainable Management of the Boreal Forest*. Ottawa: NRC Research Press, 65–112.

Stone, C. D. (2010). *Should Trees Have Standing? Law, Morality, and the Environment*. 3rd edn. Oxford: Oxford University Press. 264 pp.

Stout, L. (2015). "Corporations Don't Have to Maximize Profits," *New York Times*, April 15. Available online at https://www.nytimes.com/roomfordebate/2015/04/16/what-are-corporations-obligations-to-shareholders/corporations-dont-have-to-maximize-profits (accessed January 11, 2024).

Thorsen, B. J., Mavsar, R., Tyrväinen, L., Prokofieva, I., and Stenger, A. (eds). (2014). *The Provision of Forest Ecosystem Services, i. Quantifying and Valuing Non-Marketed Ecosystem Services*. Finland: European Forest Institute, Joensuu. 73 pp. Available online at https://efi.int/sites/default/files/files/publication-bank/2018/efi_wsctu5_vol1_2014.pdf (accessed February 1, 2024).

Timko, J. A., and Satterfield, T. (2008). "Seeking Social Equity in National Parks: Experiments with Evaluation in Canada and South Africa," *Conservation and Society*, 6/3: 238–54.

Tittler, R., Messier, C., and Fall, A. (2012). "Concentrating Anthropogenic Disturbance to Balance Ecological and Economic Values: Applications to Forest Management," *Ecological Applications*, 22/4: 1268–77.

Tucker, C. M. (2004). "Community Institutions and Forest Management in Mexico's Monarch Butterfly Reserve," *Society and Natural Resources*, 17/7: 569–87.

Vincent, J. R., and Binkley, C. S. (1993). "Efficient Multiple-Use Forestry May Require Land-Use Specialization," *Land Economics*, 69/4: 370–6.

Wang, S. (2019). "Managing Forests for the Greater Good: The Role of the Social License to Operate," *Forest Policy and Economics*, 107: 101920.

Ward, C., and Erdle, T. (2015). "Evaluation of Forest Management Strategies Based on Triad Zoning," *Forestry Chronicle*, 91/1: 40–51.

Watson, J. E., Darling, E. S., Venter, O., et al. (2016). "Bolder Science Needed Now for Protected Areas," *Conservation Biology*, 30/2: 243–8.

WCED (World Commission on Environment and Development). (1987). *Our Common Future*. Oxford: Oxford University Press. 383 pp.

White, D. A., Silberstein, R. P., Balocchi-Contreras, F., Quiroga, J. J., Meason, D. F., Palma, J. H., and de Arellano, P. R. (2021). "Growth, Water Use, and Water Use Efficiency of *Eucalyptus globulus* and *Pinus radiata* Plantations Compared with Natural Stands of Roble-Hualo Forest in the Coastal Mountains of Central Chile," *Forest Ecology and Management*, 501: 119676.

Williams, D. R., and Stewart, S. I. (1998). "Sense of Place: An Elusive Concept that Is Finding a Home in Ecosystem Management," *Journal of Forestry*, 96/5: 18–23.

Wilson, J. (1998). *Talk and Log: Wilderness Politics in British Columbia*. Vancouver: UBC Press. 452 pp.

Young, K. A. (2000). "Riparian Zone Management in the Pacific Northwest: Who's Cutting What?" *Environmental Management*, 26/2: 131–44.

Zhang, H. F., Ouyang, Z. Y., and Zheng, H. (2007). "Spatial Scale Characteristics of Ecosystem Services," *Chinese Journal of Ecology*, 26/9: 1432–7.

Zhang, Z., Sherman, R., Yang, Z., et al. (2013). "Integrating a Participatory Process with a GIS-Based Multi-Criteria Decision Analysis for Protected Area Zoning in China," *Journal for Nature Conservation*, 21/4: 225–40.

CHAPTER 3

In Pursuit of Sustainability

The idea of sustainability is surprisingly simple: resource consumption cannot exceed resource production over time.

Donald W. Floyd (2002)

3.1 Sustainability as a core human aspiration

It is human nature to desire assured access to resources, well-being in all its dimensions, and some level of confidence that those requirements will be met into the future and for one's offspring. Today these aspirations are encompassed in the term sustainability, which basically denotes persistence and endurance, but can be interpreted in many different ways. Differences of opinion often reflect alternative priorities for what should be sustained, or over what scale. This chapter explores some of the complications and implications of planning for sustainability, and the challenge of moving from simple sustained-yield forestry to the more complex paradigm of sustaining all forest values in a changing world. It can be argued that all human decisions and actions now need to be questioned through the revolutionary lens of sustainability (Edwards 2005) as we approach or surpass many planetary limits. But there are many complex interacting drivers of change in forests and in natural-resource sectors, often responding to global factors (Sesser et al. 2019; see Figure 3.1). Combined with the usual uncertainties in predicting the future, the rapid pace of global change now suggests that planning for sustainability may not be a sufficiently robust strategy to assure the persistence of the world's many forest ecosystems and forest ecosystem services.

For most of human existence (and well before "sustainability" and "resilience" became management buzzwords), our species generally equated success with survival and the ability to leave descendants. This perception of success equates well with the biological concept of fitness, which can be achieved by as many diverse strategies and adaptations as there are species on this planet. But, somewhere along the line, the bipedal, tool-making, and flexible behavior of one line of tribal apes emerged as an exceptional strategy for survival and fitness. Moving forward over the millennia, the human species has mastered fire, domesticated plants and animals, cut down forests, diverted rivers, built cities and nation states, and spread throughout the planet with the concomitant evolution of thousands of languages and cultures. Recent centuries have seen the harnessing of fossil fuels and nuclear power, and the manufacture of plastics and persistent organic compounds, which all have the potential to alter drastically the Earth system. Indeed, it is now widely recognized that humankind has become too fit and too "successful": the legacy of billions of people, our consumption patterns, our waste and toxic byproducts, and our domination of the natural world means that we have exceeded several of the earth's life-support systems or are at risk of doing so (Rockström et al. 2009; Costanza et al. 2015; Steffen et al. 2015; Wackernagel and Beyers 2019). The United Nations proposed seventeen sustainable development goals (SDGs; see Box 1.2) in response to these challenges of the modern world, including the need to reduce marked discrepancies in human well-being around the planet. Sustainable and resilient forest

Resilient Forest Management. Philip J. Burton, Oxford University Press. © Philip J. Burton (2025). DOI: 10.1093/oso/9780198832997.003.0003

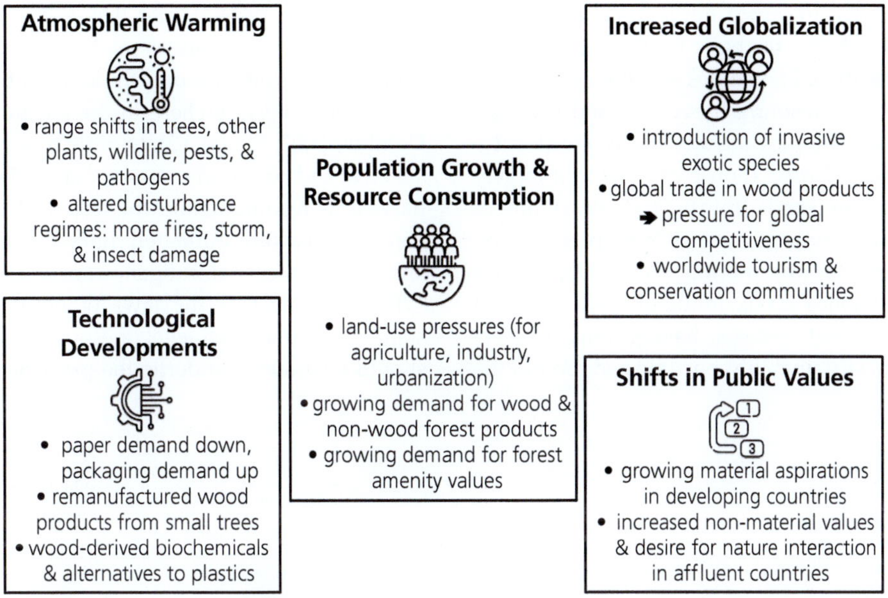

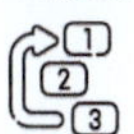

Figure 3.1 Some mechanisms of global change and their potential effects on forests, sustainable forest management, the forest products sector, and forest planning (graphics from www.flaticon.com, used with permission)

stewardship can help attain all those goals for planetary sustainability.

Even before planetary boundaries were approached, population growth and territorial expansion by *Homo sapiens* were not universally beneficial. Colonial enterprises and nationalist or religious missions of conquest were often violent affairs, accompanied by genocide or cultural suppression of Indigenous peoples, the transcontinental transmission of disease, and the wholesale destruction of ecosystems. With wealth, power, and the writing of history accruing to the victors, territorial expansion and competitive domination were undertaken repeatedly over the millennia and were viewed as heroic. One might say that today's corporate empires are the natural descendants of history's conquistadors, with growth and the pursuit of wealth underlying their raison d'être (see Box 2.4). While economic growth can occur through the development of new goods, services, and transaction pathways (Costanza et al. 2015), growth in human populations and increasing levels of resource consumption cannot continue indefinitely; that is, they are unsustainable. At the personal and household level, many people would agree that the

reliability of a standard of living and other non-material values are ultimately more important than the acquisition of more material goods and exotic experiences, because increasing wealth beyond a certain level does not generate greater well-being or happiness (Ahuvia 2008; Popescu 2016). The recognition of limits—whether planetary, territorial, ecological, financial, or social—is usually a required precursor to discussions about sustainability and how to attain it.

3.2 Lessons from human experience

From a global perspective, one might think that resource limitations to human well-being are a recent phenomenon. But local and regional "limits to growth" were encountered in many societies well before global limits were drawn to our attention in the Meadows et al. (1972) book by that name. Societies dependent on hunting, fishing, and subsistence agriculture have regularly implemented conservation measures when food stocks diminished (Gadgil et al. 1993; Johannes 2002). Failure to do so often led to ecological degradation, human suffering, conflict, and strife (Hardin 1968;

Diamond 2005). Resource limitations can also be interpreted as being the basis for the spatial delineation of traditional territories and their associated jurisdiction over resource harvesting and management rights. Examples include the *laxyip* (territories) of Gitxsan *huwilp* (house groups) that largely conform to watersheds (Burda et al. 1999), one of the many land management systems practiced by multiple First Nations in British Columbia, Canada (Turner and Jones 2000). Innumerable traditional land management systems, having similar objectives of local sustainability, once existed around the world (Davidson-Hunt and Berkes 2001). Another example is the *ahupua'a* sea-to-mountain traditional tenure system found throughout the Hawaiian Islands and elsewhere in the Pacific (Mueller-Dombois 2007; Figure 3.2). In order to assure group survival and continuity, a unit of land, the rights to harvest its resources, and the responsibility for its stewardship were intimately linked to the culture and belief systems of Indigenous peoples.

Expectations that the laws of nature are constant and that a given level of economic well-being should be assured underlie the practices, customs,

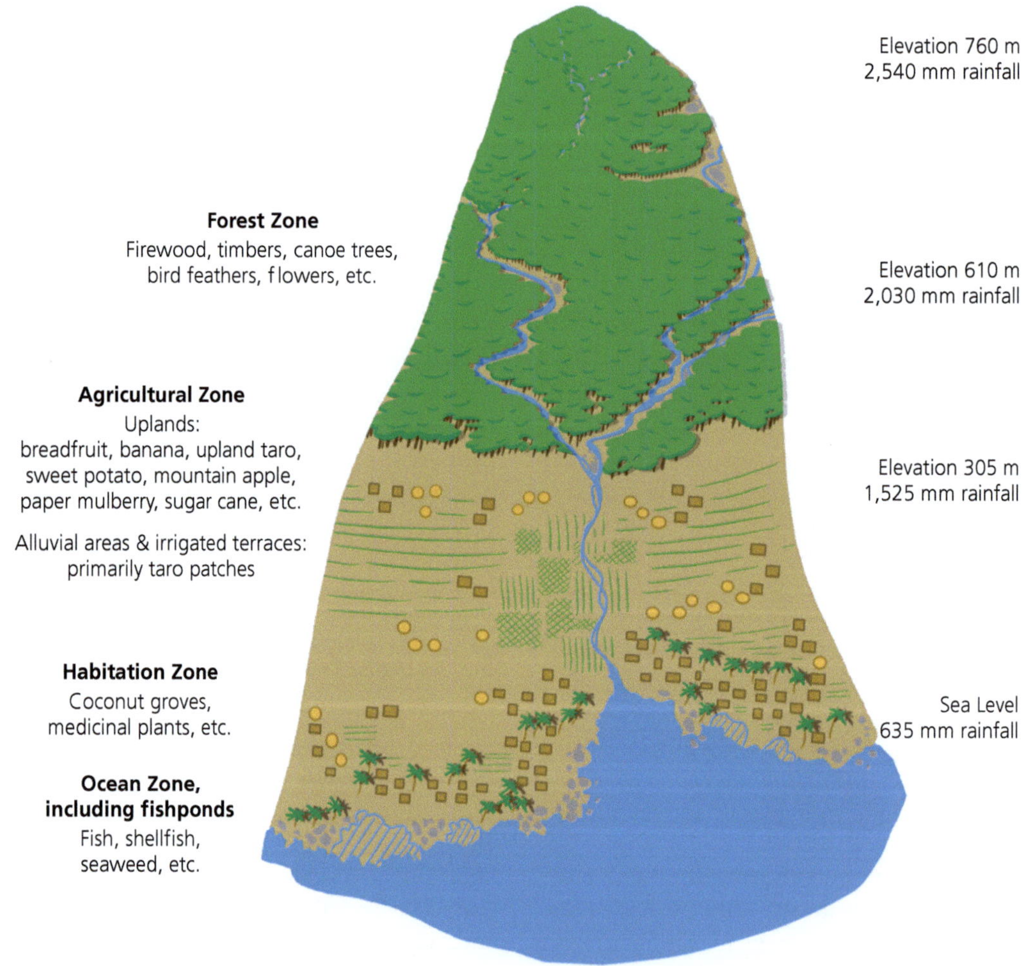

Figure 3.2 Spatial portrayal of the *ahupua'a* land management system in traditional Hawai'i

Note: Generally following catchment boundaries, each *ahupua'a* reaches from offshore coral reefs to mountain peaks, spanning a range of climates and ecosystem types, to ensure access to a sustainable supply of all the food and material needs of a village (graphic by Petroglyph Studios (www.petroglyphstudios.org), based on Mueller-Dombois (2007) and County of Maui (2010)).

and laws of all societies. Relatively stable and predictable conditions—peace and tranquility, if you will—are typically associated with a safe and fulfilling life, while upheaval and uncertainty lead to psychological stress (Nolan et al. 2000; Reser 2007). By definition, all renewable resources—timber, fish, hunted wildlife, field crops—should be infinitely sustainable under constant conditions. However, nature and the planet's physical environment have always been dynamic, such that the only "balance of nature" we can maintain is in humankind balancing its impacts within the bounds of natural variability and adaptability (Botkin 1990).

Efforts to maintain resource productivity and minimize uncertainty over long periods have met with success in many parts of the world and in many sectors. There are notable examples demonstrating that long-term sustainability is possible: traditional practices of organic fertilization and rotational cropping are responsible for millennia of sustainable farming in both the Old World and the Americas (e.g. Ellis and Wang 1997; Pulido and Bocco 2003). On the other hand, we have seen widespread collapses of fisheries in the wake of poorly regulated industrial harvesting (Pauly et al. 2005). In addition to the threat of short-term overexploitation of resources, there are always surprises that challenge stability and can upset the status quo. This wisdom is expressed in many cultures through the character of a "trickster," a zoomorphism of events and conditions that do not conform to normal expectations (Box 3.1). Trickster and trickster events may seem to challenge sustainability, but encourage preparedness and resilience, so can facilitate sustainability in the long run.

In its simplest form, concern for sustainability can be equated with long-term thinking or showing consideration for the needs of future generations. Social scientists use the term "intergenerational equity" to describe this principle, an attempt to give rights and options to people not yet born so not at the negotiating table. Continuing the theme of insights from ancient cultures, we might invoke the consideration of seven generations as advocated by Haudenosaunee (Iroquois), Anishinaabe, and other Indigenous peoples of eastern North America. One interpretation of the "Seven Generations Principle" is that one should learn from and respect the wishes

of grandparents and also consider the needs of one's grandchildren when making decisions. Alternatively, it can simply mean planning for seven generations hence, which would average out to something like 175 years! This perspective invites more thoughtful and far-reaching thinking and planning than merely meeting the demands of the present.

3.3 Sustainability can be complicated

Long-term thinking is a good start to achieving sustainability, but it requires more than that. When formalized, as in the case of sustainable forestry or when adopted as government policy, sustainability requires a number of associated parameters and assumptions. Whether planning is for seven generations or a time horizon of "in perpetuity," issues of the sustainability of what, for whom, and over what geographical or jurisdictional area always emerge. Despite the quoted epigraph opening this chapter, there can be many different interpretations of sustainability and many obstructions, risks, and uncertainties in its pursuit.

The Brundtland Commission's call (WCED 1987) for "sustainable development" spawned a discourse on sustainability that continues to this day. Whether this is considered an oxymoron, a negotiated compromise between environmental and economic interests, or an enlightened way forward, every institution that has endorsed sustainable development probably has its own interpretation of what that means. Sustainability has become a term adopted in business plans, institutional mission statements, and political campaigns around the world. Nonetheless, the Brundtland Report helped usher in an era of public consciousness that accepted and expected sustainability as a criterion for investment, government policy, and development plans. Sustainability was no longer a concept limited to the world of agronomy and soil fertility nor to foresters and timber yields, but extended to the management of human endeavors in general. Because the Brundtland Report entwined sustainability (that is, environmental conservation) with development (that is, economic growth), there is considerable latitude in weighting those two extremes. Social values are invoked as a means of weighting or rationalizing the emphasis to be

Box 3.1 Enter the trickster

The potential for—or even widespread prevalence of—surprise in worldly affairs is recognized and respected in the folklore of many peoples around the world. Often taking the form of an animal (see Figure 3.3) or a being that can travel in both the material and supernatural realms, "trickster" characters defy social norms, play pranks on unsuspecting protagonists, or serve as strong creative forces of nature, often turning things upside down in the process. Ranging from the shape-shifter Loki of Nordic legend and the clever fox of Aesop's Fables, to the hare in several African cultures, and from the badger in Japan to the coyote and raven in North America Indigenous mythology, such characters invite derision owing to foolish actions or respect because of their wily nature and creative powers.

It has been said that tricksters indulge in "creative negation" (Babcock 1975), a strong parallel to the "creative destruction" recognized as recurring in the adaptive cycle of many ecological and social systems (Holling 2001; see Section 4.4). The trickster may precipitate unexpected disturbance, destruction, and initiation of an "omega" phase of chaotic release in the adaptive cycle ... and/or may stimulate the creative reorganization of new ways of being in an "alpha" phase (see Figure 4.4 for phases of the adaptive cycle). Sometimes their actions serve as a morality tale of what not to do, or as a creation tale of how certain aspects of the natural world, technological achievement, or cultural tradition came into being.

Shared experience and ancient wisdom alert us to the fact that the world is sometimes populated with episodes of unexpected change having both destructive and creative elements. Folklore reminds leaders and heroes as well as everyday people that they are not always in control, that we should not be surprised if a wild card (for example, the joker, a jester—a trickster) shows up. While not always the main point of original Indigenous stories (Fagan 2010), the risk of surprise should inspire humility and a degree of contingency in all planning. On the other hand, trickster always reminds us that creative ways forward may emerge if we are willing to ignore traditions and societal norms.

Figure 3.3 Raven and Coyote are two powerful tricksters in the Indigenous traditions of Northwest Coast and North American Plains cultures, respectively (artwork by Petroglyph Studios (www.petroglyphstudios.org)).

placed on environmental or economic priorities, hence the emphasis in Chapter 2 on consultation as an important component of forest management and planning. This three-way compromise results in the metaphor by which sustainability is portrayed as three pillars, a three-legged stool, or the intersection of a Venn diagram of environmental, economic, and social acceptability (Figure 3.4a). The relative weights applied to these bundles of values and the resulting social endorsement ("social license to operate") can differ considerably among local, regional, national, or international arenas. Furthermore, it is being increasingly acknowledged that economic considerations are a subset of social values (Figure 3.4b), both of which rest on a foundation of environmental services and capacity (Figure 3.4c).

Another source of disagreement over the interpretation of sustainability centers on whether one prioritizes specific individual values (for example, wildlife populations, jobs, or financial returns on investment) or net condition of the whole socioecological system. In general, policy tries to maximize the net "wealth" or "welfare" (well-being) of whole systems (Adamowicz and Burton 2003). When advocating for the ideal of "strong sustainability," the goal is for all individual environmental goods and services to continue in perpetuity. On the other hand, "weak sustainability" may include the conversion of some natural (environmental) assets into human infrastructure and capacity (Kuhlman and Farrington 2010). A fundamental, if unspoken, assumption

of weak sustainability is that economic, ecological, and social values are interchangeable or substitutable, and that they can be reasonably weighed against each other for the sake of overall system-wide sustainability. This rationale is the basis for using the revenue from large timber volumes harvested from old-growth forests to build infrastructure such as forest roads and bridges in support of conversion to managed second-growth forest. While environmentalists generally support strong sustainability, they recognize the arbitrary nature of benchmarks in space and time, whereby many current conditions should not be sustained, and many desirable conditions of the past require restoration, not just sustainability (Ekins et al. 2003). Industrial developers and promoters of economic growth likewise recognize that weak sustainability (as the degree to which one can accept the conversion of natural assets to human assets) still has natural limits and is contingent on social acceptability that varies from place to place and over time (Parsons et al. 2014). In practice, forest policy and management decisions usually fall somewhere intermediate along the gradient from strong to weak sustainability, reflecting the socioeconomic pressures of the day.

Sustainability assumes a degree of constancy and predictability in the natural and socioeconomic worlds, yet such constancy is usually short lived if experienced at all (Botkin 1990). Natural disturbances (fires, floods, earthquakes, volcanic eruptions, landslides, and so on), erratic weather, disease, and war have disrupted farms, forests, fac-

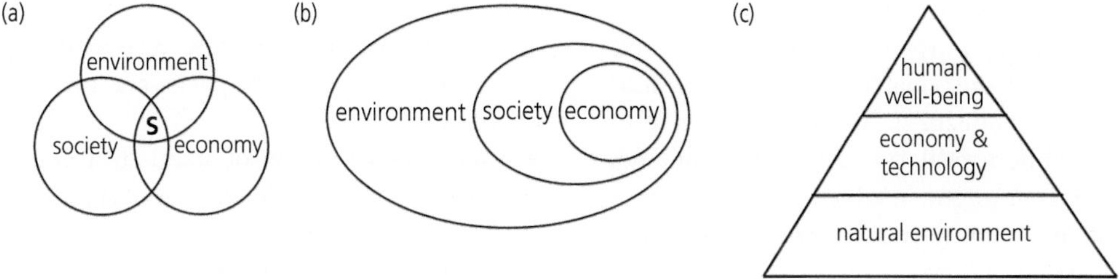

Figure 3.4 Alternative schematic portrayals of the requirements for sustainability and human wellbeing: (a) the feasible intersection, S, of economic, social, and environmental values; (b) recognition of the economy as a subset of the social system, which in turn is subsumed within the biophysical environment; (c) a pyramid in which well-being is underlain by economic and technological processes that refine and distribute ecosystem services provided by the earth's natural environment (from Burton (2020), reproduced with permission of Burleigh Dodds Science Publishing).

tories, and the agendas of nation states for all of history. When faced with business and investment decisions, additional disruptions to careful planning for a predictable, reliable, sustainable future include technological developments, the emergence of competitors, the opening and closing of trade routes and political alliances, and gradual shifts in consumer preferences and public opinion. Of course, many of those disruptions have been positive, not just negative or destructive, in empowering and enlightening the human endeavor. With more than eight billion people on the planet, instantaneous communications, widespread intercontinental trade and travel, and a turbulent biosphere in the process of adjusting to a warming atmosphere, changes and disruptions are occurring at a rate and scale unprecedented in recent human history (Morris 2019; Sage 2020; see Figure 3.1). Occasional longings for a slower world and "the good old days" now represent more than nostalgia, but may reflect a deeper craving for a more constant, predictable, and reliably sustainable world. Yet, unless we experience a precipitous collapse of population, social order, and its technological supports, such a slow-down is unlikely.

It is worth noting that sustainability has become such a widely accepted goal and management principle that it has essentially lost much of its meaning. The term has been co-opted to mean "good" or "responsible" or "feasible" management and policy in general. Gradually creeping into the vision statements of every institution and corporation, we now see sustainability invoked even by the most demonstrably exploitative and unsustainable sectors. Such "greenwashing" can be seen in the appearance of a *Journal of Sustainable Mining* (https://jsm.gig.eu/), claims of sustainability in the petroleum sector (API 2021), and sustainable logging of primary forests (Tammemagi 2021). Although sustainability has broad public appeal, it is clearly time for a new management paradigm, in which sustainability can be aspirational but is not sufficient to guide human endeavors into the future.

3.4 Sustained-yield forestry and its requirements

Sustainability has been the goal, indeed the hallmark, of responsible forestry for centuries (Burton

2020). The key feature of sustained-yield forestry is that the amount of wood harvested from a defined area and period of time should not exceed the amount of wood grown over that same area and time. If practiced carefully, timber extraction regulated in this manner, in theory, should be able to proceed indefinitely. Sustained yield forestry can be likened to living off the interest or dividends of an investment, never drawing down the capital to levels that cannot be assuredly rebuilt over a defined period of time.

Those seemingly simple requirements for sustainable forests and a sustainable forest products industry remain aspirational goals around much of the world. While the desire to protect, conserve, or sustain other (non-wood) forest values requires a set of even more demanding considerations (as discussed in the Section 3.4), the information, regulation, assumptions, and repeated assessments needed for sustained fiber production alone can be daunting. Some of those needs and challenges include the following:

- clear delineation of the area of forestland available for timber production and harvesting, excluding all wetlands, croplands and other untreed areas, non-commercial forests, protected areas, conservancies, riparian or roadside buffers, sacred groves, and other areas not available for logging;
- accurate inventories (typically volumetric, m^3) of the timber available on that land base, including its species composition and the age-class breakdown for each tree species or forest type;
- estimates of commercially recoverable proportions of wood for each species and age-class combination;
- reliable regeneration of stands to full or known levels of stocking after timber harvesting;
- accurate assessments of tree growth and volume increment, coupled with reliable yield projections and determinations of the time it will take for newly regenerated stands/trees to be worth harvesting in the future;
- exclusion, control, or reliable estimates of timber losses to damaging agents such as fire, insect outbreaks, wind storms, and pathogens;

- where the starting point for forest management is bare land, recognition that several decades or more are required to plant, nurture, and protect trees before the promise of a sustained wood supply can be met;
- where forest management starts with a wild forest of irregular age-class structure, recognition that it may also take several decades (often harvesting suboptimal wood, trees that are too small, or trees that have high levels of waste) to regulate the age-class structure so it will generate a reliable and continuous stream of wood.

Viewed from an economics and policy perspective, the rationale and requirements for sustained yield become even more constrained. Sustained-yield timber management, with its dependence on multi-decadal projections and forest protection, requires a high degree of socio-environmental stability and certainty. Will a forested land-use and a forest management plan (including harvesting constraints and silvicultural requirements) be respected for many decades or even a century? Furthermore, the incentive for forest conservation and long-term management for reliable access to timber as a commodity assumes a scarcity of land and a closed economy, conditions that may have applied to medieval European villages but not to a modern world characterized by efficient transportation networks and widespread trade (Behan 1978). Alternative models for responsible resource stewardship have long existed that do not have the rigid assumptions and requirements of traditional sustained-yield forestry. Raup (1964: 26) invited us to "think in terms of massive uncertainty, flexibility and adjustability," and Behan (1978) likewise advocated a "hang-loose approach" to sustainable forestry. Work et al. (2003) saw the development of sustainable forest management as an invitation for innovation and experimentation, not rigid adherence to ensconced policies and practices. These authors recognized and wrestled with many of the socio-ecological dynamics that led to the emergence of resilience thinking in the twenty-first century (Walker and Salt 2006; Gunderson and Allen 2010).

The scale at which forests are expected to produce a sustained yield of timber is an important consideration (Box 3.2), and one that has many implications regarding the range of options available in the forest management toolbox. Entwined with the scale issue is a debate that has raged for more than a century: whether trees should be grown and harvested as even-aged crops or as complex stands consisting of different ages and sizes of trees from which individual trees are harvested. Most dramatically expressed as the contrast between plantation forestry with clearcut logging versus various forms of continuous cover forestry (e.g. Mason et al. 2004; Pukkala and von Gadow 2012; Seedre et al. 2018), this debate invokes deep emotions on behalf of managers committed to efficiency and stakeholders committed to environmental protection. Yet both models have many natural precedents, and both can deliver sustainable supplies of timber and non-timber values and ecosystem services, though at different scales of space and time. Box 3.2 illustrates how the sustainable age-class structure of landscapes managed under the even-aged model is expressed in different terms from the sustainable age-class structure of stands managed under the uneven-aged model.

Location-specific comparisons of even-aged and uneven-aged management have been made from time to time (e.g. Guldin and Baker 1988; Laiho et al. 2011). The superiority of one approach over the other usually depends on management direction for the retention of canopy cover, the size of the forest estate over which sustainability is planned, and the silvics (notably shade tolerance) of the tree species being managed. Where it is an objective to manage forests in a close-to-nature manner, logging activities can be customized to emulate the natural disturbance regime. Observations of natural forests and their response to natural disturbances then guide the selection of forest opening sizes, criteria for retained trees, and the frequency of harvesting (Bergeron et al. 1999; Seymour et al. 2002; Long 2009; Kuuluvainen and Grenfell 2012; Palik et al. 2021). This means that many patterns of timber harvesting and silviculture intermediate between even- and uneven-aged management can be seen on the ground (for example, in the form of two- or three-cohort stands), undertaken with various degrees of commercial thinning or partial cutting (Palik et al. 2002; Kuttner et al. 2013).

Box 3.2 Sustainable forest age-class distributions

Careful consideration and manipulation of the forest age- or size-class structure is a central requirement to assure an even flow of harvestable timber under the auspices of sustained-yield forestry. This is especially true where the age and size structure of the forest is not already "normalized" or "regulated," such that a steady stream of sub-mature trees will always be ready to replace the mature trees being harvested. Natural forests tend to have irregular age structures (e.g. Figure 3.5) reflecting the stochastic incidence of disturbance events, tree seed production, and weather conditions suitable for forest regeneration. Forest stands that have developed on agricultural land or after periods of economic exploitation or political disruption (for example, wars) likewise tend to have been initiated at irregular intervals in history (Pan et al. 2011; Vilén et al. 2012). Consequently, a long period of age-class adjustments can be required to bring a forest estate or landscape under sustainable management.

A regulated forest of even-aged stands is expected to consist of equal areas of each age class (often defined in 5-, 10-, or 20-year bins, depending on growth rates) over large areas (see Figure 3.6). With trees growing under mostly open conditions, even-aged management systems lend themselves to the management of fast-growing, early successional, shade-intolerant tree species. In practice, the objective of achieving the theoretically optimal age-class structure is rarely achieved. Witness, for example, a deficit of 5–9-year-old trees in Figure 3.6(b), despite a large stable land base in which primarily one crop tree species is being managed on relatively short rotations. Such irregularities in a managed forest estate can represent historical lulls in harvesting operations (owing to market or weather constraints), or losses due to pests and other disruptions. Variation over the oldest twenty years shown in Figure 3.6(b) may represent differences in growing sites, some additional, slower-growing species (e.g. Douglas-fir, *Pseudotsuga menziesii*), or decisions to produce some larger trees. When a stand is harvested, it must be promptly regenerated to initiate a new youngest age class, thereby ensuring production of a new crop of timber.

The alternative to a uniform age-class distribution of even-aged stands over a large area is a negative exponential abundance of tree sizes (which undergo thinning as the stand matures) within each stand (see Figure 3.7). Note that tree size (as portrayed in Figure 3.7(b)) is a good proxy for tree age in closely managed forests, but this is not true of wild forests, where many trees can undergo long periods of suppressed growth and trees of all ages may be small in size. This model for sustained-yield forestry is often employed where sustainability is desired over a small area (for example, a farm woodlot), and lends itself to the growth of slower-growing, late successional, shade-tolerant tree species. Such uneven-aged stands are characterized by multiple canopy layers and a patchy mosaic of recently harvested gaps, regeneration clusters, and older maturing trees. Most of the forest area thus remains under tree cover even after the harvesting of single trees or small groups of trees, which can be an important consideration where watershed protection or the conservation of biodiversity dependent on mature or old forest are priorities.

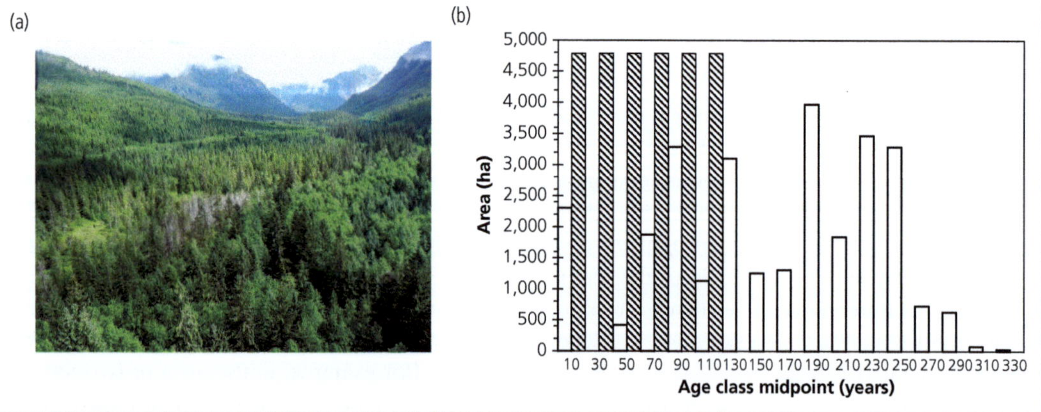

Figure 3.5 (a) A wild, undeveloped forest valley in northwestern British Columbia, Canada. (b) The natural age structure of such a landscape, compared to the idealized age structure of a fully regulated forest, in which all stands are harvested at 120 years of age. Open bars: natural age structure; cross-hatched bars: fully regulated forest (redrawn from Burton et al. (1999)).

Box 3.2 *Continued*

Figure 3.6(a) A landscape characterized by several different age classes of even-aged stands of radiata pine (*Pinus radiata*) on the Kaingaroa Forest Estate in New Zealand. This represents a sustainable supply of timber that can be harvested and renewed at regular intervals. The forest estate for sustainable wood production consists of thousands of hectares (photo by Oscar Garcia).

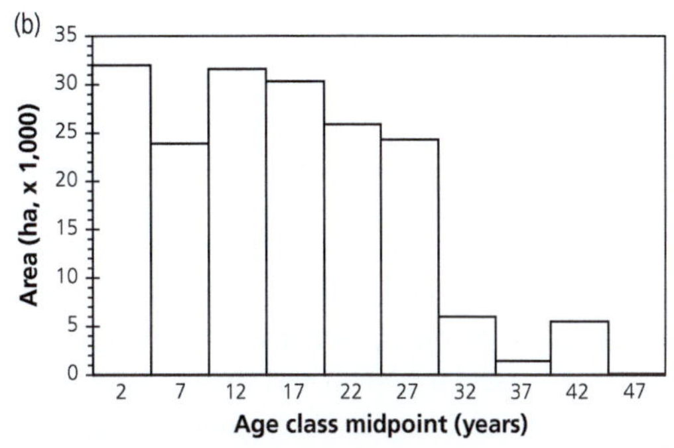

Figure 3.6(b) Graphical portrayal of the age-class structure, in hectares, of the largely normalized forest shown in Figure 3.6(a). Most trees are harvested at 25–30 years of age, showing approximately equal areas of land in each age class of forest (data from Timberlands Limited (2018)).

Figure 3.7(a) An uneven-aged stand of European beech (*Fagus sylvatica*) in northern Germany, managed using the *plenterwald* or single-tree selection system. All age- or size-classes of trees can be found in a single stand (typically in clusters where new trees establish in the gaps left by the harvest of mature trees), which is managed for sustainable wood production on the scale of a few or tens of hectares.

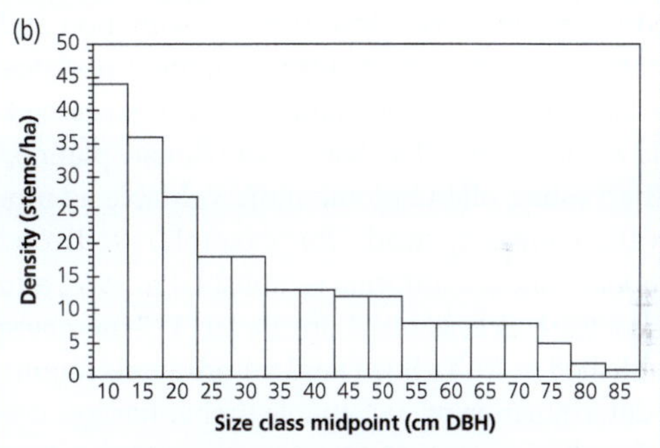

Figure 3.7(b) Graphical portrayal of the size-class structure, in stems/hectare, of trees (those >7.5 cm diameter at breast height (DBH)) in a sustainable forest stand like that shown in Figure 3.7(a). This shows declining numbers of trees in successive size classes, as many small trees are naturally or purposely thinned out to support fewer but larger trees (data from Cancino and von Gadow (2002)).

The choice of silvicultural system and preferred tree species (in harvesting, when thinning and spacing, or in establishing new crop trees) depends on factors such as growth rates, market opportunities, and species resistance to local pests and environmental constraints (Nyland 1996). Shade-intolerant species are often the fast-growing, early successional colonizers (for example, many *Acacia*, *Betula*, *Eucalyptus*, *Larix*, *Pinus* spp.) that establish after large-scale disturbances such as wildfires, so lend themselves to growing in even-aged plantations. Those animal and plant species (including intolerant tree species) that prefer open conditions do better under even-aged management in which clearcut, seed tree, or shelterwood silvicultural systems are practiced. By way of contrast, more shade-tolerant tree species with desirable woods (for example, *Acer, Fagus, Picea, Swietenia* spp.) are slower growing and have evolved under disturbance regimes characterized by small-scale disturbances such as windthrow gaps. Forms of biodiversity more dependent on old-growth conditions or continuous canopy cover (for example, some species of birds, lichens, bryophytes, and understory vascular plants) and shade-tolerant tree species are more likely to thrive under single-tree or group selection (gap-based) silvicultural system and other forms of partial cutting. As often as not, however, the choice of forest management system and its many operational variants are an expression of the management philosophy of the landowner.

All the above conditions for sustained timber production are subject to errors in estimation and variation in execution, requiring frequent updates, amendments, and recalculations in any sustained-yield management plan. Forests previously planned for harvesting often become more valuable as conservation areas instead (for example, Redwood National Park in California established in 1968, and *Nationalpark Schwarzwald* in Baden-Württemberg established in 2014). Such land withdrawals require a redetermination of sustainable timber harvest levels for the forest that remains open to logging. Conversely, some lands currently used for agriculture or mining might be planted to trees and added to the forest estate. The definition of what constitutes "non-commercial" forest (often surpassing a threshold volume of m^3/ha) depends on the market price of various forest products, the cost of labor and diesel fuel, its distance from processing facilities, and other factors. Similarly, inventories are often done for "commercial" species only, but which species are considered commercial and most profitable varies with time and place (Figure 3.8). Likewise, changing technologies and economic considerations determine how much fiber is considered recoverable and marketable. Even in an age of LiDAR inventories and monumental computing power, forest inventories, growth rates, and yield projections remain estimates that are subject both to uncertainty and to the vagaries of weather and disturbance events.

The rate and quality of fiber production vary among sites and among tree species, which sometimes can be improved with management practices such as thinning, fertilization, and pruning. However, the certainty of being able to apply such practices in the future depends on the availability and cost of labor and fertilizer. What proportion of the fiber supply can be considered recoverable and available for different uses (over the course of harvesting and processing) depends on economic conditions, fiber abundance or scarcity in the market, and advances in tree breeding and wood-product utilization. Those recovery rates also differ substantially if forests are subject to natural disturbances such as wildfires, storm damage, and insect outbreaks. Even if salvage logging after natural disturbances is feasible and is conducted carefully (see Section 8.6), there will undoubtedly be greater amounts of damaged and unutilizable fiber to deal with, and potential impacts on future stand renewal and growth (Lindenmayer et al. 2008). Compounding the problem of losses to disturbance and associated uncertainties is the likelihood that we will continue to see increasing rates of extreme events (droughts, storms, wildfires) worldwide in response to a warming atmosphere (Dale et al. 2001; Seidl et al. 2014; IPCC 2022), as further discussed in Chapter 6.

As a result of all those uncertainties and unforeseen events, few sustainable forest management plans can be relied upon for an entire rotation, especially in these rapidly changing times (see Figure 3.1). Rather, a process of constant adjustment and replanning is required to re-estimate what should constitute a sustainable yield of timber from a given forest land base. In the process, we may see alternating episodes of over-cutting and under-cutting of potential yields. This is not unusual in the world of nature nor in the world of humans: animal populations (especially insects) often overshoot their carrying capacity, crash, and then take some time to recover before overshooting sustainable levels again (Sibly et al. 2007). This is analogous to a car's global positioning system repeatedly recalculating the optimal route when the driver makes a wrong turn and goes off track.

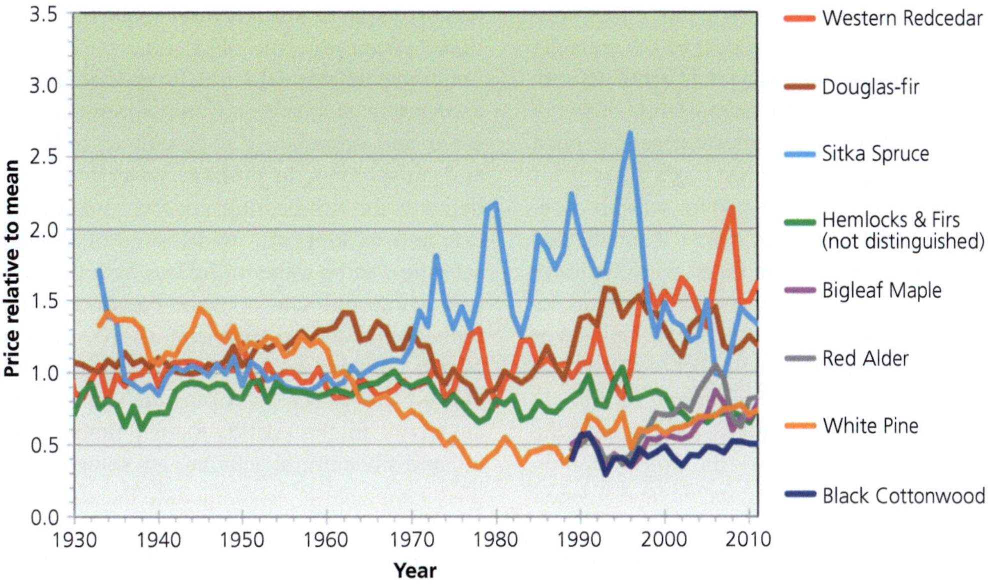

Figure 3.8 Variation over time in the relative value of logs from different British Columbia tree species, based on eighty years (an average forest rotation length in north-temperate forests) of data from the Vancouver log market (Filotas et al. (2014)).

3.5 Sustainable forest management: Sustaining all forest values

The so-called scientific approach to forest management, with its calculations of sustainable yields and its adherence to cutting cycles, has always had timber production as its primary goal. This not only reflects the historical incentive to regulate and assure reliable supplies of wood, but quite naturally reflects the dominant role by which the influence of trees (their species, density, and stature or maturity) determines other forest attributes. Nonetheless, it has always been known that forests support many non-timber resources and uses (Chapter 2): as sources of edible and medicinal plants (Emery and O'Halek 2001; van Andel 2006), as habitat for the deer and swine hunted by nobility and commoners alike (Innes 2017), and as havens of peace, quiet, and spiritual rejuvenation (Williams and Harvey 2001; Bulkan 2017), for example. While forests have been home, larder, and identity for many Indigenous peoples, there has often been an uneasy relationship with forests on the part of settled agricultural societies, because forests stood in the way of cultivation and had the potential to hide bandits and marauders. Nonetheless, Europeans have always appreciated forests for their provisioning of firewood, timbers, mushrooms, and wild game. Reflecting an agricultural mindset and a history of re-establishing forests after deforestation, the historical European view of "properly managed forests" follows an agricultural model for the cropping of trees rather than ecosystem management. This worldview pervades many forestry schools and agencies to this day (Puettmann et al. 2009) but is now experiencing a widespread pushback in western and central Europe and elsewhere (Pukkala and von Gadow 2012; Remeš 2018). To supplement the brief overview provided in Chapter 1, very readable histories of the diversity of forest uses and the evolution of forest management perspectives and policies are provided by Vogt et al. (2007), Sands (2013), Ghazoul (2015), and Innes (2017).

Traditionally trained forest managers typically recognized but rarely had opportunities to provide active support for non-wood forest uses. Such uses and users were tolerated, but rarely planned for, so long as they did not conflict with timber-management operations. It was widely asserted that other forest uses, such as hydrological regulation and opportunities for livestock grazing, could largely be supported by the normal cycle of forest harvesting and renewal (e.g. Davis 1954). This principle came to the fore with respect to the management of public forest and range lands in the US with approval of the *Multiple-Use and Sustained-Yield Act* (US Congress 1960). Such legislation gave explicit direction to manage federal forest lands not only for sustained yield of wood, but also for wildlife, livestock fodder, recreation, and water. Nonetheless, initiatives to intervene in the forest were typically undertaken and paid for by those seeking to utilize timber resources, and other uses—if recognized at all—would be allocated to small pockets of land.

Following the 1992 Earth Summit in Rio de Janeiro, a number of documents that became the basis for international treaties—the Convention on Biological Diversity, the Forest Accord, and Agenda 21—further emphasized a multidimensional perspective on sustainability for the management of forests (Burton et al. 2003). Governments around the world adopted general principles of sustainable forest management (SFM) that protect biological diversity, conserve soil and water resources, respect the rights of Indigenous peoples, and share the benefits of forest development with local communities. Faced with the threat of consumer boycotts and recognizing the public distrust of governments that regulated but also profited from forest exploitation, independent third-party auditing processes such as the Forest Stewardship Council (FSC) and the Programme for Endorsement of Forest Certification (PEFC) were developed to assure that those multiple values were being respected in the harvesting, processing, and marketing of forest products (see Section 10.2). Many observers place faith in the ability of these suites of criteria and indicators of sustainable forest management to promote, or at least track progress toward, responsible forest stewardship (Cashore et al. 2006; MacDicken et al. 2015).

The now-formalized systems of SFM criteria, indicators, and forest products certification help assure that industrial forest practices abide by the rules of the land, and that damage to the natural world is in some way minimized or mitigated. If instead we define SFM as the extension of sustained-yield forestry to all forest values (Brown and Carder 1977; Adamowicz and Burton 2003; Higman et al. 2005), we would then see active management to sustain more than fiber production. Specifically, true SFM could be expected to undertake intentional actions to enhance habitat for a diversity of plant and animal species, to conserve and stimulate the growth of wild berries and mushrooms, to promote water retention and delivery for fish-bearing streams and for human use, and to support a wide range of recreational activities and aesthetic values. Yet, because the harvesting and sale of wood-based commodities remain the only mainstream source of revenue associated with most forest management activities, we rarely see such purposeful interventions anywhere in the world. Rather, timber harvesting and silvicultural activities are sometimes modified somewhat to accommodate these other forest values.

Bernard Fernow was the father of several forestry schools in the US and Canada, and clearly articulated that "the first and foremost purpose of a forest ... is to supply us with wood material" (Fernow 1902: 85). Forestry textbooks written in the middle of the twentieth century often characterized the purpose of forest management as building up, setting in order, and keeping in order a forest business (Allen 1950; Davis 1954). This utilitarian, business-oriented "ligniculture" emphasis continues to constrain the adoption of SFM as a guiding principle in forestry schools, corporate boardrooms, and government agencies around the world. Only where there are landowners, private or public, devoted to non-timber values (sometimes called "social values" to describe the many subjective and non-market benefits of forests, as described in Chapter 2; Lawrence 2004) and with the financial resources to promote them, do we see a true expression of the real potential of multipurpose forests and SFM. Some examples do exist, for example:

- creating small forest openings to promote snow-pack accumulation and delayed surface runoff in support of water delivery (Golding and Swanson 1978);
- pre-commercial forest thinning to improve forage production for moose, *Alces alces* (McLaren et al. 2000);

- planting berry-producing shrubs for the benefit of brown bears, *Ursus arctos* (Braid et al. 2016); and

- private landowners undertaking forest operations (trail construction, thinning) to improve forest aesthetics and scenic views (Klessig 2002).

Such initiatives need to be expanded and become much more widely adopted if we are to see SFM as the prevailing forest management paradigm.

Let us imagine that a forest manager embraces the multi-value perspective of SFM, of which sustainable timber production is one goal of many. Planning and decision-making are still inevitably constrained by the wishes of the landowner and the financial resources available or that can be generated by the forest. This manager might be trying to regulate the foraging for mushrooms, improve habitat for both hunted and endangered animal species, conserve and sequester carbon, and control invasive species in addition to undertaking the traditional forestry practices of logging, planting, and thinning for wood production. The physical placement and scheduling of such interventions need to be undertaken in a manner that minimizes detrimental impacts on other resources or values. These interventions can sometimes be accommodated on the same land base but often require zoning to separate conflicting uses (see Chapter 2). If other resource management objectives can also generate financial returns (for example, from mushroom harvesting, permitted hunting or camping, trapping for furs, or finding a market for stand thinnings), then a degree of stability may be gained through the management of a more diverse resource portfolio. On the other hand, planning based on investment in multiple resources also exposes one to additional resource-specific risks related to weather, the threat of pathogens, market shifts, and so on. Balancing these risks and rewards is not an exact science, and inevitably reflects the wishes and priorities of managers, landowners, and community stakeholders as well as the practicalities of available labor, technology, and capital. It is a challenging task indeed to assure a healthy status for all those resources, and all values and interests, while working within real-world constraints.

3.6 Sustainability comes with baggage and expectations

Even if we acknowledge that strong sustainability and true SFM are nearly impossible to achieve, the sustainability paradigm is so embedded in today's zeitgeist that the term is not easily replaced in discourse or policy. So, when engaging in research, discussion, evaluation, or planning around sustainability, it is important to conduct a reality check with all stakeholders or project participants to clarify underlying assumptions and expectations.

One of the common variants of sustainable resource management is the objective of managing for "maximum sustained yield." This paradigm follows from the objective of optimizing financial returns from a particular resource and has found applications in agriculture and fisheries as well as forestry. Often dependent on repeated inputs (for example, of fertilizers, weed control, pest management), yield optimization in agriculture and forestry usually focuses on a single crop species. Optimizing fishery and forestry operations also involves determination of optimal harvest ages, typically where the rate of growth starts to decline, or, technically, the inflection point in a logistic (S-shaped) growth curve. In forestry, if wood volume or biomass is the sole attribute being managed for, that optimal harvest time (or rotation age) can be expressed graphically as the point in time at which current (or periodic) annual increment (m^3/ha grown in an individual year) intersects with mean annual increment (m^3/ha averaged over stand age to date); see Figure 3.9. This optimality results in harvest ages much lower than natural rotation cycles in which tree death reflects a combination of natural disturbance regimes (typical return intervals of fires, wind storms, or fatal pests) and the pathological or physiological lifespan of individual tree species. A sustainable ecological rotation is the time required for soil nutrient reserves to recover after timber harvesting; repeated timber harvesting at intervals less than the ecological rotation risks site degradation, especially on shallow or coarse-textured soils (Kimmins 2004).

When optimizing timber volume production, few or no old-growth values can be supported, even

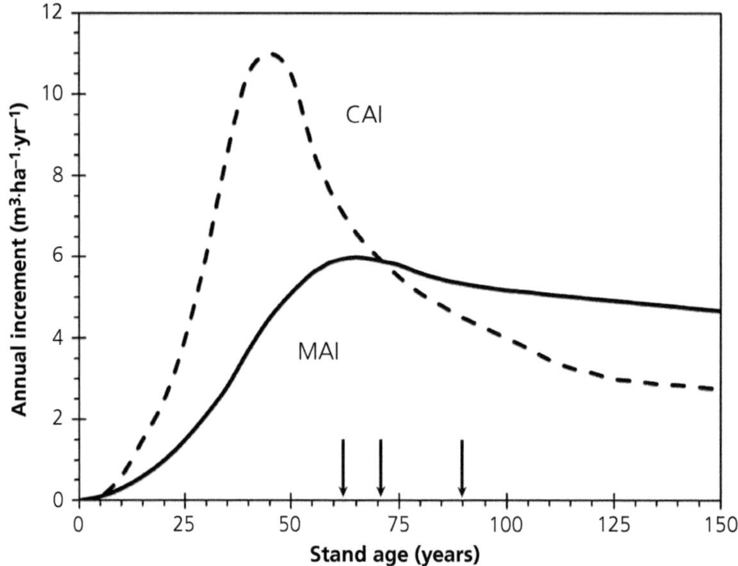

Figure 3.9 Classical timber-focused forestry calculates the optimal rotation length or harvest age as the stand age at which current annual increment (CAI) intersects mean annual increment (MAI). In this hypothetical example, merchantable wood volume production over time is maximized by harvesting stands at 71 years of age, but optimal biomass harvesting (with no minimal piece size) might be at 62 years, and the optimal rotation for sawlogs (requiring at least a 15-cm top at 6 m height) might be at 90 years.

though wood production is being managed sustainably. Stakeholders need to understand such corollaries and assumptions of sustained-yield forestry, including the intended forest products and the eventual forest age-class structure that will result. Maximum sustained-yield management involves producing biomass with the greatest efficiency possible—that is the nature of optimization. But it also means that it is vulnerable to disruptions that may occur through natural disturbances, variable weather or shifts in climate, outbreaks of pests and disease, or as a result of changing socioeconomic realities. Such consequences can be seen in fisheries collapses around the world (Pauly et al. 2005), and in the impacts of fires and insect outbreaks on timber supplies (Burton 2010; see Box 3.3). In short, no yield can ever be sustained indefinitely at a maximum level (Walker and Salt 2006), in a sense making "maximum sustained yield" a self-contradictory term.

Another common expectation of sustainability is to take it literally as a promise of constant delivery of ecosystem services. We understandably might think of sustainability like a sustained note in music,

which holds a specific pitch and volume for an extended period of time. This constancy is an important consideration in providing wood as the feedstock for manufacturing facilities (sawmills, pulp mills, pellet plants, and so on) that depend on steady inputs in order to run continuously and provide steady employment. Expressed as a requirement for an "even flow" timber supply, there nonetheless is recognition of some variability in the rate of harvesting in response to weather (for example, wet ground that cannot support heavy machinery), and to market conditions in terms of price and demand for different timber species. Those sources of variation are anticipated, and can be buffered by the ability to hold inventories of logs or finished product until conditions become more favorable. Much planning is done around average conditions, but, as the old saying goes, "no year ever experiences normal weather;" rather, so-called normal conditions are the average of what can be wildly oscillating extremes.

Where uneven-aged or continuous cover forestry is practiced, cutting cycles associated with repeated entries mean that a pulse of wood and tree cover is

---

Box 3.3 Disrupted sustainability: Bark-beetle outbreaks

Here is a dramatic example of the warming climate disrupting plans for sustainable forestry, even in a large area over which differences in forest age classes, tree species, and site types should assure a reliable timber supply. Bark beetles (Coleoptera, Curculionidae, Scolytinae) expanded their populations in the conifer forests of western North America through much of the late 1990s and early 2000s (Raffa et al. 2008; Bentz et al. 2010). Each species of beetle is co-evolved with host trees of a single species or genus, with tree defenses (for example, resin production and chemistry) similarly co-evolved to defend against beetle attack. Spruce beetles (*Ips typographus* in Europe, *Dendroctonus rufipennis* in North America) frequently invade defenseless weakened or dead windthrown trees, in which they are then able to complete their life cycle and increase local population sizes before attacking living spruce trees in high numbers (Ruel et al. 2023). The mountain pine beetle of western North America (*Dendroctonus ponderosae*) attacks only living pine trees, congregating in their hundreds and thousands for a mass attack on mature trees through pheromone signaling. Larvae eat their way through the phloem, cambium, and outer sapwood, depositing spores of fungi (*Ophiostoma* spp. and others) that eventually grow to permeate the sapwood and block tracheids, such that the tree eventually dies of moisture stress (Raffa 1988; Carroll and Safranyik 2003).

Mature trees of lodgepole pine (*Pinus contorta* var. *latifolia*) are the preferred host of the mountain pine beetle, the tree species to which it is most closely co-evolved. Large areas of central and southern British Columbia, Canada, are dominated by this fire-adapted tree species. They established from seed released by serotinous cones after widespread fires from 1880 to 1920, an era associated with railroad building and Euro-Canadian settlement. Beetle outbreaks emerged from about 1999 to 2005 at several independent outbreak centers (Aukema et al. 2006) after several decades of effective fire suppression, and (starting in the 1980s) after many years without the deep cold temperatures in autumn and winter that had limited beetle populations in the past (Burton 2010). Despite sanitation logging efforts, the mountain pine beetle outbreak quickly overcame control efforts, eventually affecting approximately twenty million ha of forest land and killing more than 720 million m^3 of timber in western Canada (Dhar et al. 2016). This unprecedented insect outbreak was cited by the Intergovernmental Panel on Climate Change (Field et al. 2007: 623) as an early example of climate change being reflected in altered disturbance regimes.

The management response to mountain pine beetle outbreaks is to harvest recently attacked trees (indicated by resin (pitch) extrusions, followed by yellowing and then reddening foliage) and to remove those trees and burn or process the logs before another batch of adult beetles emerges from the pupae. Government-regulated levels of logging (known as the allowable annual cut (AAC)) were subsequently elevated to support such sanitation logging as the outbreak continued to grow. But, when large areas of forest became affected by the growing population of beetles, much of it unroaded, forest policy shifted to salvage logging in order to capture the wood value of already-dead trees before they deteriorated. Collectively, harvesting uplifts of an extra 14.5 million m^3/yr in BC and 3.5 million m^3/yr in Alberta were approved (Dhar et al. 2016), despite the hydrological, biodiversity, and economic impacts of such accelerated forest development (Lindenmayer et al. 2008; Thorn et al. 2018). With clearcut harvesting methods predominating, many living pine, spruce, and fir trees that could have provided a mid-term timber supply in the future were nonetheless harvested as "bycatch" in the salvage logging operations (Burton 2006, 2010). While AAC levels had been more or less sustainable prior to the mountain pine beetle outbreak, uplifts essentially gave the forest product industry license to log forty-five years of timber supply in slightly more than a decade (see Figure 3.10). With the outbreak having run its course and the wood quality of standing dead trees deteriorating, AAC levels were being stepped down in the 2020s. At the time of writing, many sawmills in interior BC are experiencing the predicted (Parfitt 2005; Burton 2010) shutdowns associated with the trough in timber supply (see Figure 5.4). Harvestable timber in the region is too expensive to access, employment and communities are suffering, while the major timber companies are reinvesting their profits internationally. The large cutblocks created by salvage logging are primarily being regenerated with lodgepole pine again, setting the stage for a disturbance loop in which large expanses of even-aged, monotypic forest will once again be susceptible to mountain pine beetle attack fifty or sixty years after logging (Burton et al. 2020).

Box 3.3 *Continued*

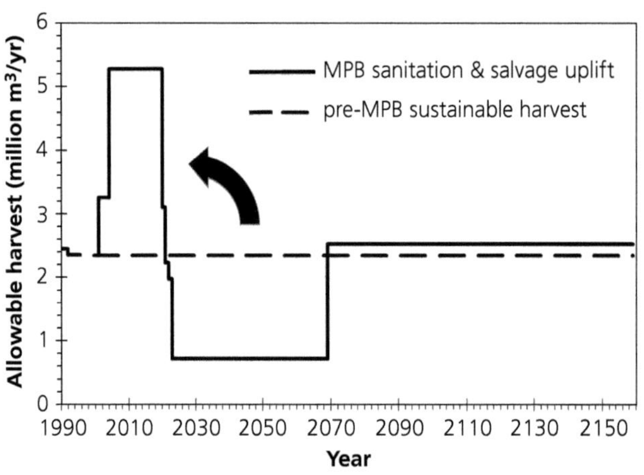

Figure 3.10 Projected timber flow for the Quesnel Timber Supply Area in British Columbia, showing the sustainable level of annual timber harvest and uplifted harvest volumes to salvage timber from stands affected by the mountain pine beetle (MPB). A "trough" of much-reduced harvesting opportunities for decades post-outbreak arises because much of the mid-term timber supply is harvested earlier than originally planned, green trees were harvested along with dead trees as a result of clearcut logging, and second-growth stands (harvested and regenerated pre-outbreak) will not yet be mature (data from BCMRF (2010)).

removed from the forest every five, ten, or maybe twenty years, followed by years of recovery and growth. At the local or stand level, single-tree or group-selection harvesting is episodic but small in footprint, so forest habitat remains minimally disrupted. But episodic timber sales are not something that the owner of a small woodlot could count on for an annual income. A sawmill supplied by numerous woodlots, each practicing continuous cover forestry, could access a continuous supply of logs if harvesting was coordinated among multiple stands. The ecological effects of this pulsing cut-and-recovery pattern are much more pronounced under even-aged management, where logging (with or without the retention of wildlife tree patches, seed trees, or a shelterwood of overstory trees) results in severe habitat alteration from which it takes many decades to recover. In such cases, forest product processing facilities or log buyers must draw from a much larger area of managed forest to achieve an even flow of feedstock. Except for wide-ranging animals, there is the danger that large areas under even-aged management will not have forest cover and habitat values sustained locally.

Furthermore, timber supply too can be disrupted from that predictable even flow by forest fires, insect outbreaks, and other large natural disturbances (see Box 3.3). A sustained flow of any forest ecosystem service or product might be achievable at large enough scales, although most forest management decisions are made at small to intermediate scales. Many ecosystem services (for example, scenic vistas from a tourist lodge, habitat suitable for a rare species, wild food harvestable within walking distance of a village) may be locally unavailable if those management decisions are not coordinated and sensitive to local needs.

Sustainable forest management can require clarifying and managing stakeholder expectations as much or more than it consists of manipulating natural resources. Is there an agreed understanding among managers and stakeholders regarding exactly what is expected to be sustained, and how? Even planning for a sustainable timber supply can denote a wide range of practices and long-term results, including different rotation lengths and management practices. Harvest intervals, stand densities, and stand-tending practices such as thinning and pruning will vary if managing for generic wood volume (for example, for pulp or pellets), aboveground biomass (for example, to feed boilers used for community heating or electricity generation), or for sawlogs to produce lumber of particular strengths and piece sizes (see Figure 3.9). Stands managed those different ways can affect habitat value, hydrology, and other ecosystem services. So all interested parties need to be on the same page regarding which ecosystem goods and services are to be protected, conserved, and actively managed and sustained under the umbrella of an SFM plan. Each of those ecosystem values has its own inherent uncertainties, as well as vulnerabilities to different threats and environmental changes, which often means that one cannot sustain everything one would like to sustain.

Sustained-yield forestry and sustainable forest management marked revolutionary advances in natural resource stewardship (Hütte 2000), and provide important aspirational goals for resource managers. But the sustainability framework can also be considered a product of an equilibrial view of the natural world, a view that has largely been discredited as we gain a better understanding of episodic change in socio-ecological systems (Botkin 1990; Levin 1999). With several of the multiple objectives of SFM being incompatible, and with global change accelerating, it is time to review the suitability of sustainable forest management as the guiding paradigm for forest policy (Prins et al. 2023). We may still hope and plan for sustainability—with all the caveats of uncertainty, expected disruption, and conflicting goals identified above—but we can also count on it being difficult or unattainable without a lot of extra work and continual readjustment. That is where planning for resilience can provide a supplemental perspective and a back-up set of considerations, and perhaps serve as an alternative paradigm for natural resource management.

Box 3.4 Key points

- Sustainability is about intergenerational equity under conditions where resources are limited and resource use can damage the environment.
- Many aspects of modern society, with current levels of population growth, resource use, and land-use change, are unsustainable and threaten biodiversity, the climate, and human health. Different ways in which we use or manage forests can increase or decrease some of those negative impacts.
- Sustained-yield forestry was historically developed to assure permanent supplies of timber, and this concept can be expanded to the conservation of all forest values under sustainable forest management (SFM).
- Claims to sustainability are so widespread that clarification is always needed. Is constancy or an even flow of benefits intended? Sustainability of what, over what area, over what time horizon, and for whom?
- With the long time periods required for trees to grow, forest plans are subject to disruption by natural disturbances, climate change, market conditions, and shifts in public values or policy direction, making sustainable management difficult.
- Frequent replanning and managing for resilience are suggested as insurance strategies against future risk and uncertainty.

Destroying rain forest for economic gain is like burning a Renaissance painting to cook a meal.

E. O. Wilson (c.1990)

References cited

Adamowicz, W. L., and Burton, P. J. (2003). "Sustainability and Sustainable Forest Management," in P. J. Burton, C. Messier, D. W. Smith, and W. L. Adamowicz (eds), *Towards Sustainable Management of the Boreal Forest*. Ottawa: NRC Research Press, 41–64.

Ahuvia, A. C. (2008). "Wealth, Consumption and Happiness," in A. Lewis (ed.), *The Cambridge Handbook of Psychology and Economic Behaviour*. New York: Cambridge University Press, 199–226.

Allen, S. W. (1950). *An Introduction to American Forestry*. 2nd edn. New York: McGraw-Hill. 413 pp.

API (2021). *The Natural Gas and Oil Industry's Commitment to Sustainability*. Washington: American Petroleum Institute. Available online at https://www.api.org/news-policy-and-issues/sustainability (accessed June 7, 2023).

Aukema, B. H., Carroll, A. L., Zhu, J., Raffa, K. F., Sickley, T. A., and Taylor, S. W. (2006). "Landscape Level Analysis of Mountain Pine Beetle in British Columbia, Canada: Spatiotemporal Development and Spatial Synchrony within the Present Outbreak," *Ecography*, 29/3: 427–41.

Babcock, B. (1975). "A Tolerated Margin of Mess: The Trickster and his Tales Reconsidered," *Journal of the Folklore Institute*, 11/3: 147–86.

BCMFR (2010). "Quesnel TSA Timber Supply Analysis Public Discussion Paper." Victoria, BC: Forest Analysis and Inventory Branch, BC Ministry Forests and Range. 17 pp. Available online at https://www.for.gov.bc.ca/ftp/hts/external/!publish/S_Vinnedge/Quesnel_TSA/26ts10pdp.pdf (accessed June 11, 2023).

Behan, R. W. (1978). "Political Popularity and Conceptual Nonsense: The Strange Case of Sustained Yield Forestry," *Environmental Law*, 8/2: 309–42.

Bentz, B. J., Régnière, J., Fettig, C. J., et al. (2010). "Climate Change and Bark Beetles of the Western United States and Canada: Direct and Indirect Effects," *BioScience*, 60/8: 602–13.

Bergeron, Y., Harvey, B., Leduc, A., and Gauthier, S. (1999). "Forest Management Guidelines Based on Natural Disturbance Dynamics: Stand-and Forest-Level Considerations," *Forestry Chronicle*, 75/1: 49–54.

Botkin, D. B. (1990.) *Discordant Harmonies: A New Ecology for the Twenty-First Century*. New York: Oxford University Press. 241 pp.

Braid, A. C., Manzer, D., and Nielsen, S. E. (2016). "Wildlife Habitat Enhancements for Grizzly Bears: Survival Rates of Planted Fruiting Shrubs in Forest Harvests," *Forest Ecology and Management*, 369: 144–54.

Brown, T. C., and Carder, D. R. (1977). "Sustained Yield of What?" *Journal of Forestry*, 75/11: 722–3.

Bulkan, J. (2017). "Social, Cultural and Spiritual (SCS) Needs and Values," in J. L. Innes and A. V. Tikina (eds), *Sustainable Forest Management: From Concept to Practice*. London: Routledge, 241–56.

Burda, C., Collier, R., and Evans, B. (1999). *The Gitxsan Model: An Alternative to the Destruction of Forests, Salmon and Gitxsan Land*. Victoria, BC: Eco-Research Chair of Environmental Law and Policy, University of Victoria. 23 pp.

Burton, P. J. (2006). "Restoration of Forests Attacked by Mountain Pine Beetle: Misnomer, Misdirected, or Must-Do?" *Journal of Ecosystems and Management*, 7/2: 1–10.

Burton, P. J. (2010). "Striving for Sustainability and Resilience in the Face of Unprecedented Change: The Case of the Mountain Pine Beetle Outbreak in British Columbia," *Sustainability*, 2: 2403–23.

Burton, P. J. (2020). "The Scope and Challenge of Sustainable Forestry," in J. A. Standurf (ed.), *Achieving Sustainable Management of Boreal and Temperate Forests*. Cambridge: Burleigh Dodds Science Publishing, 1–22.

Burton, P. J., Jentsch, A., and Walker, L. R. (2020). "The Ecology of Disturbance Interactions," *BioScience*, 70/10: 854–70.

Burton, P. J., Kneeshaw, D. D., and Coates, K. D. (1999). "Managing Forest Harvesting to Maintain Old Growth in Boreal and Sub-Boreal Forests," *Forestry Chronicle*, 75/4: 623–31.

Burton, P. J., Messier, C., Weetman, G. F., Prepas, E. E., Adamowicz, W. L., and Tittler, R. (2003). "The Current State of Boreal Forestry and the Drive for Change," in P. J. Burton, C. Messier, D. W. Smith, and W. L. Adamowicz. (eds), *Towards Sustainable Management of the Boreal Forest*. Ottawa: NRC Research Press, 1–40.

Cancino, J., and von Gadow, K. (2002). "Stem Number Guide Curves for Uneven-Aged Forests: Development and Limitations," in K. von Gadow, J. Nagel, and J. Saborowski (eds), *Continuous Cover Forestry: Assessment, Analysis, Scenarios*. Dordrecht: Kluwer Academic Publishers, 163–74.

Carroll, A. L., and Safranyik, L. (2003). "The Bionomics of the Mountain Pine Beetle in Lodgepole Pine Forests: Establishing a Context," in T. L. Shore, J. E. Brooks, and J. E. Stone (eds), *Mountain Pine Beetle Symposium: Challenges and Solutions*. Information Report BC-X-399. Victoria, BC: Canadian Forest Service, 21–32.

Cashore, B., Gale, F., Meidinger, E., and Newsom, D. (2006). "Forest Certification in Developing and Transitioning Countries: Part of a Sustainable Future?" *Environment: Science and Policy for Sustainable Development*, 48/9: 6–25.

Costanza, R., Cumberland, J. H., Daly, H., Goodland, R., Norgaard, R. B., Kubiszewski, I., and Franco, C. (2015). *An Introduction to Ecological Economics*. 2nd edn. Boca Rotan, FL: CRC Press. 356 pp.

County of Maui (2010). *2030 General Plan: Countywide Policy Plan*. Wailuku, HI: County of Maui. 82 pp. Available online at https://www.mauicounty.gov/420/Countywide-Policy-Plan (accessed June 12, 2023).

Dale, V. H., Joyce, L. A., McNulty, S., et al. (2001). "Climate Change and Forest Disturbances," *BioScience*, 51/9: 723–34.

Davidson-Hunt, I. J., and Berkes, F. (2001). "Changing Resource Management Paradigms, Traditional Ecological Knowledge, and Non-Timber Forest Products," in I. Davidson-Hunt, L. C. Duchesne, and J. C. Zasada (eds), *Forest Communities in the Third Millennium: Linking Research, Business, and Policy toward a Sustainable Non-Timber Forest Product Sector*. General Technical Report NC-217. St Paul, MN: USDA Forest Service, 78–92.

Davis, K. P. (1954). *American Forest Management*. New York: McGraw-Hill. 482 pp.

Dhar, A., Parrott, L., and Heckbert, S. (2016). "Consequences of Mountain Pine Beetle Outbreak on Forest Ecosystem Services in Western Canada," *Canadian Journal of Forest Research*, 46/8: 987–99.

Diamond, J. (2005). *Collapse: How Societies Choose to Fail or Succeed*. New York: Penguin Books. 592 pp.

Edwards, A. R. (2005). *The Sustainability Revolution: Portrait of a Paradigm Shift*. Gabriola Island, BC: New Society Publishers. 207 pp.

Ekins, P., Simon, S., Deutsch, L., Folke, C., and De Groot, R. (2003). "A Framework for the Practical Application of the Concepts of Critical Natural Capital and Strong Sustainability," *Ecological Economics*, 44/2–3: 165–85.

Ellis, E. C., and Wang, S. M. (1997). "Sustainable Traditional Agriculture in the Tai Lake Region of China," *Agriculture, Ecosystems & Environment*, 61/2–3: 177–93.

Emery, M., and O'Halek, S. L. (2001). "Brief Overview of Historical Non-Timber Forest Product Use in the US Pacific Northwest and Upper Midwest," *Journal of Sustainable Forestry*, 13/3–4: 25–30.

Fagan, K. (2010). "What's the Trouble with the Trickster? An Introduction," in D. Reder and L. M. Morra (eds), *Troubling Tricksters: Revisioning Critical Conversations*. Waterloo, ON: Wilfrid Laurier University Press, 3–20.

Fernow, B. E. (1902). *Economics of Forestry*. New York: T. Y. Crowell & Company. 520 pp.

Field, C. B., Mortsch, L. D., Brklacich, M., et al. (2007). "North America," in M. L. Parry, O. F. Canziani, J. P. Palutikof, P. J. van der Linden, and C. E. Hanson (eds), *Climate Change 2007: Impacts, Adaptation and Vulnerability: Contribution of Working Group II to the Fourth Assessment Report of the Intergovernmental Panel on Climate Change*. Cambridge: Cambridge University Press, 617–52.

Filotas, E., Parrott, L., Burton, P. J., et al. (2014). "Viewing Forests through the Lens of Complex Systems Science," *Ecosphere*, 5/1: 1.

Floyd, D. W. (2002). *Forest Sustainability: The History, the Challenge, the Promise*. Durham, NC: Forest History Society. 83 pp.

Gadgil, M., Berkes, F., and Folke, C. (1993). "Indigenous Knowledge for Biodiversity Conservation," *Ambio*, 22/2–3: 151–6.

Ghazoul, J. (2015). *Forests: A Very Short Introduction*. Oxford: Oxford University Press. 150 pp.

Golding, D. L., and Swanson, R. H. (1978). "Snow Accumulation and Melt in Small Forest Openings in Alberta," *Canadian Journal of Forest Research*, 8/4: 380–8.

Guldin, J. M., and Baker, J. B. (1988). "Yield Comparisons from Even-Aged and Uneven-Aged Loblolly-Shortleaf Pine Stands," *Southern Journal of Applied Forestry*, 12/2: 107–114.

Gunderson, L. H., and Allen, C. R. (2010). "Introduction: Why Resilience? Why Now?" in L. H. Gunderson, C. R. Allen, and C. S. Holling (eds), *Foundations of Ecological Resilience*. Washington: Island Press, pp. xiii–xxv.

Hardin, G. (1968). "The Tragedy of the Commons," *Science*, 162/3859: 1243–48.

Higman, S., Mayers, J., Bass, S., Judd, N., and Nussbaum, R. (2005). *The Sustainable Forestry Handbook: A Practical Guide for Tropical Forest Managers on Implementing New Standards*. 2nd edn. London: Earthscan. 332 pp.

Holling, C. S. (2001). "Understanding the Complexity of Economic, Ecological, and Social Systems," *Ecosystems*, 4/5: 390–405.

Hütte, G. (2000). "Perceived Images of Various Actors Engaged in Sustainability Discussions," in K. von Gadow, T. Pukkala, and M. Tomé (eds), *Sustainable Forest Management*. Dordrecht: Kluwer Academic Publishers, 193–216.

Innes, J. L. (2017). "Sustainable Forest Management: From Concept to Practice," in J. L. Innes and A. V. Tikina (eds), *Sustainable Forest Management: From Concept to Practice*. London: Routledge, 1–32.

IPCC (Intergovernmental Panel on Climate Change) (2022). "Summary for Policymakers," in H.-O. Pörtner, D. C. Roberts, E. S. Tignor, et al. (eds), *Climate Change 2022: Impacts, Adaptation and Vulnerability*. Contribution of Working Group II to the Sixth Assessment Report of the Intergovernmental Panel on Climate Change. Cambridge: Cambridge University Press, 3–33.

Johannes, R. E. (2002). "Did Indigenous Conservation Ethics Exist?" *Traditional Marine Resource Management and Knowledge Information Bulletin*, 14: 3–7 (Secretariat of the Pacific Community, Nouméa, New Caledonia). Available online at semanticscholar.org (accessed October 27, 2019).

Kimmins, J. P. (2004). *Forest Ecology: A Foundation for Sustainable Forest Management and Environmental Ethics in Forestry*. 3rd edn. San Francisco: Benjamin Cummings. 720 pp.

Klessig, L. (2002). *Woodland Visions: Appreciating and Managing Forests for Scenic Beauty*. Publication G3762. Madison, WI: University of Wisconsin Extension. 36 pp. Available online at https://woodlandinfo.org/wp-content/uploads/sites/310/2017/09/G3762.pdf (accessed June 4, 2023).

Kuhlman, T., and Farrington, J. (2010). "What Is Sustainability?" *Sustainability*, 2/11: 3436–48.

Kuttner, B., Malcolm, J. R., and Smith, S. M. (2013). "Multi-Cohort Stand Structure in Boreal Forests of Northeastern Ontario: Relationships with Forest Age, Disturbance History, and Deadwood Features," *Forestry Chronicle*, 89/3: 290–303.

Kuuluvainen, T., and Grenfell, R. (2012). "Natural Disturbance Emulation in Boreal Forest Ecosystem Management: Theories, Strategies, and a Comparison with Conventional Even-Aged Management," *Canadian Journal of Forest Research*, 42/7: 1185–203.

Laiho, O., Lähde, E., and Pukkala, T. (2011). "Uneven-vs Even-Aged Management in Finnish Boreal Forests," *Forestry*, 84/5: 547–56.

Lawrence, A. (2004) "Social Values," in J. Burley, J. Evans, and J. Youngquist (eds), *Encyclopaedia of Forest Sciences*. Oxford: Elsevier. 1126–31.

Levin, S. A. (1999). *Fragile Dominion: Complexity and the Commons*. Reading, MA: Perseus Books. 254 pp.

Lindenmayer, D. B., Burton, P. J., and Franklin, J. F. (2008). *Salvage Logging and its Ecological Consequences*. Washington: Island Press. 227 pp.

Long, J. N. (2009). "Emulating Natural Disturbance Regimes as a Basis for Forest Management: A North American View," *Forest Ecology and Management*, 257/9: 1868–73.

MacDicken, K. G., Sola, P., Hall, J. E., Sabogal, C., Tadoum, M., and de Wasseige, C. (2015). "Global Progress toward Sustainable Forest Management," *Forest Ecology and Management*, 352: 47–56.

McLaren, B. E., Mahoney, S. P., Porter, T. S., and Oosenbrug, S. M. (2000). "Spatial and Temporal Patterns of Use by Moose of Pre-Commercially Thinned, Naturally-Regenerating Stands of Balsam Fir in Central Newfoundland," *Forest Ecology and Management*, 133/3: 179–96.

Mason, B., Kerr, G., Pommerening, A., Edwards, C., Hale, S., Ireland, D., and Moore, R. (2004). "Continuous Cover Forestry in British Conifer Forests," *Forest Research Annual Report and Accounts, 2003–2004*. Edinburgh: UK Forestry Commission, 38–53. Available online at https://www.forestresearch.gov.uk/publications/archive-forest-research-annual-report-and-accounts-2003-2004/ (accessed June 7, 2023).

Meadows, D., Meadows, D., Randers, J., and Behrens, W. W. (1972). *The Limits to Growth: A Report for the Club of Rome's Project on the Predicament of Mankind*. New York: Universe Books. 206 pp.

Morris, D. W. (2019). "A Human Tragedy? The Pace of Negative Global Change Exceeds Human Progress," *Anthropocene Review*, 6/1–2: 55–70.

Mueller-Dombois, D. (2007). "The Hawaiian Ahupua'a Land Use System: Its Biological Resource Zones and the Challenge for Silvicultural Restoration," *Bishop Museum Bulletin in Cultural and Environmental Studies*, 3: 23–33.

Nolan, J. P., Wichert, I. C., and Burchell, B. J. (2000). "Job Insecurity, Psychological Well-Being and Family Life," in E. Heery and J. Salman (eds), *The Insecure Workforce*. London: Routledge, 193–221.

Nyland, R. D. (1996). *Silviculture: Concepts and Applications*. New York: McGraw-Hill. 633 pp.

Palik, B. J., D'Amato, A. W., Franklin, J. F., and Johnson, K. N. (2021). *Ecological Silviculture: Foundations and Applications*. Long Gove, IL: Waveland Press. 343 pp.

Palik, B. J., Mitchell, R. J., and Hiers, J. K. (2002). "Modelling Silviculture after Natural Disturbance to Sustain Biodiversity in the Longleaf Pine (*Pinus palustris*) Ecosystem: Balancing Complexity and Implementation," *Forest Ecology and Management*, 155/1–3: 347–56.

Pan, Y., Chen, J. M., Birdsey, R., McCullough, K., He, L., and Deng, F. (2011). "Age Structure and Disturbance Legacy of North American Forests," *Biogeosciences*, 8: 715–32

Parfitt, B. (2005). *Battling the Beetle: Taking Action to Restore British Columbia's Interior Forests*. Vancouver: Canadian Centre for Policy Alternatives. 51 pp. Available online at https://policyalternatives.ca/sites/default/files/uploads/publications/BC_Office_Pubs/bc_2005/pinebeetle.pdf (accessed June 26, 2023).

Parsons, R., Lacey, J., and Moffat, K. (2014). "Maintaining Legitimacy of a Contested Practice: How the Minerals Industry Understands its 'Social Licence to Operate,'" *Resources Policy*, 41: 83–90.

Pauly, D., Watson, R., and Alder, J. (2005). "Global Trends in World Fisheries: Impacts on Marine Ecosystems and Food Security," *Philosophical Transactions of the Royal Society B: Biological Sciences*, 360/1453: 5–12.

Popescu, G. H. (2016). "Does Economic Growth Bring about Increased Happiness?" *Journal of Self-Governance and Management Economics*, 4/4: 27–33.

Prins, K., Köhl, M., and Linser, S. (2023). "Is the Concept of Sustainable Forest Management Still Fit for Purpose?" *Forest Policy and Economics*, 157: 103072.

Puettmann, K. J., Coates, K. D., and Messier, C. (2009). *A Critique of Silviculture: Managing for Complexity*. Washington: Island Press. 189 pp.

Pukkala, T., and von Gadow, K. (eds) (2012). *Continuous Cover Forestry*. 2nd edn. Dordrecht: Springer. 296 pp.

Pulido, J. S., and Bocco, G. (2003). "The Traditional Farming System of a Mexican Indigenous Community: The Case of Nuevo San Juan Parangaricutiro, Michoacán, Mexico," *Geoderma*, 111: 249–65.

Raffa, K. F. (1988). "The Mountain Pine Beetle in Western North America," in A. A. Berryman (ed.), *Dynamics of Forest Insect Populations: Patterns, Causes, Implications*. New York: Plenum Press, 505–30.

Raffa, K. F., Aukema, B. H., Bentz, B. J., et al. (2008). "Cross-Scale Drivers of Natural Disturbances Prone to Anthropogenic Amplification: The Dynamics of Bark Beetle Eruptions," *BioScience*, 58/6: 501–17.

Raup, H. M. (1964). "Some Problems in Ecological Theory and their Relation to Conservation," *Journal of Animal Ecology*, 33: 19–28.

Remeš, J. (2018). "Development and Present State of Close-to-Nature Silviculture," *Journal of Landscape Ecology*, 11/3: 17–32.

Reser, J. P. (2007). "The Experience of Natural Disasters: Psychological Perspectives and Understandings," in J. Lidstone, L. M. Dechano, and J. P. Stoltman (eds), *International Perspectives on Natural Disasters: Occurrence, Mitigation, and Consequences*. Dordrecht: Springer, 369–84.

Rockström, J., Steffen, W., Noone, K., et al. (2009). "A Safe Operating Space for Humanity," *Nature*, 461/7263: 472–5.

Ruel, J.-C, Wermlinger, B., Gauthier, S., Burton, P. J., Waldron, K., and Shorohova, E. (2023). "Selected Examples of Interactions between Natural Disturbances," in M. M. Girona, H. Morin, S. Gauthier, and Y. Bergeron (eds), *Boreal Forests in the Face of Climate Change: Sustainable Management*. Cham, Switzerland: Springer Nature, 123–41.

Sage, R. F. (2020). "Global Change Biology: A Primer," *Global Change Biology*, 26/1: 3–30.

Sands, R. (2013). "A History of Human Interaction with Forests," in R. Sands (ed.), *Forestry in a Global Context*. 2nd edn. Wallingford, UK: CAB International, 1–36.

Seedre, M., Felton, A., and Lindbladh, M. (2018). "What Is the Impact of Continuous Cover Forestry Compared to Clearcut Forestry on Stand-Level Biodiversity in Boreal and Temperate Forests? A Systematic Review Protocol," *Environmental Evidence*, 7: 1–8.

Seidl, R., Schelhaas, M.-J, Rammer, W., and Verkerk, P. J. (2014). "Increasing Forest Disturbances in Europe and their Impact on Carbon Storage," *Nature Climate Change*, 4: 806–10.

Sesser, A. L., Rockhill, A. P., Magness, D. R., et al. (eds). (2019). *Drivers of Landscape Change in the Northwest Boreal Region*. Fairbanks: University of Alaska Press. 326 pp.

Seymour, R. S., White, A. S., and Philip, G. D. (2002). "Natural Disturbance Regimes in Northeastern North America: Evaluating Silvicultural Systems Using Natural Scales and Frequencies," *Forest Ecology and Management*, 155/1–3: 357–67.

Sibly, R. M., Barker, D., Hone, J., and Pagel, M. (2007). "On the Stability of Populations of Mammals, Birds, Fish and Insects," *Ecology Letters*, 10/10: 970–6.

Steffen, W., Richardson, K., Rockström, J., et al. (2015). "Planetary Boundaries: Guiding Human Development on a Changing Planet," *Science*, 347/6223: 1259855.

Tammemagi, H. (2021). "Certifying Old-Growth Logging as Sustainable Is Absurd," *Focus on Victoria*. Victoria, BC. Available online at https://www.focusonvictoria. ca/forests/88/ (accessed June 7, 2023).

Thorn, S., Bässler, C., Brandl, R., et al. (2018). "Impacts of Salvage Logging on Biodiversity: A Meta-Analysis," *Journal of Applied Ecology*, 55/1: 279–89.

Timberlands Limited (2018). *Public Summary: TL Standard Management Plan for Certified Forests ES231*. Rotorua, New Zealand: Timberlands Limited. 53 pp. Available online at http://www.tll.co.nz/vdb/document/ 231 (accessed October 30, 2019).

Turner, N. J., and Jones, J. T. (2000). "Occupying the Land: Traditional Patterns of Land and Resource Ownership among First Peoples of British Columbia." Presented at *Constituting the Commons: Crafting Sustainable Commons in the New Millennium, Bloomington, Indiana, May 31–June 4, 2000*. Available online at http://dlc.dlib.indiana.edu/dlc/bitstream/handle/ 10535/1952/turnern041200.pdf (accessed October 27, 2019).

US Congress (1960). *An Act to Authorize and Direct that the National Forests Be Managed under Principles of Multiple Use and To Produce a Sustained Yield of Products and Services, and for Other Purposes*. Public Law 86-517. Washington: 86th Congress of the United States of America. Available online at https://www.govinfo. gov/content/pkg/STATUTE-74/pdf/STATUTE-74- Pg215.pdf (accessed January 26, 2024).

van Andel, T. (2006). *Non-Timber Forest Products: The Value of Wild Plants*. Agrodek Series, No. 39. Wageningen, Netherlands: Agromisa Foundation. 70 pp. Available online at https://www.agromisa.org/wp-content/ uploads/Agrodok-39-Non-timber-forest-products.pdf (accessed February 3, 2024).

Vilén, T., Gunia, K., Verkerk, P. J., Seidl, R., Schelhaas, M. J., Lindner, M., and Bellassen, V. (2012). "Reconstructed Forest Age Structure in Europe 1950–2010," *Forest Ecology and Management*, 286: 203–18.

Vogt, K. A., Gara, R. I., Honea, J. M., et al. (2007). "Historical Perceptions and Uses of Forests," in K. A. Vogt, D. J. Vogt, R. L. Edmonds, et al. (eds), *Forests and Society: Sustainability and Life Cycles of Forests in Human Landscapes*. Wallingford, UK: CAB International, 1–28.

Wackernagel, M., and Beyers, B. (2019). *Ecological Footprint: Managing our Biocapacity Budget*. Gabriola Island, BC: New Society Publishers. 280 pp.

Walker, B., and Salt, D. (2006). *Resilience Thinking: Sustaining Ecosystems and People in a Changing World*. Washington: Island Press. 174 pp.

WCED (World Commission on Environment and Development). (1987). *Our Common Future*. Oxford: Oxford University Press. 383 pp.

Williams, K., and Harvey, D. (2001). "Transcendent Experience in Forest Environments," *Journal of Environmental Psychology*, 21/3: 249–60.

Work, T., Spence, J., Volney, J., and Burton, P. J. (2003). "Sustainable Forest Management as License to Think and to Try Something Different," in P. J. Burton, C. Messier, D. W. Smith, and W. L. Adamowicz (eds), *Towards Sustainable Management of the Boreal Forest*. Ottawa: NRC Research Press, 953–70.

CHAPTER 4

Resilience Concepts

Resilience is accepting your new reality, even if it's less good than the one you had before.
Elizabeth Edwards (2009)

4.1 What does resilience mean?

This book takes an expansive view of resilience as denoting the ability to persist under pressure and following various forms of stress, disruption, or disturbance. With specific reference to forests, resilience is the ability of land to continue delivery of some or all forest ecosystem services (see Section 2.2). The world's forests have fluxed and waned but ultimately persisted through geological epochs of changing climates, just as the human species has persisted through millennia of wars and plagues. How has that persistence, that resilience, been expressed and how did it come about? What are the general properties of resilient systems, so that we can learn from them and encourage those properties in forests and forest management?

The literal meaning of resilience or resiliency[1] is "the action or an act of rebounding or springing back; rebound, recoil," although that definition is considered obsolete according to the *Oxford English Dictionary*. Current usage more commonly refers to "elasticity; the power of resuming an original shape or position after compression, bending, etc." or "the quality or fact of being able to recover quickly or easily from, or resist being affected by, a misfortune, shock, illness, etc.; robustness; adaptability" (*OED* 2022). Another dictionary defines resilience as: "1. the capability of a strained body to recover its size and shape after deformation caused especially

by compressive stress, or 2. an ability to recover from or adjust easily to misfortune or change." It also includes a discussion of its more specific use in physics and engineering, whereby "resilience is the ability of an elastic material (such as rubber or animal tissue) to absorb energy (such as from a blow) and release that energy as it springs back to its original shape" (Merriam-Webster 2023). This physics-based definition encompasses a duality of meaning—absorbing energy and subsequently releasing it—that marks the expression of resilience in socio-ecological systems as well.

Derived from the Latin equivalents of *resiliens* or *resilentia*, resilience is etymologically connected to resiling, or "going back" (*OED* 2022; Merriam-Webster 2023). Like many other "re" words (repair, recover, rehabilitate, restore, and so on), resilience is somewhat shackled by interpretations that it must denote return to some previous condition (Higgs et al. 2014). The concept has continued to develop and expand, with a growing emphasis on adaptability and alternative strategies conferring persistence, rather than simply a return to previous conditions. Falk et al. (2022) equate system resistance and persistence, in contrast to a broader role for resilience in conferring a persistence of ecosystem services. Botkin (1990) and Walker (2020) insist that resilience must involve some degree of change, adaptation, and system "learning," not simply the ability to "bounce back." In addition to its use and physics and engineering, the term is used in fields as diverse as medicine, psychology, economics, politics, community planning, and ecology. Much like

---

[1] "Resilience" and "resiliency" are exact synonyms. The simpler "resilience" is used throughout this book, but literature searches included both terms.

Resilient Forest Management. Philip J. Burton, Oxford University Press. © Philip J. Burton (2025). DOI: 10.1093/oso/9780198832997.003.0004

the concept of sustainability (see Chapter 3), there is now a growing appreciation of resilience as a desirable property for people, corporations, and communities in many biophysical and social contexts (Zolli and Healy 2012; Wordsworth 2014; Wiig and Fahlbruch 2019). Through these many applications, the resilience concept has evolved to encompass several different forms of recovery, evolution, and persistence.

Resilience is ultimately the property of a system, whether that system be a biological organism, an ecosystem, a community, a business, an economy, or communications infrastructure. Across these and other systems, different aspects of resilience are recognized and can sometimes be measured. Typical behavior of a resilient system includes the ability to absorb and resist stress, the ability to minimize the degree of disruption to system functions, and the ability to return rapidly to normal (previous or acceptable) levels of performance (see Figure 4.1). What constitutes system functionality is often multidimensional (Haines 2009). In the case of human communities, for example, we look for healthy levels of employment, a wide range of available goods and services, and effective representative governance. In the case of forests, functionality takes the form of many different ecosystem services and forest values (see Chapter 2), each of which may have its own vulnerability to and response to various stressors (Carpenter et al. 2001).

This chapter presents a brief overview of the evolution of resilience theory as a formal concept and a popular mantra, and proposes a gradient or progression of approaches by which resilience can aid in the management of natural resources. Different approaches to measuring and monitoring resilience stem from the alternative definitions, mechanisms, and applications of resilience concepts that have emerged over recent decades. The greatest uptake in resilience thinking has been in its power as a metaphor, even though it was first formalized in the context of ecosystem analysis by ecological modellers who estimated numerical parameters and rate coefficients for non-linear equations solved with differential calculus (e.g. Ludwig et al. 1978; Walker et al. 1981). While some quantitative thresholds are suggested (which may be useful in monitoring and adaptive management; see Chapter 10), this and subsequent chapters rely more heavily on analogies and metaphors to convey the essence of managing for resilience.

4.2 Embracing a widespread socio-ecological concept

The mechanisms of and preconditions for resilience in multiple disciplines can inform our exploration of resilience in forests and forest management. In the world of ecology, resilience was recognized in the 1970s as one means by which ecosystems achieve

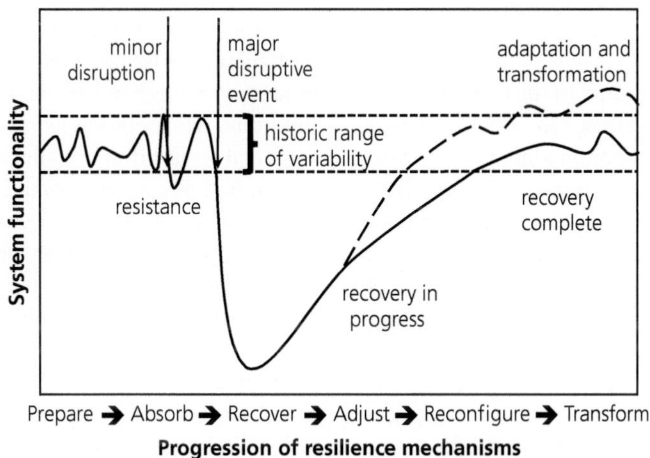

Figure 4.1 System response and progression through various mechanisms of resilience when faced with disruptive and persistent change

stability. Holling (1973: 14) defined resilience as "a measure of the persistence of systems and of their ability to absorb change and disturbance and still maintain the same relationships between populations or state variables." He went on to clarify that "resilience is the property of the system and persistence or probability of extinction is the result" (Holling 1973: 17). Other uses and implications of the concept of ecological resilience have emerged since, inspiring innovative understanding and applications in the areas of conservation biology and natural resource management (Connell et al. 2016). Walker (2020) defines resilience simply as "the ability to cope with shocks and keep functioning in much the same way."

Although terms have proliferated and are sometimes used interchangeably, a general consensus has emerged that there are three broad forms of resilience in response to disturbance or environmental change:

- robustness, through which the ecological system remains largely unchanged or recovers rapidly to exhibit the same composition, structure, and function;
- adaptation, in which the system undergoes minor or major change in composition or structure that is better suited to new conditions while processes and ecosystem services remain largely unchanged; and
- transformation, denoting substantive change in the ecosystem, but that facilitates maintenance of some ecosystem services or helps confer persistence at a larger scale.

These forms of resilience can also be interpreted as a gradient of system responses in which different degrees of change are accommodated or encouraged. Some authors equate resilience solely with some form of adaptation (e.g. Peterson St-Laurent et al. 2021), or consider transformation equivalent to regime shift and a loss of resilience, but neither of those interpretations applies universally. Seidl and Turner (2022) reserve the term "resilience" for no change in structure and composition, and use the term "restructuring" to denote structural changes without compositional change, and the term "reassembly" when composition changes but structure does not, or "replacement" if both structure

and composition change. The three-phase typology of resistance, resilience, and transformation is widely used, for which Williams (2022) interprets the gradient of management strategies as options to resist, to adapt to, or to direct system change.

Different definitions and perspectives on resilience, its modes, and component mechanisms in part reflect specific interests, which can vary from individual ecological elements (for example, current canopy trees) or populations (for example, of hunted wildlife) to more holistic interpretations of communities or ecosystems (for example, food webs), or particular ecosystem services (for example, timber supply, erosion control). It is inevitable that different disciplines will develop their own specific definitions and criteria for recognizing and quantifying resilience. For example, in the world of dendrochronology (tree ring analysis), resilience is interpreted as a return to pre-disturbance conditions of tree growth, consisting of both short-term resistance (the degree to which growth is insensitive to a disruption such as drought), and long-term or multi-year recovery after such an event (Castagneri et al. 2022). Nikinmaa et al. (2020) consider engineering resilience or resistance to be nested within ecological resilience, which in turn is nested within socio-ecological resilience, each complementary to each other in the context of forest science and management.

Combined with the many potential mechanisms of and contributions to ecological persistence, such diverse perspectives make it important to explain how the term "resilience" is being used in any particular discussion. Table 4.1 outlines the major modes and mechanisms of resilience as used in this book, while acknowledging that they represent some rather arbitrary distinctions and that other interpretations are widespread and are just as valid. For example, as elaborated upon below, one might label flexibility, substitution, and diversification all as "mechanisms" or tactics by which adjustment and adaptation can be enhanced. Some latitude in the labeling of such terms can be accepted without trying to construct a definitive taxonomy of resilience terminology.

The literature of systems analysis and management is now well populated with reviews, synthesis articles, and books concerning the concept of

Table 4.1 Different forms and mechanisms of resilience, by which the persistence of one or more values or desired ecosystem attributes is facilitated

Mode of resilience	Mechanism of resilience	Example and citation	Degree of alteration	Change in composition	Change in structure	Change in function	Ability to cope with novelty
Robustness	Resistance	Competition preventing invasion by an exotic sp.; Saccone et al. (2013)	Minor or none	None	None	None	Poor
	Recovery	Direct regeneration of serotinous-cone pines after fire; Illisson and Chen (2009)	Minor	None	Temporary	Temporary	Poor
Adaptation	Adjustment	Shift in tree species in response to browsing pressure; Côté et al. (2004)	Minor to moderate	Minor, but potentially long-term	Minor or none	Minor	Moderate
	Restructuring	Shifting from managing single to multiple tree spp. to improve ecological and economic flexibility; Knoke et al. (2005)	Moderate to major	Major, long-term	Minor to moderate	Minor	Good
Transformation	Replacement	Forest conversion to scrub or grassland after repeated fire; Whitman et al. (2019)	Substantive	May be total	Often major	Minor-major	Good short term; poor long term

resilience and its application in various management scenarios (e.g. Walker and Salt 2006, 2012; Standish et al. 2014; Linkov and Trump 2019). Xu and Kajikawa (2018) have mapped out the cross-disciplinary citations found in academic writings on resilience, with major clusters identified by research domain (see Figure 4.2). In the disciplines of medicine and human health, we recognize resilience in the resistance to disease, and in the ability to recover from pathogens or injury (Silverman and Deuster 2014; Döring et al. 2015). Reflecting our familiarity with the homeostatic nature of human physiology—for example, normal body temperature of 37°C—the emphasis in these disciplines is mostly about maintaining "healthy" or "normal" conditions, although definitions include the potential for enhanced vigor. Similarly, in the world of psychology, resilience denotes the maintenance of mental health, motivation, and direction in the face of stress, adversity, and calamity (e.g. Davydov et al. 2010; Prince-Embury 2013). Extending the scope to community health and viability, resilience typically refers to the ability to maintain local populations and access to goods and services upon the loss of core industries and employers (Beckley et al.

2002) or after natural disasters (Renschler et al. 2010; MacAskill and Guthrie 2014). In this respect there is considerable overlap with planning and policies supporting sustainable cities (Coyle 2011) and human security in general (Lautensach and Lautensach 2013), but with a greater emphasis on crisis avoidance, response, and recovery.

Much of the literature on resilience in ecological systems (e.g. Walker and Salt 2006; Gunderson et al. 2012) emphasizes the avoidance of switches ("flips" or "regime shifts") in system state, which also can be considered crises in sustainable resource management. From this perspective, much of the emphasis is on keeping a system within a desired "stability domain" (see Figure 4.3). As expanded upon below and throughout this volume, the utility of managing for resilience need not be restricted to crisis avoidance—that is, the robustness mode of resilience. Although various measures or indicators have been proposed to confirm a regime shift or predict an impending one (e.g. Brock and Carpenter 2006), such a shift cannot always be objectively defined. In part this is due to the natural range of variability (NRV) in socio-ecological systems (Perretti and Munch 2012), and the rather subjective

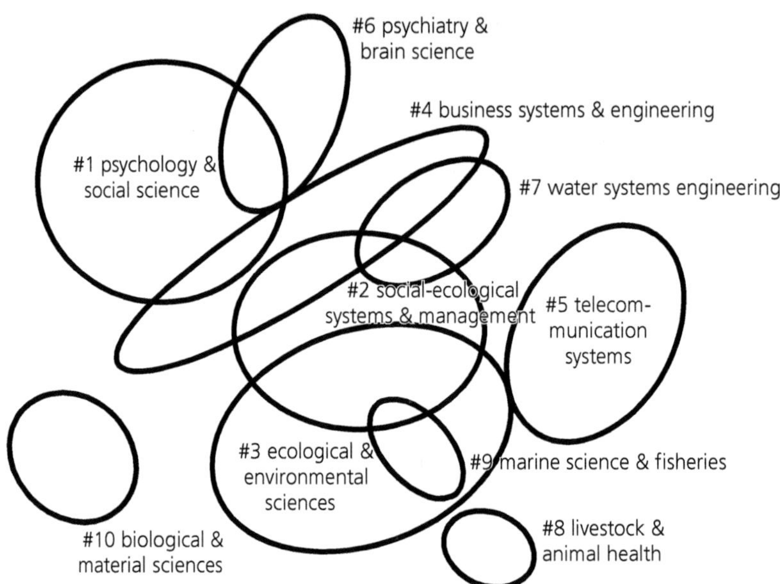

Figure 4.2 Bibliometric analysis of 86,029 citations in 19,459 papers addressing resilience (published from 1933 to 2015), showing the top ten clustering domains or disciplinary areas of research. Overlap indicates how much each referred to work in other domains. Clusters are numbered in rank order of the number of papers identified on Web of Science (redrawn from Xu and Kajikawa (2018)).

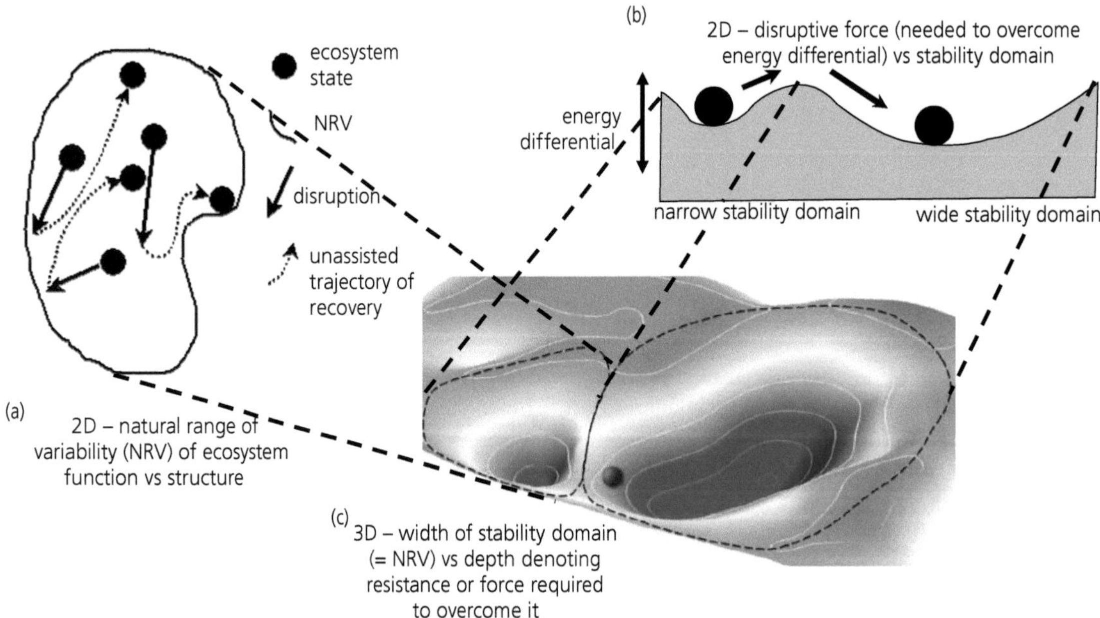

Figure 4.3 Different two-dimensional and three-dimensional portrayals of resilience within one or two alternative stability domains: (a) Most ecosystems have a natural range of variability (NRV) or stability domain (here represented as an envelope) within which natural recovery and variation constitute a healthy system (graphic from Burton 2005). (b) Disturbance or other disruption may shift the system outside of its NRV (here represented by width of a depression), into a state of degradation or an alternative stability domain (graphic redrawn from Gunderson 2000). (c) More realistically, the stability domain is 2-, 3-, or n-dimensional, representing resilience of different ecosystem attributes to different stressors (graphic from Walker et al. (2004), reproduced under the Creative Commons CC BY 4.0 license, http://creativecommons.org/licenses/by/4.0/).

decision whether to redefine the NRV envelope, or to declare that envelope or stability domain surpassed. Recognition of a regime shift or system transformation depends on one's perspective and ecosystem services of interest. For example, loss of a single keystone species or one providing critical food for people might qualify. More usually, a relatively rapid reversal of dominant species, growth forms, or functional groups qualifies as a system flip. Examples include a change from conifer dominance to broadleaf dominance (Baltzer et al. 2021), or woody plant (tree or shrub) cover prevailing over grass cover in a savanna system (Folke et al. 2004). However, both of those changes can also be considered expressions of ecosystem resilience (see Section 4.3). Such changes could also occur more gradually over time through ecological succession after disturbance rather than denoting a categorical shift to alternative stable states (Hamilton and Burton 2023).

Shifts in ecosystem state are often precipitated by disturbances such as wildfire or a severe drought. Other functional thresholds are occasionally found in natural systems without necessarily constituting a disturbance event (see Section 10.3). Perhaps more usually, such shifts also occur gradually over time ("creeping regime change") or are identified only on the basis of socioeconomic (human) criteria. This difference in the interpretation of resilience concepts is related to the debate on whether ecosystems predominantly vary continuously along environmental gradients or are largely found in identifiable categories reflecting inherent homeostatic mechanisms or "basins of attraction" (e.g. Scott 1995; Haeussler 2011). Clearly both situations exist in nature, and thresholds defining different regime attributes or stability domains may or may not emerge empirically (Connell and Ghedini 2015). It is also reasonable for regime thresholds (which may or may not differentiate stability domains) to be identified by stakeholders and managers on the basis of social or economic acceptability.

The promotion of resilience in modern human society also places considerable emphasis on the protection and recovery of physical services associated with transportation, communication, and energy infrastructure (Coyle 2011). Network analysis in support of infrastructure resilience has many commonalities with ecosystem analysis in its consideration of functional redundancies, alternative pathways, and weak versus strong linkages and dependencies (e.g. Vodopivec and Miller-Hooks 2019). With our growing dependence on the Internet and the World Wide Web, resilience of the world's interconnected digital network is of growing importance to business and governments everywhere (Kott and Linkov 2019). Indeed, it is the instantaneous redirection of digital traffic to alternative routes in the early design of the Internet that has led to its general robustness and its inspiration in addressing vulnerabilities in other engineered systems. Resilience-based management of built infrastructure in general emphasizes preparedness to avoid the loss of functionality, or to restore normal functionality as rapidly as possible (e.g. Hariri-Ardebili 2018). Although there is some use of multiple modalities in back-up systems, resilience for engineered systems has apparently placed little emphasis on plasticity or reconfiguration (evolution)—features characteristic of organic systems—as tractable strategies for ensuring continued functionality. Rather, adaptation must go through fits and starts over time, as systems go through the design and approval process . . . much as institutional policies and procedures tend to do. After a brief period of construction, followed by a long period of use and maintenance, built infrastructure undergoes a period of gradual deterioration and increasing maintenance costs, until considerations of safety, functionality, and cost dictate that the system must be replaced and rebuilt. Unfortunately, the need to "deconstruct" engineered structures contributes considerable inertia and resistance to major change and adaptation. This inertia can be broken by tragic accidents (loss of human life), partial collapse, or severe disturbances that do some of the deconstructing. There is a growing movement in the architectural and manufacturing sectors, however, to design buildings and consumer products with easy disassembly and recycling in mind (Sundin et al. 2012; Rios et al. 2015), which may somewhat reduce resistance to change and the upgrading of engineered systems.

Resilience assessment and improvement are an increasingly popular theme in the disciplines of economics and finance across all scales from micro to macro: at the level of individual households and firms, communities, countries, and regions. There is frequent emphasis on the post-facto evaluation of the role of institutional structures and policies that appear to have contributed to differences in recovery times following economic downturns (e.g. Fay and Nordhaug 2002; Amann and Jaussaud 2012). Other examples include the evaluation of individual sectors (e.g. Kofner 2014) or demographics (e.g. Béné 2013) in terms of vulnerability and resilience to chronic stressors or disruptive events. Manufacturers and retailers sometimes evaluate their vulnerability to supply-chain disruptions, and devise inventory policies to offset anticipated risks. Prudent lenders often put mortgage applications through a "shock test" to evaluate the ability of applicants to withstand an increase in interest rates. Developing a diversity of products/services and customers/clients is seen as the best buffer against disrupted and uncertain markets. Investment advice typically emphasizes the importance of portfolio diversification and responsivity to changing conditions, although high levels of diversification can also contribute to the propagation of a financial crisis (Wagner 2010). Many of these applications of resilience thinking in other disciplines and sectors provide useful analogies that can inform and inspire resilient forest management.

4.3 Modes, pathways, and mechanisms of resilience

4.3.1 Recovery

We might first remain true to the original literal definition of resilience—namely, that it denotes recovery, the reliable return to a previous state. This makes identification of and preference for the previous state a key assumption in promoting recovery. A physical metaphor for resilience as the power of recovery is a coil spring or elastic band that promptly resumes its shape after being compressed or extended (see Box 4.1, Figure 4.4a). The springs utilized in horse-drawn carriages, railway cars, and automobiles are quite literally "shock absorbers," mechanisms that reduce the impact of bumps in the road. Similarly, an ecosystem that recovers reliably and promptly after a disturbance is less likely

to experience a severe disruption in the delivery of ecosystem services. If we extend the metaphor, however, it should be pointed out that it is possible to damage an elastic band if it is pulled out too far—it can break or remain stretched if the strain is too great. In the context of forest ecosystems, even so-called fire-adapted species can indeed be killed by very intense fire, much-abbreviated fire return intervals, and disturbance interactions (Whitman et al. 2019).

Examples of "recovery pathways" in natural ecosystems are widespread, particularly where a single species or growthform has hit upon a reliable solution to persist under a predictable climate and the dominant disturbance regime such as recurrent fire or mammalian browsing. Examples include temperate forests and woodlands dominated by oaks (*Quercus* spp.) and many other broadleaf species that vegetatively resprout when seedlings are nipped off by deer or burned by cool surface fires. Many of these species also retain resprouting potential as saplings and adults (Del Tredici 2001). The shrubby chaparral of Mediterranean ecosystems typically consists of a diverse suite of species, most of which can recover vegetatively or by seed after fire (Keeley 1986). In all such cases, one can identify a somewhat reliable agent and frequency of disturbance (for example, fire every twenty to fifty years) causing mortality in the dominant vegetation, followed by prompt regeneration to the same species or set of species. The measurement of resilience in such scenarios is typically the period of time to recover, although the recovery time for ecosystem structure and services can be different from the recovery time of species composition. Where the state–space landscape is dominated by this one stability domain, recovery time is generally governed by the "steepness" of the domain cup in the sense portrayed graphically in Figure 4.3b or 4.3c (Gunderson 2000).

Many Australian eucalypt species (for example, *Eucalyptus banksii, E. diversicolor*) can be severely damaged and apparently killed by fire, but will soon sprout new foliage from charred stems and branches (Clarke et al. 2013). The same strategy of epicormic resprouting (see Figure 4.5a) in the upper tree stem and branches is true of a few pines (for example, *Pinus canariensis, P. echinata, P.*

Box 4.1 Metaphors for different mechanisms of resilience

There are many kinds of resilience in the world around us, which have analogs in forest ecosystem dynamics and management options. It is useful to consider some physical examples in communicating alternative mechanisms by which a system persists and continues to fill key roles or provide desired services.

(a) *Recovery* is here represented by a coil spring that can be extended or compressed, but then reliably returns to its original shape.
(b) *Resistance* is epitomized by a bedrock massif such as the Rock of Gibraltar, practically immutable against most meteorological or human forces in the short run.
(c) *Adjustment* is expressed in flexibility, such as an adjustable wrench suitable for use on a continuous range of nut or bolt sizes. Adjustment could also consist of a minor change in some structure or composition.
(d) *Reconfiguration* denotes a major change in structure or composition that confers success in a new situation. Adjustments are substantial, often involving addition, loss, or substitution of important components, such as when a glass window is shuttered or filled in, keeping out the weather but eliminating the ability of light to pass through.
(e) *Transformation* changes something's identity. Made of many of the same underlying components, the classic example is metamorphosis in amphibians and insects, where the adult form looks and functions as a fundamentally different organism from its larval form, while retaining its underlying genetic information.

None of these mechanisms of persistence is absolute, for springs can become permanently stretched and even mountains can be moved by earthquakes. Resistance may be successful to a certain point, after which the system may collapse, and then may or may not recover or adjust in some manner. Similarly, adjustment, reconfiguration, and transformation constitute a continuum, for which distinctions often depend on individual values of interest (see Chapter 2) rather than consideration of the system as a whole. Furthermore, different mechanisms can function at different scales in a system, with adaptation or transformation at a lower scale contributing to resistance or recovery at a higher level.

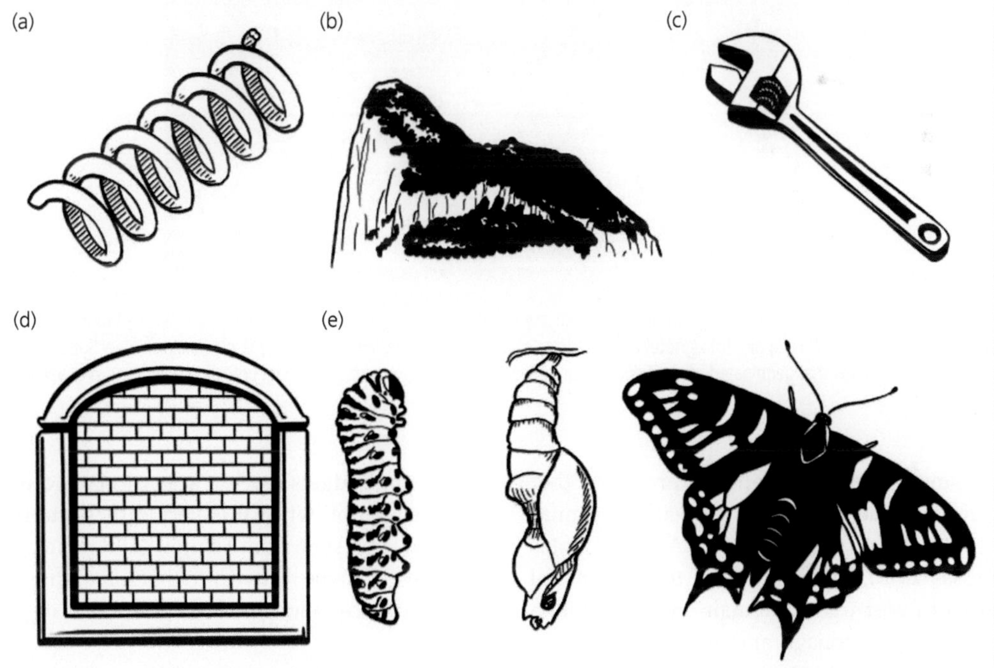

Figure 4.4 Metaphors for different mechanisms of resilience: (a) recovery; (b) resistance; (c) adjustment; (d) reconfiguration; and (e) transformation (graphics by Petroglyph Studios (www.petroglyphstudios.org)).

Figure 4.5 Evolutionary strategies conferring resistance and recovery to wildfire: (a) Canary Island pine (*Pinus canariensis*) branches and boles resprouting after wildfire, La Palma, Spain. (b) Lignotuber at the base of a toppled turpentine tree (*Syncarpia glomulifera*), New South Wales, Australia. (c) Dense seedlings that germinated after wildfire released seeds from serotinous cones of lodgepole pine (*Pinus contorta* var. *latifolia*), British Columbia, Canada.

rigida; Climent et al. 2004) and an array of other tree species characterizing the semi-arid savannas of Australia, South America, and Africa (Dantas and Pausas 2013). This ability for the substantial structure of forest trees to remain intact despite a temporary loss of foliage and fine twig biomass blurs the distinction between resistance and recovery pathways (see Table 4.1), but the persistence of most woody biomass has critical implications

to carbon budgets. Such a scenario is additionally instructive, for, when it is very difficult to disrupt such an ecosystem from a single, stable state, there is often also reliable recovery to that same state should a disturbance manage to budge it from its resting condition.

Several woody species that evolved in fire-prone habitats, as found in much of Australia and the Mediterranean region, also have swollen root crowns known as lignotubers (see Figure 4.5b), which hold carbohydrate reserves and can sprout new shoots after disturbance. Pine species with serotinous cones (for example, *Pinus banksiana, P. halepensis, P. pinaster, P. rigida*) are also very successful at releasing vast quantities of seed that can take advantage of the scorched forest floor and high light conditions after a forest fire, thereby regenerating another pine stand (see Figure 4.5c). Although both vegetative sprouts and seedlings can be effective in regenerating the dominant pre-disturbance tree species, sprouts often grow more quickly and are more drought tolerant than seedlings because they can utilize a well-developed root network.

Ecosystem recovery patterns and adaptations have been counted on and utilized by humankind in our use and management of many renewable natural resources, such as grasslands, forests, and fisheries. The evolutionary strategy of vegetative resprouting in many trees and shrubs is the basis of several silvicultural and restoration practices, including coppicing for forest stand regeneration and fuelwood production (Evans 1992). Other applications include taking small hardwood cuttings for the production of clonal nursery stock (Macdonald 2006), and the use of large cuttings planted as live stakes, wattle fences, and pole drains in the practice of soil bioengineering (Gray and Sotir 1996).

4.3.2 Resistance

Rather than responding with rapid, reliable recovery when challenged with stress or disturbance, some systems do not succumb to such disruption or disturbance in the first place. This mechanism of robustness denotes resistance to change, and is sometimes referred to as "engineering resilience" (Holling 1996; Gunderson 2000: 427). In searching for a physical metaphor, we might recognize a robust, disturbance-resistant structure such as a concrete wall or a granite massif (see Box 4.1, Figure 4.4b). Neither a concrete wall nor a mountain is infinitely robust or resistant to change, and we can recognize a scale or gradient of resistance, with a granite rock being more robust than one made of

limestone, which in turn tends to be more robust than a brick wall.

To measure robust resilience, one typically wants to know the force needed to budge, alter, or destroy the system in question or the services it provides. But, like a bridge girder or a concrete dam, one expects the system to remain structurally stable, stationary, and unchanging in order to remain functional. From a systems perspective, the state–space landscape is dominated by a single stability domain from which it does not budge, or is not expected to budge. If and when the system fails, however, it is often with catastrophic results. The engineering response to failure (for example, if a road washes out or collapses) would first be to repair the damage, perhaps strengthening the structure while doing so. If failures occur repeatedly, some sort of buffer or protection might be installed, and if that too proves ineffective, rerouting or complete reconstruction is called for.

In the realm of ecology, we can recognize some "resistance pathways" of ecosystem dynamics. For example, large, established thick-barked trees in drier, fire-dependent, open, or patchy forests and savannas are resistant to the effects of surface fires that are fueled mostly by grass between the trees (Pausas 2015). Another example is healthy conifer trees that exude resin successfully to pitch out bark beetles attempting to bore into the phloem to lay eggs (Christiansen et al. 1987). Competition and herbivory by resident species are sometimes able to resist (or at least constrain) successful establishment by invasive alien species (Levine et al. 2004; Saccone et al. 2013). Resistance strategies are counted on by managers when employing restoration burns in woodlands and savannas, and in silvicultural practices (such as thinning, pruning, or fertilization) designed to enhance the vigor of individual trees and to improve their ability to withstand stressors such as drought and pests (see Chapter 8).

4.3.3 Adjustment

Once some level of persistent change to the system is desired or inevitable, adaptation can be undertaken by different mechanisms and to different degrees. Adaptation or adaptive resilience has been proposed as the principal alternative to engineering resilience (Gunderson 2000), although here it is divided into minor shifts (adjustment) and major shifts (reconfiguration). Adaptive capacity (or adaptability) implies a high level of flexibility, expressed either through broad tolerances or through a nimble ability to change (Wagner et al. 2014).

The first option for adaptation, or the third mechanism of resilience in general, is that of adjustment, in which some components or processes are changed but the prevailing structure and function is retained. A physical metaphor for adaptive potential is the adjustable wrench (see Box 4.1, Figure 4.4c)—possibly not the best tool for a particular nut or bolt, but a more reliable choice if you are uncertain of the set of tasks with which you will be dealing. Other metaphors for adjustment typically involve some degree of reallocation or substitution, but without a reduced ability of the system to perform its essential functions. For example, sailboat crews regularly switch or realign sails to take better advantage of current wind direction and strength. A river might carve a new channel as it meanders across a floodplain. Gaps in a stone wall can be replaced with bricks with no change in function. In team sports, adjustment might consist of substituting one or two players in the middle of a game.

Analogous examples of adjustment can also be found in forest ecosystems. As individual trees develop, they often undergo adjustments of biomass allocation to shoot growth or to root growth, or to growth in height or growth in girth, often referred to as allometric plasticity (Lines et al. 2012; Jucker et al. 2015). Another example is the growth response to sediment deposition exhibited by several spruce (*Picea*) species, in which new adventitious roots are produced from the original stem, thereby taking advantage of the new rooting medium (see Figure 4.6a). Note that this physiological/anatomical adjustment mechanism confers resistance to the potentially destructive effects of disturbance at the whole-plant level, illustrating the scale-dependent nature of describing resilience mechanisms.

Ecological adjustment often occurs through the ability of some components to function under a broad range of conditions or roles. This flexible ability might reflect broad environmental tolerances, responsive phenotypic plasticity, or high genetic

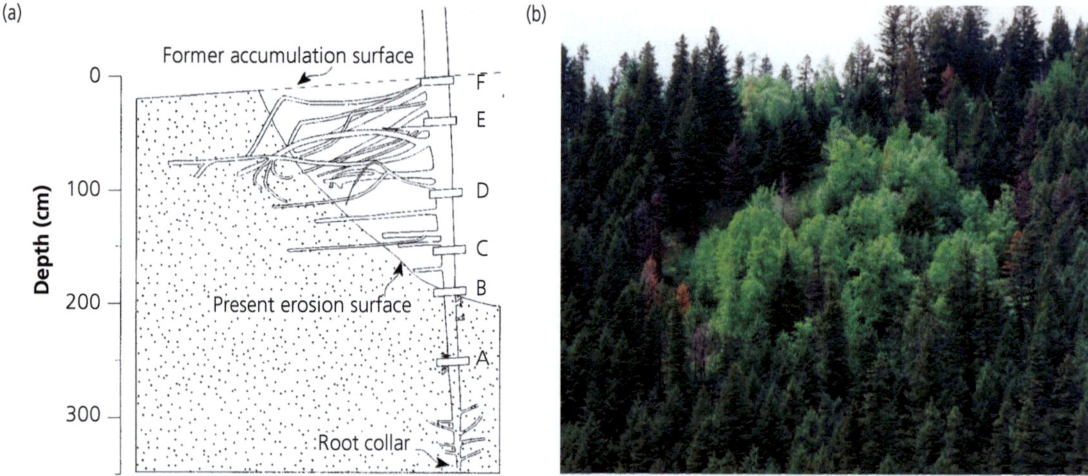

Figure 4.6 Examples of ecological adjustment mechanisms and patterns: (a) White spruce (*Picea glauca*) generates successive layers of adventitious roots from the aboveground tree trunk when the base is buried by water-borne sediments or mobile sand dunes (from Cournoyer & Filion 1994, copyright © Regents of the University of Colorado, reproduced by permission of Taylor & Francis Ltd, http://www.tandfonline.com on behalf of Regents of the University of Colorado). (b) A pocket of laminated root rot (*Phellinus sulphurescens*) in a conifer forest has resulted in the mortality of *Pseudotsuga* and *Abies* species, and their replacement by more *Phellinus*-tolerant *Betula* and *Populus* species (from Cleary et al. (2011), reproduced with permission).

diversity within or among populations. By this definition, tree species with broad niches and wide climatic ranges (for example, *Populus tremuloides* in North America) are likely to have a greater capacity to adapt to a changing climate than species with narrower tolerances and more restricted geographic ranges.

At the plant-community level, there have been several historical examples in which exotic pests have largely removed individual tree species (for example, *Castanea dentata*, *Ulmus americana*, *Fraxinus* spp.) from the forest canopy. Over time, the relative abundance of other forest tree species has increased to take advantage of the niche space and physical space vacated by the missing species, but with little change in forest structure (Brunet et al. 2014). If a minor species occupying a small portion of the forest is lost, subsequent infilling by other species can be considered a local adjustment. For example, pockets of *Phellinus* root rot in a conifer forest are gradually replaced by patches of more resistant broadleaf species (see Figure 4.6b). "Adjustment pathways" at the ecosystem level are thus sometimes characterized by a minor shift in species composition, but with forest structure, productivity, and the delivery of ecosystem services more or less maintained. The gradual changes in species composition that are found over an environmental gradient or during the course of succession could also be characterized as examples of system adjustment to changes (over space rather than over time) in the relative importance of limiting factors.

At the individual tree or canopy gap level, however, a switch in tree species—or even a switch in the genotypes replacing the previous trees—could be considered reconfiguration or transformation, once again emphasizing the importance of scale and arbitrary distinctions in all efforts to categorize the mechanisms of resilience.

4.3.4 Reconfiguration

If adjustments to changing conditions are major and long term, one might recognize the accompanying change as one of wholesale reorganization, reconfiguration, or restructuring. Unlike bricks being used to repair the gap in a stone wall, using bricks to infill a former door or window (see Box 4.1, Figure 4.4d)

confers a major change in function: light no longer passes through, although outdoor weather is still kept out. In team sports, reconfiguration would consist of an entire shift change or team changeover at one point during a game.

The adaptive response of forests to major changes is exemplified by the ingrowth of other deciduous tree species after chestnut blight (*Cryphonectria parasitica*) decimated populations of American chestnut (*Castanea dentata*) throughout eastern North America (Thompson et al. 2013). With the pathogen spreading outward from New York from 1910 through the 1940s, there were undoubtedly multiple impacts of widespread tree death (see Box 4.2), yet the forest did not degrade to scrub or grassland, but rather responded by other tree species eventually taking the place of chestnut. In this manner, the forest ecosystem reconfigured itself, adapting to the significant loss of a dominant tree species in some stands through the ecological flexibility of remaining tree species such as northern red oak (*Quercus rubra*), red maple (*Acer rubrum*), and several other species not previously recorded (see Figure 4.7). Nonetheless, at the ecosystem level, the loss of American chestnut must have had major impacts on carbon storage, nutrient cycling, and food webs (Ellison et al. 2005).

Forests regularly reconfigure themselves over the course of stand development and succession. When a cohort of trees establishes itself after a stand-level natural disturbance such as severe wildfire, it often develops into a relatively homogenous even-aged stand of shade-intolerant trees. Over time, however, as some trees die and canopy gaps form, the stand becomes more irregular in structure, with new shade-tolerant trees becoming established in gaps and eventually dominate (Oliver and Larson 1996; Kneeshaw and Burton 1997). In the absence of another stand-level disturbance, very old forest stands can become very different, reconfigured versions of their predecessors, with a sufficient continuity of tree cover at appropriate scales to support different complements of birds, understory plants, and epiphytes as well (Nordén et al. 2014; McGee 2018). Similar patterns of species replacement are evident from the pollen record, as forests reconfigured themselves in response to the world's history of glaciation and deglaciation (see Figure 4.8). In

Box 4.2 Scale and perspectives in considering forest response to chestnut blight

Forest response to the effects of invasive exotic pests such as chestnut blight (see Figure 4.7) also illustrates the continuous nature of disruptive impacts, ecosystem responses, and the importance of scale in characterizing resilience. At the scale of individual trees—often scattered individuals in the case of American chestnut in eastern forests—the physiological stress associated with the cumulative effects of pathogen attack eventually result in tree death, a disturbance event associated with canopy opening, reduced competition, and increased litterfall in the influence zone of that individual tree. At the forest stand level, infilling by surviving (non-chestnut) trees constitutes ecosystem adjustment if chestnut trees occupied less than half of the stand, or ecosystem reconfiguration if chestnuts made up more than half of the pre-disturbance stand. In some situations, where chestnut completely dominated a stand, its transformation to a stand dominated by red maple (for example) has been considered a system flip or regime shift (Shackleton et al. 2018).

Although dominance by deciduous broadleaf tree cover continues, we may never know the carbon pool impacts associated with the loss of these massive trees and their persistent root systems. At the landscape scale, the near-complete loss of the bountiful fruit production associated with adult chestnut trees has undoubtedly constituted a regime shift to the many animal species such as wild turkey (*Meleagris gallopavo*) that depended on chestnuts as an important part of their diet (Martin et al. 1961). Likewise, species adjustment occurring in the root-rot pocket described in Figure 4.6b could be considered a forest reconfiguration where such fungal effects dominate a forest stand, or a transformative response at the scale of the individual pocket of conifer mortality that then flips to dominance by broadleaf trees. In conclusion, distinctions among adjustment, reconfiguration, and transformative expressions of resilience depend on both the ecological attributes of interest and the physical scale of reference.

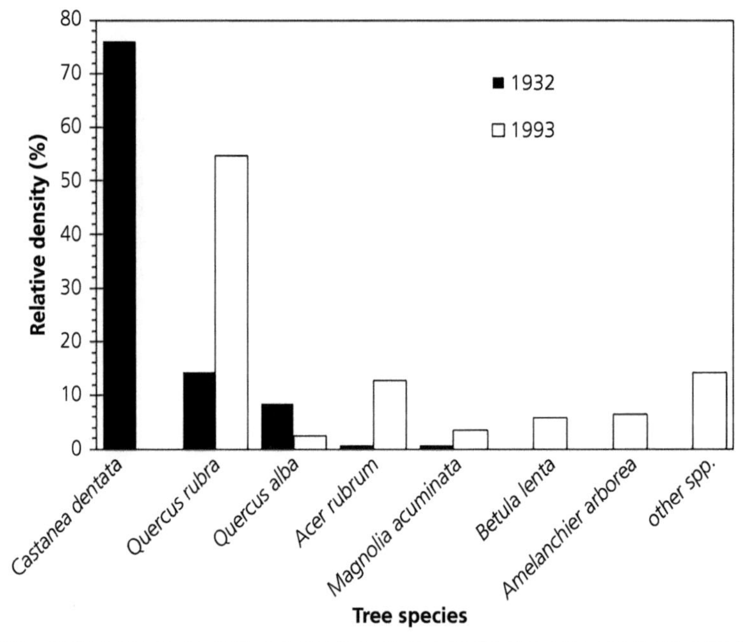

Figure 4.7 Changes in species composition of a previously chestnut-dominated forest stand in Virginia, US, before and approximately six decades after infestation by the exotic chestnut blight fungus (based on data from Agrawal and Stephenson (1995)).

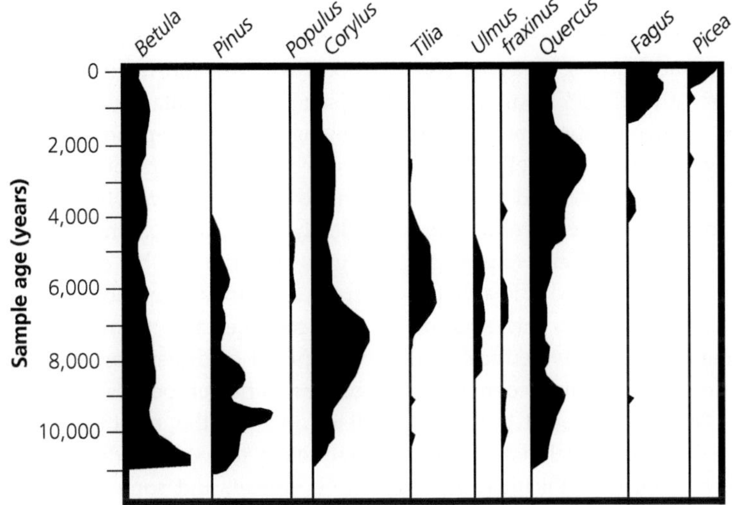

Figure 4.8 Simplified pollen diagram portraying the Holocene history of Gribskov Forest in Denmark. Early dominance by *Betula* and *Pinus* gave way to greater abundance of *Corylus*, *Tilia*, and *Ulmus*, followed by *Quercus* dominance and then *Fagus* and *Picea* most recently. Current *Picea* abundance in the pollen record represents plantations established in recent centuries (redrawn and smoothed from Overballe-Petersen et al. (2014)).

those situations, the timeline of change is longer than observed for succession, and the transitions were driven by climate fluctuations as well as disturbance regime and dispersal ability. Modern examples of forest transformations as a result of human actions exist too. The replacement of natural broadleaf forests with planted stands of Norway spruce (*Picea abies*) or other conifers represents such a transformation, as do recent efforts at reversing that trend (Hansen and Spiecker 2016). Invasive species, including trees, fungal pathogens, and insect or mammalian herbivores, can reconfigure the physical and food web structure of a forest in relatively short periods of time (Hughes and Denslow 2005; Liebhold et al. 2017).

4.3.5 Transformation

When a system changes state, this is typically regarded as the antithesis or failure of resilience. Yet much depends on the objectives of management, the role of the ecosystem in a landscape, the scale of change, and the array of alternative states available. Sometimes, conditions change so drastically that state change or transformation is the only means by which some values can be maintained locally, even if all values cannot. In other cases, local transformation facilitates resilience on a larger scale. In essence, one can think of transformation as a change in "identity," a change in the category of objects, constructs, or systems by which we recognize and manage components of the world around us (Walker and Salt 2012). So we can consider transformation, not necessarily as collapse or management failure, but as a mode of resilience by means of "shape shifting" in which most components or structures are changed but some function is retained. Classic examples from the world of biology include the many species that undergo metamorphosis as part of their life cycle: amphibians such as frogs and salamanders that have gilled aquatic and air-breathing terrestrial stages; and the many insect species that transform from voracious sedentary larvae into winged, reproductive adults (see Box 4.1, Figure 4.4e). In the world of popular culture, successful toys and movies feature cars and trucks that can transform into robots. Indigenous beliefs frequently incorporate the roles and actions of mortal humans who can transform into animals or mystical beings.

Many ecosystems have alternative stable states that can persist under a given set of environmental conditions, or under different disturbance regimes. For example, the birch-pine woodland and heathlands of Scotland are alternative stable states for natural vegetation in that part of the world (Davies

2010). Accepting *Calluna vulgaris* heath development and persistence (in response to centuries of woodcutting, grazing, and burning) may be considered a "failure" or "giving up" from the perspective of forest management or forest restoration. Yet, if the pre-existing system is not well adapted to the current regime of climate, land use, and human pressures, it may require heroic efforts and repeated investment of time and money to promote and sustain native woodlands instead of heathland. Whether to invest limited resources into such an endeavor at more than representative educational locations is a social decision. We can recognize the transformation from *Betula-Pinus* forest to *Calluna* heath as a mode of resilience by which native vegetation persists, erosion is controlled, the hills remain green, and various—though different—species of wildlife are supported.

Localized transformation (for example, to meadows or wetlands in forest) can be an important mechanism for generating landscape-level diversity. For example, there is no doubt that the ponds created by beaver (*Castor* spp.) dams flood and kill large numbers of trees and woodland plants, converting forested ecosystems into aquatic ecosystems. Yet, from the landscape perspective, those beaver ponds serve as important resources and platforms for forest biodiversity (Johnston 2017). In other words, some transformation is important to the integrity (identity and persistence) of temperate and boreal forest landscapes as a whole.

4.4 Resilience and the adaptive cycle in complex adaptive systems

Resilience is an emergent property of complex adaptive systems, and a dimension by which socio-ecological systems change in cycles of growth and reorganization. These features are outlined in the following paragraphs, with their particular relevance to forests and forest management more fully explored by Messier et al. (2013) and Filotas et al. (2014). Complexity and adaptive capacity are repeatedly invoked in offering options to promote resilience in forests and forest management in the chapters that follow.

Expressions of resilience are fundamental to the success and evolutionary potential of complex adaptive systems. Examples of complex adaptive systems include ecosystems and human organizations. An essential feature of a complex adaptive system is its modularity, with multiple components having enough independence (agency) and enough connectedness (interactions) both to absorb shocks and to change/evolve when necessary (Levin 1998). Ideally, this connectedness varies with different fluxes experienced by a system: very fluid and conducive to information flow, for example, but with modules able to disconnect in order to halt the spread of disease or other destructive disturbances. Other properties of complex adaptive systems include the responsivity of those individual actors to local conditions in a heterogeneous environment, and some mechanism (for example, genetic mutation, disturbance) that is sufficiently disruptive to promote innovation, experimentation, and novelty. The independent but interacting elements of the system then have the capacity to self-organize in a variety of persistent combinations. As a result, complex adaptive systems go through repeated cycles of development and reorganization (see Figure 4.9), with characteristic attributes at each stage. As an alternative to embracing long periods of rigid stasis, systems analysts and managers of a wide array of natural, social, and engineered systems are now advocating and experimenting with the self-organizing aspects of complex adaptive systems, for which resilience is an emergent property (Rouse 2007; Filotas et al. 2014).

Different stages of the adaptive cycle are denoted by r (the standard symbol for intrinsic rate of increase in population models) during the resource exploitation and rapid-growth phase, K (the standard symbol for carrying capacity in population models) during the mature or conservation phase, Ω (omega, the last letter of the Greek alphabet) for the disrupted conditions associated with the release phase, and α (alpha, the first letter of the Greek alphabet) for the exploratory phase of system reorganization. Phases of the adaptive cycle can vary greatly in length, although the back-loop processes of release and reorganization tend to occur rapidly. Those familiar with forest dynamics may equate the r phase with successional development, the K phase with old-growth or climax forest, and the Ω phase

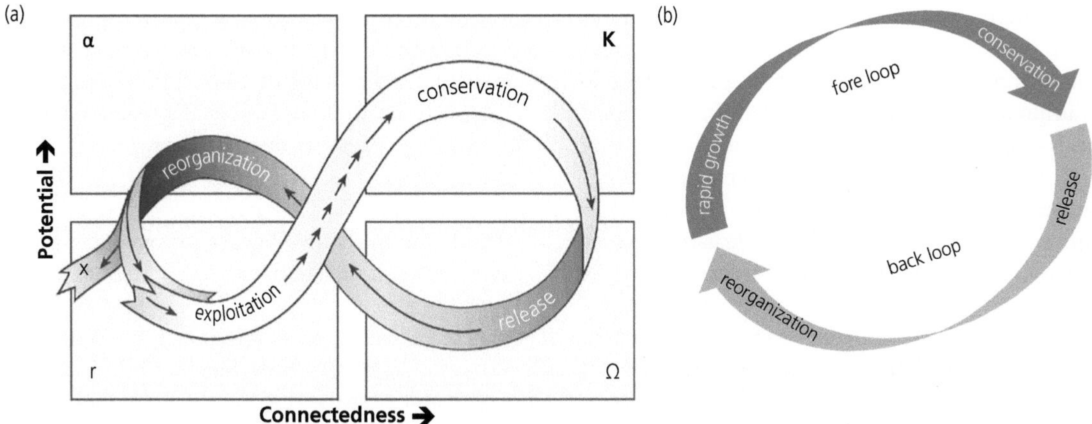

Figure 4.9 The adaptive cycle of change and rejuvenation characteristic of many ecological and historic socioeconomic (business, institutional, political) systems: (a) The four-phase cycle as originally described by Gunderson and Holling; (2002); graphic copyright © 2002 Island Press, reproduced by permission of Island Press, Washington, DC; (b) simplified as a two-phase cycle, emphasizing the development (fore loop) and reorganization (back loop) components (redrawn from Walker and Salt (2006)).

as corresponding to disturbance events. Classic portrayals of stand dynamics in the past did not pay much attention to the α phase and the potential for ecosystem reorganization under new environmental conditions. Although exits from the cycle or flipping to an alternative state (denoted by x in Figure 4.9a), can occur at any stage of the cycle, this is most likely to occur during the transition from α to r phases.

Each phase of the adaptive cycle has a distinctive set of characteristics (see Table 4.2). Holling and Gunderson (2002) posit that the adaptive cycle functions in a third dimension, that of resilience, in addition to the "potential" and "connectedness" framework portrayed in Figure 4.9a. Potential denotes productivity or future options, connectedness refers to the degree of internal control, and resilience is the opposite of vulnerability to external disruption. Resilience decreases as a system moves into the conservation (K) quadrant and is low in the immediate wake of disruptive release (the Ω stage) but then expands as the system explores reorganization options (the α stage) and exploits newly available resources (in the r quadrant). Furthermore, one can see resilience being expressed in different modes (as outlined in Table 4.1) at different stages of the adaptive cycle. Robustness prevails in the K phase, recovery dominates the r phase, and (depending on the severity of the disruption that led to the Ω phase) adjustment or reconfiguration prevail in the α phase. If the system flips to an alternative state, as denoted by the x option in Figure 4.9a, some attributes and values of the pre-disturbance system usually persist in the transformed system.

Most adaptive cycles at the scale of interest to managers (for example, forest stands and landscapes) exist within a hierarchy of space and time scales—a "panarchy"—such that smaller cycles of

Table 4.2 Characteristics of different phases of the adaptive cycle

Phase	Potential	Connectedness	Resilience	Rate of change	Competition	Rigidity	Predictability
Exploitation, r	Low	Low	High	Medium	Scramble	Low	Low
Conservation, K	Very high	High	Low	Slow	Contest	High	High
Release, Ω	Very low	High	Low	Very rapid	Not important	Very low	Low
Reorganization, α	High	Low	High	Rapid	Scramble	Very low	Very low

fast-changing variables or conditions (for example, insect populations) are constantly initiating change, while larger cycles of slow-changing variables and conditions (for example, climate, terrain, biotic legacies) constrain change and provide continuity or "system memory" over time (Holling et al. 2002). Those scales represent the duality and range of options for conferring resilience, by resisting change at one level, while experiencing potentially adaptive changes and disruptions at another level. Transformation or collapse of the system can often be avoided if there are sufficient opportunities for change within the system.

4.5 Resilience as a management approach

There are some common themes among the many interpretations and applications of resilience thinking as briefly outlined in the previous section. One common resilience-enhancing approach is the diversification of assets, activities, and connections. Another common feature is identification of some core components that are solid and reliable, and some that might expand with opportunities or contract under duress. Effective management for resilience avoids overdependence on any one component and incorporates preparation and planning based on scenarios, contingencies, buffers, and back-ups. Successful strategies for coping with crises and systemic change include good monitoring and communication, regular check-ups or report cards, and flexible responsivity to those indicators. But survival in the face of stress and disruption can often depend on intangible levels of leadership, motivation, or commitment too. Sometimes the winning strategy is to "stick with the plan," while at other times it is better to "bend with the wind." The most important aspect of managing for resilience probably revolves around situational awareness informing decisions of when and how to switch from one management mode to another (see Chapter 10).

In promoting policies and practices to support resilient socio-ecological systems in general, several commentators have devised some broad guidelines and recommendations. Lynch et al. (2022) suggest recurrent strategies of resisting, accepting, or directing ecosystem change in an adaptive framework for coping with environmental change. Walker and Salt (2012: 193–5) offer a list of eleven attributes (some with associated probing questions) to serve as goals for managing a resilient system. Six themes emerge that offer management direction, paraphrased here from Walker and Salt (2012: 195–8):

- Think multiple scales—every system experiences cross-scale interactions at local, regional, and global scales.
- Put a focus on thresholds—get to know the ecosystem services, how they are bundled, and how vulnerable they are to a change in the current stability domain.
- Celebrate change—growth and innovation are episodic and can be scary; be prepared to abandon obsolete concepts, keep effective ones, and introduce new ones.
- Embrace uncertainty—feedbacks and non-linearities that generate surprise often unlock unhealthy levels of rigidity.
- Foster innovation—encourage the free exchange of diverse ideas and perspectives; look for opportunities to experiment and to develop networks of learning.
- Don't forget governance—natural resource management is people management, so decisions must follow principles of equity and fairness in order to be successfully implemented.

Perhaps because of the diversity of definitions and applications, managing explicitly for resilience—that is, promoting resilience as a management strategy—is sometimes considered too vague a concept to operationalize (Hodgson et al. 2015). More specifically, "managing for resilience" is not sufficient as a management objective (Klein et al. 2003; Lindenmayer and Hunter 2010; Fuller and Quine 2016). On the other hand, if persistence of ecosystem services is the ultimate goal, then the apparent vagueness of "promoting resilience" can instead map a broad range of options for achieving that goal. Different management approaches are appropriate before, during, and after a crisis, for example. Similarly, different modes of resilience are seen and can be promoted at different scales, from individual trees and microsites, through to forest stands, landscapes, and regions. There also is evidence that nature can employ multiple mechanisms of resilience at any one scale, such as a combination of avoidance, resistance, and recovery in tree growth when faced with moisture limitations or fire (Lloret et al. 2011).

Although a desirable attribute, it might be most productive to view resilience as a family of approaches to guide policies and management, while not being a goal of management per se (Fuller and Quine 2016). Thus resilience can be considered an intermediate objective, facilitated by a set of principles and indicators, that guides tactics in support of the broader goal of survival and persistence. Derissen et al. (2011) consider resilience a descriptive concept, which may or may not be necessary and sufficient in support of the normative goal of sustainability. Clearly, maladaptive resistance (as one persistence tactic) will not support sustainability in changing times, while adaptation (adjustment or reconfiguration) is essential for achieving sustainability under such circumstances. Resilience might be invoked as an aid to achieving the widespread goal of socio-ecological sustainability. But the popular concept of sustainability comes with a lot of corollary assumptions and norms, some of which are unrealistic or counterproductive in a changing world (as discussed in Chapter 3). Consequently, resilience-based management is more appropriately framed (here and throughout this book) as a means of encouraging the persistence of identified ecosystem services. Managing forests for resilience supports resilient forest stewardship, and resilient forest stewardship supports resilient forests.

Box 4.3 Key points

- There is considerable interest and experience in understanding and promoting resilience in many natural and human systems, which can provide guidance in applying the concept to forest management.
- Resilience refers to persistence, but this can be achieved through either robust stability or adaptive change.
- Robustness is typically achieved through resistance to disruptive events, or relatively rapid and reliable recovery after disruption.

- Adaptability is needed when conditions change, and it is typically achieved through overall flexibility or the substitution, adjustment, restructuring, or reconfiguration of some system components.
- System transformation or a state shift can still maintain some ecosystem services.
- These modes and mechanisms exist along a continuum, and their interpretation depends on social definitions of system attributes, values of interest, and reference scale.
- Resilience further exists in a hierarchy or panarchy of space and time, with local, fast-paced factors initiating changes that tend to be constrained or averaged out at regional scales.

Nature can be far more generous than we deserve.
David Suzuki, closing episode of
The Nature of Things (2023)

References cited

Agrawal, A., and Stephenson, S. L. (1995). "Recent Successional Changes in a Former Chestnut-Dominated Forest in Southwestern Virginia," *Castanea*, 60/2: 107–13.

Amann, B., and Jaussaud, J. (2012). "Family and Non-Family Business Resilience in an Economic Downturn," *Asia Pacific Business Review*, 18/2: 203–23.

Baltzer, J. L., Day, N. J., Walker, X. J., et al. (2021). "Increasing Fire and the Decline of Fire Adapted Black Spruce in the Boreal Forest," *Proceedings of the National Academy of Sciences*, 118/45: e2024872118.

Beckley, T., Parkins, J., and Stedman, R. (2002). "Indicators of Forest-Dependent Community Sustainability: The Evolution of Research," *Forestry Chronicle*, 78/5: 626–36.

Béné, C. (2013). "Towards a Quantifiable Measure of Resilience." IDS Working Paper 434. Brighton, UK: Institute of Development Studies, University of Sussex. 27 pp. Available online at https://www.ids.ac.uk/publications/towards-a-quantifiable-measure-of-resilience/ (accessed January 5, 2020).

Botkin, D. B. (1990.) *Discordant Harmonies: A New Ecology for the Twenty-First Century*. New York: Oxford University Press. 241 pp.

Brock, W. A., and Carpenter, S. R. (2006). "Variance as a Leading Indicator of Regime Shift in Ecosystem Services," *Ecology and Society*, 11/2: 9.

Brunet, J., Bukina, Y., Hedwall, P. O., Holmström, E., and von Oheimb, G. (2014). "Pathogen Induced Disturbance and Succession in Temperate Forests: Evidence from a 100-Year Data Set in Southern Sweden," *Basic and Applied Ecology*, 15/2: 114–21.

Burton, P. J. (2005). "Ecosystem Management and Conservation Biology," in S. B. Watts and L. Tolland (eds), *Forestry Handbook for British Columbia*. 5th edn. Vancouver: Faculty of Forestry, University of British Columbia, 307–22.

Carpenter, S., Walker, B., Anderies, J. M., and Abel, N. (2001). "From Metaphor to Measurement: Resilience of What to What?" *Ecosystems*, 4/8: 765–81.

Castagneri, D., Vacchiano, G., Hacket-Pain, A., DeRose, R. J., Klein, T., and Bottero, A. (2022). "Meta-Analysis Reveals Different Competition Effects on Tree Growth Resistance and Resilience to Drought," *Ecosystems*, 25/1: 30–43.

Christiansen, E., Waring, R. H., and Berryman, A. A. (1987). "Resistance of Conifers to Bark Beetle Attack: Searching for General Relationships," *Forest Ecology and Management*, 22/1–2: 89–106.

Clarke, P. J., Lawes, M. J., Midgley, J. J., et al. (2013). "Resprouting as a Key Functional Trait: How Buds, Protection and Resources Drive Persistence after Fire," *New Phytologist*, 197/1: 19–35.

Cleary, M., Sturrock, R., and Hodge, J. (2011). "Southern Interior Forest Region: Laminated Root Disease Stand Establishment Decision Aid," *Journal of Ecosystems and Management*, 12/2: 17–20.

Climent, J., Tapias, R., Pardos, J. A., and Gil, L. (2004). "Fire Adaptations in the Canary Islands Pine (*Pinus canariensis*)," *Plant Ecology*, 171/1–2: 185–96.

Connell, S. D., and Ghedini, G. (2015). "Resisting Regime-Shifts: The Stabilizing Effect of Compensatory Processes," *Trends in Ecology & Evolution*, 30/9: 513–15.

Connell, S. D., Nimmo, D. G., Ghedini, G., MacNally, R., and Bennett, A. F. (2016). "Ecological Resistance—why Mechanisms Matter: A Reply to Sundstrom et al.," *Trends in Ecology & Evolution*, 31/6: 413–14.

Côté, S. D., Rooney, T. P., Tremblay, J.-P., Dussault, C., and Waller, D. M. (2004). Ecological impacts of deer overabundance on temperate and boreal forests. *Annual Review of Ecology, Evolution, and Systematics*, 35: 113–47.

Cournoyer, L., and Filion, L. (1994). "Variation in Wood Anatomy of White Spruce in Response to Dune Activity," *Arctic and Alpine Research*, 26/4: 412–17.

Coyle, S. J. (2011). *Sustainable and Resilient Communities: A Comprehensive Action Plan for Towns, Cities, and Regions*. Hoboken, NJ: John Wiley & Sons. 404 pp.

Dantas, V. D. L., and Pausas, J. G. (2013). "The Lanky and the Corky: Fire-Escape Strategies in Savanna Woody Species," *Journal of Ecology*, 101/5: 1265–72.

Davies, A. (2010). "Palaeoecology, Management, and Restoration in the Scottish Highlands," in M. Hall (ed.), *Restoration and History: The Search for a Useable Environmental Past*. London: Routledge, 92–104.

Davydov, D. M., Stewart, R., Ritchie, K., and Chaudieu, I. (2010). "Resilience and Mental Health," *Clinical Psychology Review*, 30/5: 479–95.

Del Tredici, P. (2001). "Sprouting in Temperate Trees: A Morphological and Ecological Review," *Botanical Review*, 67: 121–40.

Derissen, S., Quaas, M. F., and Baumgärtner, S. (2011). "The Relationship between Resilience and Sustainability of Ecological–Economic Systems," *Ecological Economics*, 70/6: 1121–8.

Döring, T. F., Vieweger, A., Pautasso, M., Vaarst, M., Finckh, M. R., and Wolfe, M. S. (2015). "Resilience as a Universal Criterion of Health," *Journal of the Science of Food and Agriculture*, 95/3: 455–65.

Ellison, A. M., Bank, M. S., Clinton, B. D., et al. (2005). "Loss of Foundation Species: Consequences for the Structure and Dynamics of Forested Ecosystems," *Frontiers in Ecology and the Environment*, 3/9: 479–86.

Evans, J. (1992). "Coppice Forestry—an Overview," in G. P. Buckley (ed.), *Ecology and Management of Coppice Woodlands*. London: Chapman & Hall, 18–27.

Falk, D. A., van Mantgem, P. J., Keeley, J. E., et al. (2022). "Mechanisms of Forest Resilience," *Forest Ecology and Management*, 512: 120129.

Fay, C. K., and Nordhaug, K. (2002). "Why are there Differences in the Resilience of Malaysia and Taiwan to Financial Crisis?" *European Journal of Development Research*, 14/1: 77–100.

Filotas, E., Parrott, L., Burton, P. J., et al. (2014). "Viewing Forests through the Lens of Complex Systems Science," *Ecosphere*, 5/1: 1.

Folke, C., Carpenter, S., Walker, B., et al. (2004). "Regime Shifts, Resilience, and Biodiversity in Ecosystem Management," *Annual Review of Ecology, Evolution, and Systematics*, 35: 557–81.

Fuller, L., and Quine, C.P. (2016). "Resilience and Tree Health: A Basis for Implementation in Sustainable Forest Management," *Forestry*, 89/1: 7–19.

Gray, D. H., and Sotir, R. B. (1996). *Biotechnical and Soil Bioengineering Slope Stabilization: A Practical Guide for Erosion Control*. New York: John Wiley & Sons. 400 pp.

Gunderson, L. H. (2000). "Ecological Resilience—in Theory and Application," *Annual Review of Ecology and Systematics*, 31: 425–39.

Gunderson, L. H., Allen, C. R., and Holling, C. S. (eds).(2012). *Foundations of Ecological Resilience*. 2nd edn. Washington: Island Press. 496 pp.

Haeussler, S. (2011). "Rethinking Biogeoclimatic Ecosystem Classification for a Changing World," *Environmental Reviews*, 19: 254–77.

Haines, Y. Y. (2009). "On the Definition of Resilience in Systems," *Risk Analysis*, 29/4: 498–501.

Hamilton, N. P., and Burton, P. J. (2023). "Wildfire Disturbance Reveals Evidence of Ecosystem Resilience and Precariousness in a Forest–Grassland Mosaic," *Ecosphere*, 14/3: e4460.

Hansen, J., and Spiecker, H. (2016). "Conversion of Norway Spruce (*Picea abies* [L.] Karst.) Forests in Europe," in J. A. Stanturf (ed.), *Restoration of Boreal and Temperate Forests*. 2nd edn. Boca Raton, FL: CRC Press, 355–64.

Hariri-Ardebili, M. A. (2018). "Risk, Reliability, Resilience (R3) and beyond in Dam Engineering: A State-of-the-Art Review," *International Journal of Disaster Risk Reduction*, 31: 806–31.

Higgs, E., Falk, D. A., Guerrini, A., et al. (2014). "The Changing Role of History in Restoration Ecology," *Frontiers in Ecology and the Environment*, 12/9: 499–506.

Hodgson, D., McDonald, J. L., and Hosken, D. J. (2015). "What Do You Mean, 'Resilient'?" *Trends in Ecology & Evolution*, 30/9: 503–6.

Holling, C. S. (1973). "Resilience and Stability of Ecological Systems," *Annual Review of Ecology and Systematics*, 4/1: 1–23.

Holling, C. S. (1996). "Engineering Resilience versus Ecological Resilience," in P. C. Schulze (ed.), *Engineering within Ecological Constraints*. Washington: National Academy Press, 31–44.

Holling, C. S., and Gunderson, L. H. (2002). "Resilience and Adaptive Cycles," in L. H. Gunderson and C. S Holling (eds), *Panarchy: Understanding Transformations in Human and Natural Systems*. Washington: Island Press, 25–62.

Holling, C. S., L. H. Gunderson, and Peterson, G. D. (2002). "Sustainability and Panarchies," in L. H. Gunderson and C. S. Holling (eds), *Panarchy: Understanding Transformations in Human and Natural Systems*. Washington: Island Press, 63–102.

Hughes, R. F., and Denslow, J. S. (2005). "Invasion by a N_2-Fixing Tree Alters Function and Structure in Wet Lowland Forests of Hawaii," *Ecological Applications*, 15/5: 1615–28.

Ilisson, T., & Chen, H. Y. (2009). The direct regeneration hypothesis in northern forests. *Journal of Vegetation Science*, 20(4): 735–44.

Johnston, C. A. (2017). *Beavers: Boreal Forest Engineers*. New York: Springer. 272 pp.

Jucker, T., Bouriaud, O., and Coomes, D. A. (2015). "Crown Plasticity Enables Trees to Optimize Canopy Packing in Mixed-Species Forests," *Functional Ecology*, 29/8: 1078–86.

Keeley, J. E. (1986). "Resilience of Mediterranean Shrub Communities to Fires," in D. Dell, Hopkins, A. J. M., and Lamont, B. B. (eds), *Resilience in Mediterranean-Type Ecosystems*. Dordrecht: Springer, 95–122.

Klein, R. J. T., Nicholls, R. J., and Thomalla, F. (2003). "Resilience to Natural Hazards: How Useful Is this Concept?" *Global Environmental Change Part B: Environmental Hazards*, 5/1–2: 35–45.

Kneeshaw, D. D., and Burton, P. J. (1997). "Canopy and Age Structures of Some Old Sub-Boreal *Picea* Stands in British Columbia," *Journal of Vegetation Science*, 8/5: 615–25.

Knoke, T., Stimm, B., Ammer, C., & Moog, M. (2005). "Mixed Forests Reconsidered: A Forest Economics Contribution on an Ecological Concept," *Forest Ecology and management*, 213(1-3): 102–16.

Kofner, S. (2014). "The German Housing System: Fundamentally Resilient?" *Journal of Housing and the Built Environment*, 29/2: 255–75.

Kott, A., and Linkov, I. (eds).(2019). *Cyber Resilience of Systems and Networks*. Cham, Switzerland: Springer. 475 pp.

Lautensach, A. K., and Lautensach, S. W. (eds). (2013). *Human Security in World Affairs: Problems and Opportunities*. Vienna: Caesarpress. 504 pp.

Levin, S. A. (1998). "Ecosystems and the Biosphere as Complex Adaptive Systems," *Ecosystems*, 1: 431–6.

Levine, J. M., Adler, P. B., and Yelenik, S. G. (2004). "A Meta-Analysis of Biotic Resistance to Exotic Plant Invasions," *Ecology Letters*, 7/10: 975–89.

Liebhold, A. M., Brockerhoff, E. G., Kalisz, S., Nuñez, M. A., Wardle, D. A., and Wingfield, M. J. (2017). "Biological Invasions in Forest Ecosystems," *Biological Invasions*, 19: 3437–58.

Lines, E. R., Zavala, M. A., Purves, D. W., and Coomes, D. A. (2012). "Predictable Changes in Aboveground Allometry of Trees along Gradients of Temperature, Aridity and Competition," *Global Ecology and Biogeography*, 21/10: 1017–28.

Linkov, I., and Trump, B. D. (2019). *The Science and Practice of Resilience*. Cham, Switzerland: Springer. 209 pp.

Lindenmayer, D., and Hunter, M. (2010). "Some Guiding Concepts for Conservation Biology," *Conservation Biology*, 24/6: 1459–68.

Lloret, F., Keeling, E. G., and Sala, A. (2011). "Components of Tree Resilience: Effects of Successive Low-Growth Episodes in Old Ponderosa Pine Forests," *Oikos*, 120/12: 1909–20.

Ludwig, D., Jones, D. D., and Holling, C. S. (1978). "Qualitative Analysis of Insect Outbreak Systems: The Spruce Budworm and Forest," *Journal of Animal Ecology*, 47/1: 315–32.

Lynch, A. J., Thompson, L. M., Morton, J. M., et al. (2022). "RAD Adaptive Management for Transforming Ecosystems," *BioScience*, 72/1: 45–56.

MacAskill, K., and Guthrie, P. (2014). "Multiple Interpretations of Resilience in Disaster Risk Management," *Procedia Economics and Finance*, 18: 667–74.

Macdonald, B. (2006). *Practical Woody Plant Propagation for Nursery Growers*. Portland, OR: Timber Press. 600 pp.

McGee, G. G. (2018). "Biological Diversity in Eastern Old Growth," in A. M. Barton and W.S. Keeton (eds), *Ecology* and *Recovery of Eastern Old-Growth Forests*. Washington: Island Press, 197–216.

Martin, A. C., Zim, H. S., and Nelson, A. L. (1961). *American Wildlife & Plants: A Guide to Wildlife Food Habits*. Dover reprint of 1951 McGraw-Hill edition. New York: Dover Publications. 504 pp.

Messier, C., Puettmann, K. J., and Coates, K. D. (eds). (2013). *Managing Forests as Complex Adaptive Systems: Building Resilience to the Challenge of Global Change*. London: Routledge. 353 pp.

Merriam-Webster (2023). "Resilience," in *The Merriam-Webster.com Dictionary*. Springfield, MA: Merriam-Webster Incorporated. Available online at https://www.merriam-webster.com/dictionary/resilience (accessed January 5, 2023).

Nikinmaa, L., Lindner, M., Cantarello, E., et al. (2020). "Reviewing the Use of Resilience Concepts in Forest Sciences," *Current Forestry Reports*, 6: 61–80.

Nordén, B., Dahlberg, A., Brandrud, T. E., Fritz, Ö., Ejrnaes, R., and Ovaskainen, O. (2014). "Effects of Ecological Continuity on Species Richness and Composition in Forests and Woodlands: A Review," *Ecoscience*, 21/1: 34–45.

OED (2022). "Resilience," in *Oxford English Dictionary: The Definitive Record of the English Language*. Oxford: Oxford University Press. Available online at https://www.oed.com/ (accessed June 14, 2023).

Oliver, C. D., and Larson, B. C. (1996). *Forest Stand Dynamics, Update Edition*. New York: John Wiley & Sons. 520 pp.

Overballe-Petersen, M. V., Raulund-Rasmussen, K., Buttenschøn, R. M., and Bradshaw, R. H. (2014). "The Forest Gribskov, Denmark: Lessons from the Past Qualify Contemporary Conservation, Restoration and Forest Management," *Biodiversity and Conservation*, 23: 23–37.

Pausas, J. G. (2015). "Bark Thickness and Fire Regime," *Functional Ecology*, 29/3: 315–27.

Perretti, C. T., and Munch, S. B. (2012). "Regime Shift Indicators Fail under Noise Levels Commonly Observed in Ecological Systems," *Ecological Applications*, 22/6: 1772–9.

Peterson St-Laurent, G., Oakes, L. E., Cross, M., and Hagerman, S. (2021). "R–R–T (Resistance–Resilience–Transformation) Typology Reveals Differential Conservation Approaches across Ecosystems and Time," *Communications Biology*, 4/1: 39.

Prince-Embury, S. (2013). "Resiliency Scales for Children and Adolescents: Theory, Research, and Clinical Application," in S. Prince-Embury and D. H. Saklofske (eds), *Resilience in Children, Adolescents, and Adults: Translating Research into Practice*. New York: Springer, 19–44.

Renschler, C. S., Frazier, A. E., Arendt, L. A., Cimellaro, G. P., Reinhorn, A. M., and Bruneau, M. (2010). *A Framework for Defining and Measuring Resilience at the Community Scale: The PEOPLES Resilience Framework*. MCEER, University at Buffalo, State University of New York. 91 pp. Available online at http://www.eng.buffalo.edu/mceer-reports/10/10-0006.pdf (accessed January 5, 2020).

Rios, F.C., Chong, W. K., and Grau, D. (2015). "Design for Disassembly and Deconstruction: Challenges and Opportunities," *Procedia Engineering*, 118: 1296–304.

Rouse, W. B. (2007). "Complex Engineered, Organizational and Natural Systems: Issues Underlying the Complexity of Systems and Fundamental Research Needed to Address These Issues," *Systems Engineering*, 10/3: 260–71.

Saccone, P., Girel, J., Pages, J. P., Brun, J. J., and Michalet, R. (2013). "Ecological Resistance to *Acer negundo* Invasion in a European Riparian Forest: Relative Importance of Environmental and Biotic Drivers," *Applied Vegetation Science*, 16/2: 184–92.

Scott, D. (1995). "Vegetation: A Mosaic of Discrete Communities, or a Continuum?" *New Zealand Journal of Ecology*, 19/1: 47–52.

Seidl, R., and Turner, M. G. (2022). "Post-Disturbance Reorganization of Forest Ecosystems in a Changing World," *Proceedings of the National Academy of Sciences*, 119/28: e2202190119.

Shackleton, R., Larson, B., and Biggs, R. (2018). "American Chestnut Dominant Forests to Red Maple Dominant Forests." Available online in the *Regime Shifts Database*, at https://www.regimeshifts.org/item/617-american-chestnut-dominant-forests-to-red-maple-dominant-forests (accessed December 31, 2019).

Silverman, M. N., and Deuster, P. A. (2014). "Biological Mechanisms Underlying the Role of Physical Fitness in Health and Resilience," *Interface Focus*, 4/5: 20140040.

Standish, R. J., Hobbs, R. J., Mayfield, M. M., et al. (2014). "Resilience in Ecology: Abstraction, Distraction, or where the Action is?" *Biological Conservation*, 177: 43–51.

Sundin, E., Elo, K., and Mien Lee, H. (2012). "Design for Automatic End-of-Life Processes," *Assembly Automation*, 32/4: 389–98.

Thompson, J. R., Carpenter, D. N., Cogbill, C. V., and Foster, D. R. (2013). "Four Centuries of Change in Northeastern United States Forests," *PloS One*, 8/9: e72540.

Vodopivec, N., and Miller-Hooks, E. (2019). "Transit System Resilience: Quantifying the Impacts of Disruptions on Diverse Populations," *Reliability Engineering & System Safety*, 191: 106561.

Wagner, S., Nocentini, S., Huth, F., and Hoogstra-Klein, M. (2014). "Forest Management Approaches for Coping with the Uncertainty of Climate Change: Trade-Offs in Service Provisioning and Adaptability," *Ecology and Society*, 19/1: 32.

Wagner, W. (2010). "Diversification at Financial Institutions and Systemic Crises," *Journal of Financial Intermediation*, 19/3: 373–86.

Walker, B. (2020). "Resilience: What It Is and Is Not," *Ecology and Society*, 25/2: 11.

Walker, B., Holling, C. S., Carpenter, S. R., and Kinzig, A. (2004). "Resilience, Adaptability and Transformability in Social–Ecological Systems," *Ecology and Society*, 9/2: 5.

Walker, B. H., Ludwig, D., Holling, C. S., and Peterman, R. M. (1981). "Stability of Semi-Arid Savanna Grazing Systems," *Journal of Ecology*, 69/2: 473–98.

Walker, B., and Salt, S. (2006). *Resilience Thinking: Sustaining Ecosystems and People in a Changing World*. Washington: Island Press. 174 pp.

Walker, B., and Salt, S. (2012). *Resilience Practice: Building Capacity to Absorb Disturbance and Maintain Function*. Washington: Island Press. 226 pp.

Whitman, E., Parisien, M.-A., Thompson, D. K., and Flannigan, M. D. (2019), "Short-Interval Wildfire and Drought Overwhelm Boreal Forest Resilience," *Scientific Reports*, 9/1: 1–12.

Wiig, S., and Fahlbruch, B. (eds). (2019). *Exploring Resilience: A Scientific Journey from Practice to Theory*. SpringerBriefs in Safety Management. Cham, Switzerland: Springer. 128 pp.

Williams, J. W. (2022). "RAD: A Paradigm, Shifting," *BioScience*, 72/1: 13–15.

Wordsworth, D. (2014). "Why Does Everything Suddenly Need 'Resilience'?" *The Spectator*. Available online at https://www.spectator.co.uk/2014/03/resilience/ (accessed January 5, 2020).

Xu, L., and Kajikawa, Y. (2018). "An Integrated Framework for Resilience Research: A Systematic Review Based on Citation Network Analysis," *Sustainability Science*, 13: 235–54.

Zolli, A., and Healy, A. M. (2012). *Resilience: Why Things Bounce Back*. New York: Simon & Shuster. 323 pp.

PART II

PROMOTING RESILIENCE IN THE MANAGEMENT OF FOREST ECOSYSTEMS

Carpenter et al. (2012) postulate that the properties of complex adaptive systems confer general, broad-spectrum resilience to unexpected events and shocks. Experience in promoting resilience in a wide range of socio-ecological systems provides a number of general principles and effective approaches with applicability in forest management. The previous chapters have outlined the broad scope of forest ecosystem services and values that we expect to persist at some appropriate level on the land, and the principles of complex adaptive systems that can be drawn upon to promote their resilience. When applying general principles for resilient forest management, different approaches or emphases are required in protected forests, multipurpose forests, and production forests.

In an earlier paper, Carpenter et al. (2001) pointed out that the measurement and promotion of resilience depend on both the value of concern and on the particular threats facing it: They astutely ask "Resilience of what, to what?" So, in addition to outlining general strategies for resilient forest management, some specific but inter-connected chal-lenges are explored in greater detail in the following chapters—particularly those stemming from a warming climate and growing risks from natural disturbances. Other issues such as unforeseen shifts in politics, trade, technology, consumer preferences, and public opinion are mentioned briefly, but are not explored in detail. Collectively, the principles, considerations, and examples presented in these four chapters provide the basis for decision-makers to pursue more resilient forest policies, and for forest managers to undertake more resilient stewardship interventions, as further developed in Part III.

References cited

Carpenter, S. R., Arrow, K. J., Barrett, S., et al. (2012). "General Resilience to Cope with Extreme Events," *Sustainability*, 4/12: 3248–59.

Carpenter, S., Walker, B., Anderies, J. M., and Abel, N. (2001). "From Metaphor to Measurement: Resilience of What to What?" *Ecosystems*, 4: 765–81.

CHAPTER 5

Enhancing General Resilience in Forests

Be prepared.

Motto of the Boy Scouts;
Robert Baden-Powell (1907)

5.1 Introduction

General resilience denotes the capacity to persist in the face of a variety of different stressors and constitutes a desirable attribute of all ecosystems and organizations. Enhanced general resilience does not substitute for vulnerability assessments and planning to address specific threats (such as wildfire, pest outbreaks, climate change, and socioeconomic disruptions). Rather, managing for general resilience can be considered a necessary—but not sufficient—aspect of managing forests, forest enterprises, and forest communities for long-term sustainability under an uncertain future. The recommended approaches are most useful in a planning context, and can be considered principles for responsible forest stewardship (Johnson et al. 1999; Chapin et al. 2009), irrespective of any particular threats or uncertainties.

A comprehensive review of the resilience literature documented the recurrence of several key terms and concepts that are indicative of their importance in the application of resilience concepts (Xu and Kajikawa 2018). Focusing on three of the disciplinary clusters identified in their multivariate analysis (see Figure 4.2), we might interpret terms used in "ecological and environmental services" as being most relevant to protected areas or conservation forests, "socio-ecological systems and management" as relating to multipurpose forests, and "business systems and engineering" concepts prevailing in production forest management. If such a cross-walk is valid, diversity emerges as being of overwhelming importance in conservation and multipurpose forests, and is the second-most important concept for production forest resilience (see Table 5.1). All resilience characteristics other than diversity become increasingly important along the sequence from conservation through multipurpose to production forests, while reference to diversity decreases. Interestingly, concepts of preparedness—including risk assessment, scenario analysis, and contingency planning—can be interpreted as being more frequently employed than diversity in production forests, which have a stronger emphasis on forest engineering and business considerations. This brief analysis suggests that the management of different forests, or different portions of a forest estate, will naturally employ different approaches in the common pursuit of resilience.

Most authors commenting on resilience have identified characteristics such as those listed in Table 5.1 or introduced in Section 4.2 as factors contributing to resilience. However, no set of factors is equally relevant to each of the five mechanisms of resilience described in Section 4.3, nor to each phase of the adaptive cycle described in Section 4.4. Depending on the hypothesis being

Resilient Forest Management. Philip J. Burton, Oxford University Press. © Philip J. Burton (2025). DOI: 10.1093/oso/9780198832997.003.0005

Table 5.1 Key characteristics of resilient systems, and the frequency with which these terms were used in three categories of published papers with perceived applicability to forest management

Characteristic	Ecological and environmental sciences (re: conservation forests)	Socio-ecological systems and management (re: multipurpose forests)	Business systems and engineering (re: production forests)
Diversity	4,904 (93.3%)	6,518 (75.5%)	485 (23.5%)
Flexibility	86 (1.6%)	714 (8.3%)	347 (16.8%)
Preparedness	34 (0.7%)	405 (4.7%)	691 (33.5%)
Connectedness	64 (1.2%)	379 (4.4%)	224 (10.9%)
Redundancy	147 (2.8%)	374 (4.3%)	154 (7.5%)
Social capital	22 (0.4%)	228 (2.6%)	149 (7.2%)
Resourcefulness	0 (0.0%)	11 (0.1%)	13 (0.6%)

Source: Summarized from Xu and Kajikawa (2018).

tested, researchers have often identified different drivers that facilitate or compromise resilience in various forest ecosystems, even when the same stressor is being examined. For example, mycorrhizal fungi facilitate resilience to drought in seasonally dry forests (Pickles and Simard 2017), while a study in European spruce-fir forests (in which fungi were not investigated) concluded that reduced stand density or more broadleaf trees hold the key to drought resilience (Bottero et al. 2021). Consequently, it can be challenging for policymakers and land managers to translate general concepts about promoting resilience into actionable initiatives (e.g. Greiner et al. 2020). Nikinmaa et al. (2023) point out the many trade-offs that must be balanced when trying to operationalize resilience in forestry, for which they suggest a framework of principles, criteria, and indicators analogous to that used in forest certification. With clear overlaps among recommendations for how to promote general resilience, some of those most useful in forest stewardship are expanded upon below. As a whole, the general properties of complex adaptive systems confer broad-spectrum resilience to unexpected disruptions (Gunderson et al. 2002; Carpenter et al. 2012), and are highly relevant to forests (Filotas et al. 2014). Nikinmaa et al. (2023) highlight the importance of diversity, connectedness, and adaptive capacity for achieving resilience in forests. The following sections explore the forestry applications of a slightly broader set of resilience attributes, including (1) multiple forms of diversity, (2) reserves and system memory, (3) capacity for flexibility and adaptation, (4) a balance of modularity and connectedness, and (5) responsive feedback mechanisms.

That is not an exhaustive list of features that characterize complex adaptive systems, and they may not be the most effective for all forests or for coping with all stressors under all management objectives. Readers may wish to consult other sources, such as Messier et al. (2013), for further ideas and inspiration for forestry applications.

5.2 Diversity, diversity, diversity!

Compositional, structural, and functional diversity at all scales confers the ability to absorb and recover from disruptions, typically through functional redundancy and the spreading of risk. In many ways, other characteristics of general resilience—reserves, flexibility, connectedness—can also be considered elaborations upon the theme of diversity. Diversity by its very nature provides the potential for experimentation and flexibility in response to novel situations, much like it constitutes the raw material upon which evolution operates through natural selection. Even in a stable and predictable environment, "a diversity of diversities" is required to sustain the many values we expect of forests (Hunter 1990).

It is easy to think that a diversity of tree species at the stand level might be the primary means of managing forests for resilience, and that approach is certainly important. But diversity in many other forms and at many other scales is just as important. Those include genetic diversity within species, the role of diverse non-tree species, multi-aged and mutilayered forest stands, compositionally and structurally diverse forest landscapes, diverse perspectives around the management table, and the use of diverse silvicultural tools. Multiple pathways for energy and for succession help maintain functional ecosystem resilience; a diverse product line, multiple fiber sources, and multiple markets help sustain a forest products industry. Sometimes diversity simply spreads the risk of unknown future stressors, analogous to the portfolio approach to financial investment (Oliver 1995; Neuner et al. 2013). In other cases, "diversity" is shorthand for networks of complex interactions that collectively are responsible for system stability, persistence, and adaptability. Table 5.2 provides a broad overview of the mechanisms by which diversity confers resilience in forests. In practice, many examples of the demonstrable benefits of diversity involve two or more of those mechanisms. Collectively, insurance effects, dilution effects, and trait complementarity are components of portfolio theory in the world of financial investment and are equally applicable in the world of forest stewardship.

In many respects, the maintenance or promotion of diversity at stand and landscape levels can be considered a risk management strategy. Threats to ecosystem integrity can include invasive or eruptive pests (both native and exotic), discrete disturbance events such as wildfires and landslides, extreme weather such as droughts and storms, and changing values and markets for the ecosystem services provided by an area under management. With forest plans encompassing timber rotations that span decades or even more than a century, this leaves their assets exposed to those many risk factors for a long period of time (Taylor and Fortson 1991; Galik and Jackson 2009), with much uncertainty as to which risks are most important to address. Risks can be spread by managing several tree species that differ in their physiological tolerances, pest resistance, and other life history traits (Aubin et al. 2016), and by maintaining a variety of plant communities (differing in composition and structure, some seral,

Table 5.2 Broad mechanisms by which diversity maintains or increases the resilience of forests in the face of specific or general threats and uncertainties

Mechanism	Explanation	Example
Insurance effect	Distributes risk among multiple species or locations; functional redundancy in filling ecosystem roles means that losses to the health or value of any one element is minimized by an abundance of others	Belgium: severe levels of mortality were avoided in plantations of mixed tree species (Van de Peer et al. 2016)
Trait complementarity	Permits differential utilization of resources in space (e.g. rooting depth) or in time (phenologically within a year, or over the course of succession), and includes different thresholds in response to stressors (e.g. drought, frost); a diversity of such traits thereby enhances resource availability and retention, and facilitates community persistence under different levels of stress	Quebec: trees with architecturally different crowns (e.g. broadleaf and conifer) yielded more when grown together than in monocultures or with species having similar crowns (Williams et al 2017)
Dilution effect	Decreases the interaction of similarly susceptible elements (e.g. species exposed to a herbivorous insect or fungal pathogen, or forest type exposed to fire spread) by interspersing susceptible and non-susceptible elements	Costa Rica: *Virola koschnyi* experienced lower levels of insect defoliator damage when growing with three other species than when grown alone (Montagnini et al. 1995)
Spatial disruption	Often a dilution effect, but also includes mechanisms by which susceptible elements are structurally hidden or barred from threats, such as tree seedlings obscured from herbivores by other plants; and certain landscape locations constitute refugia from wildfires owing to slope, aspect, and the position of adjoining water bodies	British Columbia: saplings of *Picea engelmannii x glauca* experienced less attack by *Pissodes strobi* when leaders were overtopped by broadleaf trees (Taylor et al. 1996)
Higher-level ecosystem effects	Include food web complexity (multiple energy and nutrient pathways) and temporal effects by which alternative hosts or food sources minimize effects on any one species; reservoirs of predators and parasites are maintained to dampen future outbreaks of herbivores	New Brunswick: *Abies balsamea* had reduced defoliation in the presence of broadleaf trees that support alternative hosts for wasps that parasitize spruce budworm, *Choristoneura fumiferana* (Campbell et al. 2008)

some climax) that provide habitat value for different animal species. There should be a greater chance of continuity in forest ecosystem services if those different tree species and plant communities are found across an array of elevations, slope positions, and aspects, as some sites can take advantage of optimal growing conditions while other sites may have a better chance of avoiding drought, windthrow, or wildfire. Holding forest resources in multiple disjunct parcels interspersed with non-forest land can provide some protection from disturbances such as wildfire and invasive pests. When forest assets are spread across different geographic locations and topographic positions, they are less vulnerable to shifts in climate as well. Perhaps even more important is to employ a diversity of stand management techniques, including a range of silvicultural systems and regeneration practices. Those techniques may include the use of prescribed fire as well as tree-cutting and vegetation control, different microsite modification techniques, and different ways of introducing plant materials, implemented in patches or strips at a variety of scales that span the distinction between uneven-aged and even-aged management—essentially managing for complexity (Nyland 1996; Lieffers et al. 2003; Palik et al. 2021).

The response of forests to population explosions (outbreaks) of herbivorous insects provides a good illustration of the value of tree-species diversity. Capable of exponential population growth under favorable conditions, many forest insects have evolved specialized behaviors for feeding on a single genus of trees, and almost every tree species has a suite of adapted insect herbivores (Ciesla 2011). In fact, many textbooks and manuals of phytophagous forest insects are organized or indexed by their host-tree genera or species (e.g. Sathe 2009; Ciesla 2011). Especially important as the cause of widespread tree mortality are larvae of the order Lepidoptera, which feed on foliage and hence are described as defoliators, and beetles (order Coleoptera, particularly in the family Curculionidae, subfamily Scolytinae) that feed on tree cambium and phloem tissues and so are known as bark beetles (see Figure 5.1). Collectively, herbivorous insects disturbed more than eighty-five million hectares of world forest over the 2003–12 time period (van Lierop et al. 2015), and their populations and ranges are expected to expand under climate change (Pureswaran et al. 2018).

The very nature of herbivorous insect host specificity—often with a phytochemical basis—means that compositional diversity in forest trees, within and among stands, confers strong resistance to the impacts of any one outbreaking insect species. Though patently obvious, so rarely documented, forests or stands consisting of multiple species consequently have greater survival of

Figure 5.1 Two widespread insect species that can cause high levels of damage to forest trees: (a) larva of spongy moth, *Lymantria dispar*, a defoliator of broadleaf trees (public domain photo by J.E. Appleby, U.S. Fish and Wildlife Service); (b) pupae of European spruce bark beetle, *Ips typographus*, embedded in galleries eaten into the phloem of *Picea abies* when in larval form (photo by D. Kandasamy, reproduced under Creative Commons license CC BY 4.0, https://creativecommons.org/licenses/by/4.0/).

mature trees than stands consisting of a single susceptible species. For example, eighteen million hectares of forest in British Columbia, Canada, were affected by eruptive mountain pine beetle (*Dendroctonus ponderosae*) populations (as described in Box 3.3). What was missing from the newspaper headlines, however, was that most of the affected forest consisted of mixed-species stands, so mature forest cover was not lost over most of that area (see Box 5.1). Similar patterns of tree survival have been documented after several years of defoliation by spruce budworm (*Choristoneura fumiferana*) in low-diversity and high-diversity forests in New Brunswick, Canada. Stands consisting of 42 percent balsam fir (*Abies balsamea*) mixed with other conifer and broadleaf species retained three times as much standing volume after the spruce budworm outbreak as found in stands consisting of 84 percent balsam fir (Su et al. 1996). Even when faced with multiple and undefined agents of mortality, greater tree-species diversity in a plantation reduces the incidence of catastrophic (>50 percent) tree loss (Van de Peer et al. 2016; Jactel et al. 2017).

Box 5.1 Forest exposure to an insect outbreak

The mountain pine beetle (*Dendroctonus ponderosae*) is a bark beetle, indigenous to western North America, that kills trees of several pine species through pheromone-mediated mass attack (Carroll and Safranyik 2003). Facilitated by widespread stands of mature even-aged lodgepole pine (*Pinus contorta* var. *latifolia*) and several years of mild winters, the mountain pine beetle killed trees over more than 18 million hectares in British Columbia from 1999 to 2013 (Dhar et al. 2016). It is clear that forest consisting of non-susceptible trees should be more resistant to widespread mortality due to monophagous (selective) herbivores than forest characterized primarily by a single, susceptible host species, yet that response has rarely been characterized quantitatively. Figure 5.2 portrays the composition of forest cover prior to mapped mountain pine beetle outbreaks over a large area of interior British Columbia. These data illustrate that the forest region as a whole was not decimated by the outbreak (contrary to widespread media portrayals and appeals for government assistance).

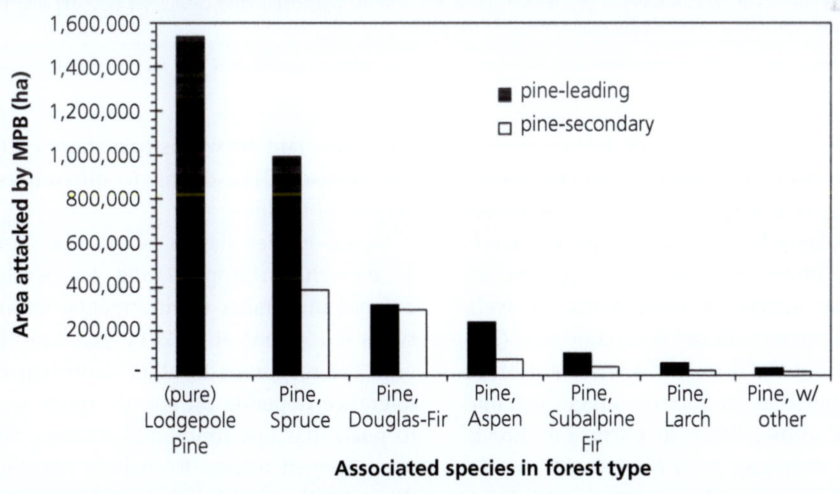

Figure 5.2 Forest inventory of 4.1 million hectares of forest types containing lodgepole pine in interior British Columbia prior to attack by mountain pine beetle (MPB) between 1959 and 2002. Approximately one-third of pine-containing forest can be considered "largely pure pine" (≥80% cover), but mixed stands predominate. Spruce is mostly *Picea engelmannii*, *P. glauca*, and their natural hybrids; Douglas-fir is *Pseudotsuga menziesii*; aspen is *Populus tremuloides*; subalpine fir is *Abies lasiocarpa*; larch is *Larix laricina* and *L. occidentalis*. Based on the same data reported in Figure 5 of Taylor et al. (2006)

> **Box 5.1** *Continued*
>
> A benefit of mixed stands containing several non-host species is further illustrated in Figure 5.3, which shows the status of forest stands in north-central British Columbia after mountain pine beetle attack. Every additional non-host tree denotes one less susceptible tree in a fully stocked forest stand, and greater living tree cover after the outbreak has run its course. Furthermore, not all lodgepole pine trees are killed by the bark beetle because it selects larger trees, promoting the development of multi-storied (multi-aged or uneven-aged) stands that are subsequently less susceptible to beetle attack (Alfaro et al. 2015).
>
> 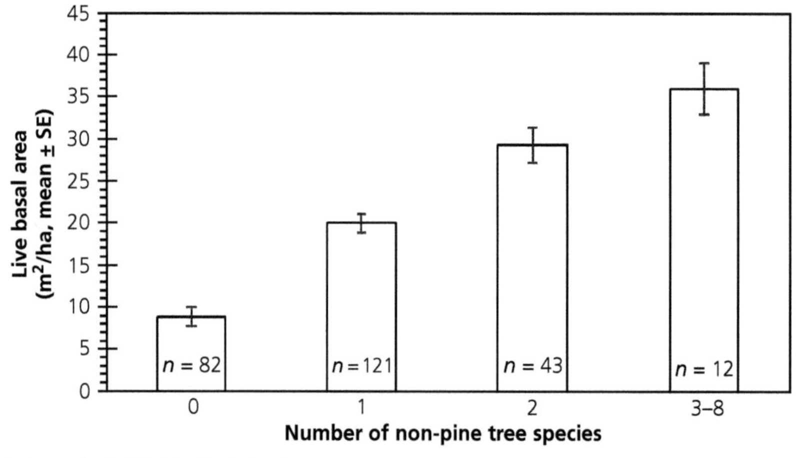
>
>
> **Figure 5.3** Persistence of living trees following the mountain pine beetle outbreak as related to plot-level overstory diversity. Sampled using 259 prism plots (all containing lodgepole pine, tallied with 4 m²/ha resolution). Total basal area, including lodgepole pine trees now dead, averaged 39.5 m²/ha (unpublished data of the author).

A broader meta-analysis of 119 earlier studies confirmed that stands of multiple tree species are particularly resistant to specialist insect herbivores (Jactel and Brockerhoff 2007). Mixed-species stands are apparently more resistant than single-species stands to other agents of disturbance as well, including small mammals and soil pathogens (Jactel et al. 2017). But the details of species attributes and site conditions are important, suggesting that assessments of vulnerability to particular threats and resilience strategies to thwart those specific risks (as discussed in subsequent chapters) are critical. For example, mixed stands of broadleaf trees and conifers can be more resistant than conifers alone to the effects of fire (e.g. Hély et al. 2000), reflecting differences in flammability (foliar moisture content and phytochemistry). Intimate mixtures of coniferous and broadleaf trees are also more resistant to windstorm damage (e.g. Griess and Knoke 2011), owing to differences in crown and root profiles.

Mixed-species stands can also be more resistant to drought in tropical and temperate climates, but not in harsher environments nor in all forest types (Grossiord et al. 2014; Jactel et al. 2017). As with the responses to other disturbances, drought tolerance depends on species traits as they relate to particular environmental stresses. So it is wise for managed forests to include species that have different thresholds of drought sensitivity, often expressed as a trade-off between drought tolerance and productivity. Greater utilization of the total soil water profile may confer resistance to a moderate drought but could result in catastrophic forest loss under severe drought. A general conclusion might be that functional diversity—assemblies of species with complementary structural and ecophysiological traits—is more important than simple species richness in conferring forest resilience (Aubin et al. 2016; Schmitt et al. 2020). However, plant community resilience often shows poor or contradictory relationships to trait diversity (Lipoma et al. 2024). Despite efforts to seek generality, species matter and site matters, so it is wise to identify and address specific threats.

Many strategies to maintain or enhance resilience in the management of natural resources entail a cost, or a trade-off of productivity or profit (depending on the timeline considered and the information available) for the sake of greater flexibility or persistence (Wagner et al. 2014; Williams and Johnson 2015; Homayounfar et al. 2022). Such an optimization versus resilience trade-off is not always the case when it comes to the benefits of increased tree diversity at the stand level. Several studies have demonstrated enhanced productivity (often called "over-yielding") in mixed-species stands compared to monocultures (e.g. Zhang et al. 2012; Liang et al. 2016). This is especially well demonstrated when there is phenological complementarity, such as deciduous (typically broadleaf) species co-mingled with evergreen (typically coniferous) species, allowing net stand-level carbon uptake to proceed with reduced inter-tree shading in early spring and late autumn (Ishii and Asano 2010; Feng et al. 2022). Again, it is the complementarity and compatibility of particular species traits—not species diversity per se—that is important (Díaz and Cabido 2001). Productivity benefits are demonstrable for managing two well-matched species compared to one, especially if they also differ in crown architecture and shade tolerance as well as phenology, but yields are not necessarily enhanced further with more than two species.

Forest structural diversity is characterized by multiple canopy layers and trees of all sizes, and can also be expressed belowground in the form of different rooting depths and patterns. Trees of multiple sizes and different branching patterns above and below ground can make a greater use of resources and are more likely to survive or recover from disturbances. Structural diversity is recognized as promoting biodiversity, especially of canopy-dwelling organisms such as birds, insects, and epiphytes (Ishii et al. 2004). Such biodiversity can further

contribute to disturbance resistance and recovery. For example, high bird populations constrain the level of damage caused by defoliating insects by preying on the caterpillars (Bereczki et al. 2014). Complex stand structure is often achieved with multiple tree species, and that structural diversity may be more influential than species diversity per se in promoting productivity (Dănescu et al. 2016). However, there are also benefits of managing diverse stand structures in single-species stands. Windthrow is typically limited to canopy trees, surface fires can kill seedlings and saplings without killing adults (especially if they have thick bark or a high crown base), and many outbreaking insect species have preferred sizes or ages of host trees. The mycorrhizal connections between adult trees and seedlings are increasingly recognized as a means by which regeneration is subsidized and environmental stresses are buffered (Simard et al. 2013). Conversely, some complex stands can be more susceptible to specific pests: uneven-aged stands of western hemlock (*Tsuga heterophylla*) have more intense infestations of parasitic dwarf mistletoe, *Arceuthobium tsugense* (Muir and Geils 2002), and uneven-aged stands of Douglas-fir (*Pseudotsuga menziesii*) are more susceptible to western spruce budworm, *Choristoneura occidentalis* (Alfaro and Maclauchlan 1992). Mixed stands of trees with different rooting patterns may overyield under normal conditions, but their more effective use of soil water can be a liability under severe drought conditions (Jacobs et al. 2021). Once again, these broadly useful approaches for promoting general resilience must often defer to the defensive strategies needed to address identifiable threats.

The value of landscape-level diversity and heterogeneity for sustaining resilient forest ecosystems is much more well established. Many species of vertebrates require different habitats for foraging and for breeding, or seasonally utilize different food resources that are found in spatially disjunct locations in a landscape. Hunter (1990) pointed out the importance of a variety of forest ages, opening sizes, and transitions (edges) between stands for a wide range of game and non-game animal species. But the benefits of landscape diversity are not limited to the maintenance of wildlife populations and the conservation of biodiversity: there are clear benefits for timber production as well. A well-balanced mosaic of different stand age classes is fundamental to sustainable forestry based on even-aged management (see Section 3.4). Breaking up large contiguous forest blocks of host species with less vulnerable age classes and mixed stands containing non-host species minimizes the impact of spruce budworm (Robert et al. 2018). Similarly, the interspersion of different vegetation types—particularly broadleaf-dominated stands with conifer-dominated stands—can slow the spread of wildfires (Fechner and Barrows 1976; Nesbit et al. 2023). A geographically diverse and widely dispersed forest estate helps ensure that some portions are more likely to escape catastrophic fires (for example, on the cooler, moister poleward side of mountains, or in the lee of lakes; Stralberg et al. 2020), and the random track of a tornado or a derecho. To the extent that different resource emphasis zones (see Chapter 2) or different tree crops (see Figure 3.8) constitute different land uses, financial analysis based on informed risk aversion and optimized land-use diversification recognizes value in a diverse forest landscape (Knoke et al. 2011).

5.3 Reserves

Keeping resources in various kinds of "reserves"—not immediately used or needed for current operations—is another broadly applicable strategy for weathering and recovering from disruptions. Natural, social, and economic reserves all assist in conferring socio-ecological resilience (Walker and Salt 2012). The concept of caching resources for times of scarcity is part of our evolutionary and cultural heritage. Seasonal food scarcities are largely predictable in many parts of the world, so evolution has favored the instincts that motivate non-migratory species to store food for use in winter: squirrels cache seed cones, jays hide acorns, and bears put on extra weight prior to hibernation. Similarly, humans developed methods to smoke fish, store grain, dry fruits, and preserve vegetables in preparation for winter. Ancient scriptures honor prophets who foresaw the need for extraordinary levels of preparation to survive flood (Atra-Hasis, Tablet III) or famine (Genesis 41: 46–57). In modern ecological parlance, we might say that those heroes (Atrahasis in ancient Sumeria and Joseph in ancient Egypt) were prompted to retain "biological legacies" in order to ensure community continuity and survival.

The world of nature has many other examples and lessons by which the persistence of organisms depends on resources held in reserve. Rhizomes are underground stems that neither photosynthesize nor absorb moisture or nutrients; they are an energetic cost to the plant in the short term. But, in the long term, they are foraging and survival organs that sprout new shoots where fire or herbivores damage aboveground stems (Chapin et al. 1990). Reserves of genetic diversity—the raw material for evolution and adaptation—exist within species in the form of recessive genes. Genetic and floristic diversity among species can also be held in reserve in the form of dormant seeds and spores. The eventual germination and emergence of different species, or ecotypes within species, in response to distinctive environmental cues can be an important means by which many plant species buffer the effects of year-to-year variation in weather, microclimate, and disturbances (Bazzaz 1996).

Natural ecological reserves exist at the landscape level too, providing important templates for biodiversity conservation, land-use planning, and forest management (Franklin et al. 2000; Gustafsson et al. 2012). For example, depending on the forest type, individual large old trees and random clusters of green trees often survive forest fires (Perry et al. 2011; Andison and McCleary 2014). Those "lifeboats" for forest-dependent organisms then play an important role in recolonizing the disturbed landscape around them (Franklin et al. 1997). Analogous islands of green can be found in the form of non-host species surviving an insect outbreak, and as rafts of intact vegetation that are relocated by a landslide. Scientists have only recently developed a unified theory of ecological refugia, recognizing that certain landscape positions and vegetation types repeatedly survive the disruptions of fires and other disturbances, and therefore tend to support old-growth forests and high biodiversity (Krawchuk et al. 2020; Stralberg et al. 2020). The expectation that climate in some locations is likely to remain more stable than elsewhere as the global climate warms is important in planning the location of ecological reserves and other protected areas (Ashcroft et al. 2009; Rose and Burton 2009). In a warming climate, shaded depressions with reliable water supply on the poleward side of hills and mountains may constitute important refugia for less drought-tolerant species. As fires become more frequent, mature forests sometimes persist on the leeward side of lakes and wetlands according to the direction of prevailing winds. Many such "relic (or relict) ecosystems" found outside their normal range exist around the world, providing important contributions to landscape diversity (Hampe and Jump 2011). The analogous management practice of retaining patches of old-growth forest when harvesting and managing forests has gained widespread adoption as a means of providing diversity within stands of managed tree crops. The expectation is that various forms of biodiversity will disperse from these reserves into the second-growth forest that develops around them (Gustafsson et al. 2012; Lindenmayer et al. 2012; Mori and Kitagawa 2014).

Reserves have a long history in forestry, from the hunting reserves of royalty (see Chapter 1), through the naval reserves designed to assure long-term supplies of masts and spars, timber tracts reserved from agricultural conversion, and the culturing of high-value large trees ("standards") in stands otherwise managed by coppicing (Nyland 1996). Today, patch retention increasingly is being used in managed second-growth stands to provide in-stand structure and protect some of the wildlife habitat and biodiversity associated with mature forests (Gustafsson et al. 2012). There is also a "conservative," not just a "conservationist," dimension that spurs the setting-aside of reserves from immediate use or exploitation. Risk-adverse financial management includes decisions to forgo high profits that are accompanied by high uncertainty, and instead to accept lower rates of return for greater surety, while also maintaining a healthy level of cash reserves and easily liquefiable assets in order to survive or respond to unforeseen disruptions.

The adage for households to "save for a rainy day" has both short-term and long-term components: urging one to have resources on hand because accidents, illness, or other trauma could strike at any time, while also pointing out the need to set aside funds for retirement—that is, when work-related income ceases. Similar principles apply in the world of timber and conservation management, as a means of buffering both foreseeable and unexpected stresses. For example, sawmills regularly stockpile logs in anticipation of feedstock delivery challenges (for example, owing to seasonal monsoons or freeze/thaw transitions that prevent heavy loads from being hauled on forest roads) and may hold on to product in anticipation of new or more favorable markets.

A related concept that invokes the wisdom of holding back from marginally sustainable resource development is "the precautionary principle." Widely associated with the promotion of sustainable development stemming from the 1992 Earth Summit in Rio de Janeiro, the precautionary principle has since emerged as an axiom to favor environmental sustainability over economic development under conditions of uncertainty. That formalized principle can be considered an extension of the maxim to "take only what you need," widely taught in many Indigenous cultures (Turner 2005; Kimmerer 2013), and admonitions against greed in the world's major religions. Managing too close to the feasibility or sustainability line—practicing insufficient caution when exploiting natural resources . . . or simply being too greedy for immediate profit—is likewise incautious. Leaving some resources and options in reserve is key to surviving turbulent times. Some empirical evidence supports the role of precautionary management in maintaining the viability of forest products enterprises and the jobs and communities that depend on them (see Box 5.2).

Box 5.2 The precautionary principle and employment sustainability

The year 2019 was a daunting one for sustainability of the forest-dependent communities in interior British Columbia (BC), Canada. The allowable annual cut (AAC) from public forests had been temporarily increased 10–15 years earlier to encourage the salvage logging of widespread lodgepole pine stands affected by an unprecedented

---

Box 5.2 *Continued*

outbreak of the mountain pine beetle (see Box 3.3 and Box 5.1), for which sawmills upgraded their facilities and hired extra shifts of workers, and contract loggers invested in new and more equipment (Burton 2010). But those temporary uplifts in AAC were ending, and large forest fires in 2017 and 2018 further limited the accessible timber supply. The reference price for standard spruce-pine-fir dimensional lumber was only US$331 per thousand board-feet (approximately €125/m^3) in mid-2019 compared to 2018 prices that reached over US$600 per thousand board-feet. On top of everything else, tariffs on Canadian softwood lumber imported into the US persisted in the absence of a renewed Softwood Lumber Agreement, making BC lumber up to 20 percent less competitive in its primary export market.

As a result of these circumstances, forest products companies were squeezed by the combination of high delivered wood costs (as available timber was far from mills, often at high elevations with high road-construction costs) and reduced demand and prices for their products. Consequently, at least twenty-one sawmills and producers of plywood and oriented strandboard (OSB) paneling were forced to curtail shifts or shut down for a few weeks, while six mills closed permanently. Those shutdowns obviously had a traumatic effect on many households and rural communities where those mills were located.

But not all sawmills in the BC Interior shut down in the face of these challenges. A cursory analysis reveals a potentially important distinction (see Figure 5.4). Those mills situated in timber supply areas (TSAs) or that held tree farm licenses (TFLs) that were being harvested at levels close to, or sometimes above, permissible and sustainable levels were more likely to cease operations permanently. In contrast, TSAs in which harvest levels were less than two-thirds of sustainable levels experienced neither temporary nor permanent shutdowns. Those mills, TSAs, and associated communities with more options to access timber "held in reserve" were better able to weather the storm.

The discipline to avoid complete exploitation of available resources can be considered an extension of "the precautionary principle." This concept gained prominence as Principle 15 of the 1992 Rio Declaration, specifically in terms of "Where there are threats of serious or irreversible damage, lack of full scientific certainty shall not be used as a reason for postponing cost-effective measures to prevent environmental degradation"(United Nations 1992). While sustainable timber harvesting need not be construed as always causing serious or irreversible damage, this principle urges us to err on the side of caution in favor of protecting environmental values and maintaining future options. Conversely, it is always risky to exploit resources too close to the theoretical limit of sustained yield.

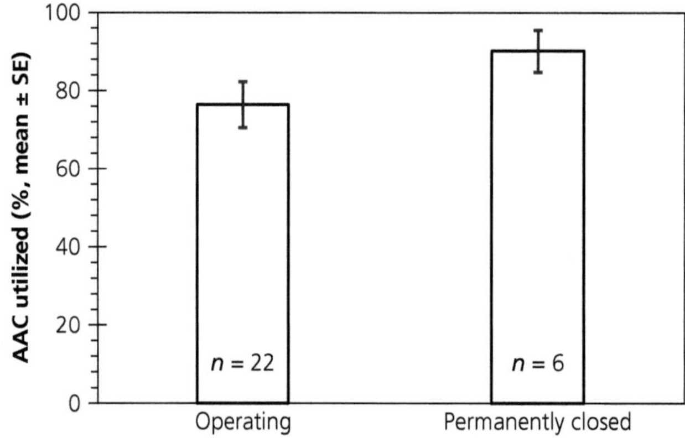

Figure 5.4 Differences in relative timber harvest levels, 2004–18, in interior British Columbia forest management units that did or did not experience lumber-mill closures in 2019. Levels of allowable annual cut (AAC) are independently determined for each forest management unit to denote permissible levels of timber harvest under a sustained yield paradigm and many levels of uncertainty. (Information summarized from billing data and timber supply reviews prepared by the British Columbia Ministry of Forests).

5.4 Flexibility

The hierarchy of biodiversity (at gene, species, community, and ecosystem levels) across variable landscapes and complex disturbance regimes results in many possible mechanisms and trajectories of species and ecosystem persistence. The combination of biodiversity, functional redundancy, and stochastic processes generates an incredible level of flexibility in how natural ecosystems respond to stress and disruption. Drawing upon one of many quotations from popular culture, we can remind ourselves that "life finds a way" (Ian Malcolm in the novel, *Jurassic Park* (1990), written by Michael Crichton). That inherent ecological flexibility and adaptability needs to be protected in conservation forests and can be harnessed in multiple-use and industrial forests.

Such ecological flexibility is well illustrated by the many conditional trajectories of forest recovery that have been observed in forests after natural disturbance. Whether in boreal forests (Burton 2013: 93–5), temperate forests (Woods 2007), or tropical forests (Norden et al. 2015), evidence suggests that no two sites, no two recovery trajectories, are identical in the pattern and timing by which composition and structure develop (see Figure 5.5). This can be frustrating to researchers and managers trying to characterize central tendencies, discretely identifiable trends, and overall predictability. On the other hand, it also suggests there is much latitude ("forgiveness") in the ability of natural systems to absorb negative human influences, and many options for their manipulation in support of sustainable resource management.

The antithesis of a rigid "command and control" management philosophy (Holling and Meffe 1996) is one that is more organically driven and distributive in the decisions made and tactics pursued. This does not mean going forward without a plan or constantly improvising, but rather planning and adjusting, followed by more planning and more adjusting. This concept of continual improvement based on changing conditions and improved information constitutes a core feature of adaptive management of natural resources (Williams 2011) and is expanded upon in Chapter 10. Ecosystems are not stationary, but rather have a number of potential futures that are unpredictable in any practical sense. This requires that management must be flexible, adaptive, and experimental at appropriate scales (Walters 1986; Holling 1996). More than a formalized program of monitoring and replanning, however, the common denominator to successful planning and managing for resilience is the willingness to be flexible, with recognition that there are always options and alternatives moving forward. This management principle has been recognized over the millennia: Publilius Syrus, in the first century BCE. wrote: "*Malum est consilium, quod mutari non potest*" or "It is a bad plan that admits of no modification." As former general and US president Dwight Eisenhower quipped in 1957: "Plans are useless, but planning is everything." Holling and Chambers (1973) conclude that an open and flexible management strategy is the only one that can accommodate conflicting views, creativity, and an unknown future. Flexibility and adaptability are core features of stewardship-focused ecosystem management (Cortner et al. 1999).

5.5 Modularity

The ability of a system to facilitate the aggregation or disaggregation of its component units in response to threats and opportunities is a key attribute of self-organizing complex adaptive systems (Levin 2005). For example, some bird species breed in well-spaced pairs, but undertake seasonal migrations in flocks; the efficiency of container shipping depends on stacking large numbers of the standardized units on ocean-going vessels, coupled with the ability to deliver individual containers from and to their destinations. Interoperable structures that can be scaled up (clustered) or scaled down provide a way for individual units to align with local environmental differences, or to share resources or distribute risk when needed. Conversely, a degree of isolation between component units can prevent failure or disaster in one unit from spreading throughout the whole system (Zolli and Healy 2012). These features suggest that there is an optimal level or balance of modularity and connectedness (Walker and Salt 2006) that might be identified under the combination of specific threats faced by a specific system.

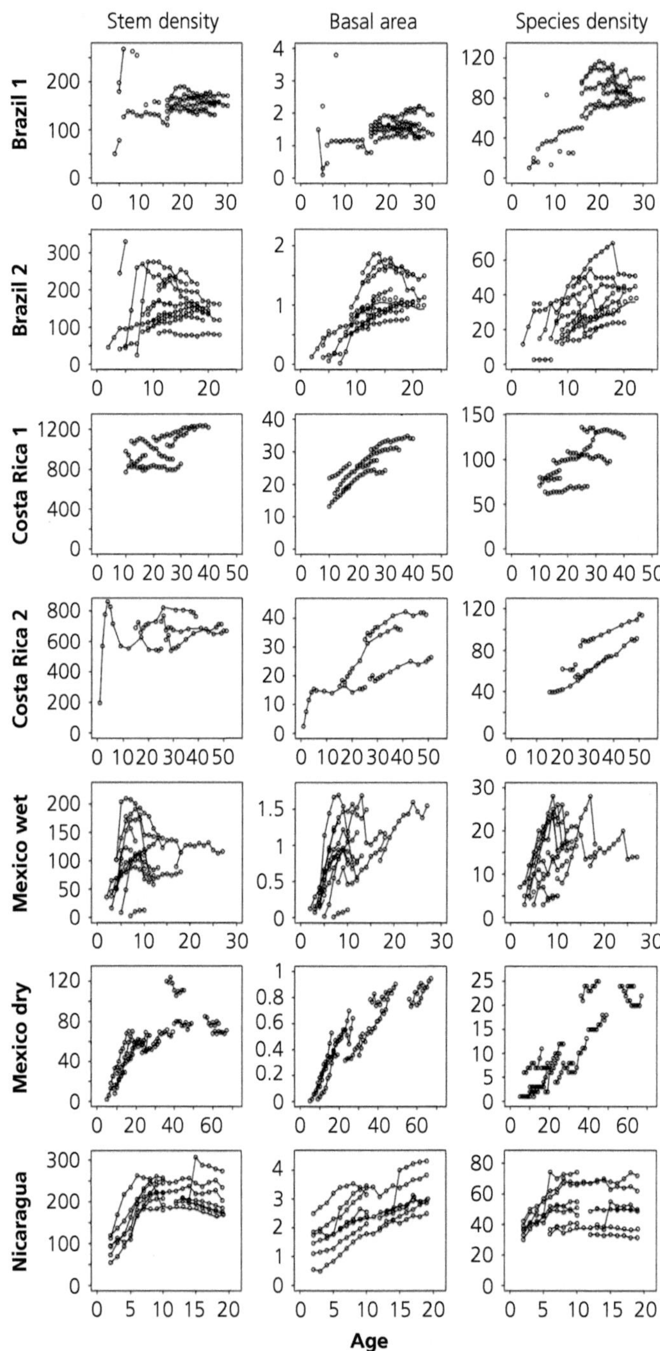

Figure 5.5 The diversity of temporal patterns in stem density, basal area, and species density in individual sample plots after logging or agriculture in seven neotropical forests. Considerable variation in species composition within and among locations further underlies these variable trajectories (Norden et al. (2015), copyright 2015 National Academy of Sciences, reproduced with permission)

Several examples related to forest and land management illustrate the value of modular organization. Central to sustainable forest management is the delineation of a "sustained yield unit" or a defined "forest management area" over which the sustainable delivery of timber and other ecosystem services is expected. Yet logs may be transported to or from other management units, and wildlife may have to migrate in or out of the area in order to make up for damage due to wildfire or other natural disturbance. Similarly, all forest management units are often expected to contribute to broader corporate or state objectives and are subject to those policies. In this manner, the broader network of forest management units can offset threats to local sustainability and remain resilient at that broader scale.

The concept of localized sustainability and calls for more local control over forest management have much in common with calls for local self-sufficiency and bioregionalism (Sale 2000; Curtis 2003), in which socio-ecological sustainability is defined over relatively small geographic units. This localism used to be the norm for social organization and resource use for most of human history, leading to the evolution of a wide variety of cultures and languages, and also some tragic (though localized) collapses (Diamond 2005, 2012). It can be argued that forest sustainability—whether for timber, amenity, or multiple values—should be planned for over a relatively small landscape unit defined on the basis of ownership, jurisdictional, or natural physiographic boundaries. Sustainable management then does not depend on compatible decisions made by external agents, and no forest animals or human forest users should have to travel long distances or cross physical boundaries (such as watershed divides, or large bodies of water) in order to access resources. Broader discussions of the value of community-based forest management in a forest governance context can be found in Sections 2.7 and 12.5.

The policies and practices of designating relatively small units of land for sustained protection/production of forest values can be well matched to the local environment and local values. For example, careful attention to maintaining scenic vistas, fuelwood supplies, and botanical forest products within easy access of the people that

count on them is important for the viability of many small businesses and subsistence economies. The alternative, with management of timber harvesting (for example) balanced only over large areas, often ignores the needs of local forest users. The presence of mushrooms, berries, wildlife, and even timber tens or hundreds of kilometers away (nominally persisting at sustainable levels over a large area) makes them essentially unavailable to forest users fixed in place. This is the case for cottagers, commercial-lodge owners, and others such as fur-trappers with small defined tenure holdings, as well as for many Indigenous peoples and those without affordable access to mechanized transport.

Another benefit of numerous independent management units is that creative experimentation is more likely. This can be beneficial to the overall system and the resilience of worldwide forest management so long as there are sufficient connections (see next section) for the sharing of information. Novel approaches typically emerge in response to unique landscape contexts and history, and the particular background and priorities of local stakeholders and managers. In this manner, multiple woodlot owners, community forests, or Indigenous forest tenures are much more likely to be managed differently from each other than if the same area of forest were managed by a single state or corporate entity. For the system as whole, modular management can represent simultaneous experimentation or an application of "active adaptive management" (Williams 2011; see Chapter 10) to compare and learn from alternative management strategies under a mandate of continuous improvement. Exploring novel solutions to sustainable forest management need not be a risky undertaking if there is sufficient flexibility for resources to be shared among units and for them to be aggregated when needed. Seasonal modularity into dispersed family units during seasonal food shortages has been a time-proven survival strategy for hunter-gatherer societies, complemented by larger gatherings and community bonding during times of plenty (Bettinger 1980; Crowther 2018).

Probably the most important value of modularity is the ability to isolate large natural disturbances. Landscape-scale fuel breaks (e.g. Figure 5.6) are widely instituted to minimize the spread of wildfires (Green 1977; Agee et al. 2000). Such

Figure 5.6 A landscape-level fuel break maintained in the fire-prone forests of *Pinus canariensis* on the island of La Palma, Canary Islands, Spain

fuel breaks—which modify vegetation to consist of widely spaced trees, less flammable tree species, herbaceous ground cover, or exposed mineral soil—typically extend or accentuate natural barriers to the spread of fire, such as lakes, rivers, rock outcrops, and ridge crests. Preferred species of host trees also constitute the food ("fuel") for population explosions of herbivorous insects that mostly disperse to adjacent trees and stands (Nenzen et al. 2017). Consequently, the interspersion of forest stands with different species and age classes of trees is proposed as a means of providing a barrier to the spread of insect outbreaks (Baskerville 1975; Fettig et al. 2014; Robert et al. 2018; see Section 5.2). It helps to know which herbivorous insect species are most likely to strike (for which one can undertake resilience planning for specific threats; see Chapter 8), but a landscape mosaic of different tree species is a generalizable defense against a wide array of oligophagous (host-specific, typically at the genus level) insect herbivores.

Another extension of the modularity concept is the spatial dispersion of risk. Having valued forest elements broadly distributed across a landscape, whether interspersed by farmland or consisting of continuous forest, makes them less vulnerable to unforeseen disturbances and disruptions. In other words, dispersed modularity equates to spreading risk spatially. This is one of the arguments supporting the designation of several, widely dispersed protected areas rather than an equivalent area in a single large reserve (Albers et al. 2016). Whether managing campgrounds in a national park or timber plantations on private land, it is usually more efficient to maintain a cluster of such assets at a single centralized location. But, if considering the risk of disruptive and destructive floods, fires, and windstorms, widely distributed assets are more likely to survive. The need to conserve multiple populations of a species at multiple locations is a fundamental tenet of conservation biology (Soulé 1987), a form of "bet hedging" (Boyce et al. 2002) analogous to risk-adverse supply-chain management in industry (Russell and Smith 2009).

5.6 Connectedness

Paradoxically, too much modularity and isolation of valued system components can also threaten overall system resilience. While multiple populations of a rare species are likely to avoid catastrophic loss to a single disturbance event, their long-term viability depends on functioning as a metapopulation, with local extirpations in suitable habitat replenished by dispersal from surviving populations and with sufficient gene flow to maintain genetic vigor (Hanski 1998). This is just one example in which connections and networks are important to long-term resilience.

Ecology, as the science of interconnections among organisms and their environment, has been documenting mutualistic networks and complex food webs for over a century. The prevalence of widespread and sometimes unexpected connections among ecosystem components and between humans and our environment prompted Commoner (1971) to coin the First Law of Ecology as "Everything is connected to everything else." Research since then has refined our understanding of food webs and other ecological networks, revealing that not all components are equally important to their stability or resilience. Dunne et al. (2002) compared the simulated response of sixteen well-documented terrestrial and aquatic food webs to removal of individual species (which might happen upon exposure to a novel pathogen, for example). They found that secondary extirpations were more likely to occur in ecosystems having fewer trophic links per species. While overall networks of tropical tree species and the many frugivorous birds that depend on them and disperse their seeds, for example, are typically resilient, they can nonetheless be vulnerable to the loss of a few key, highly connected, or generalist species (Vidal et al. 2014; Palacio et al. 2016). Building on graph theory, quantitative network analysis has developed into a sophisticated methodology for vulnerability assessment, with applications now extending from ecosystems to power grids, transportation planning, manufacturing, and organizational structures (Latora et al. 2017).

One of the most exciting demonstrations of connectness in forests is the growing appreciation of the widespread nature and importance of underground connections among plant and fungal species. The clonal structure of many woody species such as poplars (*Populus* spp.) in the northern hemisphere and the tropical woody bamboos (Bambuseae) results from vegetative reproduction and spread by means of root suckers or rhizomes (Peterson and Jones 1997). The connectedness of individual

stems and their distinction from neighboring clones is often expressed in a synchronized and clone-specific phenology, as when the autumnal foliage of trembling aspen (*Populus tremuloides*) clones exhibit different hues and timing of color change. It is apparent that resources such as water and nutrients are shared among connected stems (ramets), integrating or "averaging out" the heterogeneous resource pool available to the clone as a whole (Liu et al. 2016). Root suckers produced by *Eucryphia cordifolia* in temperate South America and by *Populus euphratica* in northwestern China are able to colonize drier, more severe microsites than those starting as seedlings (Escandón et al. 2013; Zhu et al. 2018). Natural root grafting between nonclonal trees has also been known to join the vascular systems of closely associated trees, apparently resulting in the sharing of water and nutrients and improved growth (Fraser et al. 2006; Tarroux and DesRochers 2011; Adonsou et al. 2016).

The importance of fungal (mycorrhizal) connections between and among trees, even of different species, has been well established, conceivably making many trees and other plants in a forest stand connected by fungal networks. Those connections provide pathways for the inter-plant transfer of carbon, nutrients, water, defense signals, and allelochemicals (Simard and Durall 2004; Simard et al. 2012). As with clonal plant species and those joined by root grafts, mycorrhizal networks integrate overall site and resource variability, with a joint effect on all network participants. The metaphorical power and implications of these networks—and the concept that cooperation and collaboration can prevail over competition in the natural world—have prompted references to this phenomenon as the "wood-wide web" (Simard et al. 1997; cover of *Nature*, 388/6642 (August 1997)) and "socialism in soil" (Van Der Heijden and Horton 2009). Such networks are reflected in the symbiotic relationships dominating the planet Pandora in the 2009 science-fiction movie *Avatar*, directed by James Cameron. On the other hand, inter-plant connections need not be mutually beneficial. There are many parasitic and hemiparasitic vascular plant species that confer no advantage to their host plants (Twyford 2018). Similarly, the invasive *Centaurea maculosa* receives a one-way transfer of carbon through mycorrhizal connections with the native bunchgrass, *Festuca idahoensis* (Carey et al. 2004).

The resilience provided by mycorrhizal connections and other soil-based symbiotic networks has immediate implications to forest management. Clearcut logging is ill advised in harsh climates, where adult trees can aid the establishment of associated tree seedlings (Teste and Simard 2008; Bingham and Simard 2012) and where mature trees are needed to sustain commercially valuable matsutake (*Tricholoma matsutake*) mushrooms and many other commercially harvested fungi (Palm and Chapela 1997). Broadleaf species that had been routinely brushed from young conifer plantations may actually be facilitating the growth of those crop trees (Simard and Vyse 2006) and can reduce the incidence of *Armillaria* root disease (Simard et al. 2005). The reintroduction of healthy soil biota can be key to the establishment of trees in efforts to regenerate or reclaim severely disturbed sites (Perry et al. 1989; Allen et al. 1997). Indeed, mycorrhizal networks of trees are just one part of functional meta-networks that include bacteria, mammals, and birds as well (Simard et al. 2013).

Another important mechanism to promote resilience in conservation and multiple-use forest landscapes is the maintenance of continuous expanses or linkages of mature and old forest cover, or of natural habitat free from obstructions and land uses inhospitable to native flora and fauna. Many species of plants, fungi, invertebrates, and vertebrates depend on the shade and heterogeneous canopy structure of old-growth forests, or may be slow in dispersing, growing, and reproducing, and

so are considered late successional or old-growth-dependent species. But it is not enough to protect isolated patches of old forest for species that have different habitat requirements in different seasons if we are to maintain metapopulation dynamics of inter-patch recolonization (in case of local extirpation) and healthy gene flow. Consequently, corridors of natural land—sometimes referred to as biodiversity corridors or landscape networks (see Box 5.3)—have become a key component of conservation planning and restoration (Noss and Harris 1986; Bennett 1999; Lindenmayer et al. 2006). The need for connectivity corridors is greatest where forest fragmentation by logging, agriculture, or urbanization prevails in the landscape or will do so if development proceeds without restrictions. The importance of landscape connectivity is most obvious for mobile wildlife with large home ranges (for example, tigers; Sharma et al. 2013) and seasonal migration patterns (for example, ungulates that summer in high-elevation forests, but winter in valley bottoms). But landscape corridors help maintain the viability of forest-dependent insect (Hill 1995; Várkonyi et al. 2003) and forest-dwelling plant (Wehling and Diekmann 2009; Liira and Paal 2013) populations as well. One type of habitat connectivity can be provided by riparian buffer strips that assure continuous forest cover along streams and rivers (Naiman et al. 1993), with the added benefit of maintaining shade and organic matter inputs into those flowing water bodies. In turn, those streams and rivers need to be protected from dams and diversions if the connectivity of fish populations is to be maintained.

Box 5.3 Landscape Ecology 101 for forest planners and managers

A number of terms related to the spatial arrangement of trees and other forest components across scales have technical meanings that may not coincide with their use in everyday language. In forest conservation and management, the "landscape" generally refers to a multitude of forest stands and ecosystems across a catchment basin, a forest estate, or a management area over which sustainability is expected, which is also referred to as the "forest level." A forest "stand" is a mappable unit of forest more or less

distinct in terms of its composition and structure. A uniform set of stand management (silvicultural) activities are typically applied at the stand level, although there can be some variation in recognition of within-stand site differences.

The dominant landcover type of a landscape is referred to as the landscape "matrix." Distinct polygons mapped out within that matrix are referred to as "patches." In an agricultural landscape, cropland or pasture can be expected to constitute the matrix, with woodlots, ponds,

---

Box 5.3 *Continued*

---

and farmyards as patches. In a matrix of old-growth forest, clearcuts, wetlands, and distinctive forest stands are patches. A forest matrix may be made up exclusively of contiguous forest stands—distinct at one level, but more or less equivalent in their role of providing forest cover when considered at another level. Long narrow patches are known as "corridors," which include roads, streams, and hedgerows. An interconnected set of patches and corridors of the same general land cover make up a landscape "network." One might find an aquatic network, a road network, a network of rocky ridges, and a forest cover network all in one landscape.

The landscape ecology of a forest depends on how species and ecological processes in one component of a landscape affect and are affected by those in other components. Those attributes affected by a contrast in adjacent patches are referred to as "edge effects," such as elevated levels of light and wind that reach into an uncut forest stand when a clearcut is created next door. Conversely, the uncut forest may provide shade and seeds for some distance into that clearcut. In this manner, edge effects provide a link between stand-level and landscape-level ecology. Other influences depend on the spatial continuity provided by corridors and networks, such as the ability for aquatic organisms to migrate along streams and for mammals to move between forest patches under the cover of windbreaks or hedgerows in an agricultural landscape. People and the seeds of invasive weeds likewise disperse throughout the forest landscape along roads and trails. On the other hand, corridors can be barriers to other kinds of movement, such as a river or ridgeline that stops a wildfire from spreading, a vegetated riparian zone that traps sediment,

or a road that inhibits some wildlife from crossing to the other side.

The principles of landscape ecology, in which spatial arrangement and patterns matter, can apply at many scales. For example, a single treefall "gap" in a forest stand also constitutes a patch within the matrix of the stand, or a large clearcut can constitute a "gap" in the broader matrix of uncut forest. If a stand is harvested, but a cluster of mature trees is retained, that group of trees is a patch within the matrix of the clearcut. The retention patch is sort of an island or an "inverse gap," with both the gap and the retention patch subject to edge effects and having effects on their respective matrices. At a finer level, one can even consider the spatial arrangement of individual trees, which tend to be "random" or "clustered" in natural forests, but "regular" in planted forests. Such patterns can be measured, as can the levels of "connectivity" (connections between patches) or "fragmentation" (level of isolation and edge effects) in a forest landscape. Measures of edge density and road density (in km per km^2) are particularly useful in monitoring those aspects of biodiversity that are adverse to human disturbance.

It is often challenging for a forest planner or manager to consider the landscape-level implications of actions undertaken at the stand level. Keeping landscape patterns and flows in mind at all times is a useful aid in managing forests for resilience. For more in-depth discussion and guidance in applying these concepts, see texts such as Perera et al. (2006), Bell and Apostol (2007), and Lafortezza et al. (2008). Diaz and Apostol (1992) outline a particularly accessible approach to landscape analysis and design in workbook form.

Corridors of mature native forest are especially important for biodiversity conservation in landscapes dominated by heavily modified land uses, including agriculture and production forestry (e.g. Nasi et al. 2007). Mono-specific tree crops, often of exotic origin, typically do not provide the food needed by many species of vertebrate wildlife, nor the species-specific hosts and associates for many of the invertebrates, fungi, and plants found in native ecosystems. Annual crops and grass-dominated pastures further constitute hostile habitat devoid of protection from temperature extremes and from

predators. On the other hand, altered silvicultural systems (for example, uneven-aged management, long rotations, or other forms of continuous cover forestry), agroforestry (for example, shade-grown coffee), or fruit tree horticulture can sometimes provide the desired connectivity between natural forest patches (Asare et al. 2014). Woodland connectivity can also be provided, with varying degrees of effectiveness for different species, by "greenway" walking and cycling paths designed for recreation in urban, suburban, and industrial landscapes (Hay 1991; Pirnat 2000; Li et al. 2008), and by hedgerows and windbreaks in agricultural landscapes (Wehling and Diekmann 2009). The habitat value and resilience of those planted corridors could be further enhanced by making them as wide as possible, with a rich diversity of tree, shrub, and herbaceous species (see Section 11.7).

Forested corridors are an important component to a climate-change adaptation strategy (see Chapter 7). With climate over much of the planet expected to be warmer and drier in the next several decades, continuous habitat allowing for the gradual migration of organisms to cooler and moister habitats is desirable (Townsend and Masters 2015; Keeley et al. 2018). "Escape routes" are needed for species and ecosystems to move to higher elevations and poleward, following the climatic envelopes that define their current distributions. Providing this spatial continuity is especially important for the passive and long-term migration of the undescribed or undefined biodiversity found in natural ecosystems, while managed species might be more directly and actively relocated through facilitated migration (Ste-Marie et al. 2011; Williams and Dumroese 2013). Identifying "temporal corridors" in the form of existing and likely future climate refugia (Rose and Burton 2009; Morelli et al. 2020) is an analogous adaptation strategy to the designation of spatial corridors. Connectivity in time is further provided by biological legacies (including propagules, woody material, soil biota), attributes of "ecosystem memory" that facilitate ecosystem recovery after severe disturbances (Franklin et al. 2000).

Though widely promoted and adopted, the effectiveness of landscape corridors in facilitating organism movement and survival cries out for additional research. It is likely that they are more important for the conservation of some species in some landscapes, but not others (Niemela 2001; Liira and Paal 2013). As corridors, by definition, are relatively narrow strips of habitat, they are typically dominated by edge effects, with insufficient interior habitat to protect many forest specialists (Harper et al. 2007). Some connectivity is counterproductive with respect to biodiversity conservation, such as the band of exotic and invasive plants that are typically found on roadsides (Gelbard and Belnap 2003), as roads too are landscape corridors. While forest roads may provide some barrier (modularity) to the spread of wildfire and provide enhanced access for firefighters, they are also a source of human-caused fire ignitions (Lindenmayer et al. 2009) and can constitute barriers to the movement of some interior forest species (Laurance et al. 2009). Roads are a good example of the challenges in finding the right balance, often among multiple forest values (see Chapter 2), between strong modular structure and effective connectivity.

5.7 Situational awareness

Monitoring and adaptive management are widely advocated as essential components of sustainable forest management (Duinker and Trevisan 2003; Hickey et al. 2005), especially to assure the conservation of ecological and biodiversity values (Maser et al. 1994; Lindenmayer et al. 2000; Franklin et al. 2018). Various indicators and trends can be evaluated relative to criteria considered important to sustainable management as part of a cycle of continuous management improvement (see Chapter 10). But managing for forest resilience, and the resilience of forest enterprises, requires a wider perspective than managing for sustainability, which is challenging enough in itself (Burton 2020). Formal adaptive management explicitly advocates a workshop approach to cast a wide net for the consideration of environmental and social values and a wide number of "what if" scenarios to devise resilient policies in the face of uncertainty (Holling 1978). This "wide net" and awareness of trends, patterns, and perspectives in the broader world are key to the preparedness required for resilient forest management. It is worth noting that the early roots of adaptive management (e.g. Walters 1986) and of resilience thinking (e.g. Holling 1973) have considerable overlap, with many of the foundational thinkers sharing ideas and working together.

Multiple interpretations and scales of diversity emerge as the most prevalent contributor to resilience in environmental and socio-ecological systems, but "preparedness" is most important in business and engineering applications (Xu and Kajikawa 2018; see Table 5.1). Preparedness implies anticipation of future disruptions, and appropriate planning to deal with a range of possible future scenarios. When specific normative values—such as human safety or business viability—must be assured, then we cannot count solely on the inherent resilience of nature to protect those values. Forest enterprises, forest operations, and the management of production forests thus need to place a greater emphasis (compared to the management of conservation lands or multiple-use forests) on preparedness, including attention to socioeconomic trends in markets, investment, climate, and social preferences. A long tradition of lookouts, roads, equipment caches, and trained crews ready to detect and rapidly suppress unwanted forest fires illustrates one kind of preparedness. But a broader interpretation of preparedness includes monitoring the impacts of fire suppression in the context of fuel accumulation, a warming climate, and expanding wildland–urban interfaces. Sometimes implementation of policies and practices supporting general resilience, as outlined in previous sections, are sufficient, but situational awareness is also about knowing when to prepare for specific threats (Carpenter et al. 2001). Many corporations and various levels of government are undertaking climate change vulnerability assessments (see Chapter 6), an excellent example of how situational awareness can be put into practice and communicated to decision-makers and the general public.

Risk analysis and management are considered essential components of most business plans, project proposals, and plans for the management of natural resources (Brumelle et al. 1990; Haynes and Cleaves 1999; Kangas and Kangas 2004). Yet it is impossible to anticipate all risks, to monitor all trends, or to prepare for every eventuality. One can, nonetheless, monitor specific indicators and pre-identify thresholds for shifts in policy and practices. Despite formalized monitoring systems, by the time a change is detected, confirmed, and reported, it may already be too late to avoid negative consequences. Perhaps more important is to build teams and collaborations composed of individuals and broad-reaching connections to track current trends on the land and in the outside world. There is a danger for forest planners and managers to be ensconced in urban offices, spending limited time in the forest. While more decentralized planning and management can overcome that issue somewhat, managing for general resilience also requires many eyes and ears out on the land. A good solution is the incorporation of community-based "forest guardians" into formal forest management structures. Successful examples include Indigenous guardian and watchmen programs in Canada (Ecotrust Canada 2013) and monitors of red panda (*Ailurus fulgens*) populations and habitat in Nepal (Red Panda Network 2023). Other models promote "citizen science" with technologies and recognition that reward public engagement with their environment. These approaches serve multiple roles, promoting forest patrols and monitoring for threats and trends, while also respecting traditional forest access rights, and encouraging public participation. Such programs need to be complemented with team members who come from diverse backgrounds and with diverse connections, with education not just in the biophysical and management sciences, but in the liberal arts as well. A positively engaged public, open lines of community communication, and staff members who follow politics, international markets, and popular culture collectively provide important social capital to draw upon for resilient forest management.

5.8 Closing perspectives

There is no single pathway, no widely agreed-upon checklist for how to promote resilience in natural or managed systems. Local conditions and identifiable threats will usually take precedence over general principles of managing for resilience. Yet a brief look at evolutionary strategies, cultural practices, and the success and failure of different business and political approaches offers a fairly robust set of concepts worth adhering to. The value of diversity is a common theme that permeates all resilience strategies, as are various bet-hedging approaches. Resilience also benefits from capacity, sometimes in the form of reserves and legacies, sometimes in the form of social capital and preparedness. Monitoring and awareness are important to understand and respond to changing conditions, with creativity and resourcefulness always being an asset to overcoming challenges. Some principles, notably modularity and connectivity, seem diametrically opposed, and require case-specific balancing. How do we compartmentalize the bad (undesired disturbances, disruptions, pathogens), but share the good (desired resources, gene flow, and information)? How do we engage in thoughtful planning, yet maintain adaptive flexibility? An underlying theme to building general resilience may be the conscious effort to reduce risks that threaten persistence, while still being open to beneficial opportunities.

The challenges experienced around the world during the global Covid-19 pandemic further offer some lessons with regard to these general strategies for resilience in the face of an unforeseen disruption (see Box 5.4). Although the analogy is imperfect, there is surprisingly rich food for thought when considering healthy forest ecosystems and healthy forest enterprises as sharing common features with healthy human communities.

Box 5.4 Lessons in general resilience from the COVID-19 pandemic

The worldwide pandemic resulting from the spread of a novel SARS-CoV-2 coronavirus known as COVID-19 provides an example of an unforeseen—though foreseeable—disruption to a wide array of management systems. Although it was globally disruptive, jurisdictions varied in strategies to minimize disease spread and infection, to compensate for social disruption, and to promote economic recovery. The pandemic also revealed many existing vulnerabilities in institutions and businesses, ranging from supply-chain dependencies and narrow profit margins to poor communication plans.

Several comparative analyses of these alternative strategies have now being undertaken (e.g. Kavanagh and Singh 2020; Yan et al. 2021; Popic and Moise 2022). Table 5.3 suggests there are a few analogies between some of the challenges and responses associated with the COVID-19 pandemic and potential applications in forest planning and management. Many of these features exemplify the resilience associated with complex adaptive systems and the need to transition from command-and-control management to a more responsive stewardship paradigm, as further developed in Messier et al. (2015) and in Chapter 12.

Table 5.3 Pandemic preparation and response analogies for forest planning and management

Feature	Applied in community health	Applied in forest management
Capacity	Personnel, procedures, and financial resources that can be rapidly mobilized in response to disease outbreak to minimize spread, and to test drugs and vaccines	Personnel, procedures, and financial resources that can be rapidly mobilized to respond to natural disturbances and socioeconomic disruption
Reserves	Stockpiles of personal protective equipment, testing kits, and reagents; being able to call upon vaccines and treatments that were effective or under development for similar illnesses	Strategic distribution of protected areas (natural or uncut forest); avoiding harvesting at levels too close to the sustainability break-even point; stockpiles of fire-fighting equipment, logs awaiting processing
Memory	Lessons (institutional memory) from previous outbreaks, such as the SARS and H1N1 viruses	Lessons (institutional memory) from pest outbreaks, widespread wildfires, storm events, public protest, and trade restrictions; traditional and place-based ecological knowledge; retention of biological legacies during timber harvesting
Modularity	Ability to isolate individuals and communities, to close borders	Discrete landscape units over which to plan for the sustainability of multiple forest values, employing barriers (natural or managed) to limit spread of fire or pests
Networks	Resources shared among jurisdictions as needed; information on pathology and epidemiology shared from early outbreak locales; research on disease control shared globally	Flexibility to share resources (e.g. logs, firefighting equipment) among forest management areas and political jurisdictions; information and knowledge shared widely

Box 5.4 *Continued*

Feature	Applied in community health	Applied in forest management
Monitoring and indicators	Widespread testing, reliable confirmation of disease symptoms with test results	Regular assessment of criteria and indicators for ecosystem health and for sustainable forest management
Scenario planning	Pandemic plans developed and in place in advance, including response preparation for different levels of disease severity and transmissivity	Forest management plans include contingency plans for dealing with natural disturbances, climate change, and socioeconomic disruptions
Transparency	Open sharing of information on cases, mortalities, vaccine effectiveness, and side effects; providing clear rationale for community-health decisions	Open sharing of multiple resource inventories, trends, and distributions; providing clear rationale for forest policy decisions and the forest practices undertaken

---

Box 5.5 Key points

- The specific stresses and challenges that may emerge in the future cannot be anticipated, so it is prudent to adopt forest management policies and practices that support general system resilience.
- Species, ecosystems, and human society provide many examples of resilience that can provide guidance for resilience in forest management, yet our understanding of forests as complex adaptive systems is incomplete.
- General forest resilience is promoted by diversity at multiple levels, by retaining resource reserves and legacies, by incorporating flexibility in necessary planning processes, by balancing modularity and connectedness, and by monitoring trends within and beyond the forest.
- Because exceptions to these general guidelines arise for particular species and sites, tactics to manage for specific resilience may need to supersede them when place-based vulnerability analysis reveals likely exposure to specific threats.
- Managing forests for general resilience is an important aspect of sustainable forest management, ecosystem-based management, and responsible ecosystem stewardship.

Better living through flexibility.

Dead Dog Café Comedy Hour,
written by Thomas King (1997)

References cited

Adonsou, K. E., Drobyshev, I., DesRochers, A., and Tremblay, F. (2016). "Root Connections Affect Radial Growth of Balsam Poplar Trees," *Trees*, 30/5: 1775–83.

Agee, J. K., Bahro, B., Finney, M. A., et al. (2000). "The Use of Shaded Fuelbreaks in Landscape Fire Management," *Forest Ecology and Management*, 127/13: 55–66.

Albers, H. J., Busby, G. M., Hamaide, B., Ando, A. W., and Polasky, S. (2016). "Spatially-Correlated Risk in Nature Reserve Site Selection," *PloS One*, 11/1: e0146023.

Alfaro, R. I., and Maclauchlan, L. E. (1992). "A Method to Calculate the Losses Caused by Western Spruce Budworm in Uneven-Aged Douglas Fir Forests of British Columbia," *Forest Ecology and Management*, 55/1–4: 295–313.

Alfaro, R. I., Van Akker, L., and Hawkes, B. (2015). "Characteristics of Forest Legacies Following Two Mountain Pine Beetle Outbreaks in British Columbia, Canada," *Canadian Journal of Forest Research*, 45/10: 1387–96.

Allen, E. B., Espejel, I., and Siguenza, C. (1997). "Role of Mycorrhizae in Restoration of Marginal and Derelict Land and Ecosystem Sustainability," in M. E. Palm and I. H. Chapela (eds), *Mycology in Sustainable Development: Expanding Concepts, Vanishing Borders*. Boone, NC: Parkway Publishers, 147–59.

Andison, D. W., and McCleary, K. (2014). "Detecting Regional Differences in Within-Wildfire Burn Patterns in Western Boreal Canada," *Forestry Chronicle*, 90/1: 59–69.

Asare, R., Afari-Sefa, V., Osei-Owusu, Y., and Pabi, O. (2014). "Cocoa Agroforestry for Increasing Forest

Connectivity in a Fragmented Landscape in Ghana," *Agroforestry Systems*, 88/6: 1143–56.

Ashcroft, M. B., Chisholm, L. A., and French, K. O. (2009). "Climate Change at the Landscape Scale: Predicting Fine-Grained Spatial Heterogeneity in Warming and Potential Refugia for Vegetation," *Global Change Biology*, 15/3: 656–67.

Aubin, I., Munson, A. D., Cardou, F., et al. (2016). "Traits to Stay, Traits to Move: A Review of Functional Traits to Assess Sensitivity and Adaptive Capacity of Temperate and Boreal Trees to Climate Change," *Environmental Reviews*, 24/2: 164–86.

Baskerville, G. L. (1975). "Spruce Budworm: Super Silviculturist," *Forestry Chronicle*, 51/4: 138–40.

Bazzaz, F. A. (1996). *Plants in Changing Environments: Linking Physiological, Population, and Community Ecology.* Cambridge: Cambridge University Press. 332 pp.

Bell, S., and Apostol, D. (2007). *Designing Sustainable Forest Landscapes.* London: Taylor & Francis. 368 pp.

Bennett, A. F. (1999). *Linkages in the Landscape: The Role of Corridors and Connectivity in Wildlife Conservation.* Gland, Switzerland: International Union for the Conservation of Nature (IUCN). 254 pp.

Bereczki, K., Ódor, P., Csóka, G., Mag, Z., and Báldi, A. (2014). "Effects of Forest Heterogeneity on the Efficiency of Caterpillar Control Service Provided by Birds in Temperate Oak Forests," *Forest Ecology and Management*, 327: 96–105.

Bettinger, R. L. (1980). "Explanatory/Predictive Models of Hunter–Gatherer Adaptation," in M. B. Schiffer (ed.), *Advances in Archaeological Method and Theory*, Vol. 3. New York: Academic Press, 189–255.

Bingham, M. A., and Simard, S. W. (2012). "Ectomycorrhizal Networks of old *Pseudotsuga menziesii* var. *glauca* Trees Facilitate Establishment of Conspecific Seedlings under Drought," *Ecosystems*, 15: 188–99.

Bottero, A., Forrester, D. I., Cailleret, M., et al. (2021). "Growth Resistance and Resilience of Mixed Silver Fir and Norway Spruce Forests in Central Europe: Contrasting Responses to Mild and Severe Droughts," *Global Change Biology*, 27/18: 4403–19.

Boyce, M. S., Kirsch, E. M., and Servheen, C. (2002). "Bet-Hedging Applications for Conservation," *Journal of Biosciences*, 27/4: 385–92.

Brumelle, S., Stanbury, W. T., Thompson, W. A., Vertinsky, I., and Wehrung, D. (1990). "Framework for the Analysis of Risks in Forest Management and Silvicultural Investments," *Forest Ecology and Management*, 36/2–4: 279–99.

Burton, P. J. (2010). "Striving for Sustainability and Resilience in the Face of Unprecedented Change: The Case of the Mountain Pine Beetle Outbreak in British Columbia," *Sustainability*, 2/8: 2403–23.

Burton, P. J. (2013). "Exploring Complexity in Boreal Forests," in C. Messier, K. J. Puettmann, and K. D. Coates (eds), *Managing Forests as Complex Adaptive Systems: Building Resilience to the Challenge of Global Change.* New York: Earthscan, 79–109.

Burton, P. J. (2020). "The Scope and Challenge of Sustainable Forestry," in J. A. Stanturf (ed.), *Achieving Sustainable Management of Boreal and Temperate Forests.* Cambridge: Burleigh Dodds Science Publishing, 1–22.

Campbell, E. M., MacLean, D. A., and Bergeron, Y. (2008). "The Severity of Budworm-Caused Growth Reductions in Balsam Fir/Spruce Stands Varies with the Hardwood Content of Surrounding Forest Landscapes," *Forest Science*, 54/2: 195–205.

Carey, E. V., Marler, M. J., and Callaway, R. M. (2004). "Mycorrhizae Transfer Carbon from a Native Grass to an Invasive Weed: Evidence from Stable Isotopes and Physiology," *Plant Ecology*, 172: 133–41.

Carpenter, S. R., Arrow, K. J., Barrett, S., et al. (2012). "General Resilience to Cope with Extreme Events," *Sustainability*, 4/12: 3248–59.

Carpenter, S., Walker, B., Anderies, J. M., and Abel, N. (2001). "From Metaphor to Measurement: Resilience of What to What?" *Ecosystems*, 4/8: 765–81.

Carroll, A. L., and Safranyik, L. (2003). "The Bionomics of the Mountain Pine Beetle in Lodgepole Pine Forests: Establishing a Context," in T. L. Shore, J. E. Brooks, and J. E. Stone (eds), *Mountain Pine Beetle Symposium: Challenges and Solutions.* Information Report BC-X-399. Victoria, BC: Canadian Forest Service, 21–32.

Chapin, F. S., Kofinas, G. P., and Folke, C. (eds). (2009). *Principles of Ecosystem Stewardship: Resilience-Based Natural Resource Management in a Changing World.* New York: Springer. 401 pp.

Chapin, F. S., Schulze, E.-D., and Mooney, H. A. (1990). "The Ecology and Economics of Storage in Plants," *Annual Review of Ecology and Systematics*, 21: 423–7.

Ciesla, W. M. (2011). *Forest Entomology: A Global Perspective.* Chichester, UK: Wiley-Blackwell. 441 pp.

Commoner, B. (1971). *The Closing Circle: Nature, Man, and Technology.* New York: Alfred A. Knopf. 326 pp.

Cortner, H. J., Gordon, J. C., Risser, P. G., Teeguarden, D. E., and Thomas, J. W. (1999). "Ecosystem Management: Evolving Model for Stewardship of the Nation's Natural Resources," in R. C. Szaro, N. C. Johnson, W. T. Sexton, and A. J. Malik (eds), *Ecological Stewardship: A Common Reference for Ecosystem Management*, Vol. 2. Oxford: Elsevier, 3–19.

Crowther, G. (2018). *Eating Culture: An Anthropological Guide to Food.* 2nd edn. Toronto: University of Toronto Press. 384 pp.

Curtis, F. (2003). "Eco-Localism and Sustainability," *Ecological Economics*, 46/1: 83–102.

Dănescu, A., Albrecht, A. T., and Bauhus, J. (2016). "Structural Diversity Promotes Productivity of Mixed, Uneven-Aged Forests in Southwestern Germany," *Oecologia*, 182/2: 319–33.

Dhar, A., Parrott, L., and Heckbert, S. (2016). "Consequences of Mountain Pine Beetle Outbreak on Forest Ecosystem Services in Western Canada," *Canadian Journal of Forest Research*, 46/8: 987–99.

Diamond, J. (2005). *Collapse: How Societies Choose to Succeed or Fail*. New York: Viking Penguin. 592 pp.

Diamond, J. (2012). *The World until Yesterday: What Can We Learn from Traditional Societies?* New York: Viking Press. 499 pp.

Diaz, N., and Apostol, D. (1992). *Forest Landscape Analysis and Design: A Process for Developing and Implementing Land Management Objectives for Landscape Patterns*. R6 ECO-TP-043-92. Portland, OR: USDA Forest Service. 118 pp. Available online at https://www.fs.usda.gov/treesearch/pubs/6268 (accessed May 19, 2021).

Díaz, S., and Cabido, M. (2001). "Vive la différence: Plant Functional Diversity Matters to Ecosystem Processes," *Trends in Ecology & Evolution*, 16/11: 646–55.

Duinker, P. N., and Trevisan, L. M. (2003). "Adaptive Management: Progress and Prospects for Canadian Forests," in P. J. Burton, C. Messier, D. W. Smith, and W. L. Adamowicz (eds), *Towards Sustainable Management of the Boreal Forest*. Ottawa: NRC Research Press, 857–92.

Dunne, J. A., Williams, R. J., and Martinez, N. D. (2002). "Network Structure and Biodiversity Loss in Food Webs: Robustness Increases with Connectance," *Ecology Letters*, 5/4: 558–67.

Ecotrust Canada (2013). *Aboriginal Guardian and Watchmen Programs in Canada*. Prepared for the Northeast Superior Chief's Forum. 18 pp. Available online at http://wahkohtowin.com/wp-content/uploads/2017/10/Aboriginal-Guardian-and-Watchmen-Programs-in-Canada-2013-with-recommendations.pdf (accessed July 12, 2020).

Escandón, A. B., Paula, S., Rojas, R., Corcuera, L. J., and Coopman, R. E. (2013). "Sprouting Extends the Regeneration Niche in Temperate Rain Forests: The Case of the Long-Lived Tree *Eucryphia cordifolia*," *Forest Ecology and Management*, 310: 321–6.

Fechner, G. H., and Barrows, J. S. (1976). *Aspen Stands as Wildfire Fuel Breaks*. Eisenhower Consortium Bulletin 4. Fort Collins, CO: Rocky Mountain Forest and Range Experiment Station, USDA Forest Service. 26 pp.

Feng, Y., Schmid, B., Loreau, M., et al. (2022). "Multispecies Forest Plantations Outyield Monocultures across a Broad Range of Conditions," *Science*, 376/6595: 865–8.

Fettig, C. J., Gibson, K. E., Munson, A. S., and Negrón, J. F. (2014). "Cultural Practices for Prevention and Mitigation of Mountain Pine Beetle Infestations," *Forest Science*, 60/3: 450–63.

Filotas, E., Parrott, L., Burton, P. J., et al. (2014). "Viewing Forests through the Lens of Complex Systems Science," *Ecosphere*, 5/1: 1–23.

Franklin, J. F., Berg, D. R., Thornburgh, D. A., and Tappeiner, J. C. (1997). "Alternative Silvicultural Approaches to Timber Harvesting: Variable Retention Harvest Systems," in K. A. Kohm and J. F. Franklin (eds), *Creating a Forestry for the 21st Century: The Science of Ecosystem Management*. Washington: Island Press, 111–39.

Franklin, J. F., Johnson, K. N., and Johnson, D. L. (2018). *Ecological Forest Management*. Long Grove, IL: Waveland Press. 646 pp.

Franklin, J. F., Lindenmayer, D., MacMahon, J. A., et al. (2000). "Threads of Continuity," *Conservation Biology in Practice*, 1/1: 8–16.

Fraser, E. C., Lieffers, V. J., and Landhäusser, S. M. (2006). "Carbohydrate Transfer through Root Grafts to Support Shaded Trees," *Tree Physiology*, 26/8: 1019–23.

Galik, C. S., and Jackson, R. B. (2009). "Risks to Forest Carbon Offset Projects in a Changing Climate," *Forest Ecology and Management*, 257/11: 2209–16.

Gelbard, J. L., and Belnap, J. (2003). "Roads as Conduits for Exotic Plant Invasions in A Semiarid Landscape," *Conservation Biology*, 17/2: 420–32.

Green, L. R. (1977). *Fuelbreaks and Other Fuel Modification for Wildland Fire Control*. Agricultural Handbook 499. Washington: USDA Forest Service. 79 pp.

Greiner, S. M., Grimm, K. E., and Waltz, A. E. (2020). "Managing for Resilience? Examining Management Implications of Resilience in Southwestern National Forests," *Journal of Forestry*, 118/4: 433–43.

Griess V. C., and Knoke T. (2011). "Growth Performance, Windthrow, and Insects: Meta-Analyses of Parameters Influencing Performance of Mixed-Species Stands in Boreal and Northern Temperate Biomes," *Canadian Journal of Forest Research*, 41: 1141–59.

Grossiord, C., Granier, A., Ratcliffe, S., et al. (2014). "Tree Diversity Does Not Always Improve Resistance of Forest Ecosystems to Drought," *Proceedings of the National Academy of Science*, 111: 14812–15.

Gunderson, L. H., Holling, C. S., Pritchard, L., and Peterson, G. D. (2002). "Understanding Resilience: Theory, Metaphors, and Frameworks," in L. H. Gunderson and L. Pritchard (eds), *Resilience and the Behavior of Large-Scale Resource Systems*. Washington: Island Press, 3–20.

Gustafsson, L., Baker, S. C., Bauhus, J., et al. (2012). "Retention Forestry to Maintain Multifunctional Forests: A World Perspective," *BioScience*, 62/7: 633–45.

Hampe, A., and Jump, A. S. (2011). "Climate Relicts: Past, Present, Future," *Annual Review of Ecology, Evolution, and Systematics*, 42: 313–33.

Hanski, I. (1998). "Metapopulation Dynamics," *Nature*, 396/6706: 41–9.

Harper, K. A., Mascarúa-López, L., Macdonald, S. E., and Drapeau, P. (2007). "Interaction of Edge Influence from Multiple Edges: Examples from Narrow Corridors," *Plant Ecology*, 192/1: 71–84.

Hay, K. G. (1991). "Greenways and Biodiversity," in W. E. Hudson (ed.), *Landscape Linkages and Biodiversity*. Washington: Island Press, 162–75.

Haynes, R., and Cleaves, D. (1999). "Uncertainty, Risk, and Ecosystem Management," in N. Johnson, A. J. Malk, R. C. Szaro, and W. T. Sexton (eds), *Ecological Stewardship: A Common Reference for Ecosystem Management*, Vol. 3. New York: Elsevier, 413–29.

Hély, C., Bergeron, Y., and Flannigan, M. D. (2000). "Effects of Stand Composition on Fire Hazard in Mixed-Wood Canadian Boreal Forest," *Journal of Vegetation Science*, 11: 813–24.

Hickey, G. M., Innes, J. L., Kozak, R. A., Bull, G. Q., and Vertinsky, I. (2005). "Monitoring and Information Reporting for Sustainable Forest Management: An International Multiple Case Study Analysis," *Forest Ecology and Management*, 209/3: 237–59.

Hill, C. J. (1995). "Linear Strips of Rain Forest Vegetation as Potential Dispersal Corridors for Rain Forest Insects," *Conservation Biology*, 9/6: 1559–66.

Holling, C. S. (1973). "Resilience and Stability of Ecological Systems," *Annual Review of Ecology and Systematics*, 4: 1–23.

Holling, C. S. (ed.). (1978). *Adaptive Environmental Assessment and Management*. IIASA Report 3. Chichester, UK: John Wiley & Sons. 377 pp.

Holling, C. S. (1996). "Engineering Resilience versus Ecological Resilience," in P. C. Shulze (ed.), *Engineering within Ecological Constraints*. Washington: National Academy Press, 31–44.

Holling, C. S., and Chambers, A. D. (1973). "Resource Science: The Nurture of an Infant," *BioScience*, 23/1: 13–20.

Holling, C. S., and Meffe, G. K. (1996). "Command and Control and the Pathology of Natural Resource Management," *Conservation Biology*, 10/2: 328–37.

Homayounfar, M., Muneepeerakul, R., and Anderies, J. M. (2022). "Resilience-Performance Trade-Offs in Managing Social–Ecological Systems," *Ecology and Society*, 27/1: 7.

Hunter, M. L. (1990). *Wildlife, Forests, and Forestry: Principles of Managing Forests for Biological Diversity*. Englewood Cliffs, NJ: Prentice Hall. 370 pp.

Ishii, H., and Asano, S. (2010). "The Role of Crown Architecture, Leaf Phenology and Photosynthetic Activity in Promoting Complementary Use of Light among Coexisting Species in Temperate Forests," *Ecological Research*, 25: 715–22.

Ishii, H. T., Tanabe, S. I., and Hiura, T. (2004). "Exploring the Relationships among Canopy Structure, Stand Productivity, and Biodiversity of Temperate Forest Ecosystems," *Forest Science*, 50/3: 342–55.

Jacobs, K., Bonal, D., Collet, C., Muys, B., and Ponette, Q. (2021). "Mixing Increases Drought Exposure through a Faster Growth in Beech, but not in Oak," *Forest Ecology and Management*, 479: 118593.

Jactel, H., Bauhus, J., Boberg, J., et al. (2017). "Tree Diversity Drives Forest Stand Resistance to Natural Disturbances," *Current Forestry Reports*, 3: 223–43.

Jactel, H., and Brockerhoff, E. G. (2007). "Tree Diversity Reduces Herbivory by Forest Insects," *Ecology Letters*, 10/9: 835–48.

Johnson, N., Malk, A. J., Szaro, R. C., and Sexton, W. T. (eds). (1999). *Ecological Stewardship: Key Findings*, Vol. 1. New York: Elsevier. 285 pp.

Kangas, A. S., and Kangas, J. (2004). "Probability, Possibility and Evidence: Approaches to Consider Risk and Uncertainty in Forestry Decision Analysis," *Forest Policy and Economics*, 6/2: 169–88.

Kavanagh, M. M., and Singh, R. (2020). "Democracy, Capacity, and Coercion in Pandemic Response—COVID 19 in Comparative Political Perspective," *Journal of Health Politics, Policy and Law*, 45/6: 997–1012.

Keeley, A. T., Ackerly, D. D., Cameron, D. R., et al. (2018). "New Concepts, Models, and Assessments of Climate-Wise Connectivity," *Environmental Research Letters*, 13/7: 073002.

Kimmerer, R. W. (2013). *Braiding Sweetgrass: Indigenous Wisdom, Scientific Knowledge, and the Teachings of Plants*. Minneapolis, MN: Milkweed Editions. 395 pp.

Knoke, T., Steinbeis, O. E., Bösch, M., Román-Cuesta, R. M., and Burkhardt, T. (2011). "Cost-Effective Compensation to Avoid Carbon Emissions from Forest Loss: An Approach to Consider Price–Quantity Effects and Risk-Aversion," *Ecological Economics*, 70/6: 1139–53.

Krawchuk, M. A., Meigs, G. W., Cartwright, J. M., et al. (2020). "Disturbance Refugia within Mosaics of Forest Fire, Drought, and Insect Outbreaks," *Frontiers in Ecology and the Environment*, 18/5: 235–44.

Lafortezza, R., Chen, J., Crow, T. R., and Sanesi, G. (eds). (2008). *Patterns and Processes in Forest Landscapes: Multiple Use and Sustainable Management*. Dordrecht: Springer. 425 pp.

Latora, V., Nicosia, V., and Russo, G. (2017). *Complex Networks: Principles, Methods and Applications*. Cambridge: Cambridge University Press. 559 pp.

Laurance, W. F., Goosem, M., and Laurance, S. G. (2009). "Impacts of Roads and Linear Clearings on Tropical Forests," *Trends in Ecology & Evolution*, 24/12: 659–69.

Levin, S. A. (2005). "Self-Organization and the Emergence of Complexity in Ecological Systems," *BioScience*, 55/12: 1075–9.

Li, J., Wang, Y., and Song, Y. C. (2008). "Landscape Corridors in Shanghai and their Importance in Urban Forest Planning," in M. M. Carreiro, Y.-C. Song, and J. Wu. (eds), *Ecology, Planning, and Management of Urban Forests: International Perspectives*. New York: Springer, 219–39.

Liang, J., Crowther, T. W., Picard, N., et al. (2016). "Positive Biodiversity–Productivity Relationship Predominant in Global Forests," *Science*, 354/6309: 196–208.

Lieffers, V. J., Messier, C., Burton, P. J., Ruel, J.-C., and Grover, B. E. (2003). "Nature-Based Silviculture for Sustaining a Variety of Boreal Forest Values," in P. J. Burton, C. Messier, D. W. Smith, and W. L. Adamowicz (eds), *Towards Sustainable Management of the Boreal Forest*. Ottawa: NRC Research Press, 481–530.

Liira, J., and Paal, T. (2013). "Do Forest-Dwelling Plant Species Disperse along Landscape Corridors?" *Plant Ecology*, 214/3: 455–70.

Lindenmayer, D. B., Franklin, J. F., and Fischer, J. (2006). "General Management Principles and a Checklist of Strategies to Guide Forest Biodiversity Conservation," *Biological Conservation*, 131/3: 433–45.

Lindenmayer, D. B., Franklin, J. F., Lõhmus, A., et al. (2012). "A Major Shift to the Retention Approach for Forestry Can Help Resolve Some Global Forest Sustainability Issues," *Conservation Letters*, 5/6: 421–31.

Lindenmayer, D. B., Hunter, M. L., Burton, P. J., and Gibbons. P. (2009). "Effects of Logging on Fire Regimes in Moist Forests," *Conservation Letters*, 2/6: 271–7.

Lindenmayer, D. B., Margules, C. R., and Botkin, D. B. (2000). "Indicators of Biodiversity for Ecologically Sustainable Forest Management," *Conservation Biology*, 14/4: 941–50.

Lipoma, L., Kambach, S., Díaz, S., et al. (2024). "No general support for functional diversity enhancing resilience across terrestrial plant communities," *Global Ecology and Biogeography*, 33/5: e13895.

Liu, F., Liu, J., and Dong, M. (2016). "Ecological Consequences of Clonal Integration in Plants," *Frontiers in Plant Science*, 7: 770.

Maser, C., Bormann, B. T., Brookes, M. H., Kiester, A. R., and Weigand, J. F. (1994). "Sustainable Forestry through Adaptive Ecosystem Management Is an Open-Ended Experiment," in C. Maser, *Sustainable Forestry: Philosophy, Science, and Economics*. Delray Beach, FL: St Lucie Press, 303–40.

Messier, C., Puettmann, K., Chazdon, R., et al. (2015). "From Management to Stewardship: Viewing Forests as Complex Adaptive Systems in an Uncertain World," *Conservation Letters*, 8/5: 368–77.

Messier, C., Puettmann, K. J., and Coates, K. D. (eds). (2013). *Managing Forests as Complex Adaptive Systems: Building Resilience to the Challenge of Global Change*. New York: Earthscan. 353 pp.

Montagnini, F., Gonzalez, E., Porras, C., and Rheingans, R. (1995). "Mixed and Pure Forest Plantations in the Humid Neotropics: A Comparison of Early Growth, Pest Damage and Establishment Costs," *Commonwealth Forestry Review*, 74: 306–14.

Morelli, T. L., Barrows, C. W., Ramirez, A. R., et al. (2020). "Climate-Change Refugia: Biodiversity in the Slow Lane," *Frontiers in Ecology and the Environment*, 18/5: 228–34.

Mori, A. S., and Kitagawa, R. (2014). "Retention Forestry as a Major Paradigm for Safeguarding Forest Biodiversity in Productive Landscapes: A Global Meta-Analysis," *Biological Conservation*, 175: 65–73.

Muir, J. A., and Geils, B. W. (2002). "Management Strategies for Dwarf Mistletoe: Silviculture," in B. W. Geils, J. Cibrián Tovar, and B. Moody (technical coordinators), *Mistletoes of North American Conifers*. General Technical Report RMRS-GTR-98. Ogden, UT: USDA Forest Service, 83–94.

Naiman, R. J., Decamps, H., and Pollock, M. (1993). "The Role of Riparian Corridors in Maintaining Regional Biodiversity," *Ecological Applications*, 3/2: 209–12.

Nasi, R., Koponen, P., Poulsen, J. G., Buitenzorgy, M., and Rusmantoro, W. (2007). "Impact of Landscape and Corridor Design on Primates in a Large-Scale Industrial Tropical Plantation Landscape," in E. G. Brockerhoff, H., Jactel, J. A. Parrotta, et al. (eds), *Plantation Forests and Biodiversity: Oxymoron or Opportunity?* Dordrecht: Springer, 181–202.

Nenzen, H. K., Filotas, E., Peres-Neto, P., and Gravel, D. (2017). "Epidemiological Landscape Models Reproduce Cyclic Insect Outbreaks," *Ecological Complexity*, 31: 78–87.

Nesbit, K. A., Yocom, L. L., Trudgeon, A. M., DeRose, R. J., and Rogers, P. C. (2023). "Tamm Review: Quaking Aspen's Influence on Fire Occurrence, Behavior, and Severity," *Forest Ecology and Management*, 531: 120752.

Neuner, S., Beinhofer, B., and Knoke, T. (2013). "The Optimal Tree Species Composition for A Private Forest Enterprise—Applying the Theory of Portfolio Selection," *Scandinavian Journal of Forest Research*, 28/1: 38–48.

Niemela, J. (2001). "The Utility of Movement Corridors in Forested Landscapes," *Scandinavian Journal of Forest Research*, 16(S3): 70–8.

Nikinmaa, L., Lindner, M., Cantarello, E., et al. (2023). "A Balancing Act: Principles, Criteria and Indicator

Framework to Operationalize Social–Ecological Resilience of Forests," *Journal of Environmental Management*, 331: 117039.

Norden, N., Angarita, H. A., Bongers, F., et al. (2015). "Successional Dynamics in Neotropical Forests Are as Uncertain as They Are Predictable," *Proceedings of the National Academy of Sciences*, 112/26: 8013–18.

Noss, R. F., and Harris, L. D. (1986). "Nodes, Networks, and MUMs: Preserving Diversity at All Scales," *Environmental Management*, 10/3: 299–309.

Nyland, R. D. (1996). *Silviculture: Concepts and Applications*. New York: McGraw Hill. 633 pp.

Oliver, C. D. (1995). "A Portfolio Approach to Landscape Management: An Economically, Ecologically, and Socially Sustainable Approach to Forestry," in C. R. Bamsey (ed.), *Innovative Silviculture Systems in Boreal Forests*. Edmonton: Clear Lake Ltd, 2–8. Available online at https://cfs.nrcan.gc.ca/pubwarehouse/pdfs/25732.pdf (accessed June 2, 2024).

Palik, B. J., D'Amato, A. W., Franklin, J. F., and Johnson, K. N. (2021). *Ecological Silviculture: Foundations and Applications*. Long Grove, IL: Waveland Press. 343 pp.

Palacio, R. D., Valderrama-Ardila, C., and Kattan, G. H. (2016). "Generalist Species Have a Central Role in a Highly Diverse Plant–Frugivore Network," *Biotropica*, 48/3: 349–55.

Palm, M. E., and Chapela, I. H. (eds). (1997). *Mycology in Sustainable Development: Expanding Concepts, Vanishing Borders*. Boone, NC: Parkway Publishers. 306 pp.

Perera, A. H., Buse, L. J., and Crow, T. R. (eds). (2006). *Forest Landscape Ecology: Translating Knowledge to Practice*. New York: Springer. 223 pp.

Perry, D. A., Amaranthus, M. P., Borchers, J. G., Borchers, S. L., and Brainerd, R. E. (1989). "Bootstrapping in Ecosystems," *BioScience*, 39/4: 230–7.

Perry, D. A., Hessburg, P. F., Skinner, C. N., et al. (2011). "The Ecology of Mixed Severity Fire Regimes in Washington, Oregon, and Northern California," *Forest Ecology and Management*, 262/5: 703–17.

Peterson, R. L., and Jones, R. H. (1997). "Clonality in Woody Plants: A Review and Comparison with Clonal Herbs," in H. de Kroon and J. van Groenendael (eds), *The Ecology and Evolution of Clonal Plants*. Leiden, Netherlands: Backhuys Publishers, 263–89.

Pickles, B. J., and Simard, S. W. (2017). "Mycorrhizal Networks and Forest Resilience to Drought," in N. C. Johnson, C. Gehring, and J. Jansa (eds), *Mycorrhizal Mediation of Soil: Fertility, Structure, and Carbon Storage*. Amsterdam: Elsevier, 319–39.

Pirnat, J. (2000). "Conservation and Management of Forest Patches and Corridors in Suburban Landscapes," *Landscape and Urban Planning*, 52/2–3: 135–43.

Popic, T., and Moise, A. D. (2022). "Government Responses to the COVID-19 Pandemic in Eastern and Western Europe: The Role of Health, Political and Economic Factors," *East European Politics*, 38/4: 507–28.

Pureswaran, D. S., Roques, A., and Battisti, A. (2018). "Forest Insects and Climate Change," *Current Forestry Reports*, 4/2: 35–50.

Red Panda Network (2023). *Research & Monitoring*. Kathmandu, Nepal: Red Panda Network. Available online at https://redpandanetwork.org/Research-Monitoring (accessed July 26, 2023).

Robert, L. E., Sturtevant, B. R., Cooke, B. J., et al. (2018). "Landscape Host Abundance and Configuration Regulate Periodic Outbreak Behavior in Spruce Budworm *Choristoneura fumiferana*," *Ecography*, 41/9: 1556–71.

Rose, N.-A., and Burton, P. J. (2009). "Using Bioclimatic Envelopes to Identify Temporal Corridors in Support of Conservation Planning in a Changing Climate," *Forest Ecology and Management*, 258: S64–S74.

Russell, R. S., and Smith, M. (2009). "Reeling in Outsourcing: Evaluating Supply Chain Risk and Reward under Economic Uncertainty," *Journal of Service Science*, 2/1: 57–64.

Sale, K. (2000). *Dwellers in the Land: The Bioregional Vision*. Athens, GA: University of Georgia Press. 217 pp.

Sathe, T. V. (2009). *A Textbook of Forest Entomology*. New Delhi: Daya Publishing House. 223 pp.

Schmitt, S., Maréchaux, I., Chave, J., et al. (2020). "Functional Diversity Improves Tropical Forest Resilience: Insights from a Long-Term Virtual Experiment," *Journal of Ecology*, 108/3: 831–43.

Sharma, S., Dutta, T., Maldonado, J. E., et al. (2013). "Forest Corridors Maintain Historical Gene Flow in a Tiger Metapopulation in the Highlands of Central India," *Proceedings of the Royal Society B: Biological Sciences*, 280/1767: 20131506.

Simard, S. W., Beiler, K. J., Bingham, M. A., et al. (2012). "Mycorrhizal Networks: Mechanisms, Ecology and Modelling," *Fungal Biology Reviews*, 26/1: 39–60.

Simard, S. W., and Durall, D. M. (2004). "Mycorrhizal Networks: A Review of their Extent, Function, and Importance," *Canadian Journal of Botany*, 82/8: 1140–65.

Simard, S. W., Hagerman, S. M., Sachs, D. L., Heineman, J. L., and Mather, W. J. (2005). "Conifer Growth, *Armillaria ostoyae* Root Disease and Plant Diversity Responses to Broadleaf Competition Reduction in Temperate Mixed Forests of Southern Interior British Columbia," *Canadian Journal of Forest Research*, 35/4: 843–59.

Simard, S., Martin, K., Vyse, A., and Larson, B. (2013). "Meta-Networks of Fungi, Fauna and Flora as Agents of Complex Adaptive System," in C. Messier, K. J. Puettmann, and K. D. Coates (eds), *Managing Forests as Complex Adaptive Systems: Building Resilience to the Challenge of Global Change*. New York: Earthscan, 133–64.

Simard, S. W., Perry, D. A., Jones, M. D., Myrold, D. D., Durall, D. M., and Molina, R. (1997). "Net Transfer of Carbon between Ectomycorrhizal Tree Species in the Field," *Nature*, 388/6642: 579–82.

Simard, S., and Vyse, A. (2006). "Trade-Offs between Competition and Facilitation: A Case Study of Vegetation Management in the Interior Cedar–Hemlock Forests of Southern British Columbia," *Canadian Journal of Forest Research*, 36/10: 2486–96.

Soulé, M. E. (ed.). (1987). *Viable Populations for Conservation.* Cambridge: Cambridge University Press. 189 pp.

Ste-Marie, C., Nelson, E. A., Dabros, A., and Bonneau, M. E. (2011). "Assisted Migration: Introduction to a Multifaceted Concept," *Forestry Chronicle*, 87/6: 724–30.

Stralberg, D., Arseneault, D., Baltzer, J. L., et al. (2020). "Climate-Change Refugia in Boreal North America: What, Where, and for How Long?" *Frontiers in Ecology and the Environment*, 18/5: 261–70.

Su, Q., MacLean, D. A., and Needham, T. D. (1996). "The Influence of Hardwood Content on Balsam Fir Defoliation by Spruce Budworm," *Canadian Journal of Forest Research*, 26/9: 1620–8.

Tarroux, E., and DesRochers, A. (2011). "Effect of Natural Root Grafting on Growth Response of Jack Pine (*Pinus banksiana*; Pinaceae)," *American Journal of Botany*, 98/6: 967–74.

Taylor, R. G., and Fortson, J. C. (1991). "Optimum Plantation Planting Density and Rotation Age Based on Financial Risk and Return," *Forest Science*, 37/3: 886–902.

Taylor, S. P., Delong, C., Alfaro, R. I., and Rankin, L. (1996). "The Effects of Overstory Shading on White Pine Weevil Damage to White Spruce and its Effects on Spruce Growth Rates," *Canadian Journal of Forest Research*, 26/2: 306–12.

Taylor, S. W., Carroll, A. L., Alfaro, R. I., & Safranyik, L. (2006). "Forest, climate and mountain pine beetle outbreak dynamics in western Canada," in L. Safranyik and W. R. Wilson (eds), *The Mountain Pine Beetle: A Synthesis of Biology, Management, and Impacts on Lodgepole Pine*. Victoria, BC: Canadian Forest Service. pp. 67–94.

Teste, F. P., and Simard, S. W. (2008). "Mycorrhizal Networks and Distance from Mature Trees Alter Patterns of Competition and Facilitation in Dry Douglas-Fir Forests," *Oecologia*, 158/2: 193–203.

Townsend, P. A., and Masters, K. L. (2015). "Lattice-Work Corridors for Climate Change: A Conceptual Framework for Biodiversity Conservation and Social–Ecological Resilience in a Tropical Elevational Gradient," *Ecology and Society*, 20/2: 1.

Turner, N. J. (2005). *The Earth's Blanket: Traditional Teachings for Sustainable Living*. Vancouver: Douglas & McIntyre. 298 pp.

Twyford, A. D. (2018). "Parasitic Plants," *Current Biology*, 28/16: R857–R859.

United Nations (1992). *Report of the United Nations Conference on Environment and Development (Rio de Janeiro, 3–14 June 1992), Annex I: Rio Declaration on Environment and Development*. New York: United Nations General Assembly. 5 pp. Available online at https://www.un.org/en/development/desa/population/migration/generalassembly/docs/globalcompact/A_CONF.151_26_Vol.I_Declaration.pdf (accessed July 24, 2023).

Van de Peer, T., Verheyen, K., Baeten, L., Ponette, Q., and Muys, B. (2016). "Biodiversity as Insurance for Sapling Survival in Experimental Tree Plantations," *Journal of Applied Ecology*, 53/6: 1777–86.

Van Der Heijden, M. G., and Horton, T. R. (2009). "Socialism in Soil? The Importance of Mycorrhizal Fungal Networks for Facilitation in Natural Ecosystems," *Journal of Ecology*, 97/6: 1139–50.

van Lierop, P., Lindquist, E., Sathyapala, S., and Franceschini, G. (2015). "Global Forest Area Disturbance from Fire, Insect Pests, Diseases and Severe Weather Events," *Forest Ecology and Management*, 352: 78–88.

Várkonyi, G., Kuussaari, M., and Lappalainen, H. (2003). "Use of Forest Corridors by Boreal *Xestia* Moths," *Oecologia*, 137/3: 466–74.

Vidal, M. M., Hasui, E., Pizo, M. A., et al. (2014). "Frugivores at Higher Risk of Extinction are the Key Elements of a Mutualistic Network," *Ecology*, 95: 3440–7.

Wagner, S., Nocentini, S., Huth, F., and Hoogstra-Klein, M. (2014). "Forest Management Approaches for Coping with the Uncertainty of Climate Change: Trade-Offs in Service Provisioning and Adaptability," *Ecology and Society*, 19/1: 32.

Walker, B., and Salt, S. (2006). *Resilience Thinking: Sustaining Ecosystems and People in a Changing World*. Washington: Island Press. 174 pp.

Walker, B., and Salt, S. (2012). *Resilience Practice: Building Capacity to Absorb Disturbance and Maintain Function*. Washington: Island Press. 226 pp.

Walters, C. J. (1986). *Adaptive Management of Renewable Resources*. New York: McGraw-Hill. 374 pp.

Wehling, S., and Diekmann, M. (2009). "Importance of Hedgerows as Habitat Corridors for Forest Plants in Agricultural Landscapes," *Biological Conservation*, 142/11: 2522–30.

Williams, B. K. (2011). "Passive and Active Adaptive Management: Approaches and an Example," *Journal of Environmental Management*, 92/5: 1371–8.

Williams, B. K., and Johnson, F. A. (2015). "Optimization and Resilience in Natural Resources Management," in C. Allen and A. Garmestani (eds), *Adaptive Management of Social– Ecological Systems*. Dordrecht: Springer, 217–33.

Williams, L. J., Paquette, A., Cavender-Bares, J., Messier, C., and Reich, P. B. (2017). "Spatial Complementarity in Tree Crowns Explains Overyielding in Species Mixtures," *Nature Ecology & Evolution*, 1/4: 1–7.

Williams, M. I., and Dumroese, R. K. (2013). "Preparing for Climate Change: Forestry and Assisted Migration," *Journal of Forestry*, 111/4: 287–97.

Woods, K. D. (2007). "Predictability, Contingency, and Convergence in Late Succession: Slow Systems and Complex Data-Sets," *Journal of Vegetation Science*, 18/4: 543–60.

Xu, L., and Kajikawa, Y. (2018). "An Integrated Framework for Resilience Research: A Systematic Review Based on Citation Network Analysis," *Sustainability Science*, 13: 235–54.

Yan, B., Chen, B., Wu, L., Zhang, X., and Zhu, H. (2021). "Culture, Institution, and COVID-19 First-Response Policy: A Qualitative Comparative Analysis of Thirty-One Countries," *Journal of Comparative Policy Analysis: Research and Practice*, 23/2: 219–33.

Zolli, A., and Healy, A. M. (2012). *Resilience: Why Things Bounce Back*. New York: Simon & Schuster. 323 pp.

Zhang Y., Chen, H. Y. H., and Reich, P. B. (2012). "Forest Productivity Increases with Evenness, Species Richness and Trait Variation: A Global Meta-Analysis," *Journal of Ecology*, 100: 742–9.

Zhu, C. G., Li, W. H., Chen, Y. N., and Chen, Y. P. (2018). "Characteristics of Water Physiological Integration and its Ecological Significance for *Populus euphratica* Young Ramets in an Extremely Drought Environment," *Journal of Geophysical Research: Atmospheres*, 123/10: 5657–66.

CHAPTER 6

Climatic Influences and Forest Vulnerabilities

Forewarned is forearmed.

Abraham Tucker (1768)

6.1 Atmospheric effects on forests

Most of the temperature indicators that are being monitored around the world are showing evidence of a directional warming over the last several decades, and those trends are expected to continue for the foreseeable future (IPCC 2022). We know that the distribution, productivity, and persistence of many plants, animals, and ecosystems is strongly related to the balance of temperature and precipitation. This is the basis for various systems by which biomes and ecozones are classified (e.g. Figure 6.1). Calibrating those relationships constitutes a dominant theme in the disciplines of biogeography, geographic ecology, and vegetation classification (MacArthur 1984; Lomolino et al. 2016). Regional meso-climates reflect the role of nearby oceans or large lakes, and of mountain ranges and prevailing winds. Climatic and ecological differences over space or time can be especially apparent in mountainous terrain, because temperature and available moisture are further modified by slope position and aspect through their effects on insolation and drainage (MacKenzie and Meidinger 2018). These environmental parameters had collectively been considered largely static, as the dominant geographic features surely persist, but atmospheric changes over the last century are now changing the regimes of temperature, precipitation, and air movement that forests and people experience. This chapter reviews some of the observed and anticipated effects of this changing climate, setting the context for Chapter 7, in which suggested forest management responses are outlined.

How temperature and precipitation regimes govern the distribution and productivity of ecosystems is complex and can be described in several different ways. For example, the "temperature response" of species can be expressed in quite different sets of optima and tolerances as related to mean annual temperature, minimum winter temperature, maximum summer temperature, cumulative heat sums above a threshold such as 5 °C, or length of the frost-free period. Those relative differences in response to weather conditions can result in some coexisting species having growth and competitive advantages in some years, while other species have the advantage under the weather conditions of other years. Over the long run, those differences are eventually expressed as species range limits, which may also reflect biogeographic dispersal history, and constraints owing to predation and disease.

The availability of water is a strong determinant of plant growth, ecosystem productivity, and the range limits of many species. The climatic availability or absence of water can be portrayed in terms of mean annual precipitation, rainfall during the hottest month, days between periods of significant rainfall events, or various indices relating temperature to precipitation to indicate evaporative losses. Those climatic factors are further modified by slope aspect and slope position, and by

Resilient Forest Management. Philip J. Burton, Oxford University Press. © Philip J. Burton (2025). DOI: 10.1093/oso/9780198832997.003.0006

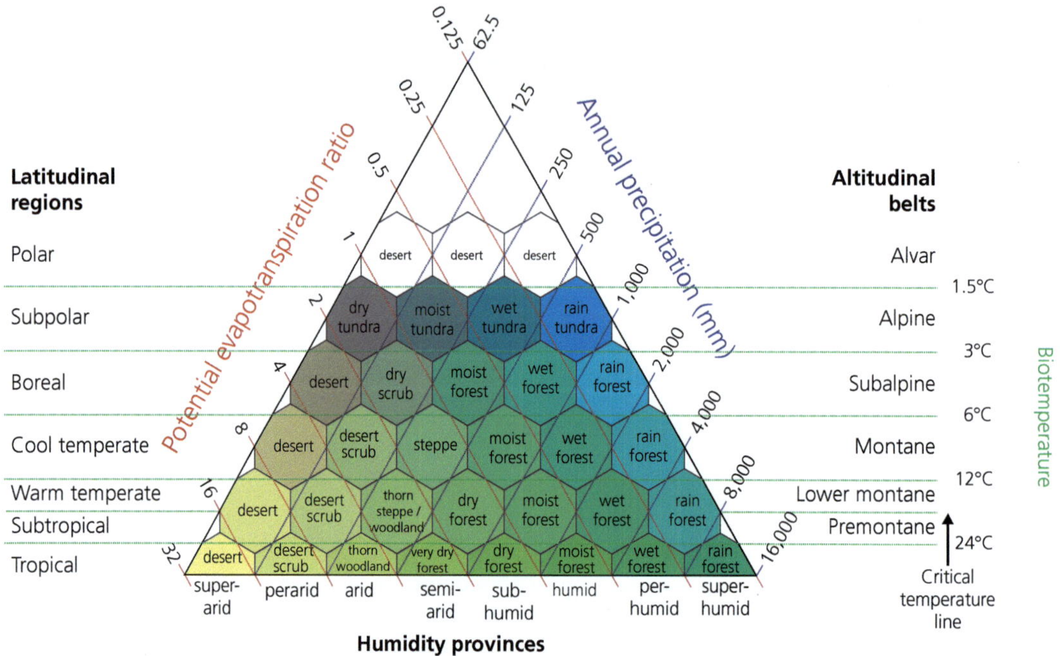

Figure 6.1 Life zones of the earth, following the classification scheme of L. R. Holdridge (Holdridge (1947, 1967); graphic by Peter Halasz, reproduced under CC BY-SA 2.5, https://creativecommons.org/licenses/by-sa/2.5, via Wikimedia Commons).

soil depth, coarse fragment content, and texture to determine a site's water-holding capacity. Plant species differ in their ability to limit water loss through stomatal closure, in metabolic differences permitting survival over long periods of drought, and in morphological adaptations allowing access to deep soil water. Many different physiological and phenological adaptations to extremes of temperature and precipitation govern the survival, growth, and reproduction of different species. So, even though a wide array of climate indices can be calculated, they are best considered simplistic proxies for describing the complex of climatic drivers and constraints that govern the distribution and productivity of species, communities, and ecosystems (see Box 6.1).

There are many components and processes of forest ecosystems that are directly or indirectly affected by conditions and trends in the earth's atmosphere. In any given year, drought, heat, soil saturation, growing-season frost, windstorms, and snow press can all limit the growth and survival of trees and other plants. While industrial air pollution, ground-level ozone, and acid rain have been subjects of concern with respect to the health of forest ecosystems—and continue to be so in some areas—the worldwide focus is now on the direct and indirect effects of atmospheric carbon dioxide (CO_2) concentrations (Hamelin and Innes 2017). Elevated CO_2 levels, primarily due to the burning of fossil fuels, were originally expected to have a fertilization effect on plant growth, with more feedstock available for sunlight to turn into sugars and cellulose (Wittwer and Strain 1985). While this effect is evident in some annual agricultural crops and juvenile trees, it has generally not been found to prevail in mature trees and other perennial plants in the wild, where CO_2 is not usually the limiting factor to plant growth (Jiang et al. 2020). Early increases in photosynthesis and growth are usually followed by physiological and carbon-allocation adjustments, including enhanced water-use efficiency, but elevated respiratory losses typically offset any gains (Bazzaz et al. 1993; Urban 2003; Jiang et al. 2020). Over the lifetime of trees and stands, growth limitations due to shading, nutrient deficiencies, and

Box 6.1 Climatic constraints, bio-envelope modelling, and the distribution of species and ecosystems

The current geographic range occupied by a species, a recognized vegetation type, or a particular ecosystem is not always limited by climatic constraints. Nonetheless, the climatic conditions prevailing within a current range indicate a set of conditions that are demonstrably acceptable to the biological or ecological feature of interest (e.g. Figure 6.2). Using climate interpolation methods (e.g. ClimateNA, Wang et al. 2016; https://climatena.ca/) to identify the various climatic attributes associated with such distributional limits is the basis for bioclimatic envelope or environmental niche modelling (Heikkinen et al. 2006). Once the environmental conditions prevailing within the range of a given species or feature are determined, the parameters of this "climatic niche" can be used to project potential future distributions as the climate changes. It is important to include the global (not just jurisdictional) range data for defining this climatic niche or its limits (bioclimatic envelope). However, many researchers work with the 90th or 95th percentiles—rather than extreme upper and lower limits—of various climate metrics to delineate core habitats, and in recognition that there may be errors in the calibrating database and climate interpolation algorithms (Franklin 2010; Booth et al. 2014).

The precise environmental conditions that determine the distributional limits of a species or forest type are not always obvious and can vary considerably around the perimeter of a bioclimatic envelope or set of range limits. Many of those variables are highly correlated, with those available for mapping and modelling not necessarily being the operative ones responsible for the repeated mortalities or reproductive failures that define the limits of a species' range. For example, projections of constraints to the westward distribution of the Coastal Douglas-Fir biogeoclimatic zone in British Columbia identify frost-free days and extreme minimum temperature (temperature regime) constraints at its northern limits, compared to low levels of heat:moisture ratios and the climatic moisture deficit (moisture regime) constraints at the south end of Vancouver Island (see Figure 6.3). Future constraints on species or ecosystem distributions may or may not be the same as those prevailing today, but again are likely to vary at different parts of their ranges.

Although a mapped ecosystem or forest type might be treated as a discrete biophysical entry, it is likely that each species will die out or spread individually, according

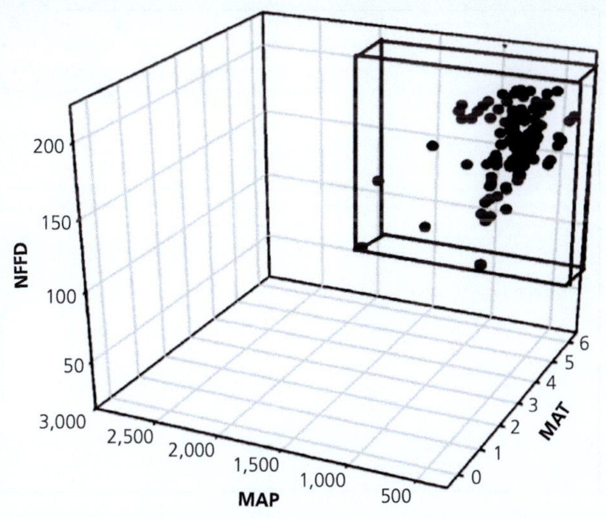

Figure 6.2 A simple bioclimatic envelope, in this case for the distribution of *Botrychium crenulatum* within the historic climate space (1961–90) found across the province of British Columbia, Canada. MAP is mean annual precipitation (mm), MAT is mean annual temperature (°C), and NFFD is number of frost-free days. In practice, additional climate attributes and indices are further used to narrow the conditions represented by the *n*-dimensional hypervolume of observed habitat locations (reproduced from Rose and Burton (2009).

Box 6.1 *Continued*

to its autecological tolerances. If climate stabilizes, familiar biological communities may reassemble themselves. But the bioclimatic envelopes that supposedly characterize an ecosystem or forest type are more correctly viewed as defining acceptable conditions for just a few dominant species.

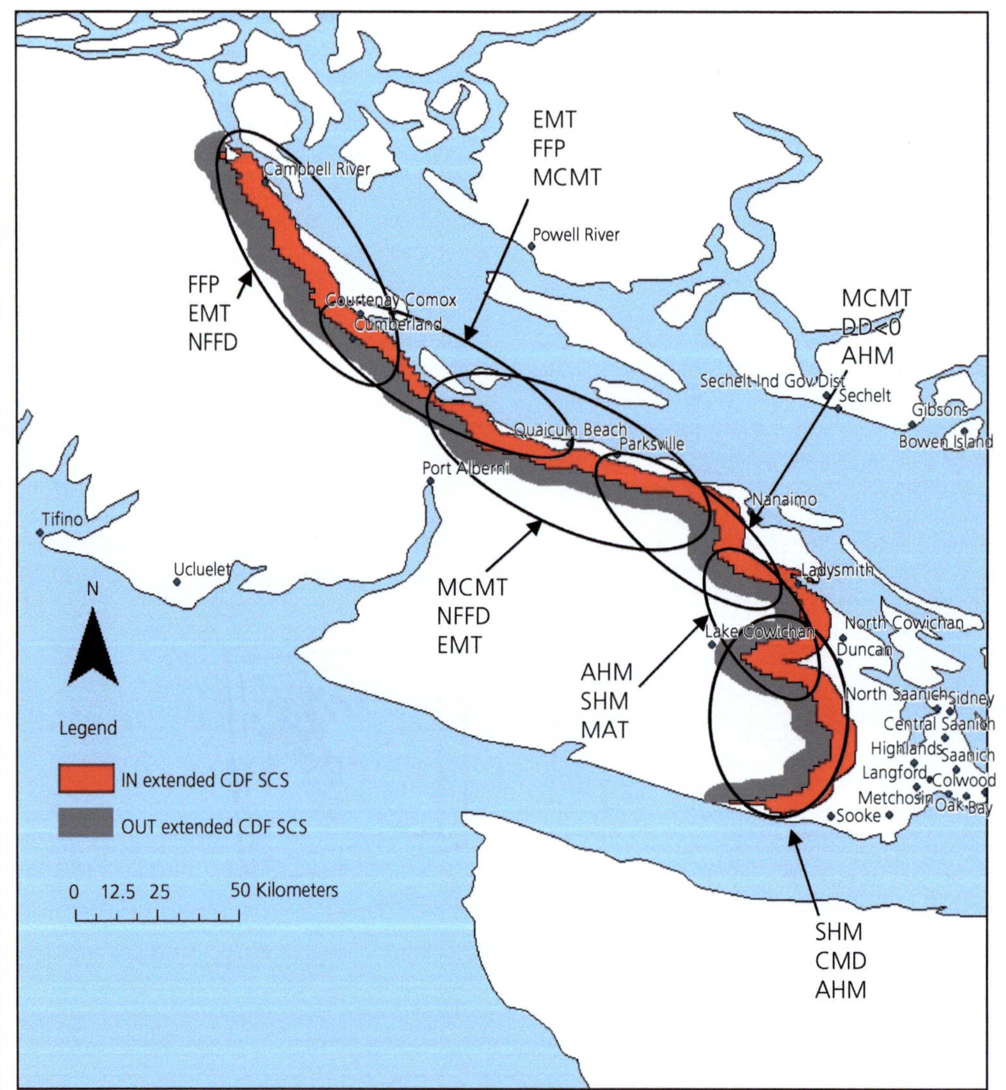

Figure 6.3 The three most significant climate variable differences from the south to the north end of the suitable climate space (SCS) for the Coastal Douglas-fir (CDF) biogeoclimatic zone on Vancouver Island, Canada (as determined by Klassen and Burton 2015). Predictive significance of climate variables is indicated by their listed order from highest to lowest. EMT = extreme minimum temperature, FFP = uninterrupted frost-free period, MCMT = mean cold monthly temperature, NFFD = number of frost-free days, DD<0 = degree-days less than 0 °C, AHM = annual heat:moisture index, SHM = summer heat:moisture index, MAT = mean annual temperature, and CMD = climatic moisture deficit (reproduced from Klassen (2012) with permission).

temperature extremes can often overwhelm any beneficial effect of CO_2 enrichment. Subtle interspecific differences related to tree age, shade tolerance, and water-use efficiency may be expressed in differential tree growth and ecosystem dynamics, especially in semi-arid forests (Huang et al. 2007). Ongoing research may yet confirm a CO_2 fertilization effect in fast-growing plantations and juvenile stands, but not mature, mixed-species forests. In general, we cannot expect to see widespread enhancement of forest productivity in response to CO_2 enrichment (van der Sleen et al. 2015; Girardin et al. 2016; Xiao et al. 2021; Norby et al. 2022).

With focal species such as trees and desired wildlife all having different tolerances to various expressions of temperature and available moisture, this presents a monumental challenge to the prediction of likely mortality events, growth differences, and range shifts. For example, the phenological cycle of tree growth, dormancy, and cold tolerance in temperate and boreal trees reflects a complex set of ecophysiological triggers related to heat sums (cumulative degree-days above thresholds that vary from species to species), chilling requirements in the autumn, and frequency of extreme cold events relative to average temperatures (Fuchigami et al. 1982; Burton and Cumming 1995; Menezes-Silva et al. 2019). Successful seed production by many insect-pollinated species (including many berry-producing species valued by birds, rodents, bears, and humans) depends on the alignment of warm rain-free periods during the critical periods of flowering (Lawson and Rands 2019), while species differ quite individualistically with regard to the particular weather conditions that best determine fruit production (Krebs et al. 2009). Similarly, the timing of many animal migrations and breeding seasons, often prompted by seasonal photoperiod changes (which, of course, remain constant regardless of climate change), depend on necessary levels of plant or insect development (which respond to weather cues) in their breeding grounds in order for offspring to thrive (Beard et al. 2019). As seasonal weather extremes become less predictable under climate change, misalignments of these several ecosystem components and relationships are already being observed (Renner and Zohner 2018; IPCC 2022). On top of these many direct and indirect effects on organisms, ecosystems are also subject to climate influences on natural disturbances such as wildfires, windstorms, and outbreaks of herbivorous insects (Seidl et al. 2017), as explored further in Chapter 8.

6.2 Forest change in the past

The world's forests have repeatedly responded to changing climates in the past. During periods of warmth and drought, species composition at a site has shifted to more heat- and drought-tolerant species, which we know from the composition and relative abundance of pollen extracted from the sediments of lakes and ponds (see Figure 4.8). Similarly, the width of tree rings in living trees—which can be correlated with contemporary meteorological records—and in well-preserved dead woody material show tree growth responses to changes in temperature and precipitation. Using pollen records, tree rings, and a variety of other proxy indicators (including corals, ice cores, and isotopic differences in sediments) to reconstruct paleoclimates, it is apparent the world last experienced conditions analogous to today's warmer climate in the mid-Holocene, six thousand to eight thousand years ago (Marcott et al. 2013). This means that many of the world's forests may already be maladapted to current conditions. Moderate scenarios for conditions projected to prevail by the end of the twenty-first century also have prehistoric precedents, but we have to reach back to the Pliocene Epoch, three million to six million years ago, to find those climate analogs (Burke et al. 2018). Forests and their myriad components have numerous mechanisms of resilience in response to disruptions and a changing environment (Falk et al. 2022), but the rate of current climate change is much more rapid than the shifts to which ecosystems adapted in the past.

In terms of biogeography and species conservation, forest species persisted by shifting their distribution to track climatic conditions over the course of hundreds of years. Such transitions occurred in fits and starts, with long lags related to the broad tolerances of mature trees and interruptions due to wildfire that could respectively delay or accelerate composition change (Ali et al. 2008). In this manner, entire biomes gradually shifted toward the equator in response to Pleistocene glaciation, and then gradually repopulated the landscape as glaciers receded, although often in unique species combinations that we do not see today (Delcourt and Delcourt 1987). Whether the global climate was warming or cooling, many species persisted in isolated refugia in which the local climate remained relatively stable or where the terrain (slope position and direction, site drainage) dampened some of the effects of temperature change. As the climate changed again, plant and animal species dispersed out of those refugia and recolonized their former ranges and took advantage of a newly suitable habitat (Pielou 1991). But, unlike conditions thousands of years ago, much of the world is now dominated by farms and cities and is criss-crossed by highways, greatly fragmenting the landscape and constraining the potential for species to find a suitable habitat safely and effectively. Furthermore, there are now eight billion people on the planet, who have great expectations of the goods and services provided by trees and forests, especially those centered around our existing communities. So, while the flux and wane of forest ecosystems in the past was natural and value free, the human species is now faced with the responsibility of having caused this latest round of climate disruption, and of trying to manage its effects on the world's forests.

6.3 Understanding and predicting a changing climate

Foremost among the agents of global change is our warming atmosphere, a planetary response to the release of CO_2 from burning fossil fuels, as accentuated by human population growth, resource consumption, and deforestation. With the harnessing and rapid exploitation of coal, petroleum, and natural gas over the last two centuries, we have upset the balance of the earth's biogeochemical systems. Elevated levels of CO_2, methane (CH_4), and other greenhouse gases (such as nitrous oxides, NO and NO_2) are now driving a difficult-to-stop trajectory of global atmospheric warming and widespread disruption of the planet's weather patterns. Addressing the climate-change crisis is the

challenge of the century. Unless humankind keeps average global air temperatures from rising more than 1.5 °C from preindustrial norms, we will be placing many unique and threatened ecosystems at very high risk, and temperatures over 2.0 °C will result in severe impacts from extreme weather events such as storms, droughts, and wildfires (IPCC 2022). Forests are clearly experiencing this warming climate already. Around the planet as a whole, 2024 experienced the hottest average surface air temperatures ever recorded; furthermore, each year from 2015 to 2024 has been warmer than any recorded since 1850 (NOAA 2025).

Early communications about "global warming" have been revised to refer instead to more nuanced but urgent messaging about "climate change," reflecting the fact that disrupted global ocean and atmospheric circulation systems can result in regional cooling as well as overall warming. For example, with accelerated release of cold water from the Greenland ice sheet and the Arctic ice cap, the Gulf Stream, the Atlantic meridional overturning circulation, and their associated weather systems are now (or soon will be) exerting a cooling trend on northern Europe (Palter 2015; Ditlevsen and Ditlevsen 2023). With polar regions warming about three times faster than temperate regions, the atmospheric temperature gradients that have historically constrained the jet stream are now less pronounced and predictable. It now seems to wobble and stall more than in the past, making it a matter of chance whether a location is north or south of the jet stream and whether prolonged weather systems are dominated by hot dry conditions or cool wet conditions (Trouet et al. 2018). While drought is generally associated with warmer climates, there is also greater evaporation from the world's oceans and a greater incidence of extreme tropical cyclones and "atmospheric rivers" (Kamae et al. 2021; IPCC 2022), which often dump massive amounts of rain and result in widespread flooding. Even where precipitation is greater, higher air temperatures can result in greater evaporation and a greater ability of the air to hold moisture in the form of humidity rather than falling as rain. Higher temperatures generate a non-linear increase in vapor pressure deficit (VPD, the drying power of air), resulting in a net reduction in water available to and retained by plants

(Jung et al. 2010; Allen et al. 2015). Collectively, these contrasting trends and the greater incidence of extreme weather have prompted some commentators to use the terms "climate wilding" or "climate chaos" to describe the worldwide phenomenon we are now experiencing. The take-home message is to be prepared for generally more extreme and unpredictable meteorological conditions and their associated effects.

There is considerable variability in the projections for future temperature and precipitation regimes owing to the uncertainty around the implementation of CO_2 emission reductions, natural climate variability, and how future climates are simulated. This is the reason they are termed projections rather than predictions. Many climate research labs around the world have developed their own version of global circulation models (GCMs), which use basic physical principles to describe the dynamics of air movement at multiple levels in the atmosphere and water movement at multiple depths of the world's oceans (Satoh 2016). Simulations incorporate thermodynamic atmospheric variables such as wind, temperature, humidity, and air pressure, and consider cloud and precipitation formation, plus the dynamics of shortwave and longwave radiation. Those drivers are modeled in a three-dimensional grid that couples the dynamics of the atmosphere with that of the oceans, land surface, and biosphere. Models have become increasingly sophisticated and reliable in the three or four decades over which they have been used and refined. They now address the complex interaction of factors such as multiple greenhouse gases (GHGs), volcanic aerosols and other emissions, variations in solar irradiance, tropospheric ozone, and changes in surface albedo (Arias et al. 2021). They have been tested and fine-tuned accurately to "backcast" climate conditions over the last 100–150 years for comparison with empirical observations. Backcasts and forecasts are often presented in a manner that contrasts natural and human drivers (for example, with or without the role of fossil-fuel emissions), and thus can evaluate different scenarios of future emission controls. It is largely the assumptions about the degree and timing of any global reductions in fossil-fuel emissions that are responsible for the wide range of potential climate futures that scientists are presenting.

The Intergovernmental Panel on Climate Change (IPCC) is a joint initiative of the World Meteorological Organization (WMO) and the United Nations Environment Programme (UNEP). It serves as the compiler of scientific information on global climate-change projections and impacts. The IPCC first met in 1988 and released its First Assessment Report (AR1) in 1990. Subsequent updates and special reports have been released intermittently, with the Sixth Assessment Report (AR6) completed in 2023; the IPCC website (https://www.ipcc.ch/) is the definitive source for updated global climate-change information. The terminology for future scenarios as described by climate scientists, ecologists, and the IPCC has changed over time. These have included double or triple the preindustrial CO_2 concentrations (estimated to have been 275–300 parts per million), nominal storylines for future GHG emission controls described according to the IPCC Special Report on Emissions Scenarios (SRES), and converting all potential natural and human effects on climates in terms of radiation forcing in W/m^2, described as representative concentration pathways (RCPs). The radiative forcing approach allows a direct comparison of the climate change effects of different GHGs without converting their concentrations to CO_2 equivalence. Most recently, the trajectory of these radiative forcings is described as shared socioeconomic pathways (SSPs) in the AR6 report. Although the assumptions and trajectories of change preclude SRES, RCP, and SSP scenarios from being analogous, Rogelj et al. (2012) provide a comparison of SRES (AR4) and RCP (AR5) scenario projections for the range of temperatures expected at the end of the century, and Lee et al. (2021) compare the projected temperature trajectories for RCP (AR5) and SSP (AR6) scenarios. In most cases, these various scenarios can be summarized as a business-as-usual (no emissions control, so accompanied by large and rapid changes in global temperature) scenario, a low-emissions scenario, and one or more intermediate scenarios, as illustrated in Figure 6.4. Common to all projections are that temperature and precipitation are both expected to increase, that precipitation projections are more variable than those for temperature, and the alternative scenarios generate little difference among them in the near term though they diverge strongly over time.

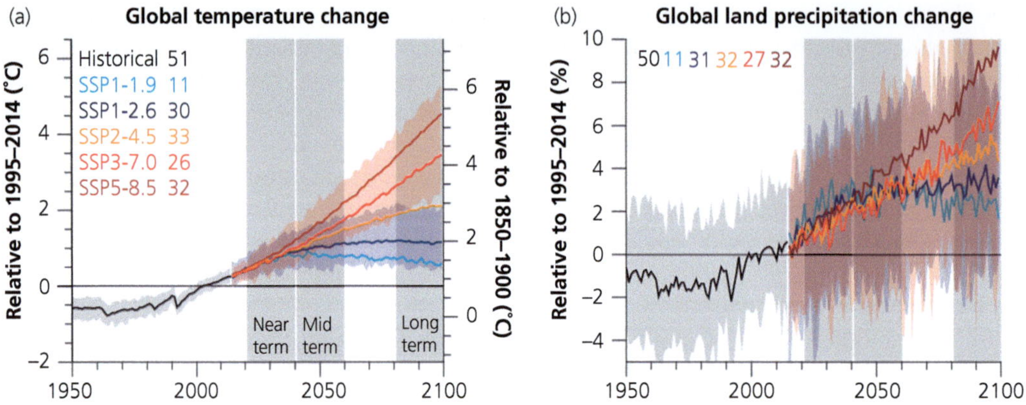

Figure 6.4 Selected indicators of global climate change from CMIP6 (IPCC's Coupled Model Intercomparison Project Phase 6) historical and scenario simulations: (a) Global surface air temperature changes relative to the 1995–2014 average (left axis) and relative to the 1850–1900 average (right axis; offset by 0.82 °C). (b) Global land precipitation changes relative to the 1995–2014 average. The number after the dash (panel a) for each of five shared socioeconomic pathways (SSPs) denotes the assumed level of radiative forcing in W/m^2, analogous to RCP pathways. The shadings around the SSP1-2.6 and SSP3-7.0 curves show 5–95% ranges. The number of simulations using each scenario is specified by the separate colored number in the upper left of each panel (Lee et al. (2021: figure 4.2), part of the IPCC AR6 Working Group I report; reproduced with permission).

Because of the stochastic nature of GCM models, trajectories of climate change can vary somewhat with each simulation run. There are also slight structural differences and priorities for model parameterization and testing reflected in the GCMs developed by different research groups. For example, the CSIRO model does a good job of explaining climatic patterns in Australia and some other southern hemisphere locations but is less successful at reconstructing historic climates in the northern hemisphere. Conversely, the Hadley model performs well for the British Isles, western Europe, and northwestern North America (northern hemisphere regions strongly affected by maritime influences). Those differences are generally averaged out by presenting "ensemble" model outputs, averaging the many models and model runs that use any given scenario to highlight the trends in projected climate space on which they agree.

The coarse horizontal resolution (1° to 4° of latitude and longitude, *c*.100–400 km) of most GCMs limits the reliability of their projections to any given forest management area. For local predictions, best practices have moved from selecting one GCM model over another and instead employ regional circulation models (RCMs) that take GCM model output to inform the simulation of air-mass movements and other dynamics as influenced by terrain and maritime/continental gradients not represented in GCMs (Feser et al. 2011). Other approaches involve statistical downscaling of GCM output, in which local weather-station correlations and elevation differences are used to generate scale-free, site-level estimates of historical, present, or future conditions (e.g. Wang et al. 2016).

Many analysts recommend working with the widest range of possible futures as projected by different models and socioeconomic scenarios (Farjad et al. 2019). Other researchers emphasize the middle-of-the-road scenarios, which are generally realistic for the near- and mid-term periods. But it can also be argued that many modeled scenarios (including those with immediate and widespread reductions to global GHG emissions) are now clearly unrealistic, as it is unlikely that we will meet the 2015 Paris Agreement goals of limiting temperature rise to 2.0 °C (or preferably 1.5 °C) above preindustrial levels (Raftery et al. 2017). In other words, we are likely to experience average global temperature rise at least 3–5 °C by the end of the twenty-first century (see Figure 6.4), and temperatures will not level off there but can be expected to rise for another century or two (Lee et al. 2021). So practitioners should probably emphasize the moderate-to-severe range of changes in global temperature and other climate changes when planning for the future forest.

6.4 Assessing impacts and vulnerabilities

In addition to regionally specific effects of atmospheric warming in different parts of the world, countries and regions differ in the degree to which they are dominated by natural ecosystems, agricultural lands, and built infrastructure. Consequently, we have seen the emergence of national, regional, and state- or province-level assessments of the expected effects of climate change; some of these are listed in Table 6.1. As well as reporting regionally relevant projections of anticipated changes in basic meteorological indicators (for example, mean annual temperature and precipitation, the frequency of extreme events), these reports often document indications of climate-change impacts on environmental and socioeconomic values that are already evident. The presence of trends in various indicators helps inform vulnerability assessments, which identify key components of the natural and built environment for which a changing climate is likely to affect the viability and sustainability of human activities. Such warnings frequently include the likelihood of greater heat and drought stress (to be experienced by trees, crops, wildlife, livestock, and humans), expanding ranges of invasive pests and pathogens, more wildfires, and other extreme events such as storms and flooding. With such forewarnings, citizens, corporations, and public agencies and institutions are urged to prepare with various adaptation strategies. National adaptation programs often refer to the needs and roles of forests and the forestry sector as contributing to overall environmental and economic policy adaptation direction (Roberts et al. 2009; Locatelli et al. 2010). Some of the more likely and widespread climate-change impacts and vulnerabilities as they relate to forests and forest management have already been referred to, while Chapter 7 summarizes some suggestions for how forests and forest management can respond as best they can.

When faced with an identifiable risk or threat, one of the first steps one should take is an assessment of one's vulnerabilities to that threat, where "vulnerability" denotes a combination of susceptibility and ability to cope. Climate change is a multidimensional threat, with several interacting mechanisms by which a rapid change in atmospheric conditions can disrupt current forest ecosystem processes and current forest management activities with broad levels of uncertainty (Patt et al. 2005). Nonetheless, some locations and some forest management policies and practices are more sensitive to climatic disruption than others, so place-based vulnerability assessments are widely recommended as the first step in implementing climate-change adaptation strategies for forestry (e.g. Upgupta et al. 2015; Thorne et al. 2018). The results of vulnerability assessments then help direct what kinds of resilience-enhancing actions should be prioritized for what elements or locations of the system.

Forest vulnerability assessments can be undertaken using a variety of methods, usually consisting of a combination of quantitative measurements and qualitative rankings. In most applications, some measure of "climate exposure" and "climatic sensitivity" is used to derive anticipated climate-change impacts. The theoretical basis for climate vulnerability assessment maps nicely onto the conceptual basis for system resilience: the geographic proximity of a species, ecosystem, or process to the trailing (retracting) boundary of its climatic niche is analogous to measuring system precariousness based on its proximity to the edge of a stability domain (represented by the ball near the topographic ridge in Figure 4.3(c)). Similarly, ecological vulnerability can be inferred from multivariate proximity of an ecosystem's composition and structure to that of an alternative state (Lexer and Seidl 2009).

Downscaled climate-change projections are typically employed to evaluate whether future possible temperature and precipitation regimes are likely to be outside the current bioclimatic envelopes for species or ecosystems of interest (Rose and Burton 2009; Araújo and Peterson 2012). The magnitude or speed of anticipated changes can be expressed as "climate change velocity" (Loarie et al. 2009), which implies that there is greater potential for climate stress and less ability to adapt when environmental change is rapid. The degree to which other stressors, such as wildfires or storm damage (see Chapter 8) are associated with the expected future climate can

Table 6.1 Some examples of climate change effects and vulnerabilities identified at national or regional levels

Nation or region	Report title	Notes on forests or forestry	Reference[*]
Australia	*State of the Climate 2022*	Continued increase in the number of dangerous fire weather days and a longer fire season for southern and eastern Australia; more time in drought, but with short-duration heavy-rainfall events and flooding; fewer but more intense tropical cyclones	Australian Government (2022)
Brazil	*Climate Risk Country Profile: Brazil* [†]	Intensive deforestation and forest fires in the Amazon are altering precipitation patterns; increasing aridity threatens forest cover and unique biodiversity; sea-level rise threatens mangrove ecosystems	World Bank Group (2021)
Canada	*Canada in a Changing Climate: Synthesis Report*	Increased severity of heatwaves, which contribute to droughts and wildfires; northern forests are already experiencing permafrost thaw, ground subsidence, and loss of winter roads; the frequency, severity, and size of wildfires and pest infestations are increasing, with impacts on carbon storage and biodiversity	Lulham et al. (2023)
Europe	*Europe's Changing Climate Hazards: An Index-Based Interactive EEA Report*	Mean air temperature, growing degree-days, and humid heatwaves will continue to increase; changes in precipitation and wild weather will vary regionally, with southern Europe most at risk of droughts and wildfire, while central Europe is expected to experience more flooding; increasing frequency and intensity of storms is projected for northern and central Europe	European Environmental Agency (2021)
France	*Le Plan national d'adaptation au changement climatique 2018–2022 (PNACC-2)*	Expectations of more intense floods and droughts, plus increased risk of forest and brush fires; forests, soils, biodiversity, and natural hazards are among the priorities for research	République Française (2018)
Germany	*Klimawirkungs- und Risikoanalyse 2021 für Deutschland (Kurzfassung)*	Severe damage by dry and hot years in combination with storm and bark beetle outbreaks constitute the greatest risks, but forest fires are increasingly of concern	Kahlenborn et al. (2021)
New Zealand	*National Climate Change Risk Assessment for New Zealand: Main Report*	Changes in temperature, rainfall, wind, and drought threaten the long-term composition and stability of indigenous forest ecosystems	NZME (2020)
Russia	*Russian Forests and Climate Change*	Rising temperatures, permafrost thaw, and large forest fires already being experienced, and are expected to increase	Leskinen et al. (2020)
Southern Africa	*Climate Risk and Vulnerability: A Handbook for Southern Africa, Second Edition*	Higher temperatures and more severe droughts expected to result in tree mortality and species range shifts in a region already water limited; forest fires on the rise, including in plantations; indigenous forests and biodiversity already threatened by deforestation, which also accentuates flooding	Davis-Reddy and Vincent (2017)
United Kingdom	*Independent Assessment of UK Climate Risk: Advice to Government for the UK's Third Climate Change Risk Assessment (CCRA3)*	Hotter, drier conditions reduce the functioning of peatlands and forests and threaten their existence; they face erosion from wind and rain, increased risk of fire damage; severe drought and moisture deficits will have major impacts on biodiversity, agriculture, and forestry; new species and longer growing seasons are possible	UKCCC (2021)
United States of America	*Fifth National Climate Assessment: Overview (Chapter 1) and Forests (Chapter 7)*	Extreme heat events, drought, flooding, wildfire, and hurricanes are becoming more frequent and/or severe; more frequent and intense precipitation events, especially in the Northeast and Midwest; some coastal forests are dying owing to sea-level rise; degradation and extinction of local flora and fauna is expected in vulnerable ecosystems such as montane rainforests	Jay et al. (2023)

[*] All references are available on the World Wide Web, but locations may change over time, so may require web browser searches to locate. Note that the homepage hosting each of these publications often includes a number of related reports and guidelines—e.g. www.dcceew.gov.au/climate-change, www.changingclimate.ca, www.globalchange.gov, and www.eea.europa.eu/en/topics/at-a-glance/climate [†] The climateknowledgeportal.worldbank.org website provides concise independent overviews of climate change risks and vulnerabilities by country and major watersheds, with an emphasis on economic sectors; Brazil is one of seventy climate-risk country profiles that were available as of February 2024.

add another level of risk to forests, forest communities, or the forest sector (Thorne et al. 2018). Exposure to high levels of change in the climate does not alone denote susceptibility, but rather depends on the particular sensitivities or robustness of the system, as indicated by documented response functions to changes in temperature, precipitation, or disturbance.

Anticipated climate-change impacts are then weighed against "adaptive capacity," which denotes the ability to accommodate or cope with potential climate change impacts with minimal disruption (Turner et al. 2003; Engle 2011). A commonly accepted framework for vulnerability assessments is to score vulnerability as the product of exposure and sensitivity, minus adaptive capacity. Although indicators of each consideration are difficult to measure and cannot be arithmetically combined, simple scoring systems (ranking each as low, medium or high; 1, 2, or 3) are feasible and can be useful (Brandt et al. 2017; Janowiak et al. 2018). The abilities of a system to resist, recover, adjust, reconfigure, or transform (per Chapter 4) may all vary in their sensitivity to a given battery of potential disruptions. Collectively, the various modes of inherent system resilience also determine its overall adaptive capacity. For species and ecosystems, adaptive capacity can be

indicated by the climatic range of their geographic distribution, by species and genetic diversity, by tolerance to disturbance, and by population or landscape connectivity (Brandt et al. 2017). At the species level, various life-history traits that reflect evolutionary adaptation to the direct and indirect stressors expected to be associated with future climates are also indicative of adaptive capacity (Aubin et al. 2016).

The combination of adaptive capacity versus impacts denotes a system's vulnerability (see Figure 6.5(a)). Potential effects and impacts can be informed by technical assessments, statistical relationships, and the simulation modelling of socioecological systems. The ranking of sensitivities, adaptive capacities, and the uncertainties associated with each can be more problematic, but need not depend on ranking by experts. Whether impacts are expected to be high or low, local stakeholders may be best positioned to judge whether such anticipated changes are tolerable to a community, institution, agency, or sector. Combined with other biophysical and geopolitical uncertainties, variation in expert and stakeholder opinion contributes to the overall uncertainty in vulnerability assessment, which should also be expressed as the degree of confidence in the assessment (Tonmoy et al. 2014; see Figure 6.5(b)).

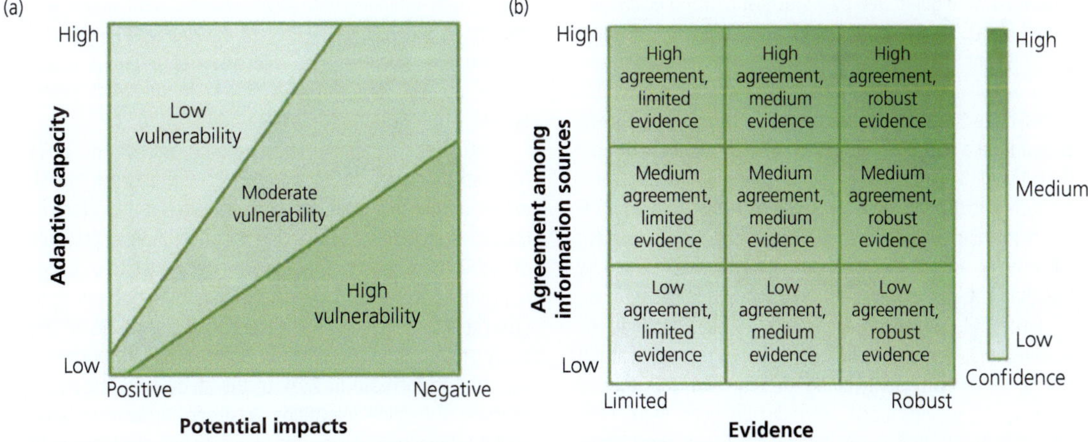

Figure 6.5 (a) Vulnerability assessments, often based on judgment calls by expert panelists, balance potential impacts against the adaptive capacity of the system. (b) Confidence in a vulnerability assessment depends on both the robustness of evidence available and the level of agreement among multiple sources of information, and should be reported as part of the assessment (reproduced from Janowiak et al. (2018); public domain, courtesy of US Department of Agriculture).

Vulnerability assessments of forests to climate change have typically employed a fairly standard set of indicators, often related to the criteria and indicators of sustainable forest management (Innes 2017). For example, Seidl et al. (2011) evaluated simulated levels of future primary productivity, timber stocks, carbon stocks, biodiversity, and disturbances as informed by downscaled climate-change scenarios to indicate the relative exposure and sensitivity of Austrian forests to climate change. Those impacts were assessed in the context of adaptive capacity as indicated by the relative position of tree species in their bioclimatic niche, silvicultural flexibility, and the cost of management, with weightings and thresholds of detection and tolerance determined in multidisciplinary workshops (see Box 6.2). Similarly,

Halofsky et al. (2011) simulated runoff, streamflow, water temperature, soil moisture, vegetation, and wildfire under different climate model scenarios for the Olympic Peninsula of Washington, US. Authors worked with regional scientists and specialists to review the literature and model outputs to identify ecosystem components and locations expected to undergo the greatest change in the light of current management activities. Analysis included the assignment of climate-sensitivity scores for several species of wildlife, which were usually considered more sensitive when rated by expert opinion than by quantitative modelling (Halofsky et al. 2011). In general, climate-change indicators should be part of an overall program of monitoring and adaptive management (discussed further in Chapter 10), in which selected features of the socio-ecological

---

Box 6.2 A climate change vulnerability assessment for Austrian forests

The Austrian Federal Forests (AFF) constitute some 591,000 hectares of public lands, of which some 368,120 hectares are managed commercially. A 164,550-hectare subset of this land base was evaluated for its vulnerability to climate change by Seidl et al. (2011). Viewed through a sustainable forest management (SFM) lens, but with timber production a priority, eight indicators were identified and weighted by stakeholders. Expected changes in those indicators were generated by a stochastic hybrid ecosystem model (PICUS v1.41), with growth, disturbance, and stand development (including silvicultural manipulations) simulated. Simulations responded to climate-change projections under three SRES scenarios (A1B, A2, and B1) as downscaled from a regionally calibrated global circulation model (ECHAM 5). By 2100, it is projected that mean annual temperatures across the study area will increase by 2.8 to 3.9 °C (depending on SRES scenario, and more in alpine areas), while mean annual precipitation is projected to change by less than 2%, but with significantly less rain in the summer.

Five SFM "sensitivity indicators" were identified by stakeholders: forest productivity, timber stocks, carbon stocks, biodiversity (with snag density and tree-species diversity as subindicators), and disturbances (bark beetle, windthrow, and snow breakage). Those sensitivities are balanced against adaptive capacity, denoted by three "state indicators:" tree-species position within its fundamental niche, silvicultural flexibility (with adapted species

composition and production goal subindicators), and cost intensity (silvicultural and harvesting costs). It should be noted that the selection of indictors, detection tolerances, acceptability tolerances, and their relative weightings in indicator aggregation were determined at a series of workshops. Those successive workshops were peopled by internal AFF staff (strategic management team, nature conservation experts, and field managers) supplemented with input from World Wildlife Fund (WWF) Austria staff and scientists from Vienna's University of Natural Resources and Applied Life Sciences; i.e., this was not a public stakeholder process, but rather a series of mediated expert panels.

Simulated aspatially for fifty-two assessment units (defined by ecoregion, elevation belt, and bedrock types), model output was evaluated against agreed-upon thresholds for each indicator. Following the precautionary principle of highlighting system responses most sensitive to a changing climate, forest productivity is in danger of decreasing to unacceptable levels (>15% reduction) across 62% of the study area by 2100. Conversely, forest productivity can be expected to increase in 23% of the study area, especially at lower subalpine elevations. A strong increase in bark-beetle activity can be expected, which (when combined with compulsory timber-salvage practices) will result in reduced standing deadwood and consequent impacts on biodiversity. Though not evaluated in a sophisticated manner, wind

---

Box 6.2 *Continued*

and snow damage is expected to be minimal. Tree niche occupancy is satisfactory for 93% of the study area, and silvicultural flexibility exists for promoting beech (*Fagus sylvatica*) instead of spruce (*Picea abies*) at lower elevations, but costs are anticipated to increase. Overall (aggregated) vulnerability is already high in 6% of the study area but is projected to become high in 40% of the study area in the second half of the twenty-first century, particularly on the half of the area underlain by calcareous geology.

With a few decades to prepare before profound vulnerability is expressed, there is time to develop and implement climate-change adaptation strategies, such as the alleviation of drought stress and bark-beetle susceptibility of spruce forests on calcareous soils in the montane elevation belt. While spruce productivity should increase at higher elevations, this opportunity will have to be balanced against a concurrent increase in the risk of bark-beetle attack. The authors conclude that there is a strong need for climate-change adaptation over much of the area (Seidl et al. 2011).

---

system are followed on the basis of their sensitivity, measurability, spatiotemporal scope, and relevance.

Specific to the tracking and assessment of climate impacts and vulnerability to inform adaptation in managed forests, Gauthier et al. (2014) suggest monitoring indicators selected from among the following categories (further discussed in Box 10.3), all of which can be subdivided further, each composed of several subindicators:

- climate drivers, e.g. peak summer temperature, climate moisture index
- the forest system
  - landforms and hydrology—e.g. area of subsidence due to permafrost thaw
  - natural disturbances—e.g. annual frequency and area of wildfires
  - species phenology—e.g. timing of tree budbreak and insect developmental stages
  - species distribution and abundance—e.g. alpine and polar treeline advancement
  - forest stand dynamics—e.g. tree regeneration, radial growth
  - edaphic conditions and processes—e.g. decomposition rates
- the human dimension—e.g. seasonal road access, hunting success

Change velocity in such indicators, combined with system susceptibility and adaptive capacity, can inform general strategies to promote resistance, recovery, adjustment, reconfiguration, or transformation. The status of those indicators then needs to be assessed (not just reported and filed away!) with respect to baseline conditions, thresholds denoting likely state transitions or tipping points, and undesired socioeconomic impacts. Where undesirable trends or precarious conditions are identified, adaptation responses may include biophysical adjustments in the forest system (for example, preferred species for forest regeneration) or changes to the policy and regulatory environment in which forests are managed (for example, with respect to the determination of sustainable harvest levels, logging methods, and logging season).

Some analysts have primarily taken a spatial overlay approach of summing vulnerability indicators. For example, Upgupta et al. (2015) used geospatial data on biological richness, disturbance, canopy cover, slope, and forest use by rural communities (all scored broadly 1, 2, or 3 in mapped grid cells) to rank the climate vulnerability of forest districts in the western Himalayan region of India. Notably, they identified important vulnerabilities at low elevations under current conditions, with high elevations being more vulnerable under future climate scenarios (Upgutpa et al. 2015). Other investigators and management teams have undertaken climate-change vulnerability assessments without the use of ecosystem simulation models or other computer tools. For example, Le Goff and Bergeron (2014) evaluated the climate vulnerability of three ecosystem-based forest management projects in

Quebec, Canada. Once climate-change scenarios and some of their ecological implications were summarized in a local context, they tabulated perceptions of management committees and those suggested by the literature and by a resource specialist, with a focus on nine biophysical processes and six social or management considerations. Le Goff and Bergeron (2014) conclude that ecological forestry initiatives are well positioned to cope with climate change because of their regional scope and an adaptive management framework. The use of many indicators and subindicators can provide a broad and robust means of assessing vulnerability, but it is common to aggregate those components into a single vulnerability index for purposes of communication and prioritizing adaptation (Choi et al. 2011; Seidl et al. 2011; Upgutpa et al. 2015). Stakeholder committees and public consultation are used to arrive at an acceptable consensus on how to weight the respective components of such aggregate indices and can be very important in prioritizing any follow-up response to vulnerability assessments.

The stressors or indicators considered in most climate-change vulnerability assessments include those widely recognized in the climate-change literature for decades: higher temperatures (especially in winter, triggering maladaptive phenologies in some cases); more variable and unpredictable precipitation (higher in some regions and seasons, lower in others); more drought stress; and more disturbance (floods, fires, pests, and storm damage). Some impacts and vulnerabilities due to the knock-on effects of heat, drought, and wildfires may be unexpected, but are evident during contemporary drought events (see Box 6.3).

Box 6.3 The 2023 drought in western Canada

A severe and extended period of low precipitation across much of western Canada during the spring and summer of 2023 (see Figure 6.6) is a precursor of things to come under a warming climate. Although in the long term we may see tree mortalities and range shifts, the immediate impacts of a single drought event can be dramatic.

Some forest-related impacts of the 2023 drought observed in the province of British Columbia included the following:

- plant and fungus dieback and reduced reproductive success;
  - ranchers demanding greater use of public rangelands to access sufficient forage for cattle, forced to purchase hay that was highly priced and in short supply;
  - very poor wild berry and mushroom crops;
    - reduced availability of food and medicinal plant materials used by Indigenous people and other subsistence and commercial foragers;
    - high numbers of black bears (*Ursus americana*) looking for food in towns, interacting with humans, owing to poor berry production;
- reduced stream flows and stream continuity, elevated water temperatures, reduced oxygen levels;
  - reduced salmon-spawning success;
- drought-stressed trees were more susceptible to insect attack and wildfire;

- potential forest regeneration failures owing to mortality of recently planted seedlings (yet to be confirmed); and
- the greatest annual area burned since records started being kept a century ago;
  - reductions in the sustainable timber supply;
    - destruction of severely burned timber;
    - less severely burned timber in remote areas is not recoverable by industry;
    - timber destroyed in the construction of fuel breaks;
  - some suspended forest operations owing to localized high wildfire risk;
  - widespread campfire bans owing to high fire risk;
  - evacuees from threatening wildfires using forest campgrounds elsewhere as temporary residences;
  - wildlife suffering from direct mortalities, food shortages, and habitat degradation;
  - potential loss of plant and animal populations having restricted distributions (yet to be confirmed);
  - health and activities of workers, recreationists, citizens, and wildlife restricted by smoke from local and distant wildfires; and
  - forest road, highway, and railway closures owing to wildfire and smoke, restricting supply chains for the forest products sector, curtailing recreational access and socioeconomic activity.

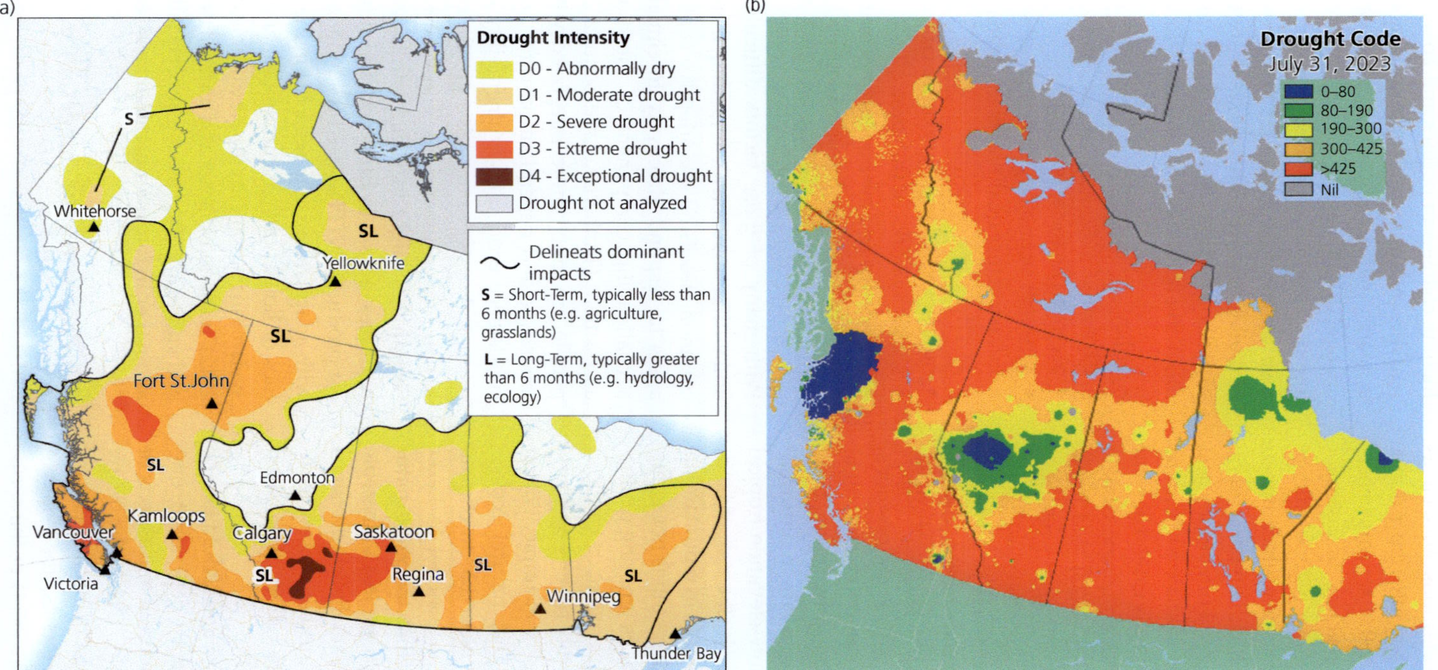

Figure 6.6 Maps portraying drought conditions across western Canada at the end of July 2023. (a) The "Canadian Drought Monitor" is based on ten different drought indicators, including precipitation percentiles, evaporation indices, satellite data, and the Palmer Modified Drought Index (from AAFC (2023), reproduced with permission of Agriculture and Agri-Food Canada); (b) the "drought code" is a representation of the cumulative and long-term moisture deficit in deep organic layers and large woody material as part of the Canadian Forest Fire Danger Rating System ((Hanes et al. 2023), Map by Canadian Forest Service, Natural Resources Canada, reproduced with the permission of the Department of Natural Resources, 2023). It is notable that a five-point scale of relative drought severity portrays more widespread moisture deficits for organic fuels (b) than for soil moisture (a).

Vulnerability assessments often evaluate current or near-term (0–30 years) conditions as well as medium-term (*c.*20/30–60 years into the future), or moderately long-term (*c.*50–80/100 years hence) projections. Many critical transitions in potential forest composition or disturbance regime emerge only in the long-term projections, for which uncertainty is greatest (Seidl et al. 2011; Duveneck et al. 2014; Wang et al. 2020). Owing to both biophysical and socioeconomic unknowns, climate projections and vulnerability assessments rarely extend beyond 2100. Yet most scenarios do not anticipate climate to have stabilized by then (IPCC 2023), and forest planners have to consider conditions beyond that date.

Even when prepared for a relatively small region, vulnerability assessments recognize the distinctive sensitivities and adaptive capacity of different forest types and different landscape positions. For example, assessment of sub-polar and high-elevation forests may need to consider freeze/thaw events and the balance of snow to rain in future climates, features irrelevant to low-elevation tropical and subtropical locales. Similarly, the risk of landslides and debris flows is important in mountainous landscapes and along river valleys but is not significant on level terrain. Generalities ignoring distinctive place-based socio-ecological contexts can lead to erroneous assessments of vulnerability and misguided adaptation responses. For example, Wilson and Turton (2011) conclude that natural ecosystems in Australia are more vulnerable to climate change than managed ones, in contrast to Lexer and Seidl (2009), who assert that the more natural forests of the Austrian Alps are less vulnerable than intensively managed forests. Policy decisions to promote one approach to resilient forest management over another need to reflect place-based assessments.

Vulnerability assessments can and should also be done for the forest sector and forest-dependent communities, focusing on the socioeconomic aspects of sustainable forest management. Driven by many of the same biophysical exposure components as vulnerability assessments of forests, the "human dimension" (per Gauthier et al. 2014) of the socio-ecological system can also be affected by climate impact pathways that have little effect on forest ecosystems. Those "external" considerations can include changes in politics and governance, global markets and trade barriers, and technological developments that may or may not respond to climate change. Consequently, there can be great variability among countries, communities, corporations, and institutions in the adaptive capacity side of the vulnerability equation. This makes the overall adaptive capacity of forest management a function of the following features (Johnston and Williamson 2007):

- policy or institutional flexibility, efficiency, or barriers
- access to financial resources
- access to technology and innovation
- infrastructure quality and maintenance (e.g. road and drainage systems)
- human capital (e.g. workforce education)
- social networks or isolation
- knowledge and awareness of potential impacts
- trust in scientists, forest managers, institutions
- perception of climate risk

With many of those factors difficult to measure, and with disagreements regarding their importance common, assessing the adaptive capacity of managers to respond effectively to climate change is a challenging component of forest vulnerability assessment (Timberlake and Schultz 2019). Where people are much more integrated with the forest landscape and dependent on it for both cash and subsistence economies, vulnerability of the socio-ecological system must give greater consideration to non-timber dimensions. Those forest values (see Section 2.2) might include hydrology and aquatic resources, access to fuelwood and non-timber forest products (NTFPs) used for food and medicine, with adaptive capacity reflected in the relative abilities of households, communities, and governance systems to respond to climate extremes (Sonwa et al. 2012). Such non-market and unregulated forest uses and values are often treated superficially or omitted altogether in forest vulnerability assessments.

6.5 Preparing for uncertainty

Perhaps the biggest challenge in responding to the known and forecast changes to our climate is deciding how to deal with the many uncertainties involved (Millar et al. 2007; Nocentini et al. 2017). Yes, we know that, at a global level, temperatures are rising and extreme weather events are becoming more frequent and intense. We have also gained a broad understanding that ecosystem shifts and discernible landscape change reflect disturbances and extreme weather events more often than average conditions. Furthermore, the directional trend (or "press") of a changing climate often sets up conditions for an extreme event (or "pulse") to be more likely and to be especially disruptive (Harris et al. 2018). When and where some disturbances and extreme events will emerge must still be considered largely random in nature, although risk and hazard mapping can assign relative probabilities. We can anticipate that floods and landslides are more likely on some terrains than others, and that the likelihood of both disturbances is accentuated by forest fires followed by extreme rainfall (de la Barrera et al. 2018, Burton et al. 2020). That knowledge can help us identify when and where forests and forest operations are more vulnerable to atmospheric rivers (high-precipitation events) than to heat domes (heat and drought stress; Allen et al. 2015), for example.

Some parts of the world may experience wetter rather than drier conditions, and some may even become cooler (at least for a period of time) in a warming world. Different species, ecosystem processes, and the sustainability of various forest values all have different thresholds for transitioning to profoundly altered states. Likewise, human communities of stakeholders have different tolerances to change, reflecting both cultural and economic differences. Such thresholds might be only weakly linked to the GCM output variables, and those projections vary among GCM models and SSP scenarios, further compounding uncertainty. Greenhouse-gas-reduction policies on behalf of the international community seem to have only sporadic implementation or effectiveness, yet nation state and corporate commitments are there: so which SSP scenario or ensemble of SSPs should be used for planning purposes? Or, indeed, should all possible futures, no matter how unlikely, be considered and planned for? Answers to such questions inevitably reflect the risk-tolerance profiles of stakeholders and the management philosophies of decision-makers. But,

if one endorses the precautionary principles for biodiversity conservation and resilience planning in general (see Section 5.3), it would be prudent to follow the Seidl et al. (2011) example (see Box 6.1) of highlighting the potential for the greatest possible changes. If we extrapolate current trends of international differences in GHG mitigation, the growing use of fossil fuels in developing countries, and the growth of regional rivalries (Yang et al. 2018), we are likely to see the moderately severe to severe impacts of SSP3 or SSP5 scenarios (see Figure 6.4). In other words, we seem to be on track to experience the most dystopian future envisioned in the Millennium Ecosystem Assessment scenario storylines (see Figure 1.5).

So how can a forest planner or manager reasonably respond to such uncertainty and to the wide range of possible futures associated with management goals that span decades or more than a century? Is it possible to plan for both drought and floods? Can forests be made more resilient to both wildfires and windstorms? Indeed, as introduced in Chapter 5, there are options for promoting general resilience; various expressions of diversification and flexibility are foremost among those options (Millar et al. 2007). Once locally focused vulnerability assessments have been conducted, then preparations can be made to cope with direct effects of climate change (see Chapter 7), and specific practices can be employed to address the risk of specific forest disturbances (see Chapter 8).

6.6 Climate change as part of global change

The magnitude and pace of climate change since the year 2000 has provided strong impetus to think in terms of forest resilience and not just sustainability. But climate trends are just some of the cumulative effects faced by forests and forest managers (Mery and Alfaro 2010), reflecting the sheer dominance of the human species on this planet. That human dominance is not just a matter of population numbers, but is also a product of our resource-consumption patterns and our collective effects on land cover and ocean food chains as we convert natural ecosystems into managed food, fiber, and fuel production systems (Vitousek et al. 1997). With human population growth continuing, human-dominated and human-modified ecosystems already constitute more than 53 percent of the world's ice-free land cover (Hooke et al. 2012). Although population growth is leveling off in developed countries, reflecting a combination of affluence, education, and reproductive rights for women, it continues unabated in much of the world.

Forest ecosystems are being modified or threatened by a plethora of human-induced changes, with the particular agents and mechanics of change, forest loss, and forest degradation varying from place to place. Primary and indigenous forests are being replaced by simplified plantation tree crops in many parts of the world, while tree cover is being replaced by agricultural crops and human infrastructure elsewhere. Tropical deforestation has slowed and continued in fits and starts under different political regimes: supporting the expansion of beef and soybean production in places such as Brazil, of oil-palm plantations in countries such as Indonesia. Likewise, primary forests in Canada and Russia continue to be logged, threatening their legacy of carbon storage and biodiversity. Much forested land in the United States is now being parcelized for ex-urban residential development and real-estate speculation rather than timber production or nature conservation.

Often facilitated by a warming climate and increased levels of disturbance, invasive exotic species pose ongoing threats to forest ecosystem integrity. Global trade and international travel are transporting unwanted weeds, destructive mammals, insect pests, and tree pathogens around the world. Strategic forest management planning is inherently difficult even under stable conditions, given the long timeframes considered, inherent inertia in forest ecosystem dynamics, and the spatial variability in most forest management areas (Davis et al. 2001; Franklin et al. 2018). Now we must overlay those long-term plans with a changing climate, the appearance of exotic species, and unexpected shifts in socioeconomic values. Not only do markets for wood and non-wood products and forest ecosystem services tend to fluctuate, but new markets (for example, for carbon retention and sequestration) can appear. Rapid shifts in public values and expectations with regard to forests (for

example, often prioritizing wildlife, recreation, or carbon storage over fiber production) can drive community, governmental, and regulatory decisions governing forests. The cumulative impact of these many twirling factors is that sustainable management plans are more aspirational rather than realistic, and tend to be inadequate soon after they are released. Forest managers may have no control over these external drivers (Farcy et al. 2016; Sesser et al. 2019), but can weather their impacts by incorporating a variety of resilience strategies. For example, to keep up with the changing world, it is inevitable that forest management plans need to be "rolling plans," scoping out goals and the vision of a desired future forest a century or more hence, but undergoing essential revision every few years. A changing climate and the other elements of global change make such updates and revisions an important approach for resilient forest management, now more than ever.

Box 6.4 Key points

- A warming atmosphere, primarily due to the burning of fossil fuels but exacerbated by deforestation and other factors, is leading to demonstrable increases in air temperatures and greater variability in precipitation. Increased incidence of droughts, storms, and wildfires is already being observed in many forests around the world.
- Numerous tools exist to project future climate conditions and associated effects on socio-ecological values, largely dependent on complex global circulation models (GCMs) and different scenarios of global trends in the emission and control of greenhouse gases (GHGs).
- To improve relevance to particular forest management areas, GCM projections can be refined through the use of statistical downscaling or regional climate models (RCMs) that account for mountainous terrain and ocean proximity.
- The vulnerability of any particular feature or process to climate change is a function of exposure, sensitivity, and adaptive capacity. Vulnerability assessments are fundamentally local in nature and can incorporate place-based thresholds and risk tolerances.
- Climate change, as part of a complex set of interacting global changes, now constitutes a lens through which all forest management plans and goals must be viewed.

Climate is what we expect, weather is what we get.

Robert Heinlein (1973)

References cited

AAFC (2023). *Drought Conditions as of July 31, 2023.* Ottawa: Canadian Drought Monitor, Agriculture and Agri-Food Canada. Available online at https:// agriculture.canada.ca/en/agricultural-production/ weather/canadian-drought-monitor (accessed August 25, 2023).

Ali, A. A., Asselin, H., Larouche, A. C., Bergeron, Y., Carcaillet, C., and Richard, P. J. H. (2008). "Changes in Fire Regime Explain the Holocene Rise and Fall of *Abies balsamea* in the Coniferous Forests of Western Québec, Canada," *The Holocene*, 18/5: 693–703.

Allen, C. D., Breshears, D. D., and McDowell, N. G. (2015). "On Underestimation of Global Vulnerability to Tree Mortality and Forest Die-Off from Hotter Drought in the Anthropocene," *Ecosphere*, 6/8: 1–55.

Arias, P. A., Bellouin, N., Coppola, E., et al. (2021). "Technical Summary," in V. Masson-Delmotte, P. Zhai, A. Pirani, et al. (eds), *Climate Change 2021: The Physical Science Basis. Contribution of Working Group I to the Sixth Assessment Report of the Intergovernmental Panel on Climate Change.* Cambridge: Cambridge University Press, 33–144.

Araújo, M. B., and Peterson, A. T. (2012). "Uses and Misuses of Bioclimatic Envelope Modelling," *Ecology*, 93/7: 1527–39.

Aubin, I., Munson, A. D., Cardou, F., et al. (2016). "Traits to Stay, Traits to Move: A Review of Functional Traits to Assess Sensitivity and Adaptive Capacity of Temperate and Boreal Forests to Climate Change," *Environmental Reviews*, 24/2: 164–86.

Australian Government (2022). *State of the Climate 2022.* Canberra: CSIRO and Bureau of Meteorology, Commonwealth of Australia. 27 pp. Available online at http://www.bom.gov.au/state-of-the-climate/index. shtml (accessed February 12, 2024).

Bazzaz, F. A., Miao, S. L., and Wayne, P. M. (1993). "CO_2-Induced Growth Enhancements of Co-Occurring Tree Species Decline at Different Rates," *Oecologia*, 96/4: 478–82.

Beard, K. H., Kelsey, K. C., Leffler, A. J., and Welker, J. M. (2019). "The Missing Angle: Ecosystem Consequences of Phenological Mismatch," *Trends in Ecology & Evolution*, 34/10: 885–8.

Booth, T. H., Nix, H. A., Busby, J. R., and Hutchinson, M. F. (2014). "BIOCLIM: The First Species Distribution Modelling Package, its Early Applications and Relevance to Most Current MAXENT Studies," *Diversity and Distributions*, 20/1: 1–9.

Brandt, L. A., Butler, P. R., Handler, S. D., Janowiak, M. K., Shannon, P. D., and Swanston, C. W. (2017). "Integrating Science and Management to Assess Forest Ecosystem Vulnerability to Climate Change," *Journal of Forestry*, 115/3: 212–21.

Burke, K. D., Williams, J. W., Chandler, M. A., Haywood, A. M., Lunt, D. J., and Otto-Bliesner, B. L. (2018). "Pliocene and Eocene Provide Best Analogs for Near-Future Climates," *Proceedings of the National Academy of Sciences*, 115/52: 13288–93.

Burton, P. J., and Cumming, S. G. (1995). "Potential Effects of Climatic Change on Some Western Canadian Forests, Based on Phenological Enhancements to a Patch Model of Forest Succession," *Water, Air, and Soil Pollution*, 82/1–2: 401–14.

Burton, P. J., Jentsch, A., and Walker, L. R. (2020). "The Ecology of Disturbance Interactions," *BioScience*, 70/10: 854–70.

Choi, S., Lee, W. K., Kwak, H., et al. (2011). "Vulnerability Assessment of Forest Ecosystem to Climate Change in Korea Using MC1 Model," *Journal of Forest Planning*, 16: 149–61.

Davis, L. S., Johnson, K. N., Bettinger, P., and Howard, T. E. (2001). *Forest Management to Sustain Ecological, Economic, and Social Values.* 4th edn. Long Grove, IL: Waveland Press. 804 pp.

Davis-Reddy, C. L., and Vincent, K. (2017). *Climate Risk and Vulnerability: A Handbook for Southern Africa.* 2nd edn. Pretoria, South Africa: Council for Scientific and Industrial Research. 193 pp. Available online at https:// www.csir.co.za/sites/default/files/Documents/ SADC%20Handbook_Second%20Edition_full%20report. pdf (accessed February 12, 2024).

de la Barrera, F., Barraza, F., Favier, P., Ruiz, V., and Quense, J. (2018). "Megafires in Chile 2017: Monitoring Multiscale Environmental Impacts of Burned Ecosystems," *Science of the Total Environment*, 637/638: 1526–36.

Delcourt, P. A., and Delcourt, H. R. (1987). "Late-Quaternary Dynamics of Temperate Forests: Applications of Paleoecology to Issues of Global Environmental Change," *Quaternary Science Reviews*, 6/2: 129–46.

Ditlevsen, P., and Ditlevsen, S. (2023). "Warning of a forthcoming collapse of the Atlantic meridional overturning circulation," *Nature Communications*, 14/1: 4254.

Duveneck, M. J., Scheller, R. M., White, M. A., Handler, S. D., and Ravenscroft, C. (2014). "Climate Change Effects on Northern Great Lake (USA) Forests: A Case for Preserving Diversity," *Ecosphere*, 5/2: 1–26.

Engle, N. L. (2011). "Adaptive Capacity and its Assessment," *Global Environmental Change*, 21/2: 647–56.

European Environmental Agency (2021). *Europe's Changing Climate Hazards: An Index-Based Interactive EEA Report.* Copenhagen: European Environmental Agency,

European Union. Available online at https://www.eea.europa.eu/publications/europes-changing-climate-hazards-1 (accessed February 12, 2024).

Falk, D. A., van Mantgem, P. J., Keeley, J. E., et al. (2022). "Mechanisms of Forest Resilience," *Forest Ecology and Management*, 512: 120129.

Farcy, C., de Camino, R., de Arano, I. M., and Briales, E. R. (2016). "External Drivers of Changes Challenging Forestry: Political and Social Issues at Stake," in G. R. Larocque (ed.), *Ecological Forest Management Handbook*. Boca Raton, FL: CRC Press, 87–105.

Farjad, B., Gupta, A., Sartipizadeh, H., and Cannon, A. J. (2019). "A Novel Approach for Selecting Extreme Climate Change Scenarios for Climate Change Impact Studies," *Science of the Total Environment*, 678: 476–85.

Feser, F., Rockel, B., von Storch, H., Winterfeldt, J., and Zahn, M. (2011). "Regional Climate Models Add Value to Global Model Data: A Review and Selected Examples," *Bulletin of the American Meteorological Society*, 92/9: 1181–92.

Franklin, J. (2010). *Mapping Species Distributions: Spatial Inference and Prediction*. New York: Cambridge University Press. 319 pp.

Franklin, J. F., Johnson, K. N., and Johnson, D. L. (2018). *Ecological Forest Management*. Long Grove, IL: Waveland Press. 646 pp.

Fuchigami, L. H., Weiser, C. J., Kobayashi, K., Timmis, R., and Gusta, L. V. (1982). "A Degree Growth Stage (°GS) Model and Cold Acclimation in Temperate Woody Plants," in P. H. Li and A. Sakai (eds), *Plant Cold Hardiness and Freezing Stress*, ii. *Mechanisms and Crop Implications*. New York: Academic Press, 93–116.

Gauthier, S., Lorente, M., Kremsater, L., et al. (2014). *Tracking Climate Change Effects: Potential Indicators for Canada's Forests and Forest Sector*. Ottawa: Canadian Forest Service, Natural Resources Canada. 86 pp. Available online at http://cfs.nrcan.gc.ca/publications?id=35231 (accessed February 13, 2024).

Girardin, M. P., Bouriaud, O., Hogg, E. H., et al. (2016). "No Growth Stimulation of Canada's Boreal Forest under Half-Century of Combined Warming and CO_2 Fertilization," *Proceedings of the National Academy of Sciences*, 113/52: E8406–E8414.

Halofsky, J. E., Peterson, D. L., O'Halloran, K. A., and Hawkins-Hoffman, C. (eds) (2011). *Adapting to Climate Change at Olympic National Forest and Olympic National Park*. General Technical Report PNW-GTR-844. Portland, OR: USDA Forest Service. 130 pp.

Hamelin, R., and Innes, J. L. (2017). "Forest Ecosystem Health and Vitality," in J. L. Innes and A. V. Tikina (eds), *Sustainable Forest Management: From Concept to Practice*. New York: Routledge, 101–33.

Hanes, C. C., Wotton, M., Bourgeau-Chavez, L., et al. (2023). "Evaluation of New Methods for Drought Estimation in the Canadian Forest Fire Danger Rating System," *International Journal of Wildland Fire*, 32/6: 836–53.

Harris, R. M., Beaumont, L. J., Vance, T. R., et al. (2018). "Biological Responses to the Press and Pulse of Climate Trends and Extreme Events," *Nature Climate Change*, 8/7: 579–87.

Heikkinen, R. K., Luoto, M., Araújo, M. B., Virkkala, R., Thuiller, W., and Sykes, M. T. (2006). "Methods and Uncertainties in Bioclimatic Envelope Modelling under Climate Change," *Progress in Physical Geography*, 30/6: 751–77.

Heinlein, R. A. (1973). *Time Enough for Love*. New York: G. P. Putnam's Sons. 605 pp.

Holdridge, L. R. (1947). "Determination of World Plant Formations from Simple Climatic Data," *Science*, 105/2727: 367–8.

Holdridge, L. R. (1967). *Life Zone Ecology*. San Jose, Costa Rica: Tropical Science Center. 206 pp.

Hooke, R. L., Martín Duque, J. F., and Pedraza Gilsanz, J. D. (2012). "Land Transformation by Humans: A Review," *GSA Today*, 22/12: 4–10.

Huang, J. G., Bergeron, Y., Denneler, B., Berninger, F., and Tardif, J. (2007). "Response of Forest Trees to Increased Atmospheric CO_2," *Critical Reviews in Plant Sciences*, 26/5–6: 265–83.

Innes, J. L. (2017). "Criteria and Indicators of Sustainable Forest Management," in J. L. Innes and A. V. Tikina (eds), *Sustainable Forest Management: From Concept to Practice*. New York: Routledge, 33–43.

IPCC (Intergovernmental Panel on Climate Change). (2022). "Summary for Policymakers," in H.-O. Pörtner, D.C. Roberts, M. Tignor, et al., *Climate Change 2022: Impacts, Adaptation and Vulnerability*. Contribution of Working Group II to the Sixth Assessment Report of the Intergovernmental Panel on Climate Change. Cambridge: Cambridge University Press, 3–33. Available online at https://www.ipcc.ch/report/ar6/wg2/ (accessed August 2, 2023).

IPCC (Intergovernmental Panel on Climate Change). (2023). *Climate Change 2023: Synthesis Report*. Contribution of Working Groups I, II, and III to the Sixth Assessment Report of the Intergovernmental Panel on Climate Change. Geneva: Intergovernmental Panel on Climate Change. 184 pp. Available online at https://www.ipcc.ch/report/sixth-assessment-report-cycle/ (accessed February 13, 2024).

Janowiak, M. K., D'Amato, A. W., Swanston, C. W., et al. (2018). *New England and Northern New York Forest Ecosystem Vulnerability Assessment and Synthesis: A Report from the New England Climate Change Response Framework*

Project. General Technical Report NRS-173. Newtown Square, PA: USDA Forest Service. 234 pp.

Jay, A. K., Crimmins, A. R., Avery, C. W., et al. (2023). "Overview: Understanding Risks, Impacts, and Responses," in A. R. Crimmins, C. W. Avery, D. R. Easterling, K. E. Kunkel, B. C. Stewart, and T. K. Maycock (eds), *Fifth National Climate Assessment*. Washington: US Global Change Research Program. 47 pp. Available online at https://nca2023.globalchange.gov/downloads/NCA5_Ch1_Overview.pdf (accessed February 12, 2024).

Jiang, M., Medlyn, B. E., Drake, J. E., et al. (2020). "The Fate of Carbon in a Mature Forest under Carbon Dioxide Enrichment," *Nature*, 580/7802: 227–31.

Johnston, M., and Williamson, T. (2007). "A Framework for Assessing Climate Change Vulnerability of the Canadian Forest Sector," *Forestry Chronicle*, 83/3: 358–61.

Jung, M., Reichstein, M., Ciais, P., et al. (2010). "Recent Decline in the Global Land Evapotranspiration Trend due to Limited Moisture Supply," *Nature*, 467/7318: 951–4.

Kahlenborn, W., Porst, L., Voß, M., et al. (2021). *Klimawirkungs- und Risikoanalyse 2021 für Deutschland (Kurzfassung)*. Berlin: Umweltbundesamt. 121 pp. Available online at https://adelphi.de/en/publications/climate-impact-and-risk-assessment-2021-for-germany-summary (accessed February 12, 2024).

Kamae, Y., Imada, Y., Kawase, H., and Mei, W. (2021). "Atmospheric Rivers Bring More Frequent and Intense Extreme Rainfall Events over East Asia under Global Warming," *Geophysical Research Letters*, 48/24: e2021GL096030.

Klassen, H. A. (2012). "Potential Impacts of Climate Change in Dry Coastal Ecosystems of British Columbia." M.Sc. Thesis, University of Northern British Columbia, Prince George, BC. 253 pp.

Klassen, H. A., and Burton, P. J. (2015). "Climatic Characterization of Forest Zones across Administrative Boundaries Improves Conservation Planning," *Applied Vegetation Science*, 18/2: 343–56.

Krebs, C. J., Boonstra, R., Cowcill, K., and Kenney, A. J. (2009). "Climatic Determinants of Berry Crops in the Boreal Forest of the Southwestern Yukon," *Botany*, 87/4: 401–8.

Lawson, D. A., and Rands, S. A. (2019). "The Effects of Rainfall on Plant–Pollinator Interactions," *Arthropod–Plant Interactions*, 13/4: 561–9.

Le Goff, H., and Bergeron, Y. (2014). "Vulnerability Assessment to Climate Change of Three Ecosystem-Based Forest Management Projects in Quebec," *Forestry Chronicle*, 90/2: 214–27.

Lee, J.-Y., Marotzke, J., Bala, G., et al. (2021). "Future Global Climate: Scenario-Based Projections and Near-Term Information," in V. Masson-Delmotte, P. Zhai, A. Pirani, et al. (eds), *Climate Change 2021: The Physical Science Basis. Contribution of Working Group I to the Sixth Assessment Report of the Intergovernmental Panel on Climate Change*. Cambridge: Cambridge University Press, 553–672.

Leskinen, P., Lindner, M., Verkerk, P. J., et al. (eds). (2020). *Russian Forests and Climate Change*. What Science Can Tell Us 11. Joensuu, Finland: European Forest Institute. 137 pp. Available online at https://efi.int/sites/default/files/files/publication-bank/2020/efi_wsctu_11_2020.pdf (accessed February 12, 2024).

Lexer, M. J., and Seidl, R. (2009). "Addressing Biodiversity in a Stakeholder-Driven Climate Change Vulnerability Assessment of Forest Management," *Forest Ecology and Management*, 258: S158–S167.

Loarie, S. R., Duffy, P. B., Hamilton, H., Asner, G. P., Field, C. B., and Ackerly, D. D. (2009). "The Velocity of Climate Change," *Nature*, 462/7276: 1052–5.

Locatelli, B., Brockhaus, M., Buck, A., et al. (2010). "Forests and Adaptation to Climate Change: Challenges and Opportunities," in M. Mery, P. Katila, G. Galloway, et al. (eds), *Forests and Society: Responding to Global Drivers of Change*. IUFRO World Series 25. Vienna: International Union of Forest Research Organizations, 21–42.

Lomolino, M. V., Riddle, B. R., and Whittaker, R. J. (2016). *Biogeography: Biological Diversity across Space and Time*. 5th edn. Oxford: Oxford University Press. 784 pp.

Lulham, N., Warren, F. J., Walsh, K. A., and Szwarc, J. (2023). *Canada in a Changing Climate: Synthesis Report*. Ottawa: Government of Canada. 71 pp. Available online at https://changingclimate.ca/synthesis/ (accessed February 12, 2024).

MacArthur, R. H. (1984). *Geographical Ecology: Patterns in the Distribution of Species*. Princeton: Princeton University Press. 288 pp.

MacKenzie, W. H., and Meidinger, D. V. (2018). "The Biogeoclimatic Ecosystem Classification Approach: An Ecological Framework for Vegetation Classification," *Phytocoenologia*, 48/2: 203–13.

Marcott, S. A., Shakun, J. D., Clark, P. U., and Mix, A. C. (2013). "A Reconstruction of Regional and Global Temperature for the Past 11,300 Years," *Science*, 339/6124: 1198–1201.

Menezes-Silva, P. E., Loram-Lourenço, L., Alves, R. D. F. B., Sousa, L. F., Almeida, S. E. D. S., and Farnese, F. S. (2019). "Different Ways to Die in a Changing World: Consequences of Climate Change for Tree Species Performance and Survival through an Ecophysiological Perspective," *Ecology and Evolution*, 9/20: 11979–99.

Mery, G., and Alfaro, R. I. (2010). "Forests in a Changing World," in M. Mery, P. Katila, G. Galloway, et al. (eds), *Forests and Society: Responding to Global Drivers of Change*. International Union Forest Research Organizatioins World Series 25. Vienna: The International Union of Forest Research Organizations, 13–17.

Millar, C. I., Stephenson, N. L., and Stephens, S. L. (2007). "Climate Change and Forests of the Future: Managing in the Face of Uncertainty," *Ecological Applications*, 17: 2145–51.

NOAA (2025). *Monthly Global Climate Report for Annual 2024*. Washington: National Centers for Environmental Information, National Oceanic and Atmospheric Administration, US Department of Commerce. Available online at https://www.ncei.noaa.gov/access/monitoring/monthly-report/global/202413/supplemental/page-1 (accessed January 15, 2025).

Nocentini, S., Buttoud, G., Ciancio, O., and Corona, P. (2017). "Managing Forests in a Changing World: The Need for a Systemic Approach: A Review," *Forest Systems*, 26/1: eR01.

Norby, R. J., Warren, J. M., Iversen, C. M., Childs, J., Jawdy, S. S., and Walker, A. P. (2022). "Forest Stand and Canopy Development Unaltered by 12 Years of CO_2 Enrichment," *Tree Physiology*, 42/3: 428–40.

NZME (2020). *National Climate Change Risk Assessment for Aotearoa New Zealand: Main Report—Arotakenga Tūraru mō te Huringa Āhuarangi o Āotearoa: Pūrongo whakatōpū*. Wellington: New Zealand Ministry for the Environment. 133 pp. Available online at https://environment.govt.nz/assets/Publications/Files/national-climate-change-risk-assessment-main-report.pdf (accessed February 12, 2024).

Palter, J. B. (2015). "The Role of the Gulf Stream in European Climate," *Annual Review of Marine Science*, 7: 113–37.

Patt, A., Klein, R. J., and de la Vega-Leinert, A. (2005). "Taking the Uncertainty in Climate-Change Vulnerability Assessment Seriously," *Comptes Rendus Geoscience*, 337/4: 411–24.

Pielou, E. C. (1991). *After the Ice Age: The Return of Life to Glaciated North America*. Chicago: University of Chicago Press. 367 pp.

Raftery, A. E., Zimmer, A., Frierson, D. M., Startz, R., and Liu, P. (2017). "Less than 2 °C Warming by 2100 Unlikely," *Nature Climate Change*, 7/9: 637–41.

Reidmiller, D. R., Avery, C. W., Easterling, D. R., et al. (eds). (2018). *Fourth National Climate Assessment*, ii. *Impacts, Risks, and Adaptation in the United States*. Washington: US Global Change Research Program. 1515 pp. Available online at https://nca2018.globalchange.gov/downloads/NCA4_2018_FullReport.pdf (accessed August 23, 2023).

Renner, S. S., and Zohner, C. M. (2018). "Climate Change and Phenological Mismatch in Trophic Interactions among Plants, Insects, and Vertebrates," *Annual Review of Ecology, Evolution, and Systematics*, 49: 165–82.

République Française (2018). *Le Plan National D'Adaptation au Changement Climatique 2018–2022 (PNACC-2)*. Paris: Ministère de la Transition Écologique et Solidaire. 24 pp. Available online at https://www.ecologie.gouv.fr/sites/default/files/2018.12.20_PNACC2.pdf (accessed February 12, 2024).

Roberts, G., Parrotta, J., and Wreford, A. (2009). "Current Adaptation Measures and Policies," in R. Seppälä, A. Buck, and P. Katila (eds), *Adaptation of Forests and People to Climate Change: A Global Assessment Report*. IUFRO World Series 22. Vienna: International Union of Forest Research Organizations, 123–35.

Rogelj, J., Meinshausen, M., and Knutti, R. (2012). "Global Warming under Old and New Scenarios Using IPCC Climate Sensitivity Range Estimates," *Nature Climate Change*, 2/4: 248–53.

Rose, N.-A., and Burton, P. J. (2009). "Using Bioclimatic Envelopes to Identify Temporal Corridors in Support of Conservation Planning in a Changing Climate," *Forest Ecology and Management*, 258: S64–S74.

Satoh, M. (2016). *Atmospheric Circulation Dynamics and General Circulation Models*. 2nd edn. Chichester, UK: Springer-Praxis. 730 pp.

Seidl, R., Rammer, W., and Lexer, M. J. (2011). "Climate Change Vulnerability of Sustainable Forest Management in the Eastern Alps," *Climatic Change*, 106/2: 225–54.

Seidl, R., Thom, D., Kautz, M., et al. (2017). "Forest Disturbances under Climate Change," *Nature Climate Change*, 7/6: 395–402.

Sesser, A. L., Rockhill, A. P., Magness, D. R., et al. (eds). (2019). *Drivers of Landscape Change in the Northwest Boreal Region*. Fairbanks: University of Alaska Press. 326 pp.

Sonwa, D. J., Somorin, O. A., Jum, C., Bele, M. Y., and Nkem, J. N. (2012). "Vulnerability, Forest-Related Sectors and Climate Change Adaptation: The Case of Cameroon," *Forest Policy and Economics*, 23: 1–9.

Thorne, J. H., Choe, H., Stine, P.A., et al. (2018). "Climate Change Vulnerability Assessment of Forests in the Southwest USA," *Climatic Change*, 148/3: 387–402.

Timberlake, T. J., and Schultz, C. A. (2019). "Climate Change Vulnerability Assessment for Forest Management: The Case of the US Forest Service," *Forests*, 10/11: 1030.

Tonmoy, F. N., El-Zein, A., and Hinkel, J. (2014). "Assessment of Vulnerability to Climate Change Using Indicators: A Meta-Analysis of the Literature," *Wiley Interdisciplinary Reviews: Climate Change*, 5/6: 775–92.

Trouet, V., Babst, F., and Meko, M. (2018). "Recent Enhanced High-Summer North Atlantic Jet Variability

Emerges from Three-Century Context," *Nature Communications*, 9/1: 180.

Turner, B., Kasperson, R., Matsone, P., et al. (2003). "A Framework for Vulnerability Analysis in Sustainability Science," *Proceedings of the National Academy of Science*, 100/14: 8074–9.

UKCCC (2021). *Independent Assessment of UK Climate Risk: Advice to Government for the UK's Third Climate Change Risk Assessment (CCRA3)*. London: United Kingdom Climate Change Committee. 141 pp. Available online at https://www.theccc.org.uk/wp-content/uploads/2021/07/Independent-Assessment-of-UK-Climate-Risk-Advice-to-Govt-for-CCRA3-CCC.pdf (accessed February 12, 2024).

Upgupta, S., Sharma, J., Jayaraman, M., Kumar, V., and Ravindranath, N. H. (2015). "Climate Change Impact and Vulnerability Assessment of Forests in the Indian Western Himalayan Region: A Case Study of Himachal Pradesh, India," *Climate Risk Management*, 10: 63–76.

Urban, O. (2003). "Physiological Impacts of Elevated CO_2 Concentration Ranging from Molecular to Whole Plant Responses," *Photosynthetica*, 41/1: 9–20.

van der Sleen, P., Groenendijk, P., Vlam, M., et al. (2015). "No Growth Stimulation of Tropical Trees by 150 Years of CO_2 Fertilization but Water-Use Efficiency Increased," *Nature Geoscience*, 8/1: 24–8.

Vitousek, P. M., Mooney, H. A., Lubchenco, J., and Melillo, J. M. (1997). "Human Domination of Earth's Ecosystems," *Science*, 277/5325: 494–9.

Wang, T., Hamann, A., Spittlehouse, D., and Carroll, C. (2016). "Locally Downscaled and Spatially Customizable Climate Data for Historical and Future Periods for North America," *PloS One*, 11/6: e0156720.

Wang, X., Studens, K., Parisien, M.-A., et al. (2020). "Projected Changes in Fire Size from Daily Spread Potential in Canada over the 21st Century," *Environmental Research Letters*, 15/10: 104048.

Wilson, R., and Turton, S. (2011). *An Assessment of the Vulnerability of Australian Forests to the Impacts of Climate Change*, iv. *Climate Change Adaptation Options, Tools and Vulnerability*. Gold Coast, Queensland, Australia: National Climate Change Adaptation Research Facility. 119 pp. Available online at https://researchonline.jcu.edu.au/15794/ (accessed August 5, 2023).

Wittwer, S. H., and Strain, B. R. (1985). "Carbon Dioxide Levels in the Biosphere: Effects on Plant Productivity," *Critical Reviews in Plant Sciences*, 2/3: 171–98.

World Bank Group (2021). *Climate Risk Profile: Brazil*. Washington: World Bank Publications. 31 pp. Available online at https://climateknowledgeportal.worldbank.org/country-profiles (accessed February 12, 2024).

Xiao, J. L., Zeng, F., He, Q.L., Yao, Y.X., Han, X., and Shi, W.Y. (2021). "Responses of Forest Carbon Cycle to Drought and Elevated CO_2," *Atmosphere*, 12/2: 212.

Yang, P., Yao, Y.-F., Mi, Z., et al. (2018). "Social Cost of Carbon under Shared Socioeconomic Pathways," *Global Environmental Change*, 53: 225–32.

CHAPTER 7

Fostering Climate Change Resilience

Though we cannot control the wind, we can adjust our sails . . . .

Grip magazine, Toronto (March 4, 1882)

7.1 Responding to climate change

As briefly outlined in Chapter 6, Planet Earth is undergoing a dramatic shift in climate owing to the increase in carbon dioxide (CO_2) and other greenhouse gases (GHGs) associated with the burning of fossil fuels and other human influences on the atmosphere. An underlying theme is that greater heat and drought will be experienced in the coming decades. Equally important considerations are that: (1) extreme temperature, precipitation, and storm events will become more frequent; and (2) global trends and extremes will be expressed differently around the world (IPCC 2022). Most of those trends will be disruptive and will have largely negative effects on natural ecosystems and on human health and socioeconomic systems in general.

Not only will forests and forest management have to cope with such changes; they can also play important roles in mitigating the effects of anthropogenic climate change. Nature-based solutions such as forest conservation and restoration can keep additional CO_2 from being leaked into the atmosphere, and sequester more carbon in long-lasting solid reservoirs (see Figure 7.1). Closed forest cover can provide important shade and habitat to moderate the effects of global warming and can provide refuge from some of the other cumulative effects of human activities. Trees and forest ecosystems provide important reservoirs of carbon, and tree growth extracts some of that elevated CO_2 from the atmosphere and sequesters it in long-lasting cellulose and the organic matter of forest soils. In addition to those mitigative effects, the sustainable use of forest biomass (and that of other crops) has the potential to displace the burning of fossil fuels. Collectively, these roles and services of trees and forests in keeping excess CO_2 out of the atmosphere are some of the most important natural solutions to climate change (Griscom et al. 2017; Marvin et al. 2023). Ideally, resilient forest management to address the identifiable threat of climate change also promotes a wide range of other ecosystem services (see Section 2.2) as well.

The first step for a land manager to address climate change is to understand the place-based implications of a warming planet to the specific area being managed. This requires some interpretation of downscaled climate-change projections, with an understanding of the most likely scenarios and the level of uncertainty around temperature and precipitation projections. Not only does exposure to the changing climate vary around the planet, but so does the sensitivity of different species, ecosystems, ecological processes, and forest management systems, as does their adaptive or mitigative capacity for coping with such change. In other words, a full assessment of climate impacts and vulnerabilities (see Section 6.4) should serve as a precursor to developing place-based climate-adaptive strategies. Such strategies, sometimes referred to as "climate-smart forestry" (Bowditch et al. 2020), must be viable under current conditions as well as those expected to prevail in the near, mid, and distant future. Furthermore, they must be undertaken in the context of the many other cumulative

Resilient Forest Management. Philip J. Burton, Oxford University Press. © Philip J. Burton (2025). DOI: 10.1093/oso/9780198832997.003.0007

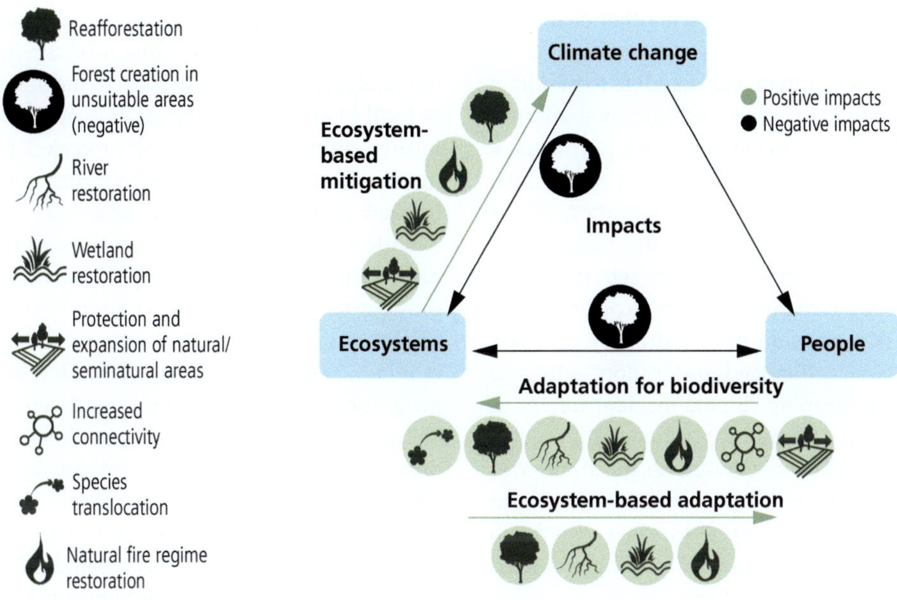

Figure 7.1 Nature-based solutions for climate-change mitigation and adaptation (from Morecroft et al. (2019), reproduced with permission).

and interactive stressors associated with the role of forests in our world, ranging from land-use conflicts to invasive species.

There is general agreement on the most promising strategies for promoting climate-change resilience in the world's forests. Different approaches are needed in different parts of the world and in different forest types, however, and to support different management objectives. "Climate resilience" and climate-smart forestry embrace the continuum of resilience concepts, promoting resistance and recovery in some cases, adjustment and reconfiguration in other situations, and sometimes a degree of system transformation (see Chapter 4). Selected approaches from that continuum are summarized here, organized by the broad triad categories (see Box 2.3) of protected, multipurpose, and production forests, but with widespread crossover of recommended strategies among land-use emphasis categories as well.

7.2 Strategies for protected forests

The world's natural ecosystems have repeatedly adjusted, evolved, and adapted to climatic changes over the many eras and epochs of prehistory. The

difference they are facing in the Anthropocene is that the cumulative effects of human activities greatly constrain the ability of species, communities, and ecosystems to persist unmodified as the climate changes. Land conversion to agricultural, urban, industrial, and transportation uses has left natural forests much reduced in area, with remaining forest ecosystems fragmented, often dominated by edge effects (Harper et al. 2005), experiencing pressure from invasive species, and isolated from each other (Lindenmayer et al. 2010). The health of many forests at high elevations and downwind from highly industrialized and urbanized areas of the planet is also compromised by atmospheric pollution in the form of acid rain and excess levels of sulfur and nitrogen (Taylor et al. 1994). Populations of many forest species (especially predators and those harvested for food) have been decimated, which greatly constrains their ability to play their full potential roles in current and future ecosystems.

7.2.1 Assure continuity in space and time

With the above constraints in mind, managers of protected forests are fighting an uphill battle under current conditions, to which a changing climate and extreme weather events pose yet another challenge. Management options in parks and protected areas can be limited when there is a mandate to maintain natural ecosystem processes with minimal interference. It is widely recognized that efforts to anticipate and respond to future conditions must supplement actions to strengthen current conservation efforts (Schmitz et al. 2015). Because of the widespread, degrading effects of many human activities, protected area management often focuses on ecological restoration (Keenleyside et al. 2012). A common theme to conservation and restoration, and also important in promoting climate-resilient forests, is the need to assure continuity: genetic and ecosystem continuity and connectivity over space and over time. That strategy needs to be implemented in both the designation of new protected areas, and as much as possible in the management of existing protected forests.

Most jurisdictions around the world are far from achieving the widely promoted goal of protecting 30 percent of their land area by the year 2030 (the "30 by 30" target endorsed by the Kunming–Montreal Global Biodiversity Framework; Corson and Campbell 2023). At the same time, accessible areas of many protected forests, adjacent areas, and the front country of many parks are facing pressures from or for greater human use and modification. These pressures and interventions may have the objective of generating revenue (for example, through resource extraction or commercial recreation) or of minimizing risks from disturbances such as wildfire. The cumulative effect of these pressures and land uses threatens to whittle away on forest ecosystem integrity and ecological continuity.

The very nature of conservation, restoration, and protected-area management has a focus on trying to preserve or rebuild the past, usually defined as conditions prevailing in some precolonial or preindustrial era. This means that the prevailing strategy for resilient forest management of protected forests is to prioritize resistance strategies. To resist and hedge the effects of climate change as much as possible, planners and managers can identify locations most likely to serve as climate refugia, landscapes and sites where truly enduring features (geographic location, topography, site drainage) are more likely to ameliorate the effects of heat and

drought (Anderson and Ferree 2010; Schmitz et al. 2015). Such locations may already be serving as spatial catalysts (Pressey et al. 2007) to the development of local biodiversity hotspots. Potential refugia are ideally connected as part of protected (or lower human impact) networks within and among designated parks and reserves (Anderson et al. 2023; Saunders et al. 2023). Map overlays of future climate indicators (for example, projected length of the growing season, various drought indices) can identify locations where the velocity of climate change is minimized (Loarie et al. 2009). Overlays of the geographic extent of the bioclimatic envelopes or habitat suitability projections (see Box 6.1) for prioritized species, populations, or ecosystem types can identify locations for "temporal corridors" or refugia most likely to provide habitat continuity (Rose and Burton 2009; Morelli et al. 2020). With an emphasis on local and regional prioritization exercises, Table 7.1 outlines approaches for identifying such refugia, and for incorporating climate change into conservation planning at the ecosystem level.

7.2.2 Protect high-carbon reservoirs

Existing old-growth forests often indicate stable environments that have persisted and avoided natural disturbances such as fire for centuries (Krawchuk et al. 2020), so those areas should be prioritized for enhanced protection. Such locations often have a more secure moisture supply or reduced moisture demand—that is, dominated by maritime weather patterns, poleward slope aspects, or lower slope positions (Rogeau et al. 2018; Krawchuk et al. 2020). Older forests and the ecosystem services they provide can be less sensitive to climate change than younger forests (Thom et al. 2019; Stritih et al. 2021). Whether related to the sites on which they are found or to their complex composition and structure, this suggests that protected old forests are imbued with greater resilience to climate change than if they were managed on short even-aged rotations optimal for timber production.

Under appropriate climates and disturbance regimes, protected areas can serve an important climate-mitigation role by retaining large amounts of carbon (see Box 7.1). By this argument, however, conservation areas should also be protected from

Table 7.1 Complementary approaches for promoting resilience in conservation areas, and the scientific assessments needed to support such actions

Approach	Assessments needed	Management examples
1. **Protect current patterns of biodiversity**	• Map species occurrences and biodiversity hotspots • Map ecosystems and their associated services	• Extend greater protection status to include representative, rare, and high-diversity locales
2. **Protect large, intact, natural landscapes**	• Forecast climate-change effects on selected important species, including potential pests and invasives • Map potential future patterns of fire, hydrology, and carbon sequestration • Map locations where ecosystem services operate and provide human value	• Devise recovery or control plans as needed • Designate resource emphasis zones, with separate management guidelines • E.g. identify areas most resilient to heavy recreation use and zone accordingly
3. **Protect the physical setting**	• Map areas of high ecological integrity • Map land facets in relation to current climate patterns • Map areas of high topographic complexity	• Designate as core conservation areas, buffered with lower-integrity areas • Project climate change impacts and prepare management guidelines by facet class • Inventory and monitor biodiversity for potential changes
4. **Maintain and restore ecological connectivity**	• Identify crucial habitats and potential movement corridors for species • Map connections between current and future locations • Anticipate species invasions around current and future human infrastructure	• Extend greater protection status to existing key habitats and corridors • Project species distribution models under climate change and assure protection to connect current and future climate space • Concentrate human infrastructure; monitor for and control invasives
5. **Identify and manage areas to provide future climate space for species expected to be displaced**	• Map potential range of selected important species under likely future climate • Forecast ecosystem vulnerability to climate change • Map locations that would support shifts in vegetation types and biomes	• Project species distribution models under climate change and assure protection to future climate space • Devise place- or species-specific adaptation plans for high-vulnerability conservation elements • Designate areas where resistance strategies can give way to adjustment, restructuring, or transformation
6. **Identify and protect climate refugia**	• Identify areas where selected important species can be expected to persist • Map habitats with high natural resistance to climate change (e.g. spring-fed streams) • Map areas projected to experience little change in vegetation	• Extend greater protection status according to species importance/rarity • Enhance the level of protection, with appropriate buffers • May be of least concern if already protected

Based on and extended from Schmitz et al. (2015).

wildfire and other such disturbances, which are important natural processes in many of the world's forests. Balancing such conservation priorities—carbon sequestration and old-growth forests versus some degree of natural disturbance processes—has to be done on a place-specific basis with input from stakeholders. That balance might also be achieved over a network of protected areas and other forests; protected area planning needs to be part of a regional landscape planning process.

The identification of existing or potential refugia and emphasizing the spatiotemporal continuity of ecosystems represent an emphasis on "resistance"

Box 7.1 The role of protected areas in climate-change mitigation

Protected forests play a climate regulation role by retaining large reservoirs of carbon, offering further incentive for the expansion of protected areas as a climate mitigation strategy. The United Nations program promoting Reduction in Emissions from Deforestation and Forest Degradation (REDD+) provides financial incentives for the protection of forests in developing nations, particularly those in the global south (Dudley 2008; Kanninen et al. 2010). Devised from 2005 to 2015, REDD+ provides a framework to reduce forest loss to agriculture and other land-use changes, to restore degraded forests (see Chapter 9), and to promote more sustainable timber harvesting activities. Its goal is to offset CO_2 emissions from the burning of fossil fuels by promoting more carbon storage in living trees and forest ecosystems, especially in the tropics where deforestation had been rampant but trees can grow fast. Following documentation and auditing, countries undertaking such initiatives are eligible to receive transfer payments from the Green Climate Fund, set up under the United Nations Framework Convention on Climate Change (UNFCCC 2014). There may also be opportunities in developed nations for enhanced forest protection through carbon offset programs (Gresh 2018; Kadam 2018). However, for true mitigation benefits, such protection needs to occur without shifting wood production to the cutting of primary forests in other nations. Collectively, these mechanisms for channeling governmental and private-sector carbon market investments provide an underutilized means of underwriting and justifying the expansion of parks and reserves, so essential for the natural adjustment of plant and animal communities to a changing climate.

Where indigenous trees are able to achieve great ages and sizes, old forests can continue to sequester high levels of CO_2 from the atmosphere (Luyssaert et al. 2008). Where climates are likely to remain wet and/or cold enough to avoid wildfire and rapid decomposition (namely in some of the world's rare temperate, boreal, and tropical montane rainforests), such carbon-dense primary forests can serve as important reservoirs of solid carbon for centuries to come (Carter and Buma 2024; see Figure 7.2).

Figure 7.2 Long-lasting carbon reservoirs in the temperate rainforest of coastal British Columbia, Canada, where wildfires are very rare: (a) Massive western redcedar (*Thuja plicata*) trees have tapered boles and so are windfirm, can live more than 1,000 years, and have very rot-resistant wood that persists for centuries after death (photo by Darwyn Coxson). (b) Soil profile from a hypermaritime coastal forest, dominated by organic layers more than one meter deep; Lv denotes recently deposited litter undergoing initial decomposition, Fm is a matted layer of partially decomposed organic matter with abundant fungal mycelia, and Hh layers represent highly decomposed humus with few recognizable plant residues (photo by Paul Sanborn).

Box 7.1 *Continued*

An often-neglected carbon sink is the broad expanse of peatlands around the world, from the swamp forests of Indonesia and the wet temperate rainforests of Chile and western North America to the broad expanses of subarctic forest and tundra. Saturated with water most of the time, peatland ecosystems are often refugia from wildfires (Kuntzemann et al. 2023). Collectively, it is estimated that global peatlands—which in many ways are young fossil fuels and are sometimes exploited as such—contribute 6×10^{11} Mg (metric tons) of carbon storage in the biosphere (Yu et al. 2011). Those carbon-rich and nutrient-poor habitats support distinctive flora and fauna, including insectivorous and ericaceous plants and wide-ranging *Rangifer* (caribou or reindeer) species. But those massive deposits of dead organic matter are also susceptible to wildfire under extreme hot and dry conditions, in which flaming and smoldering fire (which can burn deeply and is particularly difficult to extinguish) can release globally significant levels of carbon dioxide and other greenhouse gasses (Zoltai et al. 1998; Lin et al. 2021).

With ongoing global warming and widespread drought threatening to lower water tables and increase fire frequency and severity, peatlands are especially vulnerable to climate change. It therefore behooves land managers actively to protect these 3.4 million km² worldwide (Zoltai et al. 1998), prioritizing them for the exclusion of fire and extractive activities that might facilitate carbon loss through combustion or accelerated decomposition. If not protected from disturbance, the carbon dynamics of peatlands, organic soils, and deep forest floors could accentuate climate change in a positive feedback loop. This means that timber harvesting from forested wetlands, muskeg, and mires needs to be avoided (see Section 11.7.3); forest regeneration is sparse and difficult to promote on such sites anyway. Historic efforts to drain peatlands, plant trees, and fertilize them (as widely practiced in Finland, Indonesia, and elsewhere) need to be reversed through the reinstitution of prior water levels and ongoing ecosystem restoration (see Chapter 9). In managed forests, the potentially conflicting goals of promoting high carbon sequestration rates (promoting densely stocked stands of fast-growing trees) and retaining high carbon stocks (in large living trees) can sometimes be accommodated by using multi-aged silvicultural systems that retain high levels of mature forest structure (Palik et al. 2021).

---

strategies, reflecting the focus on preservation and stasis inherent to protected area management. Whether such a strategy is feasible in the long run remains to be seen on a case-by-case basis, depending on the cumulative impacts of climate change and other stressors relative to the sensitivity of the natural features and processes designated for protection. In the short term, resilience after extreme weather events and natural disturbances may be expressed or promoted through "recovery" strategies that also target preexisting conditions. It is generally expected that ecosystem recovery in protected forests will proceed by natural processes with minimal human interference. There are instances, however, where weather-mediated die-offs or disturbances clear the way for invasive species to get a foothold, thereby preventing or delaying recovery by indigenous species (Hiremath and Sundaram 2005; Caldeira et al. 2015). In such cases, the control of exotic plants, animals, or diseases may be needed to facilitate the recovery of ecosystems.

7.2.3 Accommodate transitions without losing ecological integrity

If we recognize the need to conserve natural processes in protected areas, and not just current composition and pattern, then changes in natural vegetation and ecosystem structure can be expected and considered acceptable in parks and reserves (Pressey et al. 2007). Early levels of "adjustment" to a changing climate may be expressed through changes in the relative allocation of resources to reproduction and growth in plants already found there. Such evidence is found in high-resolution paleoecological data, in which pollen abundance by different species changes on an annual or decadal scale, but is not accompanied by significant changes in the floristic complement (Carcaillet et al. 2010; Minckley et al. 2012). Where protected ecosystems are sufficiently diverse and protected landscapes are sufficiently large, "reconfiguration" in terms of directional changes in genotypic or species abundance and the appearance or disappearance of species can be expected under sustained levels of climate change. Such changes typically require longer periods of time (centuries to millennia, at least in the past) to take place or to be distinguished from post-disturbance successional dynamics (Webb 1986). For such directional adaptation to occur naturally, spatial continuity or connectivity is required between the location of donor ecosystems (or their component species) and the location of future suitable climate space. Moving external species from outside a park or ecological reserve to establish in anticipation of future climatic conditions is generally frowned upon, although may be considered as part of a strategy of forward-looking restoration after instances of disturbance or degradation. Guiding ecosystem restoration to match new bioclimatic envelopes might also be considered in protected areas (see Box 9.3).

The designation and management of protected areas have historically depended on assumptions of static conditions and the desirability of avoiding change, especially human-caused change. Consequently, complete "transformation" of a protected forest in response to a changing climate—for example, from forest to grassland—may be considered a failure of conservation. On the other hand, in the presence of enduring features (terrain, rock outcrops, and so on), a completely transformed protected area may still provide important watershed protection roles, habitat for native species, and distinctive recreational experiences (Anderson and Ferree 2010). Conservation planners may decide that new examples of representative forest ecosystems need to be identified and protected elsewhere, but most existing protected areas deserve to be retained as well, even if their species composition and physiognomic structure changes radically. The potential dominance of invasive exotic species (foreign to the continent or biogeographic province) may be the greatest threat to the ability of many ecosystems to adjust, reconfigure, or transform successfully while still delivering desired ecosystem services (Pyšek et al. 2020). Invasive exotic species often capture the ground after disturbances, not only preventing recovery and modest change, but also preventing transformation into another native-dominated ecosystem that would be adapted to the new climate and would be acceptable to land managers and stakeholders.

7.3 Strategies for multipurpose forests

Extensively managed forests that are maintained and nurtured for the promotion of multiple values have a wide selection of climate-coping strategies available, including a flexible emphasis on approaches geared to protection (see Section 7.2) and to production (see Section 7.4). As in protected forests, the challenge is to promote spatiotemporal continuity, diversity, and natural processes so forests can continue to deliver desired ecosystem services. Where timber extraction is a management goal, guiding principles can generally follow those of "sustainable forest management" (Burton et al. 2003; Innes and Tikina 2017), "ecosystem management" (Kohm and Franklin 1997), "ecological forest management" (Larocque 2016; Franklin et al. 2018), or "ecosystem stewardship" (Chapin et al. 2009; Messier et al. 2015). These forest management paradigms have often embraced the emulation of natural disturbances in planning the size, frequency, and retention patterns of timber-harvesting operations (Bergeron et al. 1999; Hunter 2007). If those natural processes and patterns are to guide forest management going forward, however, the challenge is to apply lessons from historic trends—without being able to mimic them under a changing climate—based on their environmental associations.

Lessons from past natural and anthropogenic disturbances, and how forests have responded to them, can inspire effective ecological approaches to resilient forest management. Disturbances have always removed the inertia of dominance by trees that were established under conditions prevailing many decades or centuries ago. Such disturbances typically initiate secondary succession with fast-growing, shade-intolerant, wide-niched, short-lived species that gradually give way to the slower-growing, shade-tolerant, long-lived, and often specialized species that reproduce and persist indefinitely in their own shade (Finegan 1984). Climate-adapted silviculture may likewise emphasize wide-niched tree species (for example, red maple, *Acer rubrum*, in eastern North America), or fast-growing species under relatively short rotations (for example, loblolly pine, *Pinus taeda*, in the US southeast), allowing frequent readjustment to species selection and silvicultural practices in concert with climate shifts. Where emphasis is placed on the retention of old forest cover and late-successional species, timber harvesting might proceed on the scale of canopy gaps (created by the collapse of single trees or small groups of trees) that prevail in old-growth forest dynamics (Coates and Burton 1997; Kern et al. 2017). It is expected that the microclimate and structural diversity of such pockets and retained old forest cover will help resist or moderate the effects of a changing climate.

Guidance and inspiration can come from the forest composition, structure, and forestry practices of "future climate analogs" found elsewhere on the continent that currently experience climatic conditions similar to those expected in the future for the area under management (MacKenzie and Mahony 2021; Mette et al. 2021; Esperon-Rodriguez et al. 2022). Particular attention should be paid to the transition zones that support the species, structures, and processes common to both current and future conditions in order to be feasible now and in the future (Rehm et al. 2015). However, the advisability of borrowing species and silvicultural practices from elsewhere can also depend on soil and terrain attributes, as well as infrastructure and community values (Kellett et al. 2015). Where "no-analog" conditions are projected, a traits-based approach to tree species selection is advisable (Park et al. 2014; Aubin et al. 2016).

Actively managing multipurpose forests for resilience in the face of climate change requires more of an anticipatory approach than is needed for protected forests. It is still assumed that most multipurpose forests have a mandate to protect indigenous biodiversity, which requires a focus on sustaining native tree cover and largely natural ecosystems. Accordingly, all harvesting and silvicultural activities should conserve those natural compositions, structures, and processes that confer ability for a forest to self-adjust to changing conditions. In other words, conditions supportive of key ecological processes must continue: energy capture, water and nutrient cycling, the functional interaction of diverse species at all trophic levels, effective

dispersal and migration, successful vegetative and sexual reproduction, and evolution through natural selection. Many other forest ecosystem services, including the provision of clean and reliable water, non-timber forest products, and recreational opportunities, are also supported by assuring as much ecological integrity of forests as possible. Those considerations are also important in managing multipurpose forests that may have been planted or have a blend of native and exotic tree species, especially where there is an agenda of long-term naturalization.

7.3.1 Promote tree-species diversity

One of the most effective ways to facilitate climate change resilience is to protect, promote, and establish forest stands consisting of multiple tree species (Bauhus et al. 2017). As introduced in Chapter 5, diversity is a key risk-reduction and bet-hedging strategy for persisting in an uncertain future. Forest stands with two or more tree species have a greater chance of maintaining forest cover and the delivery of ecosystem services if exposed to drought, growing-season frost, storm damage, or the expansion of invasive pests (Pretzsch et al. 2013; Jactel et al. 2017; Kneeshaw et al. 2021). Mechanisms vary, but complementary resource utilization niches (Kabrick et al. 2017), different stress tolerances, and a variety of adaptive traits (Aubin et al. 2016) play a role in explaining the benefits of diversity.

Increased resilience can be conferred to some degree by adding more species to a forest stand, in which multiple species provide insurance against species-specific disruptors such as herbivores and pathogens (see, e.g., Figure 5.3; see also Section 8.5). Complementary crown architecture (especially coniferous and broadleaf, or evergreen and deciduous) between at least two tree species is especially important, as illustrated in Figure 7.3(a) (Pretzsch 2014; Williams et al. 2017). Such stands can be intimately mixed even-aged forests, stratified mixtures consisting (for example) of slow-growing conifers under fast-growing broadleaves, or complex stands with high vertical and horizontal variability in the distribution of two or more tree species (see Figure 7.3(b)). Because of differences in shade tolerance, phenology, and growth form, multi-species

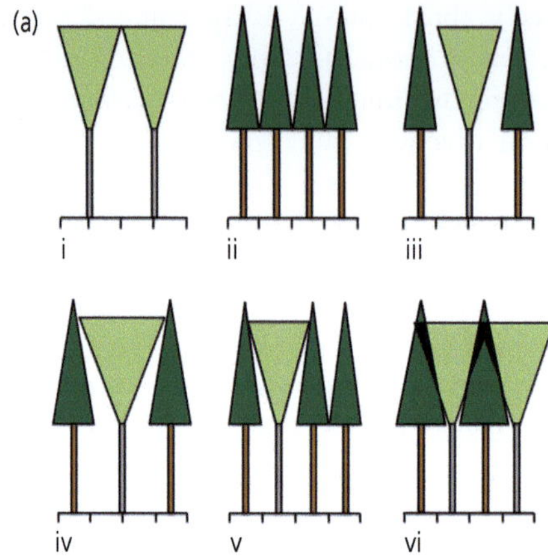

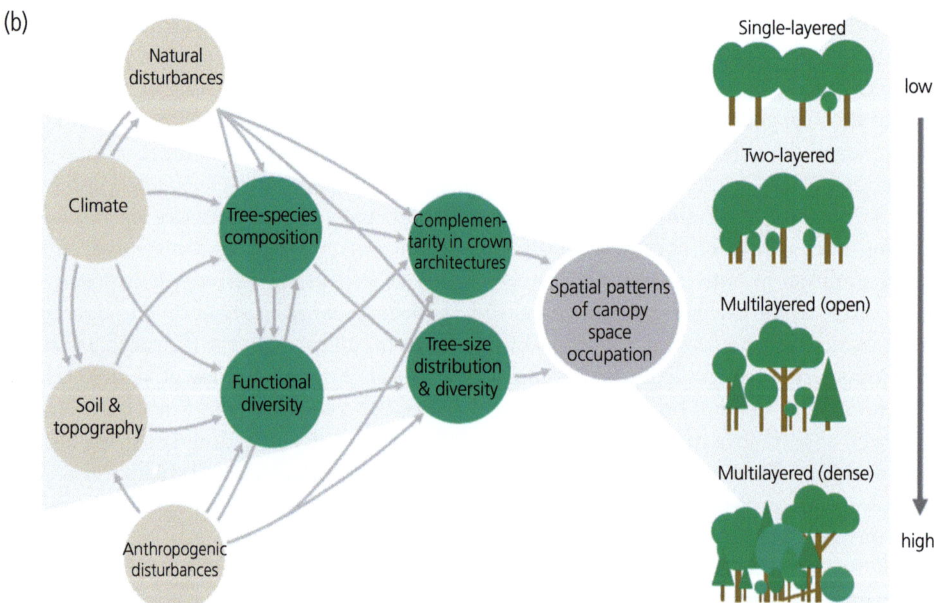

Figure 7.3 Schematic representations of some canopy structures possible in multi-species forest stands, compared to those in even-aged single-species stands: (a) Potential effects of species mixing in even-aged stands when two species with complementary crown shapes (i and ii) are mixed with shape and density remaining constant to result in an additive effect (iii), or when wider crown expansion (iv) or higher density (v) or both (vi) facilitate a multiplicative mixing effect with denser canopy space filling (Pretzsch (2014), reproduced under the open access CC BY-NC-ND license, http://creativecommons.org/licenses/by-nc-nd/3.0/); (b) abiotic and biotic controls on forest structural complexity, which increases with a greater diversity of tree sizes and greater complementarity in crown architectures (Ehbrecht et al. (2021), reproduced under the open access CC-BY-4.0 license, http://creativecommons.org/licenses/by/4.0/).

stands often have more structural complexity than single-species stands. Many of these forest structures are superior at delivering non-timber forest ecosystem services (for example, providing habitat for wildlife and berry-producing shrubs) than uniform monocultures (Brockerhoff et al. 2017).

There is increasing evidence for "overyielding" in multi-species forest stands as well: the fact that two species with complementary traits can produce more wood and sequester more carbon if grown together than if grown separately (Kelty 1992; Williams et al. 2017). However, where mixed stands are more effective at fully utilizing resources such as light or water (which helps support greater yields), there is also danger of greater resource depletion. One species may facilitate the growth and survival of another through mechanisms such as hydraulic lift, extracting water from deep soil strata and releasing it closer to the surface (Pretzsch et al. 2013). Depending on drought severity, the tree species, and their degree of stomatal control (to limit water loss during drought), there is a risk of longer-lasting damage and mortality if all available soil moisture is depleted rather than conserved (Allen et al. 2010; Pretzsch et al. 2013). So stand diversification may assist ecological resistance in the case of drought, but may compromise recovery potential (which is discussed further and illustrated in Subsection 7.4.1).

7.3.2 Reduce stand densities

Where drought or recurrent seasonal moisture deficits are expected, the spacing or thinning of a stand can allow the remaining trees to access more soil resources to survive and grow (Allen et al. 2010; Sohn et al. 2016; Figure 7.4). Thinning is also widely advocated as a fire-risk reduction treatment (see Chapter 8), in which "ladder fuels" are removed to reduce the chance of surface fire becoming a crown fire, and the overstory is thinned to reduce canopy bulk density and crown fire intensity (Agee and Skinner 2005). Thinning operations provide the opportunity to enhance the relative abundance of large living and dead trees, rare and future-adapted tree species, and generally to increase stand complexity by applying techniques such as variable density thinning (Carey 2003; Palik et al. 2021). A pattern of gaps (open sunlight) and skips (leaving clusters of trees) increases in-stand heterogeneity (Brodie and Harrington 2020). This approach improves forage, browse, or berry production by understory species in most years, while the shade reduces heat stress and supports some level of shrub growth and berry production in dry years. Sunlit gaps may also be needed for the introduction of future-adapted species (see Subsections 7.3.3 and 7.4.3) that may not be very shade tolerant. Where non-crop vegetation significantly constrains the establishment of trees, vegetation control is generally less acceptable than it is in production forests, especially if chemical herbicides are to be employed.

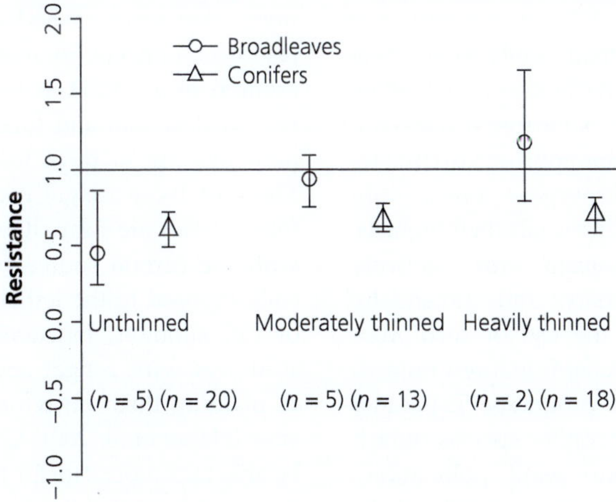

Figure 7.4 Results of a meta-analysis of twenty-three forest-thinning experiments from around the world. Short-term resistance is expressed as an index denoting radial stem growth one year after a drought event relative to one year before the drought. Values of 1 mean that no change in growth has taken place, whereas index values of 0.5 mean that a 50% reduction occurred and 1.5 means a 50% relative increase in growth. Sample sizes (*n*) denote the number of individual studies, and fiducial lines represent 95% bootstrapped confidence limits. Note that broadleaved species generally respond to thinning more favorably than conifers (Sohn et al. (2016), reproduced with permission)

7.3.3 Introduce adapted genetic material

A direct way of gradually preparing forests for a future climate is the assisted migration of selected provenances (genotypes, ecotypes) or species from populations evolved in locations where climates better match those expected to prevail in the future (Ste-Marie et al. 2011). Assisted migration trials are now being actively undertaken and evaluated (Sáenz-Romero et al. 2021), although without complete understanding about potential phenological, competitive, herbivory, or symbiotic mismatches (Park and Talbot 2018), and with some concerns about potential invasiveness and changing the character of local forests (Pelai et al. 2021). Particularly applicable to trees used in artificial forest regeneration (tree seeding or planting), assisted migration is discussed further in Section 7.4.

Artificial regeneration can be considered a form of intensive forest management, but the planting of native tree species is widely practiced in extensively managed, multiple-use public forests, and in forest restoration (Lieffers et al. 2003; Lamb 2015). Even species not indigenous to the region can be safely introduced to forests managed by ecological principles if found nearby on the same continent, in the same floristic province, or with paleoecological evidence of previous association. Although many plant-animal associations have evolved at the genus level (e.g. Laurindo et al. 2019), intercontinental transfers of tree species are inadvisable where there is an objective to sustain native biodiversity (Pedley et al. 2014). Introductions of exotic tree species cannot be considered a responsible approach to climate-change adaptation in multipurpose forests until ecological research more fully reveals their impacts.

An alternative mechanism for actively introducing genetic diversity and potentially climate-adaptive genes is the use of seed from open-pollinated tree seed orchards in forest nursery and artificial regeneration programs. Especially relevant to widely planted conifer species (which depend on outcrossing and wind pollination), there are tree-improvement programs around the world that have typically focused on selecting trees—with seeds or cuttings collected from multiple locations—that have superior growth or pest resistance. Those selected trees or their offspring are grown together in specialized seed-production stands (seed orchards) where pollen is shared among the trees brought together from diverse geographic locations. Despite narrower genetic diversity in the selected trait or traits, seed produced in seed orchards nonetheless—and perhaps counter-intuitively—can have wider genetic variability (heterozygosity) than found in any one local population of trees (El-Kassaby and Askew 1998; Hosius et al. 2006). Immersed in that genetic variability are mechanisms for a wide range of tolerances to drought, heat, frost, and so on, although the possibility of genetic linkages between some traits (for example, rapid growth and high water use or low drought tolerance) has not been widely researched. The seed orchard or "polycross approach" of developing widely adapted seed sources can also be applied to plants other than trees, such as grasses and legumes for revegetating roadsides (Burton and Burton 2002).

7.3.4 Maintain viable fungal connections

One ecological relationship that is being increasingly understood, appreciated, and promoted is the symbiotic association between fungal species and the roots of vascular plants (see Box 7.2). Research is increasingly emphasizing the need to protect forest soil health, rhizosphere biodiversity, mycorrhizal networks, and the "mother trees" that serve as nodes for the distribution of resources and environmental cues below ground (Simard et al. 2021). It remains to be determined how widespread and functionally important such networks are in most forests (Karst et al. 2023). Many of those fungal connections are lost when forest stands are clearcut logged (Durall et al. 1999), with the carbon source from host trees lost and soils exposed to the lethal heat and drying effects of full sunlight. Inoculating nursery-grown tree seedlings with fungal spores or fragments prior to planting them on reforestation and restoration sites (Habte et al. 2001; Quoreshi et al. 2009) may be able to compensate for that loss, though without the benefit of subsidies from mature trees. In order to maintain the stress-resilient network of fungal connections in a changing climate, it would seem prudent to employ systems of uniform overstory tree retention, or to limit the size and shape of harvested openings so no regeneration site is more than one or two tree-heights from mature trees.

Box 7.2 Mycorrhizal networks leverage climate-change resilience

A key component of healthy forests is the fungal network that extends the soil foraging ability of trees and connects them in dynamic interactive networks. It has long been known that the symbiotic association of some fungi with tree root tips (see Figure 7.5(a)) effectively extends the reach of fine roots through even narrower hyphae. Those are fed carbohydrates by the tree and are able to access nutrients (such as ammonium or nitrate, but especially the more immobile nutrients such as phosphorus), some of which it subsequently passes to its host tree (Voigt 1969; Nehls and Plassard 2018). More recently, it has been found that mycorrhizae can also transport water from tightly bound soil particle surfaces to trees (especially small seedlings) and other vascular plants (Parke et al. 1983; Lehto and Zwiazek 2011; Ruiz-Lozano et al. 2012). In addition, mycorrhizal networks allow the transfer of nutrients and carbohydrates between trees of different species, typically from those with greater access to resources (light, moisture, nutrients) to those that are suffering a deficit (Ek et al. 1997; Simard et al. 1997).

There is some evidence that established trees use mycorrhizal connections (see Figure 7.5(b)) to subsidize the carbon balance of their nearby offspring until those seedlings become successfully established (Teste et al. 2009). Trees can also transmit distress signals through their fungal connections to indicate the presence of insect herbivores, prompting their neighbors to initiate the production of defensive secondary compounds. Dying trees may even shunt carbohydrates and valuable nutrients to their surviving neighbors via those underground connections (Song et al. 2015) although this may be a passive process. Fungal connections and processes of resource sharing and signal-ing are not restricted to ectomycorrhizae or to trees, but also consist of other types of fungi and are found connecting trees to shrubs (e.g. Horton et al. 1999) and to herbaceous plants (e.g. Lerat et al. 2002; Dickie et al. 2004). Mediated by a healthy and diverse fungal community, a diverse understory plant community may thus be supportive of the trees and overall forest ecosystem structure and functions that we typically wish to retain or promote. Small mammals and invertebrates can be important for the dispersal of fungal spores, especially into disturbed areas (Vašutová et al. 2019; Stephens et al. 2021), emphasizing the value of maintaining the full biotic community.

Fungal mycelia and their ability to connect trees with resources and with each other have many attributes that we recognize as contributing to resilience, particularly to the vicissitudes of weather and the foreseeable stresses associated with a changing climate (Mohan et al. 2014). The "resistance" of trees to drought is ameliorated by mycorrhizal fungi tapping into deeper or more tightly retained soil water, sometimes through osmotic adjustments associated with the generation of sugar alcohols (Lehto and Zwiazek 2011). Nutrient availability is constrained by a lack of soil water in which anions and cations can be transported, but fungal hyphae offer a cytoplasmic pathway for nutrient transport under drought conditions. "Recovery" from the effects of drought, wildfire, or insect outbreaks (as well as logging) can be facilitated by surviving trees, and perhaps non-tree species, that convey photosynthate and other key resources through shared mycorrhizal connections. As the climate shifts more discernably and reliably, we can expect to see "adjustment" through shifts in the relative

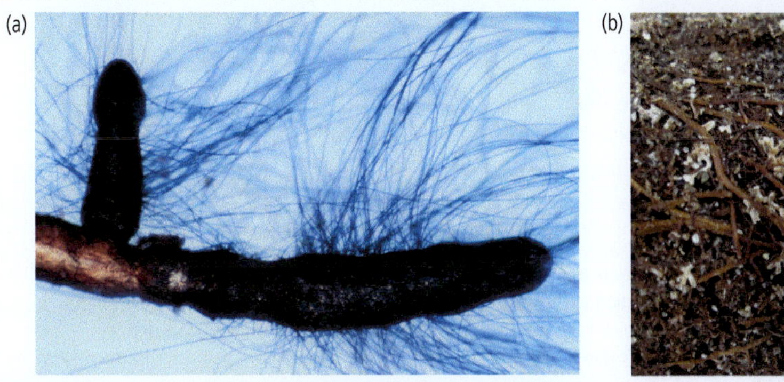

Figure 7.5 Ectomycorrhizal fungi: (a) *Cenococcum* sp. mycelia and fungal sheath on *Abies lasiocarpa* root tips (photo courtesy of Hugues Massicotte, reproduced from Peterson et al. (2004) with the permission of NRC Research Press, Ottawa); **(b)** *Suillus* sp. or *Rhizopogon* sp. mycorrhizal clusters (white) on *Pinus albicaulis* root tips, with ropey rhizomorphs (brown) potentially providing connectivity among trees (photo by Hugues Massicotte).

Box 7.2 *Continued*

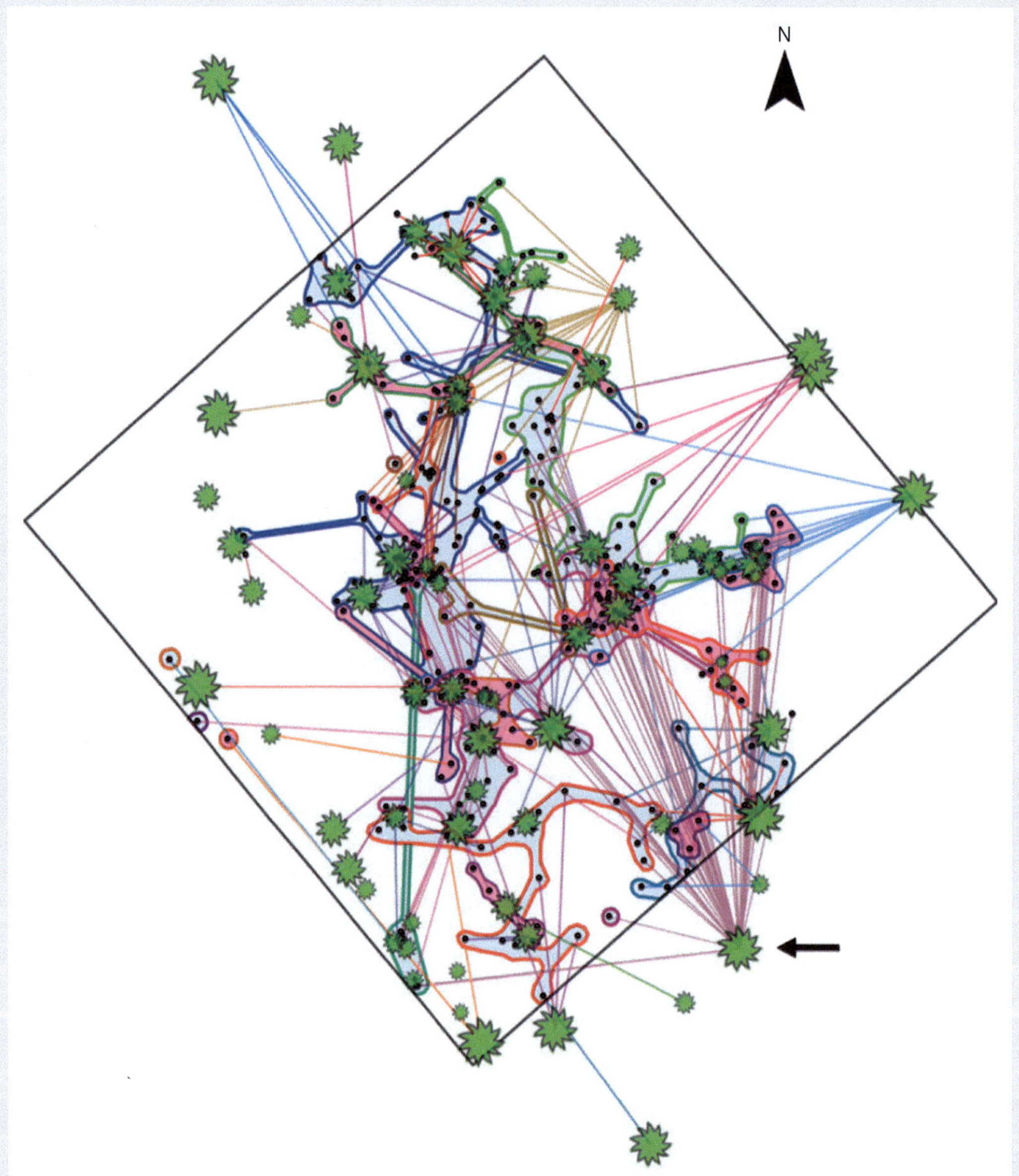

Figure 7.6 Mycorrhizal connections among sixty-seven trees in a 30m × 30m plot in a Douglas-fir (*Pseudotsuga menziesii*) forest. Green sunbursts represent Douglas-fir trees, scaled in proportion to their diameter; black dots denote sites sampled for mycorrhizal fungi; blue lines illustrate *Rhizopogon vesiculosus* linkages; pink lines illustrate *R. vinicolor* linkages; blue or pink shaded areas denote single genets, distinguished by different outline colors. The arrow points to the most highly connected tree, which was linked to forty-seven other trees (Beiler et al. (2010), reproduced with permission).

Box 7.2 *Continued*

---

dominance of different genotypes, species, or functional groups of fungi in the mycorrhizal community (Shi et al. 2002; Izzo et al. 2005). Furthermore, "reconfiguration" is likely to emerge more rapidly by means of evolution through natural selection in fungi that have much shorter life cycles than trees. In temperate and boreal forests, "transformation" of both the vascular plant flora and the fungal community is often evident in shifts to grassland, shrubland, or invasive species that are accentuated by shifts from forest-based ectomycorrhizal to vesicular-arbuscular mycorrhizal (VAM) communities that more often characterize grasslands. Conversely, restoration back to healthy forests often requires

inoculation with the appropriate fungal community (Perry et al. 1989).

Whether following source-sink gradients, constituting coevolved patterns of mutualistic adaptation, or reflecting a super-organismal level of biotic organization, it is clear that the symbioses of trees and fungi in forests constitute modular but networked self-organizing complex adaptive systems that confer resilience (Filotas et al. 2014). The fungal network that connects trees to each other (see, e.g., Figure 7.6) literally constitutes a safety net for their collective persistence in the face of environmental stress and perturbations.

7.3.5 Multiple options, multiple expectations

Place-based planning or zoning within the multiple-use forest can be an important tool in preparing for the accelerating effects of a changing climate. Prioritizing any existing old-growth forest stands and other likely climate refugia (for example, cooler, moister micro/meso-sites; see Section 7.2) for protection, temporary canopy retention, or longer rotations can promote some ecological resistance. Resilience through ecological recovery (for example, after severe weather or disturbance events) can be augmented through artificial regeneration with future-adapted species and genotypes, usually by planting nursery-grown tree seedlings, but potentially using shrub shoot or rhizome cuttings and herbaceous plant seed too. Adjustment strategies can be incorporated into resistance and recovery treatments as well, selecting maladapted trees (individuals, or species) for removal during spacing, and substituting high-diversity seed or seedling sources in regeneration programs. Actively promoting reconfiguration will probably require facilitated migration (see Section 7.4), with the purposeful introduction of better climate-adapted tree provenances and species selected from locations that currently reflect future climate expectations for the areas under management.

Because multipurpose forests have a wide latitude in management objectives and approaches, some forest transformation may be completely acceptable in multi-valued landscapes. For example, in some areas and at some stage, recurrent droughts and/or wildfires are likely to make forest protection and restoration efforts ineffective. This means that the prevailing disturbance regime may then align better with the promotion of grasslands and livestock grazing than with efforts to retain forest cover. Native grasslands and meadows produce more forage and sustain important biodiversity quite distinct from that of forests (Feurdean et al. 2018). Especially in forests having supportive stakeholders, it can make sense to embrace the inevitable and allow a transition to herb-dominated rather than tree-dominated native vegetation.

Along with a wide range of options for promoting several modes of resilience, multipurpose forests also tend to support multiple and diverse public expectations. It can often be difficult to negotiate and agree to an acceptable balance of interventions to address climatic stressors, natural disturbances, and other unforeseen disruptions. This makes effective and equitable systems of stakeholder engagement and forest governance especially important (Timberlake and Schultz 2017). Once again, zoning within a forest landscape (see Section 2.5), with place-based responses to climate change vulnerability (see Section 6.4), can help sustain as many values and supportive ecosystem attributes as possible.

7.4 Strategies for production forests

Forests dedicated to timber production, as the component of forest landscapes most directly manipulated by humans, offer opportunities for direct interventions to address climate change. Forest plantation establishment on abandoned farmland and industrial sites (for example, mining waste) is a particularly attractive option for increasing CO_2 sequestration for climate mitigation (Gagné et al. 2016). Climate-resilient management of most existing plantations is likely to emphasize resistance, adjustment, and reconfiguration. Where existing timber stocks are not yet at the preferred rotation age or are not dominated by trees of harvestable size, it can be assumed that emphasis will be placed on implementing practices, such as thinning, that will promote the resistance of current forest stands to climate variability and weather extremes. Identifying the need for resistance strategies and which ones to implement will depend on place-based vulnerabilities and capacity of the socio-ecological system to respond (see Section 6.4). On the other hand, the freedom of an inevitable timber harvest conducted under a sustainable forest management framework allows managers to ask, "What attributes do we want in the future forest?" It is in the design of the future forest that targeted shifts in plant materials and silvicultural practices can be implemented to adjust and reconfigure the forest to thrive under expected changes in the climate.

7.4.1 Reduce competition and moisture stress

A widespread near-term approach to enhance forest resistance is to conduct forest thinning, with the intention of removing competitive constraints on the growth and vigor of crop trees, leaving them more widely spaced (as introduced in Subsection 7.3.2). This is an especially important technique in forest types, climates, and sites subject to recurrent drought, where competition for soil water is a dominant constraint to tree growth and survival (Allen et al. 2015; Sohn et al. 2016). Commercial thinning operations (where the cut trees can be used for some sort of wood product) can be done with mechanized harvesters if the stand has dedicated machine trails, but pre-commercial thinning is

usually done manually with motorized brush saws or chainsaws. Spacing trees for stand resistance is a "low thinning" or "thin from below" treatment (Nyland 1996) in which suppressed or subdominant trees are removed. The traditional intent of low thinnings is to accelerate the growth of the remaining crop trees to merchantable size, maintaining stands at densities less than those that would result in imminent mortality owing to competition (self-thinning) while still ensuring that trees occupy as much growing space as possible (Newton 2021). However, thinning intensities to improve drought resistance (or to reduce fire spread or insect herbivory (see Chapter 8)) may be much greater than traditionally practiced, and need to be determined on a case-by-case basis with consideration of site attributes and tree-species traits, along with the economic costs, benefits, and risks.

An important consideration when implementing thinning programs is the fate of removed trees: ideally this material could be used and sold to recover at least the costs of removing it. If dropped on the ground to rot (typically the case for pre-commercial thinning), this dead material will dry out and can be a fire hazard for a year or more. Whether such material burns or rots, it will release CO_2 into the atmosphere. On the other hand, utilization of the removed trees may depend on labor availability (if thinnings are small enough to be dragged to roadside by hand) and on markets for small boles (for example, fencing materials) or biomass (for example, for institutional/municipal heating or electricity generation).

The effectiveness of stand thinning in promoting forest resilience is somewhat ambiguous. The potential both to resist and to recover from droughts inherently differs among tree species, varies with tree age, and appears to be enhanced by stand thinning, but with widespread tradeoffs between recovery and resistance abilities (see Figure 7.7). Although thinning can alleviate growth declines during drought, the effects on growth after moisture stress are uncertain (Castagneri et al. 2022). Where the remaining trees exhibit reduced post-drought recovery potential, this may be due to the loss of carbon sharing through root grafts and mycorrhizal networks. In other cases, thinning may not detectably avoid drought-induced

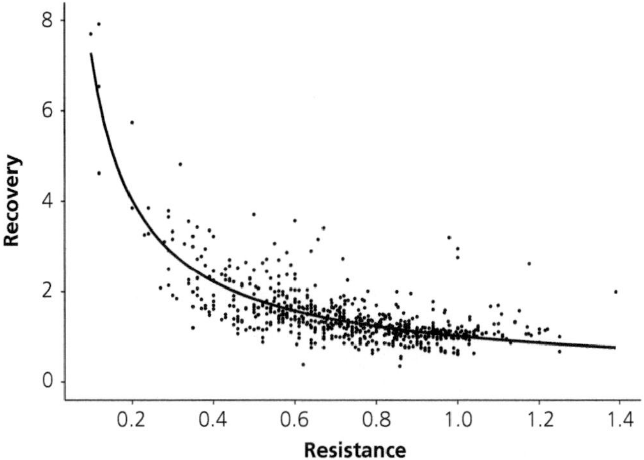

Figure 7.7 The empirical trade-off in two forms of tree growth resilience to drought, inferred from tree ring data from sixty-six published studies. Resistance describes a tree's capacity to absorb changes in radial growth during an extreme event by comparing radial growth during the year of disturbance to mean radial growth in a pre-disturbance period; recovery is calculated as average growth in the year(s) following the extreme event in relation to the growth level during the drought event year (simplified from Schwarz et al. (2020), courtesy of Julia Schwarz, reproduced with permission).

growth reduction or mortality, but may still promote more rapid growth recovery after a drought (Kohler et al. 2010). Other research suggests thinning can promote drought resistance and recovery in juvenile stands, but the remaining crop trees may have greater tree-level moisture demands when older (D'Amato et al. 2013). Such variable responses indicate that more site-specific and generalizable research needs to be done on the efficacy of thinning as a drought resilience treatment. Finally, it must be realized that any stand thinning results in decreased levels of total stand-level carbon sequestration and biomass production, and that such treatments may still fail to prevent tree mortality owing to accelerating climate change (Allen et al. 2015).

In the past, high stocking densities of trees were considered desirable in order fully to utilize the land value of plantations, quickly close cover to shade out competition from the understory, promote self-pruning and low taper, and leave open a range of stand density options. But, if droughts are likely to increase in frequency and severity, lower tree densities may be desirable at the outset of stand establishment. Lower densities may also induce a degree of wildfire resistance (see Section 8.3) but could also

increase the cover and growth of understory shrubs and grasses. Efforts to reduce competition for moisture can be extended to managing herbs, shrubs, and undesired tree species. Vegetation management (often using chemical herbicides) is frequently employed on timber-producing lands where non-crop vegetation threatens the establishment and early growth of planted trees. Tree water relations and growth can be significantly improved when nearby grasses or shrubs are suppressed on sites subject to severe belowground competition (Coll et al. 2003; Watt et al. 2003). A useful alternative to chemical herbicides that can also directly help with the conservation of soil water is the use of various plastic or biodegradable mulches or brush mats (Thomas et al. 2001; Van Lerberghes 2004). Water supplementation (irrigation) is a less widely used technique to address drought stress in forests, but managers may wish to consider this option in locating and supporting high-value plantations. Other techniques practiced in fruit-orchard management may also be worth considering where vulnerability assessments warrant and infrastructure is available, such as the use of sprinklers to alleviate drought or reduce damage from growing season frosts.

7.4.2 Favor vegetative reproduction

Forest recovery after harvesting or other sources of tree mortality in stands managed primarily for timber typically includes tree planting or is achieved by coppicing (vegetative sprouting from stumps or root suckers). Coppice production was traditionally practiced in Europe for local fuelwood production, taking advantage of the fact that most temperate broadleaf species can regenerate from stump sprouts, although this ability varies with stump diameter (Matula et al. 2012). The resulting sprouts typically exhibit growth much superior to seedlings, as they draw resources from well-established root systems. Large areas of northern Alberta, Canada, are managed for pulpwood production, with natural regeneration of *Populus tremuloides* from root suckers, typically requiring little or no silvicultural outlay. Likewise, in Mediterranean Europe and in South America, other *Populus* and *Eucalyptus* species (often hybrids selected for rapid early growth) are coppiced for two or three short-rotation pulpwood harvests before new genetic material is replanted (Mead 2001; Freer-Smith et al. 2019).

Most gymnosperm species do not have the ability to resprout from harvested adult stumps, although there are several notable exceptions, such as coast redwood (*Sequoia sempervirens*) in California and bald-cypress (*Taxodium distichum*) in the southeastern US (O'Hara et al. 2017; Burrows 2021). Other exceptions include ginkgo (*Ginkgo biloba*), Chinese fir (*Cunninghamia lanceolata*), sugi (*Cryptomeria japonica*) in eastern Asia, and several species of the Araucariaceae (e.g. *Araucaria angustifolia* and *A. cunninghamii*) and Podocarpaceae families in the southern hemisphere (Burrows 2021). Stump sprouters typically resprout after fire as well, as fire was probably the dominant agent of natural selection leading to that coppicing behavior. Managing tree species that quickly and reliably establish a new stand by vegetative reproduction after harvesting or wildfire is a low-cost and low-risk recovery strategy. However, tree form may not be optimal for solid-wood products, and the practice hinders opportunities for adjusting the genetic composition of the next crop of trees (Nyland 1996; Buckley and Mills 2015). Tree planting (or, more rarely, tree seeding) presents the option of implementing species-level and genetic-level forest adjustment and forest reconfiguration strategies for better-adapted long-term resilience following every harvest or disturbance event.

7.4.3 Assisted migration

Most widespread tree species have considerable genetic variability in drought tolerance, frost tolerance, optimal temperature for photosynthesis, and other traits that developed over millennia of natural selection in diverse geographic locations. Consequently, using plant materials (seeds, seedlings, cuttings) taken from forest populations growing in future climate analog locations in forest regeneration programs is one of the most direct and promising means of addressing climate change in managed forests (Millar et al. 2007; Aitken and Whitlock 2013). Where suitable genetic variability does not exist naturally in a tree species, there are increasing opportunities for gene editing and transgenic gene introduction using CRISPR technology (Wang et al. 2021). With generally fewer concerns over the introduction of new or modified tree genes into a landscape than in multi-value forests, a wide range of assisted migration approaches are possible for production forests (see Figure 7.8):

(1) assisted population migration or assisted gene flow moves germplasm within the existing range of a species, based on intra-specific (provenance, or population to population) differences in species traits; this might be termed a climate-adjustment strategy;

(2) assisted range expansion anticipates the development of new suitable habitat adjacent to or just beyond the current range of a species, and moves germplasm into such locations; this might be termed a reconfiguration strategy;

(3) assisted species migration moves germplasm well beyond the current range of a species into potentially suitable habitat, overcoming dispersal and migration barriers that had not been breached historically; this could be a reconfiguration strategy, or (depending on changes to overall forest physiognomy or the product output profile, perhaps from softwood to hardwood species) could constitute a system transformation; and

(4) introducing exotic species from other continents, which have no evolutionary history of

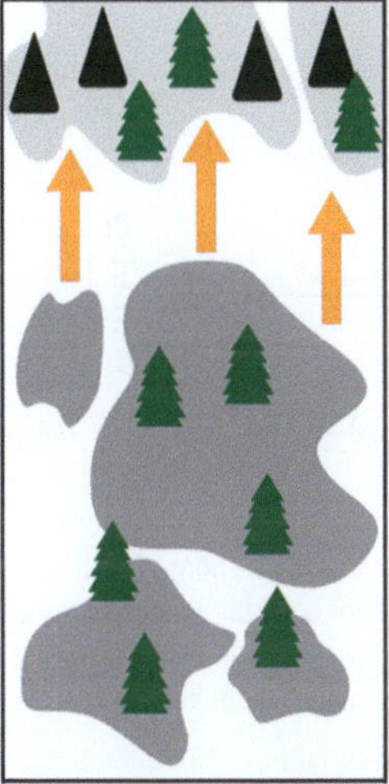
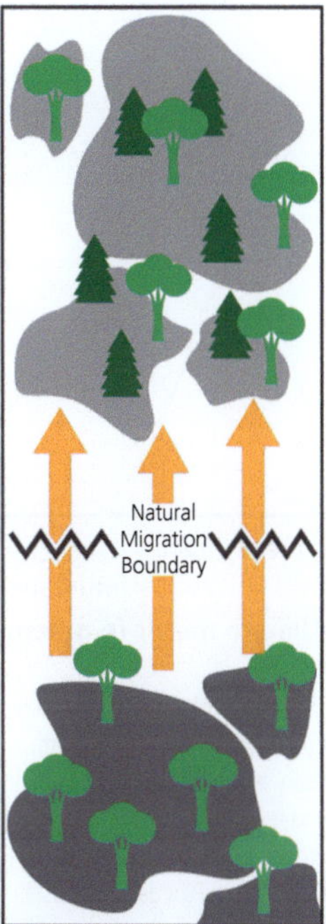

Figure 7.8 The gradient of assisted migration options for adapting managed forests to climate change. Assisted population migration moves seeds or other genetic material within the current range of a species; assisted range expansion moves genetic material just beyond the current range of a species (but potentially within natural dispersal distances or within paleo-historical ranges); and assisted species migration moves genetic material far beyond the current range of a species, potentially into different parts of the continent, or even among continents (graphic from Handler et al. (2018); public domain, courtesy of US Department of Agriculture).

interacting with other components of a region's flora and fauna; this could result in significant transformation of the system from the perspective of non-timber values yet may be perfectly acceptable to sustain ongoing or improved timber production on land designated for that purpose.

There are many considerations in undertaking assisted migration, even for intensively managed tree crops. Concerns include the risk of invasive spread beyond managed stands, the concurrent expansion of associated pests and diseases, contamination of local gene pools (Aubin et al. 2011; Pedlar et al. 2012), and inadequate compatibility with supportive soil fungi and microflora (Kranabetter et al. 2012). Long rotation lengths and lifespans of forest trees—especially in high-latitude and

high-elevation locations—also mean that plant materials can be expected to experience a wide range of climatic conditions between establishment and harvest. With climate-change trajectories and global carbon emissions increasingly uncertain for distant periods of time, it is important to adopt a low-risk, "no regrets" policy (Heltberg et al. 2009). For silviculture, this means utilizing provenances and species with broad tolerances, even if they might be less productive under current or expected future conditions (see Figure 7.9(a)). Trees and stands at all stages of development will need to tolerate the full range of conditions that will constitute their environment during stand development and must be able to thrive under current conditions as well as those expected in the future. For example, if a tree species or seed lot is being moved polewards or to a higher elevation, tree seedlings will still

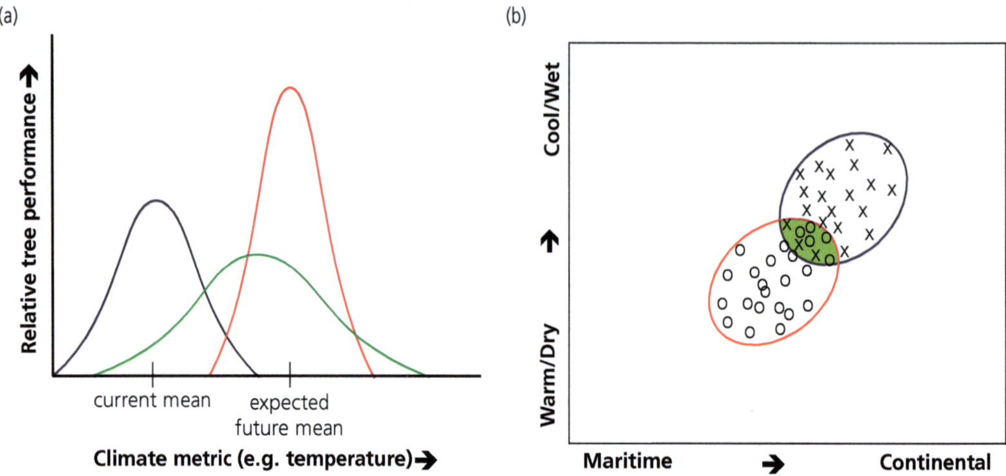

Figure 7.9 Considerations for the assisted migration of tree populations and species. (a) Resilient adaptation strategies favor use of broad-niched plant material (green curve), with wide environmental response functions or tolerances (even if performance is sub-optimal), considering a necessary establishment and transition period, or uncertainty about future conditions. (b) Plant materials must successfully establish, grow, and survive under current climatic conditions (blue ellipse) as well as those expected in the future (red ellipse), suggesting that plant material should be sourced from populations at locations (green-shaded polygon) in which current and expected future bioclimatic conditions overlap (based on O'Neill and Gómez-Pineda (2021)).

need to survive the occasional incidence of growing season frost, even if they may be well adapted to a longer frost-free growing season fifty years from now (see Figure 7.9(b)).

Fortunately, there are several online bioclimatic envelope and plant hardiness models and decision support tools available to aid in coarse-level evaluation of tree migration options. Some examples include:

- the US Forest Service "Seedlot Selection Tool" (CCRC, n.d.);
- the US Forest Service "Climate Change Tree Atlas" (Peters et al. 2020);
- Canada's "SeedWhere" climate similarity mapping tool (McKenney et al. 1999);
- Canada's "Plant Hardiness Site" (Natural Resources Canada 2022);
- The UK's "Climate Matching Tool" (Forest Research 2023); and
- the international "D4R, Diversity for Restoration" tool (Fremout et al. 2022).

Those papers and the websites hosting those tools also refer to many other useful publications and related online decision support tools.

7.4.4 Rotation and species selection

Forests being managed primarily for timber production are managed with shorter rotation lengths than those practiced in multiple-use forests. As introduced in Chapter 3, optimal rotation lengths—exclusive of any other management constraints and land-use objectives—reflect the culmination of mean annual increment (MAI), after which it would be more profitable for the landowner to harvest the existing crop and plant a new one. Those timber-optimizing rotations (see Table 7.2; see also additional plantation yields and comparisons with non-intensive forest productivity in Tables 2.5 and 2.6) are considerably shorter than the potential life span of the component species, or the age at which pathogens or other natural agents of mortality become important. Although short rotations clearly compromise other forest values (such as biodiversity and carbon sequestration) in multipurpose forests, their use in production forests allows managers to reevaluate tree species and provenance options on a frequent basis. Indeed, holding the investment as standing timber for as short a period a time as possible is a reasonable risk-reduction strategy for coping with unforeseen extreme weather

Table 7.2 Representative rotation lengths and productivity (mean annual increment, MAI) for some of the world's most important plantation tree species

Tree species	Country or region	Rotation length (years)	MAI ($m^3 \cdot ha^{-1} \cdot yr^{-1}$)
Broadleaf species			
Acacia mearnsii	Indonesia	7–10	10–25
Betula pendula or *B. pubescens*	Sweden, Finland	25–40	4–5.5
Eucalyptus x (hybrids)	Brazil	7	40
Eucalyptus camaldulensis	Australia	10–20	5–10
Eucalyptus grandis	South Africa	8–10	20
Eucalyptus globulus	Chile, Portugal	10–15	12–20
Populus x (hybrids)	Italy	10–12	17–20
Tectona grandis	India	60–80	4–8
Conifer species			
Picea abies	Sweden	70–80	4.5–7
Picea abies	France, Germany	75	13–15
Picea glauca	central Canada	55	2.5
Picea mariana	eastern Canada	90	2
Picea sitchensis	Scotland, UK	35–50	10
Pinus elliottii, P. taeda	Southeast USA	25	22
Pinus radiata	Chile, New Zealand	25	22
Pinus spp.	Brazil	15–20	16
Pseudotsuga menziesii	BC, Canada	45	6
Pseudotsuga menziesii	France	40	6
Pseudotsuga menziesii	OR & WA, USA	40–50	13

Sources: Mead (2001); Talbert and Marshall (2005); Landsberg and Waring (2014); Cameron (2015); Freer-Smith et al. (2019); Spiecker et al. (2019).

events, market shifts, and other disruptions (Lien et al. 2007). Selecting tree species or germplasm to match the climate expected to prevail over the next fifteen to twenty-five years, for example, is much more feasible and subject to less uncertainty than planning for seventy-five- or one-hundred-year rotations.

Production forests need not consist of single-species, even-aged stands. As mentioned in Section 7.3, there are important opportunities for boosting total stand yield by growing two or more complementary species (such as a conifer and a broadleaf) together. With long-term soil health and nutrition paramount, there can also be advantages to having a significant deciduous or nitrogen-fixing (herb, shrub, or companion tree) component to softwood (conifer) production forests. Likewise, maintaining a multi-species plantation is a risk-reduction strategy where vulnerability to insect pests and diseases are a consideration, as most such pests are restricted to particular tree species or genus (see Section 8.5). In this manner, tree species diversity can support timber-production objectives

while also providing ancillary benefits to wildlife such as songbirds.

7.5 Navigating change

Managing forests with climate change in mind is a daunting task, not only because of the long lifespan of trees, but because of the many uncertainties involved. As introduced in Section 6.4, regionally specific vulnerability analyses can help managers anticipate the gross direction of change—drier or wetter, warmer or cooler—that can be expected for the area of interest. Perhaps the most relevant message is that forests are simply going to be exposed to greater variability, greater extremes of weather, than experienced in the recent past. On the other hand, well-established healthy trees can have great tolerances to the vicissitudes of nature (Falk et al. 2022), while tree reproduction and young seedlings are more sensitive. Along with the cumulative effects of widespread deforestation and timber harvesting, natural disturbances are expected to become more prevalent, challenging the inherent resistance ability and (potentially, depending on disturbance severity and landscape context) recovery potential of existing forests, as explored further in Chapter 8. Conversely, those disturbances clear the way for adjusting and reconfiguring the future forest with better suited genetic material and species. Some forests may have to transform into non-wooded ecosystems, while still conserving and delivering selected ecosystem services.

Recent years have seen many publications listing an array of strategies, tactics, practices, and tools that might be employed in managing forests under a changing climate (e.g. Table 7.3). Such advice typically starts with an emphasis on the importance of place-based vulnerability assessments, evaluating potential climate (or extreme weather event) impacts, and reviewing adaptive capacity—not just of specific forest ecosystems, species, and locations, but also of the managers, companies, agencies, policies, and communities invested in the forest. Resistance strategies are generally the first priority, but, when they fail or when disturbances (natural or anthropogenic) occur, opportunities for adjustment and better adaptation to future climates present themselves. Even production forests can benefit from naturalistic approaches to climate change adaptation, however, rather than undertaking repeated and intensive interventions (Fenton et al. 2024). Hopefully, ecosystems with enough biodiversity and connectivity will have the adaptive capacity for genetic changes and species range shifts to accommodate changing conditions, but this

Table 7.3 Climate change adaptation strategies and approaches for managed forests

Strategy 1: Sustain fundamental ecological functions
- Maintain or restore soil quality and nutrient cycling
- Maintain or restore hydrology
- Maintain or restore riparian areas

Strategy 2: Reduce the impact of existing biological stressors
- Maintain or improve the ability of forests to resist pests and pathogens
- Prevent the introduction and establishment of invasive species and remove existing invasives
- Manage herbivory to protect or promote regeneration

Strategy 3: Protect forests from severe fire and wind disturbance
- Alter forest structure or composition to reduce risk or severity of fire
- Establish fuel breaks to slow the spread of catastrophic fire
- Alter forest structure to reduce severity or extent of wind and ice damage

Strategy 4: Maintain or create refugia
- Prioritize and protect existing populations on unique sites
- Prioritize and protect sensitive or at-risk species or communities
- Establish artificial reserves for at-risk and displaced species

Strategy 5: Maintain and enhance species and structural diversity
- Promote diverse age classes
- Maintain and restore diversity of native tree species
- Retain biological legacies
- Restore fire to fire-adapted ecosystems
- Establish reserves to protect ecosystem diversity

Strategy 6: Increase ecosystem redundancy across the landscape
- Manage habitats over a range of sites and conditions
- Expand the boundaries of reserves to increase diversity

Strategy 7: Promote landscape connectivity
- Use landscape-scale planning and partnerships to reduce fragmentation and enhance connectivity
- Establish and expand reserves and reserve networks to link habitats and protect key communities
- Maintain and create habitat corridors through reforestation or restoration

Strategy 8: Enhance genetic diversity
- Use seeds, germplasm, and other genetic material from across a greater geographic range
- Favor existing genotypes that are better adapted to future conditions
- Increase diversity of nursery stock to provide those species or genotypes likely to succeed

---

Strategy 9: Facilitate community adjustments through species transitions
- Anticipate and respond to species decline
- Favor or restore native species that are expected to be better adapted to future conditions
- Manage for species and genotypes with wide moisture and temperature tolerances
- Emphasize drought- and heat-tolerant species and populations
- Guide species composition at early stages of stand development
- Protect future-adapted regeneration from herbivory
- Establish or encourage new mixes of native species
- Identify and move species to sites that are likely to provide future habitat

Strategy 10: Plan for and respond to disturbance
- Prepare for more frequent and more severe disturbances
- Prepare to realign significantly altered ecosystems to meet future environmental conditions
- Promptly revegetate sites after disturbance
- Allow for areas of natural regeneration after disturbance
- Maintain seed or nursery stock of desired species for use after severe disturbance
- Remove or prevent establishment of invasives and other competitors after disturbance

---

Note: Note the analogous advice for climate change adaptation in protected areas (Table 7.1), and general advice for implementing ecological restoration (Chapter 9) and forest stewardship (Chapter 12).
Source: From Janowiak et al. (2014).

can be challenging in fragmented and degraded landscapes.

It is a widely accepted principal of ecological stewardship that the recovery or restoration of ecosystems disturbed by humans is imperative (see Chapter 10). But the promotion of simple recovery, or some kind of historical reconstruction, is not usually sufficient to assure forest sustainability under directional climate change. Hence we see widespread discussion about the purposeful movement of climate-adapted populations within a species' range, the movement of species beyond their natural ranges, and the overall reconfiguration of forest composition and structure in efforts to promote persistence and productivity in future climates. Various stakeholders may have different criteria for transitioning from resistance and recovery to adjustment and reconfiguration, and in what they consider an acceptable or unacceptable transformation of forests (see Box 7.3). Yet the value of removing other stressors on a forest landscape—pollution, land-use conversion, clearcut logging, siltation, over-hunting, invasive species, and so on—is also understood as important in facilitating its ability to

---

Box 7.3 A gradient of climate-coping strategies and options

Peterson St-Laurent et al. (2021) evaluated 104 USA conservation projects funded since 2011 for purposes of climate-change adaptation. Those projects were scored on a six-point "R–R–T" (resistance—resilience—transformation) scale (see Figure 7.10), which is analogous to the robustness—adaptation—transformation modes of resilience in Table 4.1 and the associated five-point continuum of resilience mechanisms outlined in Chapter 4. In addition to documenting ecosystem differences in the proportion of different R–R–T strategies being supported, they also report a shift from active resistance, passive resistance, and resilience strategies to autonomous, directed, and accelerated transformation strategies starting in 2016. Transition strategies often promote the development of more open woodland conditions, sometimes in response to undesirable thresholds being crossed, such as the loss of canopy species to an invasive insect (Ontl et al. 2018). Hagerman et al. (2021) report various demographic differences among 1,490 North American residents surveyed, and some noteworthy shifts in opinion upon the provision of specific information,

Box 7.3 *Continued*

regarding their comfort with interventionist approaches to conservation planning under climate change. Kolström et al. (2011) point out that there are many different limiting factors, expected climate changes, and socioeconomic conditions within Europe, which collectively influence vulnerability and preferred adaptation options on a regional basis. This is true of all continents. Such patterns affirm the need for ecosystem-specific, community-focused approaches to be developed through stakeholder engagement. Like all forest management decisions, directions selected in managing for climate-change resilience have long-term implications, and always risk being out of sync with evolving public opinion and values.

A coordinated program of place-based adaptations for stand-level forest management is being tested at more than a dozen research locations in the US and Canada (https://www.adaptivesilviculture.org/). This "Adaptive Silviculture for Climate Change" (ASCC) network embraces place-based manager–scientist collaboration to evaluate forest management options for an uncertain future (Millar et al. 2007). Building on a 3-point R–R–T framework (see Figure 7.10(b)), regionally based scientists and forest resource managers were invited to co-devise appropriate silvicultural interventions to manage forests for resistance, "resilience," or transition under a changing climate (Nagel et al. 2017). Workshops were used to review downscaled, locally relevant

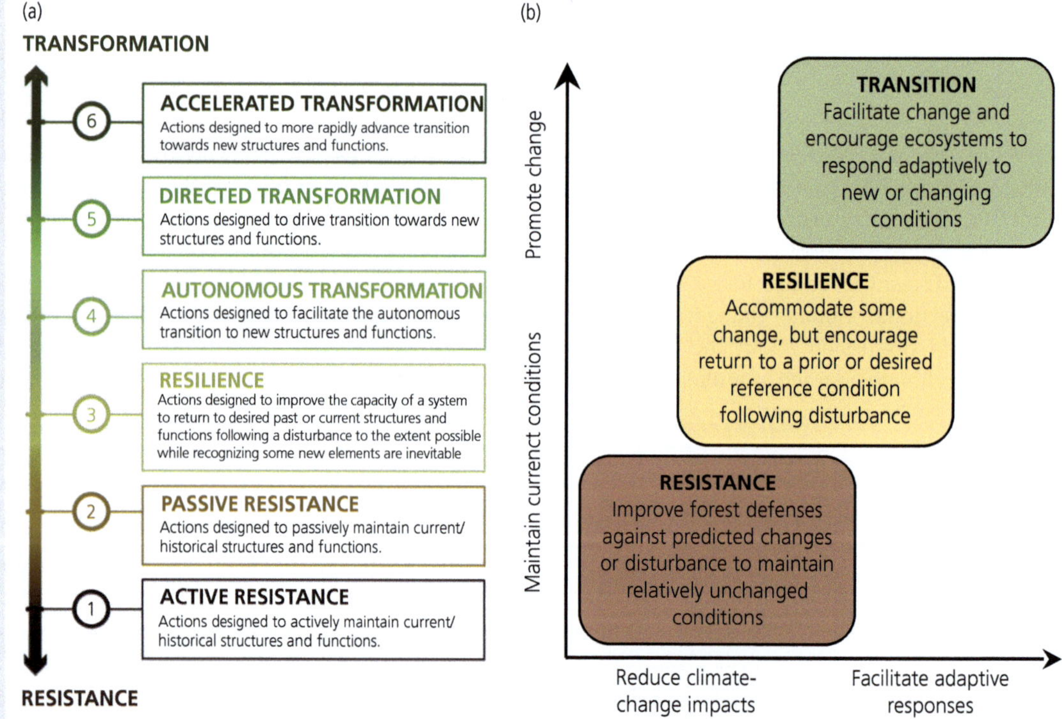

Figure 7.10 The R–R–T gradient of adapting to climate change. Interpreted in (a) an evaluation of conservation projects (Peterson St-Laurent et al. (2021), reproduced under the Creative Commons license CC BY 4.0, http://creativecommons.org/licenses/by/4.0/) and (b) adaptive silviculture for climate change (redrawn from Nagel et al. (2017) and Hoepting et al. (2019)).

Box 7.3 *Continued*

climate change projections, and to elicit local knowledge and values to identify suitable treatments, tree species, and levels of tolerable change. Typical approaches being implemented include density management (spacing, reduced stocking) to enhance resistance, manipulating stand composition to promote species with broad tolerances from among those already found in the area to improve resilience, and assisted migration of tree genotypes and species from outside the area to facilitate transition to stands better adapted to the changing climate. All treatments are being replicated in multiple stands (8–10 ha in size) at each location along with "no treatment" controls. Species composition, health, and productivity of the ground vegetation, mid-story, and overstory layers are being monitored before, immediately after, and three, five, and ten years after treatment. Overall project design facilitates valid statistical comparisons within and among locations, as well as important local learning and knowledge development. Similar operational trials and R–R–T scenarios are being modeled, installed, and monitored elsewhere (e.g. Buma and Wessman 2013).

---

cope with a changing climate and cumulative effects (Ogden and Innes 2007; Janowiak et al. 2014).

Options that forest managers may wish to consider in preparing for a changing climate are further described in publications such as Spittlehouse and Stewart (2003), Freer-Smith et al. (2007), Ogden and Innes (2007), Guariguata et al. (2008), Seppälä et al. (2009), Lindenmayer et al. (2010), Mery et al. (2010), Pérez et al. (2010), Spies et al. (2010), Kolström et al. (2011), Seidl et al. (2011), Messier et al. (2013), Gauthier et al. (2014), Janowiak et al. (2014), Keenan (2015), Schmitz et al. (2015), Gagné et al. (2016), Palik et al. (2021: ch. 16), and Montoro Girona et al. (2023), all of which represent different perspectives. Antwi et al. (2024a) list several specific forest practices that have been implemented or are being tested in different jurisdictions to date. Some potential tactics not described in detail above include restoring populations of overexploited species, reduced-impact logging, minimizing other stressors such as air pollution and over-harvesting, introducing greater structural complexity (vertical and horizontal diversity in stand structure), updating growth and yield models, and moving to continuous-cover forestry. Practitioners often refer to the need for altered harvesting seasons, improved stream crossing standards, and other adjustments to forest operations and forest management policies. Many approaches also address ways to reduce the risk of damage from natural disturbances (wildfire, insect outbreaks, wind and ice storms; see Chapter 8), to protect or enhance carbon sinks, or to shorten rotations or diversify tree species as conscious hedges against the risk of potential changes in the future value of forest goods and services. In practice, many forest managers develop contingency plans, opting for a planned reactive approach rather than proactive interventions, with response options identified a priori in the event of elevated tree mortality or other ecosystem shifts (Janowiak et al. 2014). Antwi et al. (2024b) report that the most important factors contributing to the success of climate-change adaptation implementation practices in Canadian forestry have been collaboration and adequate funding support for research and program delivery, while barriers include policy conflicts, forest tenure arrangements, and gaps in technical capacity.

Box 7.4 Key points

- Adapting to climate change is primarily about preparing for greater variability and uncertainty in the many expressions of a warming climate and the associated biophysical processes that it affects.
- Place-based climate vulnerability assessments, which also consider local priorities and acceptable thresholds for the delivery of ecosystem services, can help prioritize management responses from the continuum of resistance, recovery, adjustment, reconfiguration, and transformation options.
- Conservation-oriented approaches include the protection of high-carbon reservoirs, high-biodiversity reserves, soils and their associated biota, locations likely to escape disturbances and the most drastic impacts of a warming climate, and overall landscape connectivity.

Change is the only constant in life.

Heraclitus (c.500 BCE)

References cited

Agee, J. K., and Skinner, C. N. (2005). "Basic Principles of Forest Fuel Reduction Treatments," *Forest Ecology and Management*, 211/1–2: 83–96.

Aitken, S. N., and Whitlock, M. C. (2013). "Assisted Gene Flow to Facilitate Local Adaptation to Climate Change," *Annual Review of Ecology, Evolution, and Systematics*, 44: 367–88.

Allen, C. D., Breshears, D. D., and McDowell, N. G. (2015). "On Underestimation of Global Vulnerability to Tree Mortality and Forest Die-off from Hotter Drought in the Anthropocene," *Ecosphere*, 6/8: 1–55.

Allen, C. D., Macalady, A. K., Chenchouni, H., et al. (2010). "A Global Overview of Drought and Heat-Induced Tree Mortality Reveals Emerging Climate Change Risks for Forests," *Forest Ecology and Management*, 259/4: 660–84.

Anderson, M. G., Clark, M., Olivero, A. P., et al. (2023). "A Resilient and Connected Network of Sites to Sustain Biodiversity under a Changing Climate," *Proceedings of the National Academy of Sciences*, 120/7: e2204434119.

Anderson, M. G., and Ferree, C. E. (2010). "Conserving the Stage: Climate Change and the Geophysical Underpinnings of Species Diversity," *PloS One*, 5/7: e11554.

Antwi, E. K., Burkhardt, H., Boakye-Danquah, J., Doucet, T., and Abolina, E. (2024a). "Review of Climate Change Adaptation and Mitigation Implementation in Canada's Forest Ecosystems Part I: Reporting, Science, and Institutional/Governance Supporting Practices in Canada," *Environmental Reviews*, 32: 16–41.

Antwi, E. K., Burkhardt, H., Boakye-Danquah, J., Doucet, T., and Abolina, E. (2024b). "Review of Climate Change Adaptation and Mitigation Implementation in Canada's Forest Ecosystems Part II: Successes and Barriers to Effective Implementation," *Environmental Reviews*, 32: 42–67.

Aubin, I., Garbe, C. M., Colombo, S., et al. (2011). "Why We Disagree about Assisted Migration: Ethical Implications of a Key Debate Regarding the Future of Canada's Forests," *Forestry Chronicle*, 87/6: 755–65.

Aubin, I., Munson, A. D., Cardou, F., et al. (2016). "Traits to Stay, Traits to Move: A Review of Functional Traits to Assess Sensitivity and Adaptive Capacity of Temperate and Boreal Forests to Climate Change," *Environmental Reviews*, 24/2: 164–86.

Bauhus, J., Forrester, D. I., and Pretzsch, H. (2017). "Mixed-Species Forests: The Development of a Forest Management Paradigm," in H. Pretzsch, D. I. Forrester, and J. Bauhus (eds), *Mixed-Species Forests: Ecology and Management*. Berlin: Springer, 1–25.

Beiler, K. J., Durall, D. M., Simard, S. W., Maxwell, S. A., and Kretzer, A. M. (2010). "Architecture of the Wood-Wide Web: *Rhizopogon* spp. Genets Link Multiple Douglas-Fir Cohorts," *New Phytologist*, 185/2: 543–53.

Bergeron, Y., Harvey, B., Leduc, A., and Gauthier, S. (1999). "Forest Management Guidelines Based on Natural Disturbance Dynamics: Stand- and Forest-Level Considerations," *Forestry Chronicle*, 75/1: 49–54.

Bowditch, E., Santopuoli, G., Binder, F., et al. (2020). "What is Climate-Smart Forestry? A Definition from a Multinational Collaborative Process Focused on Mountain Regions of Europe," *Ecosystem Services*, 43: 101113.

Brockerhoff, E. G., Barbaro, L., Castagneyrol, B., et al. (2017). "Forest Biodiversity, Ecosystem Functioning and the Provision of Ecosystem Services," *Biodiversity and Conservation*, 26: 3005–35.

Brodie, L. C., and Harrington, C. A. (2020). *Guide to Variable-Density Thinning Using Skips and Gaps*. General Technical Report PNW-GTR-989. Portland, OR: USDA Forest Service. 37 pp. Available online at https://www.fs.usda.gov/pnw/pubs/pnw_gtr989.pdf (accessed February 13, 2024).

Buckley, P., and Mills, J. (2015). "Coppice Silviculture: From the Mesolithic to the 21st Century," in K. Kirby and C. Watkins (eds), *Europe's Changing Woods and Forests: From Wildwood to Managed Landscapes*. Wallingford, UK: CAB International, 77–92.

Buma, B., and Wessman, C. A. (2013). "Forest Resilience, Climate Change, and Opportunities for Adaptation: A Specific Case of a General Problem," *Forest Ecology and Management*, 306: 216–25.

Burrows, G. E. (2021). "Gymnosperm Resprouting—a Review," *Plants*, 10/12: 2551.

Burton, P. J., and Burton, C. M. (2002). "Promoting Genetic Diversity in the Production of Large Quantities of Native Plant Seed," *Ecological Restoration*, 20/2: 116–22.

Burton, P. J., Messier, C., Smith, D. W., and Adamowicz, W. L. (eds). (2003). *Towards Sustainable Management of the Boreal Forest*. Ottawa: NRC Research Press. 1039 pp.

Caldeira, M. C., Lecomte, X., David, T. S., Pinto, J. G., Bugalho, M. N., and Werner, C. (2015). "Synergy of Extreme Drought and Shrub Invasion Reduce Ecosystem Functioning and Resilience in Water-Limited Climates," *Scientific Reports*, 5/1: 1–9.

Cameron, A. D. (2015). "Building Resilience into Sitka Spruce (*Picea sitchensis* (Bong.) Carr.) Forests in Scotland in Response to the Threat of Climate Change," *Forests*, 6: 398–415.

Carcaillet, C., Richard, P. J., Bergeron, Y., Fréchette, B., and Ali, A. A. (2010). "Resilience of the Boreal Forest in Response to Holocene Fire-Frequency Changes Assessed by Pollen Diversity and Population Dynamics," *International Journal of Wildland Fire*, 19/8: 1026–39.

Carey, A. B. (2003). "Biocomplexity and Restoration of Biodiversity in Temperate Coniferous Forest: Inducing Spatial Heterogeneity with Variable-Density Thinning," *Forestry*, 76/2: 127–36.

Carter, T. A., and Buma, B. (2024). "The Distribution of Tree Biomass Carbon within the Pacific Coastal Temperate Rainforest, a Disproportionally Carbon Dense Forest," *Canadian Journal of Forest Research*, 54/9: 956–77

Castagneri, D., Vacchiano, G., Hacket-Pain, A., DeRose, R. J., Klein, T., and Bottero, A. (2022). "Meta-Analysis Reveals Different Competition Effects on Tree Growth Resistance and Resilience to Drought," *Ecosystems*, 24: 30–43.

CCRC (Climate Change Resource Center). (n.d.) *Seedlot Selection Tool*. Washington: USDA Forest Service. Available online at https://www.fs.usda.gov/ccrc/tool/seedlot-selection-toolandat https://seedlotselectiontool.org/sst/ (accessed February 13, 2024).

Chapin, F. S., Kofinas, G. P., and Folke, C. (eds) (2009). *Principles of Ecosystem Stewardship: Resilience-Based Natural Resource Management in a Changing World*. New York: Springer. 401 pp.

Coates, K. D., and Burton, P. J. (1997). "A Gap-Based Approach for Development of Silvicultural Systems to Address Ecosystem Management Objectives," *Forest Ecology and Management*, 99/3: 337–54.

Coll, L. I., Balandier, P., Picon-Cochard, C., Prevosto, B., and Curt, T. (2003). "Competition for Water between Beech Seedlings and Surrounding Vegetation in Different Light and Vegetation Composition Conditions," *Annals of Forest Science*, 60: 593–600.

Corson, C., and Campbell, L. M. (2023). "Conservation at a Crossroads: Governing by Global Targets, Innovative Financing, and Techno-Optimism or Radical Reform?" *Ecology and Society*, 28/2: 3.

D'Amato, A. W., Bradford, J. B., Fraver, S., and Palik, B. J. (2013). "Effects Of Thinning on Drought Vulnerability and Climate Response in North Temperate Forest Ecosystems," *Ecological Applications*, 23/8: 1735–42.

Dickie, I. A., Guza, R. C., Krazewski, S. E., and Reich, P. B. (2004). "Shared Ectomycorrhizal Fungi between a Herbaceous Perennial (*Helianthemum bicknellii*) and Oak (*Quercus* Seedlings," *New Phytologist*, 164/2: 375–82.

Dudley, N. (2008). "The Use of Protected Areas as Tools to Apply REDD Carbon Offset Schemes," *Policy Matters (IUCN Commission on Environmental, Economic and Social Policy)*, 16: 99–107.

Durall, D. M., Jones, M. D., Wright, E. F., Kroeger, P., and Coates, K. D. (1999). "Species Richness of Ectomycorrhizal Fungi in Cutblocks of Different Sizes in the Interior Cedar-Hemlock Forests of Northwestern British Columbia: Sporocarps and Ectomycorrhizae," *Canadian Journal of Forest Research*, 29/9: 1322–32.

Ehbrecht, M., Seidel, D., Annighöfer, P. et al. (2021). "Global Patterns and Climatic Controls of Forest Structural Complexity," *Nature Communications*, 12/1: 519.

El-Kassaby, Y. A., and Askew, G. R. (1998). "Seed Orchards and their Genetics," in A. K. Mandal and G. L. Gibson (eds), *Forest Genetics and Tree Breeding*. New Delhi: CBS Publishers & Distributors, 103–11.

Ek, H., Andersson, S., and Söderström, B. (1997). "Carbon and Nitrogen Flow in Silver Birch and Norway Spruce Connected by a Common Mycorrhizal Mycelium," *Mycorrhiza*, 6/6: 465–7.

Esperon-Rodriguez, M., Ordoñez, C., van Doorn, N. S., Hirons, A., and Messier, C. (2022). "Using Climate Analogues and Vulnerability Metrics to Inform Urban Tree Species Selection in a Changing Climate: The Case for Canadian Cities," *Landscape and Urban Planning*, 228: 104578.

Falk, D. A., van Mantgem, P. J., Keeley, J. E., et al. (2022). "Mechanisms of Forest Resilience," *Forest Ecology and Management*, 512: 120129.

Felton, A., Seidl, R., Lindenmayer, D. B., et al. (2024). "The choice of path to resilience is crucial to the future of production forests," *Nature Ecology & Evolution*, 8: 1561–3.

Feurdean, A., Ruprecht, E., Molnár, Z., Hutchinson, S. M., and Hickler, T. (2018). "Biodiversity-Rich European Grasslands: Ancient, Forgotten Ecosystems," *Biological Conservation*, 228: 224–32.

Filotas, E., Parrott, L., Burton, P. J., et al. (2014). "Viewing Forests through the Lens of Complex Systems Science," *Ecosphere*, 5/1: 1.

Finegan, B. (1984). "Forest Succession," *Nature*, 312/5990: 109–14.

Forest Research (2023). *Climate Matching Tool: Selecting Forest Reproductive Material Suited to Current and Future Climates to Maintain Forest Resilience.* Roslin, Scotland: UK Forest Research. Available online at https://www.forestresearch.gov.uk/tools-and-resources/fthr/climate-matching-tool/ (accessed February 13, 2024).

Franklin, J. F., Johnson, K. N., and Johnson, D. L. (2018). *Ecological Forest Management.* Long Grove, IL: Waveland Press. 646 pp.

Freer-Smith, P. H., Broadmeadow, M. S., and Lynch, J. M. (eds). (2007). *Forestry and Climate Change.* Wallingford, UK: CAB International. 252 pp.

Freer-Smith, P., Muys, B., Bozzano, M., et al. (2019). *Plantation Forests in Europe: Challenges and Opportunities.* Joensuu, Finland: European Forest Institute. 50 pp. Available online at https://doi.org/10.36333/fs09 (accessed February 13, 2024).

Fremout, T., Thomas, E., Taedoumg, H., et al. (2022). "Diversity for Restoration (D4R): Guiding the Selection of Tree Species and Seed Sources for Climate-Resilient Restoration of Tropical Forest Landscapes," *Journal of Applied Ecology*, 59/3: 664–79 (with a computer application calibrated for selected countries available online at https://www.diversityforrestoration.org/tool.php (accessed February 13, 2024)).

Gagné, L., Sirois, L., and Lavoie, L. (2016). "Forest Management and Climate Change: Adaptive Measures for the Temperate-Boreal Interface of Eastern North America," in G. R. Larocque (ed.), *Ecological Forest Management Handbook.* Boca Raton, FL: CRC Press, 561–87.

Gauthier, S., Bernier, P., Burton, P. J., et al. (2014). "Climate Change Vulnerability and Adaptation in the Managed Canadian Boreal Forest," *Environmental Reviews*, 22/3: 256–85.

Gresh, G. O. (2018). "Carbon Capture, Sequestration, and Storage in Washington State Parks: A Review of Relevant Policy and Project Feasibility." MA Thesis, University of Washington, Seattle. 69 pp. Available online at https://digital.lib.washington.edu/researchworks/handle/1773/42752 (accessed February 13, 2024).

Griscom, B. W., Adams, J., Ellis, P. W., et al. (2017). "Natural Climate Solutions," *Proceedings of the National Academy of Sciences*, 114/44: 11645–50.

Guariguata, M. R., Cornelius, J. P., Locatelli, B., Forner, C., and Sánchez-Azofeifa, G. A. (2008). "Mitigation Needs Adaptation: Tropical Forestry and Climate Change," *Mitigation and Adaptation Strategies for Global Change*, 13/8: 793–808.

Habte, M., Miyasaka, S. C., and Matsuyama, D. T. (2001). "Arbuscular Mycorrhizal Fungi Improve Early Forest-Tree Establishment," in W. J. Horst, M. K. Schenk, A. Bürkert, et al. (eds), *Plant Nutrition: Food Security and Sustainability of Agro-Ecosystems through Basic and Applied Research.* Dordrecht: Kluwer, 644–5.

Hagerman, S., Satterfield, T., Nawaz, S., Peterson St-Laurent, G., Kozak, R., and Gregory, R. (2021). "Social Comfort Zones for Transformative Conservation Decisions in a Changing Climate," *Conservation Biology*, 35/6: 1932–43.

Handler, S., Pike, C., St Clair, B., Abotts, H., and Janowiak, J. (2018). *Assisted Migration.* Climate Change Resource Center, USDA Forest Service. Available online at https://www.fs.usda.gov/ccrc/topics/assisted-migration (accessed February 13, 2024).

Harper, K. A., Macdonald, S. E., Burton, P. J., et al. (2005). "Edge Influence on Forest Structure and Composition in Fragmented Landscapes," *Conservation Biology*, 19/3: 768–82.

Heltberg, R., Siegel, P. B., and Jorgensen, S. L. (2009). "Addressing Human Vulnerability to Climate Change: Toward a 'No-Regrets' Approach," *Global Environmental Change*, 19/1: 89–99.

Hiremath, A. J., and Sundaram, B. (2005). "The Fire–Lantana Cycle Hypothesis in Indian Forests," *Conservation and Society*, 3/1: 26–42.

Hoepting, M., Jones, T., and Fera, J. (2019). "Collaborating on Climate Change at the Petawawa Research Forest," *Forestry Chronicle*, 95/2: 62–3.

Horton, T. R., Bruns, T. D., and Parker, V. T. (1999). "Ectomycorrhizal Fungi Associated with *Arctostaphylos* Contribute to *Pseudotsuga menziesii* Establishment," *Canadian Journal of Botany*, 77/1: 93–102.

Hosius, B., Leinemann, L., Konnert, M., and Bergmann, F. (2006). "Genetic Aspects of forestry in the Central Europe," *European Journal of Forest Research*, 125/4: 407–17.

Hunter, M. L. (2007). "Core Principles for Using Natural Disturbance Regimes to Inform Landscape Management," in Lindenmayer, D. B., and Hobbs, R. J. (eds), *Managing and Designing Landscapes for Conservation: Moving from Perspectives to Principles.* Oxford: Blackwell, 408–22.

Innes, J. L., and Tikina, A. V. (eds). (2017). *Sustainable Forest Management: From Concept to Practice.* New York: Routledge. 395 pp.

IPCC (Intergovernmental Panel on Climate Change). (2022). "Summary for Policymakers," in H.-O. Pörtner, D. C. Roberts, M. Tignor, et al., *Climate Change 2022: Impacts, Adaptation and Vulnerability.* Contribution of Working Group II to the Sixth Assessment Report of the Intergovernmental Panel on Climate Change. Cambridge: Cambridge University Press, 3–33. Available on-line at https://www.ipcc.ch/report/ar6/wg2/ (accessed February 13, 2024).

Izzo, A., Agbowo, J., and Bruns, T. D. (2005). "Detection of Plot-Level Changes in Ectomycorrhizal Communities across Years in an Old-Growth Mixed-Conifer Forest," *New Phytologist*, 166/2: 619–30.

Jactel, H., Bauhus, J., Boberg, J., et al. (2017). "Tree Diversity Drives Forest Stand Resistance to Natural Disturbances," *Current Forestry Reports*, 3: 223–43.

Janowiak, M. K., Swanston, C. W., Nagel, L. M., et al. (2014). "A Practical Approach for Translating Climate Change Adaptation Principles into Forest Management Actions," *Journal of Forestry*, 112/5: 424–33.

Kabrick, J. M., Clark, K. L., D'Amato, A. W., et al. (2017). "Managing Hardwood–Softwood Mixtures for Future Forests in Eastern North America: Assessing Suitability to Projected Climate Change," *Journal of Forestry*, 115/3: 190–201.

Kadam, P. (2018). "Ecosystem-Service Based Offset Program (ESBOP) and its Governance Structure for Canadian Parks and Protected Areas (PPAs)." Masters Thesis, University of Toronto, Toronto. 40 pp. Available online at https://tspace.library.utoronto.ca/handle/1807/81193 (accessed February 13, 2024).

Kanninen, M., Brackhaus, M., Murdiyarso, D., and Nabuurs, G.-J. (2010). "Harnessing Forests for Climate Change Mitigation through REDD+," in G. Mery, P. Katila, G. Galloway, et al. (eds), *Forests and Society: Responding to Global Drivers of Change*. IUFRO World Series, vol. 5. Vienna: International Union of Forest Research Organizations, 43–54.

Karst, J., Jones, M. D., and Hoeksema, J. D. (2023). "Positive Citation Bias and Overinterpreted Results Lead to Misinformation on Common Mycorrhizal Networks in Forests," *Nature Ecology & Evolution*, 7/4: 501–11.

Keenan, R. J. (2015). "Climate Change Impacts and Adaptation in Forest Management: A Review," *Annals of Forest Science*, 72/2: 145–67.

Keenleyside, K., Dudley, N., Cairns, S., Hall, C., and Stolton, S. (eds). (2012). *Ecological Restoration for Protected Areas: Principles, Guidelines and Best Practices*. Best Practices Protected Area Guidelines Series, No. 18. Gland, Switzerland: International Union for Conservation of Nature and Natural Resources. 120 pp.

Kellett, J., Hamilton, C., Ness, D., and Pullen, S. (2015). "Testing the Limits of Regional Climate Analogue Studies: An Australian Example," *Land Use Policy*, 44: 54–61.

Kelty, M. J. (1992). "Comparative Productivity of Monocultures and Mixed-Species Stands," in M. J. Kelty (ed.), *The Ecology and Silviculture of Mixed-Species Forests*. Dordrecht: Kluwer, 125–41.

Kern, C. C., Burton, J. I., Raymond, P., et al. (2017). "Challenges Facing Gap-Based Silviculture and Possible Solutions for Mesic Northern Forests in North America," *Forestry*, 90/1: 4–17.

Kneeshaw, D., Sturtevant, B., DeGrandpé, L., et al. (2021). "The Vision of Managing for Pest-Resistant Landscapes: Realistic or Utopic?" *Current Forestry Reports*, 7: 97–113.

Kohler, M., Sohn, J., Nägele, G., and Bauhus, J. (2010). "Can Drought Tolerance of Norway Spruce (*Picea abies* (L.) Karst.) Be Increased through Thinning?" *European Journal of Forest Research*, 129/6: 1109–18.

Kolström, M., Lindner, M., Vilén, T., et al. (2011). "Reviewing the Science and Implementation of Climate Change Adaptation Measures in European Forestry," *Forests*, 2/4: 961–82.

Kohm, K. A., and Franklin, J. F. (eds). (1997). *Creating a Forestry for the 21st Century: The Science of Ecosytem Management*. Washington: Island Press. 475 pp.

Kranabetter, J. M., Stoehr, M. U., and O'Neill, G. A. (2012). "Divergence in Ectomycorrhizal Communities with Foreign Douglas-Fir Populations and Implications for Assisted Migration," *Ecological Applications*, 22/2: 550–60.

Krawchuk, M. A., Meigs, G. W., Cartwright, J. M., et al. (2020). "Disturbance Refugia within Mosaics of Forest Fire, Drought, and Insect Outbreaks," *Frontiers in Ecology and the Environment*, 18/5: 235–44.

Kuntzemann, C. E., Whitman, E., Stralberg, D., Parisien, M.-A., Thompson, D. K., and Nielsen, S. E. (2023). "Peatlands Promote Fire Refugia in Boreal Forests of Northern Alberta, Canada," *Ecosphere*, 14/5: e4510.

Lamb, D. (2015). "Restoration of Forest Ecosystems," in K.S.-H. Peh, R.T. Corlett, and Y. Bergeron (eds). *Routledge Handbook of Forest Ecology*. Abingdon, UK: Routledge, 397–410.

Landsberg, J., and Waring, R. (2014). *Forests in our Changing World: New Principles for Conservation and Management*. Washington: Island Press. 209 pp.

Larocque, G. R. (ed.). (2016). *Ecological Forest Management Handbook*. Boca Raton, FL: CRC Press. 604 pp.

Laurindo, R. S., Novaes, R. L. M., Vizentin-Bugoni, J., and Gregorin, R. (2019). "The Effects of Habitat Loss on Bat-Fruit Networks," *Biodiversity and Conservation*, 28: 589–601.

Lehto, T., and Zwiazek, J. J. (2011). "Ectomycorrhizas and Water Relations of Trees: A Review," *Mycorrhiza*, 21: 71–90.

Lerat, S., Gauci, R., Catford, J. G., Vierheilig, H., Piché, Y., and Lapointe, L. (2002). "^{14}C Transfer between the Spring Ephemeral *Erythronium americanum* and Sugar Maple Saplings via Arbuscular Mycorrhizal Fungi in Natural Stands," *Oecologia*, 132/2: 181–7.

Lieffers, V. J., Messier, C., Burton, P. J., Ruel, J.-C., and Grover, B. E. (2003). "Nature-Based Silviculture for Sustaining a Variety of Boreal Forest Values," in P. J. Burton, C. Messier, D. W. Smith, and W. L. Adamowicz (eds), *Towards Sustainable Management of the Boreal Forest*. Ottawa: NRC Research Press, 481–530.

Lien, G., Størdal, S., Hardaker, J. B., and Asheim, L. J. (2007). "Risk Aversion and Optimal Forest Replanting: A Stochastic Efficiency Study," *European Journal of Operational Research*, 181/3: 1584–92.

Lin, S., Liu, Y., and Huang, X. (2021). "Climate-Induced Arctic-Boreal Peatland Fire and Carbon Loss in the 21st Century," *Science of the Total Environment*, 796: 148924.

Lindenmayer, D. B., Steffen, W., Burbidge, A. A., et al. (2010). "Conservation Strategies in Response to Rapid Climate Change: Australia as a Case Study," *Biological Conservation*, 143/7: 1587–93.

Loarie, S. R., Duffy, P. B., Hamilton, H., Asner, G. P., Field, C. B., and Ackerly, D. D. (2009). "The Velocity of Climate Change," *Nature*, 462/7276: 1052–5.

Luyssaert, S., Schulze, E. D., Börner, A., et al. (2008). "Old-Growth Forests as Global Carbon Sinks," *Nature*, 455/7210: 213–15.

McKenney, D. W., Mackey, B. G., and Joyce, D. (1999). "SeedWhere: A Computer Tool to Support Seed Transfer and Ecological Restoration Decisions," *Environmental Modelling and Software*, 14: 589–95 (with a computer application available online with several other forest change adaptation tools at https://natural-resources.canada.ca/climate-change/climate-change-impacts-forests/forest-change-adaptation-tools/17770 (accessed February 13, 2024)).

MacKenzie, W. H., and Mahony, C. R. (2021). "An Ecological Approach to Climate Change-Informed Tree Species Selection for Reforestation," *Forest Ecology and Management*, 481: 118705.

Marvin, D. C., Sleeter, B. M., Cameron, D. R., Nelson, E., and Plantinga, A. J. (2023). "Natural Climate Solutions Provide Robust Carbon Mitigation Capacity under Future Climate Change Scenarios," *Scientific Reports*, 13/1: 19008.

Matula, R., Svátek, M., Kůrová, J., Úradníček, L., Kadavý, J., and Kneifl, M. (2012). "The Sprouting Ability of the Main Tree Species in Central European Coppices: Implications for Coppice Restoration," *European Journal of Forest Research*, 131/5: 1501–11.

Mead, D. J. (ed.). (2001). *Mean Annual Volume Increment of Selected Industrial Forest Plantation Species*. Working Paper FP/1. Rome: Food and Agriculture Organization of the United Nations. 27 pp. Available online at https://www.fao.org/3/ac121e/ac121e.pdf (accessed August 15, 2023).

Mery, G., Katila, P., Galloway, G., et al. (eds). (2010). *Forests and Society: Responding to Global Drivers of Change*. World Series, vol. 25. Vienna: International Union of Forest Research Organizations. 509 pp.

Messier, C., Puettmann, K., Chazdon, R., et al. (2015). "From Management to Stewardship: Viewing Forests as Complex Adaptive Systems in an Uncertain World," *Conservation Letters*, 8/5: 368–77.

Messier, C., Puettmann, K. J., and Coates, K. D. (eds). (2013). *Managing Forests as Complex Adaptive Systems: Building Resilience to the Challenge of Global Change*. New York: Routledge. 353 pp.

Mette, T., Brandl, S., and Kölling, C. (2021). "Climate Analogues for Temperate European Forests to Raise Silvicultural Evidence Using Twin Regions," *Sustainability*, 13/12: 6522.

Millar, C. I., Stephenson, N. L., and Stephens, S. L. (2007). "Climate Change and Forests of the Future: Managing in the Face of Uncertainty," *Ecological Applications*, 17: 2145–51.

Minckley, T. A., Booth, R. K., and Jackson, S. T. (2012). "Response of Arboreal Pollen Abundance to Late-Holocene Drought Events in the Upper Midwest, USA," *The Holocene*, 22/5: 531–9.

Mohan, J. E., Cowden, C. C., Baas, P., et al. (2014). "Mycorrhizal Fungi Mediation of Terrestrial Ecosystem Responses to Global Change: Mini-Review," *Fungal Ecology*, 10: 3–19.

Montoro Girona, M., Morin, H., Gauthier, S., and Bergeron, Y. (eds). (2023). *Boreal Forests in the Face of Climate Change: Sustainable Management*. Cham, Switzerland: Springer. 837 pp.

Morecroft, M. D., Duffield, S., Harley, M., et al. (2019). "Measuring the Success of Climate Change Adaptation and Mitigation in Terrestrial Ecosystems," *Science*, 366/6471: eaaw9256.

Morelli, T. L., Barrows, C. W., Ramirez, A. R., et al. (2020). "Climate-Change Refugia: Biodiversity in the Slow Lane," *Frontiers in Ecology and the Environment*, 18/5: 228–34.

Nagel, L. M., Palik, B. J., Battaglia, M. A., et al. (2017). "Adaptive Silviculture for Climate Change: A National Experiment in Manager–Scientist Partnerships to Apply an Adaptation Framework," *Journal of Forestry*, 115/3: 167–78.

Natural Resources Canada (2022). *Canada's Plant Hardiness Site: Predicted Effect of Climate Change on the Size and Location of Climate Habitat of North American Plant Species*. Ottawa: Natural Resources Canada. Available online at http://planthardiness.gc.ca/?m=16&lang=en (accessed August 16, 2023).

Nehls, U., and Plassard, C. (2018). "Nitrogen and Phosphate Metabolism in Ectomycorrhizas," *New Phytologist*, 220/4: 1047–58.

Newton, P. F. (2021). "Stand Density Management Diagrams: Modelling Approaches, Variants, and Exemplification of their Potential Utility in Crop Planning," *Canadian Journal of Forest Research*, 51/2: 236–56.

Nyland, R. D. (1996). *Silviculture: Concepts and Applications*. New York: McGraw-Hill. 633 pp.

Ogden, A. E., and Innes, J. (2007). "Incorporating Climate Change Adaptation Considerations into Forest Management Planning in the Boreal Forest," *International Forestry Review*, 9/3: 713–33.

O'Hara, K. L., Cox, L. E., Nikolaeva, S., Bauer, J. J., and Hedges, R. (2017). "Regeneration Dynamics of Coast Redwood, a Sprouting Conifer Species: A Review with Implications for Management and Restoration," *Forests*, 8/5: 144.

O'Neill, G., and Gómez-Pineda, E. (2021). "Local *Was* Best: Sourcing Tree Seed for Future Climates," *Canadian Journal of Forest Research*, 51: 1432–9.

Ontl, T. A., Swanston, C., Brandt, L. A., et al. (2018). "Adaptation Pathways: Ecoregion and Land Ownership Influences on Climate Adaptation Decision-Making in Forest Management," *Climatic Change*, 146: 75–88.

Palik, B. J., D'Amato, A. W., Franklin, J. F., and Johnson, K. N. (2021). *Ecological Silviculture: Foundations and Applications*. Long Grove, IL: Waveland Press. 343 pp.

Park, A., Puettmann, K., Wilson, E., Messier, C., Kames, S., and Dhar, A. (2014). "Can Boreal and Temperate Forest Management Be Adapted to the Uncertainties of 21st Century Climate Change?" *Critical Reviews in Plant Sciences*, 33/4: 251–85.

Park, A., and Talbot, C. (2018). "Information Underload: Ecological Complexity, Incomplete Knowledge, and Data Deficits Create Challenges for the Assisted Migration of Forest Trees," *BioScience*, 68/4: 251–63.

Parke, E. L., Linderman, R. G., and Black, C. H. (1983). "The Role of Ectomycorrhizas in Drought Tolerance of Douglas-Fir Seedlings," *New Phytologist*, 95/1: 83–95.

Pedlar, J. H., McKenney, D. W., Aubin, I., et al. (2012). "Placing Forestry in the Assisted Migration Debate," *BioScience*, 62/9: 835–42.

Pedley, S. M., Martin, R. D., Oxbrough, A., Irwin, S., Kelly, T. C., and O'Halloran, J. (2014). "Commercial Spruce Plantations Support a Limited Canopy Fauna: Evidence from a Multi Taxa Comparison of Native and Plantation Forests," *Forest Ecology and Management*, 314: 172–82.

Pelai, R., Hagerman, S. M., and Kozak, R. (2021). "Whose Expertise Counts? Assisted Migration and the Politics of Knowledge in British Columbia's Public Forests," *Land Use Policy*, 103: 105296.

Pérez, Á. A., Fernández, B. H., and Gatti, R. C. (eds). (2010). *Building Resilience to Climate Change: Ecosystem-Based Adaptation and Lessons from the Field*. Gland, Switzerland: International Union for the Conservation of Nature. 164 pp.

Perry, D. A., Amaranthus, M. P., Borchers, J. G., Borchers, S. L., and Brainerd, R. E. (1989). "Bootstrapping in Ecosystems," *BioScience*, 39/4: 230–7.

Peters, M. P., Prasad, A. M., Matthews, S. N., and Iverson, L. R. (2020). *Climate Change Tree Atlas, Version 4.* Delaware, OH: US Forest Service, Northern Research Station and Northern Institute of Applied Climate Science. Available online at https://www.fs.usda.gov/ccrc/tool/climate-change-tree-atlas (accessed February 13, 2024).

Peterson, R. L., Massicotte, H. B., and Melville, L. H. (2004). *Mycorrhizas: Anatomy and Cell Biology*. Ottawa: NRC Research Press. 173 pp.

Peterson St-Laurent, G., Oakes, L. E., Cross, M., and Hagerman, S. (2021). "R–R–T (Resistance–Resilience–Transformation) Typology Reveals Differential Conservation Approaches across Ecosystems and Time," *Communications Biology*, 4/1: 1–9.

Pressey, R. L., Cabeza, M., Watts, M. E., Cowling, R. M., and Wilson, K. A. (2007). "Conservation Planning in a Changing World," *Trends in Ecology & Evolution*, 22/11: 583–92.

Pretzsch, H. (2014). "Canopy Space Filling and Tree Crown Morphology in Mixed-Species Stands Compared with Monocultures," *Forest Ecology and Management*, 327: 251–64.

Pretzsch, H., Schütze, G., and Uhl, E. (2013). "Resistance of European Tree Species to Drought Stress in Mixed versus Pure Forests: Evidence of Stress Release by Inter-Specific Facilitation," *Plant Biology*, 15/3: 483–95.

Pyšek, P., Hulme, P. E., Simberloff, D. et al. (2020). "Scientists' Warning on Invasive Alien Species," *Biological Reviews*, 95/6: 1511–34.

Quoreshi, A. M., Kernaghan, G., and Hunt, G. A. (2009). "Mycorrhizal Fungi in Canadian Forest Nurseries and Field Performance of Inoculated Seedlings," in D. Khasa, Y. Piché, and A. P. Coughlan (eds), *Advances in Mycorrhizal Science and Technology*, Ottawa: NRC Research Press, 115–27.

Rehm, E. M., Olivas, P., Stroud, J., and Feeley, K. J. (2015). "Losing your Edge: Climate Change and the Conservation Value of Range-Edge Populations," *Ecology and Evolution*, 5/19: 4315–26.

Rogeau, M. P., Barber, Q. E., and Parisien, M.-A. (2018). "Effect of Topography on Persistent Fire Refugia of the Canadian Rocky Mountains," *Forests*, 9/6: 285.

Rose, N.-A., and Burton, P. J. (2009). "Using Bioclimatic Envelopes to Identify Temporal Corridors in Support of Conservation Planning in a Changing Climate," *Forest Ecology and Management*, 258: S64–S74.

Ruiz-Lozano, J. M., Porcel, R., Bárzana, G., Azcón, R., and Aroca, R. (2012). "Contribution of Arbuscular Mycorrhizal Symbiosis to Plant Drought Tolerance: State of the Art," in R. Aroca (ed.), *Plant Responses to Drought Stress*. Berlin: Springer, 335–62.

Sáenz-Romero, C., O'Neill, G., Aitken, S. N., and Lindig-Cisneros, R. (2021). "Assisted Migration Field Tests in Canada and Mexico: Lessons, Limitations, and Challenges," *Forests*, 12/1: 9.

Saunders, S. P., Grand, J., Bateman, B. L., et al. (2023). "Integrating Climate-Change Refugia into 30 by 30 Conservation Planning in North America," *Frontiers in Ecology and the Environment*, 21/2: 77–84.

Schmitz, O. J., Lawler, J. J., Beier, P., et al. (2015) "Conserving Biodiversity: Practical Guidance about Climate Change Adaptation Approaches in Support of Land-Use Planning," *Natural Areas Journal*, 35/1: 190–203.

Schwarz, J., Skiadaresis, G., Kohler, M., et al. (2020). "Quantifying Growth Responses of Trees to Drought: A Critique of Commonly Used Resilience Indices and Recommendations for Future Studies," *Current Forestry Reports*, 6: 185–200.

Seidl, R., Rammer, W., and Lexer, M. J. (2011). "Adaptation Options to Reduce Climate Change Vulnerability of Sustainable Forest Management in the Austrian Alps," *Canadian Journal of Forest Research*, 41/4: 694–706.

Seppälä, R., Buck, A., and Katila, P. (eds). (2009). *Adaptation of Forests and People to Climate Change: A Global Assessment Report*. IUFRO World Series, vol. 22. Helsinki, Finland: International Union of Forest Research Organizations. 224 pp.

Shi, L., Guttenberger, M., Kottke, I., and Hampp, R. (2002). "The Effect of Drought on Mycorrhizas of Beech (*Fagus sylvatica* L.): Changes in Community Structure, and the Content of Carbohydrates and Nitrogen Storage Bodies of the Fungi," *Mycorrhiza*, 12: 303–11.

Simard, S. W., Perry, D. A., Jones, M. D., Myrold, D. D., Durall, D. M., and Molina, R. (1997). "Net Transfer of Carbon between Ectomycorrhizal Tree Species in the Field," *Nature*, 388/6642: 579–82.

Simard, S. W., Roach, W. J., Beauregard, J., et al. (2021). "Partial Retention of Legacy Trees Protect Mycorrhizal Inoculum Potential, Biodiversity, and Soil Resources while Promoting Natural Regeneration of Interior Douglas-Fir," *Frontiers in Forests and Global Change*, 3: 620436.

Sohn, J. A., Saha, S., and Bauhus, J. (2016). "Potential of Forest Thinning to Mitigate Drought Stress: A Meta-Analysis," *Forest Ecology and Management*, 380: 261–73.

Song, Y. Y., Simard, S. W., Carroll, A., Mohn, W. W., and Zeng, R. S. (2015). "Defoliation of Interior Douglas-Fir Elicits Carbon Transfer and Stress Signalling to Ponderosa Pine Neighbors through Ectomycorrhizal Networks," *Scientific Reports*, 5/1: 1–9.

Spiecker, H., Lindner, M., and Schuler, J. (eds). (2019). *Douglas-Fir: An Option for Europe*. Joensuu, Finland: European Forest Institute. 121 pp. Available online at https://efi.int/sites/default/files/files/publication-bank/2019/efi_wsctu9_2019.pdf (accessed February 13, 2024).

Spies, T. A., Giesen, T. W., Swanson, F. J., Franklin, J. F., Lach, D., and Johnson, K. N. (2010). "Climate Change Adaptation Strategies for Federal Forests of the Pacific Northwest, USA: Ecological, Policy, and Socio-Economic Perspectives," *Landscape Ecology*, 25/8: 1185–99.

Spittlehouse, D. L., and Stewart, R. B. (2003). "Adaptation to Climate Change in Forest Management," *BC Journal of Ecosystems and Management*, 4/1: 1. Available online at https://jem-online.org/index.php/jem/article/view/254 (accessed February 13, 2024).

Ste-Marie, C., Nelson, E. A., Dabros, A., and Bonneau, M. E. (2011). "Assisted Migration: Introduction to a Multifaceted Concept," *Forestry Chronicle*, 87/6: 724–30.

Stephens, R. B., Frey, S. D., D'Amato, A. W., and Rowe, R. J. (2021). "Functional, Temporal and Spatial Complementarity in Mammal–Fungal Spore Networks Enhances Mycorrhizal Dispersal Following Forest Harvesting," *Functional Ecology*, 35: 2072–83.

Stritih, A., Senf, C., Seidl, R., Grêt-Regamey, A., and Bebi, P. (2021). "The Impact of Land-Use Legacies and Recent Management on Natural Disturbance Susceptibility in Mountain Forests," *Forest Ecology and Management*, 484: 118950.

Talbert, C., and Marshall, D. (2005). "Plantation Productivity in the Douglas-Fir Region under Intensive Silvicultural Practices: Results from Research and Operations," *Journal of Forestry*, 103/2: 65–70.

Taylor, G. E., Johnson, D. W., and Andersen, C. P. (1994). "Air Pollution and Forest Ecosystems: A Regional to Global Perspective," *Ecological Applications*, 4/4: 662–89.

Teste, F. P., Simard, S. W., Durall, D. M., Guy, R. D., Jones, M. D., and Schoonmaker, A. L. (2009). "Access to Mycorrhizal Networks and Roots of Trees: Importance for Seedling Survival and Resource Transfer," *Ecology*, 90/10: 2808–22.

Thom, D., Golivets, M., Edling, L., et al. (2019). "The Climate Sensitivity of Carbon, Timber, and Species Richness Covaries with Forest Age in Boreal–Temperate North America," *Global Change Biology*, 25/7: 2446–58.

Thomas, K. D., Reid, W. J., and Comeau, P. G. (2001). "Vegetation Management Using Polyethylene Mulch Mats and Glyphosate Herbicide in a Coastal British Columbia Hybrid Poplar Plantation: Four-Year Growth Response," *Western Journal of Applied Forestry*, 16/1: 26–30.

Timberlake, T. J., and Schultz, C. A. (2017). "Policy, Practice, and Partnerships for Climate Change Adaptation on US National Forests," *Climatic Change*, 144: 257–69.

UNFCCC (United Nations Framework Convention on Climate Change). (2014). *Report of the Conference of the Parties on its Nineteenth Session, Held in Warsaw from 11 to 23 November 2013. Addendum Part Two: Action Taken by the Conference of the Parties at its Nineteenth Session*. FCCC/CP/2013/10/Add.1. Available online at https://unfccc.int/sites/default/files/resource/docs/2013/cop19/eng/10a03.pdf (accessed February 13, 2024)

Van Lerberghes, P. (2004). *Les Paillis Biodégradables en Plantation Ligneuse*. Forêt-enterprise, No. 157. Paris: Institut pour le Développement Forestier, Centre National de la Propriété Forestière. 5 pp. Available online at https://www.plantruffe.fr/Files/122919/32IdfLePaillageAlternativeAuDesherbageChimique.pdf (accessed February 13, 2024).

Vašutová, M., Mleczko, P., López-García, A., et al. (2019). "Taxi Drivers: The Role of Animals in Transporting Mycorrhizal Fungi," *Mycorrhiza*, 29: 413–34.

Voigt, G. K. (1969). "Mycorrhizae and Nutrient Mobilization," in E. Hacskaylo (ed.), *Mycorrhizae: Proceedings of the First North American Conference on Mycorrhizae*. Miscellaneous Publication 1189. Washington: USDA Forest Service, 122–31.

Wang, Z., Li, L., and Ouyang, L. (2021). "Efficient Genetic Transformation Method for *Eucalyptus* Genome Editing," *PLoS One*, 16/5: e0252011.

Watt, M. S., Whitehead, D., Mason, E. G., Richardson, B., and Kimberley, M. O. (2003). "The Influence of Weed Competition for Light and Water on Growth and Dry Matter Partitioning of Young *Pinus radiata* at a Dryland Site," *Forest Ecology and Management*, 183: 363–76.

Webb, T. (1986). "Is Vegetation in Equilibrium with Climate? How to Interpret Late-Quaternary Pollen Data," *Vegetatio*, 67/2: 75–91.

Williams, L. J., Paquette, A., Cavender-Bares, J., Messier, C., and Reich, P. B. (2017). "Spatial Complementarity in Tree Crowns Explains Overyielding in Species Mixtures," *Nature Ecology & Evolution*, 1/4: 1–7.

Yu, Z., Beilman, D. W., Frolking, S., et al. (2011). "Peatlands and their Role in the Global Carbon Cycle," *Eos*, 92: 97–108.

Zoltai, S. C., Morrissey, L. A., Livingston, G. P., and Groot, W. D. (1998). "Effects of Fires on Carbon Cycling in North American Boreal Peatlands," *Environmental Reviews*, 6/1: 13–24.

CHAPTER 8

Resilience to Forest Disturbances

I always tried to turn every disaster into an opportunity.

John D. Rockefeller (1839–1937)

8.1 Disturbance ecology

An ecological disturbance is any relatively abrupt change in biomass, resource availability, or ecological structure or function (Burton et al. 2020). Disturbance denotes a discrete event (Pickett and White 1985) relative to the lifespan of the dominant organisms (Peters et al. 2011), which is associated with a change in resource availability (Jentsch and White 2019). As such, a disturbance or "pulse event" is differentiated from gradual environmental change or "press perturbation," such as the disruption associated with ongoing changes in atmospheric chemistry or the unmitigated influence of exotic invasive species. A disturbance can be a selective agent of mortality to a single species (for example, a disease outbreak, or a particular species being targeted for harvesting), a general consumer of biomass (for example, through fire or grazing), or the mechanism by which organisms are physically displaced (for example, a landslide). With increasing human domination of the planet, forests are subject not just to recurrent wildfires, windstorms, and insect outbreaks, but to widespread anthropogenic disturbances such as logging, road-building, and land-use conversions as well. In addition, a warming planet is amplifying the frequency and severity of many "natural" disturbances (Dale et al. 2001; Seidl et al. 2017), blurring the distinction between natural and human-caused events.

A forest disturbance event may take just a few seconds (for example, a tree falling owing to root rot or wind), a few minutes (for example, a raging crown fire), a few hours (for example, a growing-season frost), or may unfold over the course of several weeks (for example, a heatwave or drought), or even years (for a forest insect outbreak). Extreme weather events, which are becoming more frequent under a changing climate—such as late frost in spring, summer heat waves, and droughts—constitute important disturbances when the physiological tolerances of dominant species are surpassed (Jentsch et al. 2007). Disturbances can be very localized and discrete in their footprint (for example, a treefall gap or cutblock, having clear boundaries), or can be widespread, diffuse, and difficult to map (for example, a region-wide windstorm or insect damage). Disturbance events often cascade to trigger other disturbances (for example, wildfires or hurricanes resulting in subsequent mass movements), with effects sometimes occurring far from the disturbance itself (for example, flooding resulting from a stream blocked by a landslide, beavers, or a new road), generating the potential for a myriad of disturbance and ecosystem interactions (Buma 2015; Burton et al. 2020).

Disturbances generally release a pulse of nutrients or growing space that sets the stage for ecosystem reorganization and ecological succession (Walker and Wardle 2014; Jentsch and White 2019). Although forest managers have long considered the need for "forest protection" from wildfires, animal damage, and fungal pathogens (Hawley 1937; Parthiban et al. 2019), it is only since the 1980s that natural disturbances have been considered important drivers of forest renewal, landscape

Resilient Forest Management. Philip J. Burton, Oxford University Press. © Philip J. Burton (2025). DOI: 10.1093/oso/9780198832997.003.0008

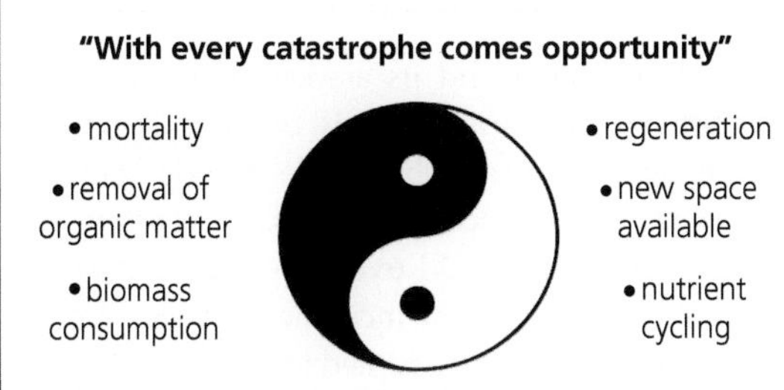

Figure 8.1 The "yin and yang" of forest disturbances. Disturbances play both destructive and regenerative roles, making every disturbance an opportunity to "build back better" when guiding the recovery and renewal process. In every catastrophe are the seeds of renewal, although the newly rebuilt system inevitably sets itself up for future disturbance.

diversity, and the evolution of biodiversity (Pickett and White 1985; Wohlgemuth et al. 2019). The component mechanisms of disturbance and succession are opposing but complementary processes (see Figure 8.1) that are collectively responsible for vegetation dynamics. Despite those positive and ecologically important roles, the disruptive and destructive nature of forest disturbances typically prompt a protective or resistance response on the part of resource managers and the general public. There has yet to be a cultural shift within the mainstream forestry community that recognizes and accepts the value and inevitability of natural disturbances. However, there is growing awareness of the opportunity for improved climate adaptation once some of the inertia of existing vegetation is reduced by disturbances (Buma and Schultz 2020). At the same time, the principles of ecological forestry and the incorporation of coarse-filter biodiversity conservation into forest management are premised on the emulation of the key attributes of natural disturbances (Perera et al. 2004; Kuuluvainen and Grenfell 2012; Kneeshaw and Bergeron 2016).

Individual disturbances can be described in terms of their size (area affected), timing (date/season initiated, and duration), intensity (energy released), severity (ecological impact), and selectivity (of effects on different species or sizes/stages of organisms). Individual disturbances, by their very nature, are largely unpredictable in terms of when and where they will occur. Disturbance regimes, on the other hand, describe historical norms or expectations with regard to the combination of disturbance agents typical of a given landscape, the return interval for such events, and the frequency distribution of features such as event size, seasonality, duration, and severity, as well as disturbance interactions (Burton et al. 2020). The disturbance regime is essentially a landscape attribute (Harvey et al. 2021), with both spatial and temporal parameters. Characterizing a landscape's disturbance regime therefore requires either the mapping of historical disturbances over large areas (as in Burton and Boulanger 2018), or a reconstruction of the history of disturbances at a site over a longer period of time (for example, analyzing tree rings and sediment cores, as in Kuosmanen et al. 2020). Simulation modelling is increasingly being used to characterize current disturbance regimes and to project their future under climate change (Sturtevant and Fortin 2021). It would be prudent for vulnerability analysis (see Section 6.4) of a forest management area—even if not considering climate change—to consider the region's disturbance regime. Such analysis can reveal the risk (probability) of a particular type of disturbance affecting a forest tract or forest type over a given period of time. Even a relative assessment of risks and vulnerability can inform a forest manager of likely types of disturbance and their potential effects for which planning is needed (Angelstam 1998).

The immediate effect of most forest disturbances is seen as death or damage to trees, with obvious impacts on the ability of the forest to continue growing fiber, sequester carbon, and sustain the same form of wildlife habitat. Less appreciated is the spatially variable nature of disturbance severity and the resulting modification of forest structure and composition (Burton et al. 2008; Turner 2010). This variability supports greater biodiversity than found in an undisturbed forest landscape. A heterogeneous mosaic of habitats, disturbance patches, successional stages, and metapopulations (multiple sources of organisms able to recolonize disturbed areas) contributes to landscape resilience (Bengtsson et al. 2003; Moritz et al. 2010). Within the spatial footprint of a disturbance event (for example, a wildfire or blowdown patch), any surviving trees, thickets of shrubs and other unburned vegetation, trees with damaged boles (that subsequently support fungal growth and the development of tree hollows and cavities), and even the standing snags, fallen logs, and dead root systems collectively contribute biological legacies of the original forest. Those legacies and their patchy distribution define unique post-disturbance habitats and support recovery of the post-disturbance ecosystem (Franklin et al. 2000; Johnstone et al. 2016). A focus on such structural and biological legacies—what is left behind, more than what is killed or removed— informs the advisability of salvage logging (see Section 8.6) and is an important consideration in designing place-appropriate harvesting and silvicultural systems in ecological forestry (Hansen et al. 1991; Palik et al. 2021; see Chapter 12).

8.2 Disturbances in a warming climate

A warming planet and its associated disruptions to weather patterns are already evident in altered disturbance regimes. Observable trends include increasing disturbance frequency or severity in many parts of the world over recent decades (see Figure 8.2), and a growing incidence of novel disturbances in ecosystems ill adapted to such disruptions (for example, fire in rainforests). Where plant, animal, and fungal species are adapted to seasonal variations in weather and to natural disturbances, mature trees and intact ecosystems can be largely resilient to the direct effects of climate change that fall within the historic range of variability (Falk et al. 2022). Indeed, it is a truism that natural vegetation and natural disturbance regimes strongly reflect the regional climate (White 1979; Jentsch and

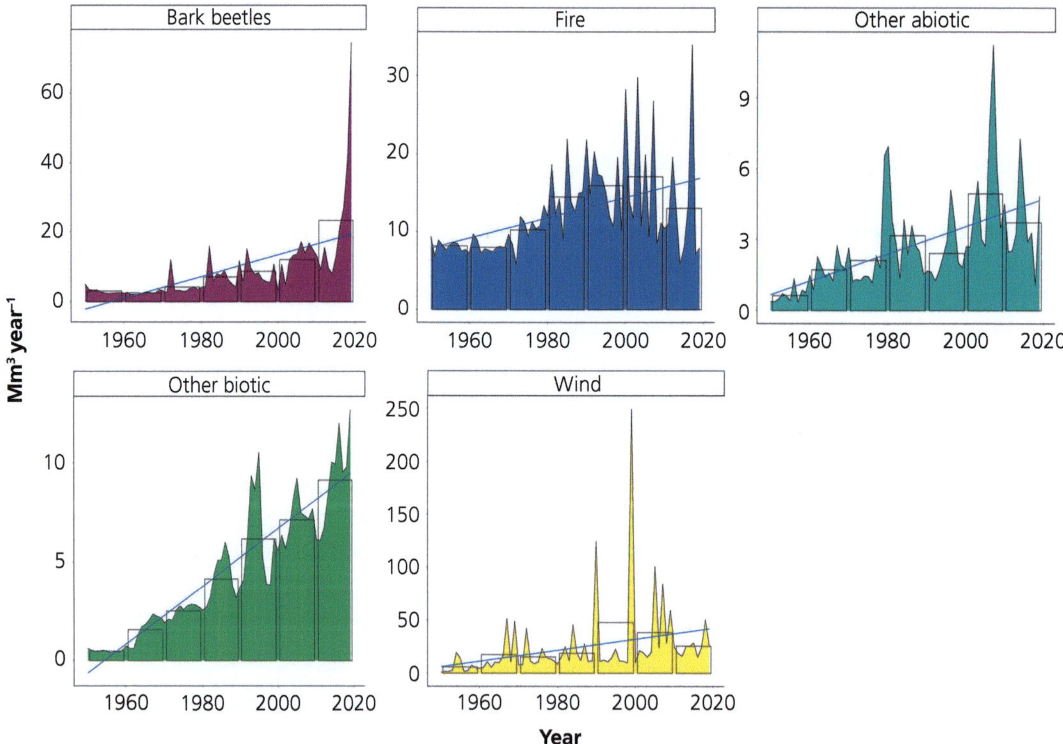

Figure 8.2 Time series of estimated volumes of timber damaged annually by natural forest disturbances in thirty-four European countries, 1950–2019. Bars represent decadal averages, and the lines represent significant linear models fit to the decadal averages. Note that the scales of vertical axes differ among panels (reproduced from Patacca et al. (2023) under Creative Commons CC BY-NC-ND 3.0 DEED (https://creativecommons.org/licenses/by-nc-nd/3.0/deed.en)).

von Heßberg 2019). Many natural disturbances—although not volcanoes, earthquakes, tsunamis, or asteroid collisions, of course—are strongly responsive to variations in weather and climate. Consequently, changes in forest disturbance regimes can become evident before gradual shifts in the range or dominance of individual species responding to altered temperature and precipitation regimes (Parry et al. 2007). In contrast, disturbances frequently catalyze rapid compositional change in forests (Johnstone et al. 2010; Turner 2010; Brice et al. 2020). Disturbances are often required to reduce the dominance (inertia) of mature trees and well-established plant communities before new genotypes, species, and communities can adjust, reconfigure, or transform forests better to match the new climate (Turner 2010; Thom et al. 2017).

We are used to describing prolonged drought, high temperatures, and gusty winds as "fire weather," and we know that prolonged rainfall and saturated soils on steep slopes constitute a susceptibility to landslides. Both of those extremes are increasingly more likely—sometimes in the same landscape—now than in the recent past (see Chapter 6). Warmer temperatures and droughts tend to result in more widespread and long-lasting outbreaks of herbivorous insects (Pureswaran et al. 2018; Lehmann et al. 2020). Furthermore, cascading effects of multiple interacting disturbances can be expected as well (see Box 8.1). For example, drought conditions frequently lead to forest fires that leave bare ground exposed to rapid runoff and downstream flooding later in the same year. After a few years of vegetation recovery and tree-root decomposition, heavy rainfall can saturate soil profiles that are no longer held in place on steep slopes by living trees, resulting in mass movements such as soil slips and debris flows (Parise and Cannon 2012). The risk of disturbance interactions and their potential to disrupt forest management and compromise forest resilience is largely absent from current forest plans and policies.

The mechanisms, projections, and implications of climate change to individual forest disturbance risks are becoming increasingly well understood, yet preparations for such inevitabilities are lagging. High air temperature, low relative humidity, and prolonged periods without rainfall are key predictors of wildfire ignition risk, while global lightning-strike frequency and density might increase (Williams 2005) or decrease (Finney et al. 2018) as the atmosphere warms. Another dimension of global change—not climate change, but increased penetration of the world's forests by infrastructure and industrial activities—is the greater risk of human-caused ignition associated with roads (Narayanaraj and Wimberly 2012) and with logging and land clearing (especially in moist forests; Lindenmayer et al. 2009). Once ignited, fire spread is strongly related to winds, another feature of the atmosphere projected to increase in some regions and seasons but not others (McInnes et al. 2011; Haughian et al. 2012). When winds are strong and fuels are dry, low forest density and fuel breaks or forest stand isolation provide little resistance to airborne firebrands.

Like wildfires, population explosions of herbivorous insects tend to be promoted by warmer air temperatures. With larval development of most insect species being strongly related to heat sums (degree-days above species-specific temperature thresholds), some insects have been documented to have two or more generations per year in locations that previously supported only one generation per year (Netherer and Schopf 2010). Also important is the pronounced reduction in overwintering mortality as winter temperatures have risen more than summer temperatures, such that the absence of killing cold has been attributed as a major factor in the population growth of several bark-beetle species in western North America and in Europe (Raffa et al. 2015). The result is historically unprecedented durations, severities, spatial extents, and range expansions associated with outbreaks of several indigenous insect species. Some species that had not previously been documented as eruptive can also emerge as serious agents of tree stress and mortality. Nonetheless, heatwaves, phenological mismatches, and improved performance of competitors, predators, and parasites mean that insect populations may be reduced in some places too (Lehmann et al. 2020), so the natural history of an outbreaking forest insect species needs to be considered in the context of available food sources under the evolving climate (Forrest 2016; Kneeshaw et al. 2021).

Box 8.1 Disturbance interactions and cumulative effects

Forest disturbances, both natural and human caused, frequently follow each other in short succession. The likelihood and distinctive effects of those disturbance combinations have long been recognized as a characteristic of a region's disturbance regime (Pickett and White 1985), but it is only in recent years that systematic considerations of these phenomena have been undertaken (Buma 2015; Johnstone et al. 2016; Burton et al. 2020; Sturtevant and Fortin 2021). In general, there is the potential for synergistic, linear, or antagonistic interaction between two disturbance events for which the first disturbance may promote, inhibit, or have a neutral effect on the likelihood or severity of a subsequent event (see Table 8.1).

Synergistic effects can include severe losses of biodiversity, or compromised ecosystem structure and function that can result in closed forests transitioning to open woodland, savanna, grassland, or shrubland states (Girard et al. 2009; Coop et al. 2020). Antagonistic interactions represent compensatory effects of the following disturbance that can somewhat offset the impacts of an initial disturbance on ecosystem integrity. Any generalizations of the examples presented in Table 8.1 depend greatly on climatic conditions, forest type, and the time periods (lags) being considered (Harvey et al. 2016; Burton et al. 2020). Promotional and synergistic disturbances that result in a cascade of multiple disturbances are typically of greatest concern. But even neutral and linear relationships between multiple disturbances can result in cumulative effects that may compromise numerous forest values, forest resilience, and plans for sustainable resource management (Venier et al. 2021). Notably, timber harvesting and associated road building can both trigger other disturbances (for example, landslides), and are a common response (in the form of salvage logging) to natural disturbances such as storm or insect damage (Lindenmayer et al. 2008).

Table 8.1 Some examples of the diverse links and cumulative effects* associated with multiple forest disturbances

Relationship	Promotional A, B ↑	Neutral** A, B =	Inhibitory A, B ↓
Synergistic ("2+2=5")	Natural disturbance → salvage logging (Lindenmayer et al. 2008; Thorn et al. 2018); see Section 8.6	Petroleum development, logging (Schneider et al. 2003; Johnson et al. 2015)	Forest fire → reduced fire probability for a while (Parks et al. 2018; Buma et al. 2020); severe impacts if short-interval fires occur
Linear ("2+2=4")	Windstorm → spruce bark-beetle outbreak (Reynolds and Holsten 1994; Wermelinger 2004)	Logging, volcanic eruption (Means and Winjun 1983; Burton et al 2020)	Severe forest fire → reduced potential for local insect outbreaks (Fleming et al. 2002; Kulakowski et al. 2003)
Antagonistic ("2+2=3")	Drought → outbreak of defoliating insects, which may reduce transpiration and moisture stress (Williams et al. 1987)	Drought ended by high rainfall associated with hurricanes (Maxwell et al. 2013)	Drought → reduced susceptibility to herbivores and to frost (Herms and Mattson 1992; Kreyling 2010)

*Disturbance A affects (increases, ↑, or decreases, ↓) or does not affect (=) the probability or severity of Disturbance B, resulting in effects that may be additive ("2+2=4") or non-additive ("2+2≠4").

**no mechanism of interaction, except in terms of context-weighted random co-occurrence, as defined by Burton et al. (2020).

It is extreme weather events that tend to constitute or drive forest disturbances, although other stand- and forest-level factors contribute to varying degrees. The challenge in addressing climate change is the greater occurrence of all kinds of weather extremes, which we might characterize as warmer and drier, warmer and wetter, cooler and drier, and cooler and wetter. Each combination seems to be occurring at different places around the planet, and sometimes in different seasons at any one place; each combination promotes a distinctive shift in the associated disturbance regime. Although the cold weather events seem to be less common and are more localized than high temperature extremes, high precipitation events seem to be just as common as drought events. With our planet dominated by oceans, the warming atmosphere is charging the air with more water vapor and cyclonic storms with more energy. So it comes as no surprise that record-breaking "heat domes," "polar vortices," "atmospheric rivers," severe hurricanes and associated events characterized by high rainfall, floods, and destructive winds are now occurring (Robinson 2021; Tradowsky et al. 2023; White et al. 2023). Storm damage from various kinds of weather events is on the increase, ranging from the effects of severe thunderstorms, tornados, and derechos to freezing rains (ice storms) and prolonged episodes of heavy wet snow that can break tree branches and compromise forest cover and tree growth. When accompanied or preceded by heavy rainfall, strong winds are more likely to result in high levels of tree windthrow (Lindner and Ramukainen 2013; Hall et al. 2020).

Other "natural" disturbances owing to climate change are on the rise too. These include landslides released as high-elevation permafrost thaws (Huggel et al. 2012; Patton et al. 2019), and the development of soil slumps and sinkholes ("thermokarst" terrain) as permafrost thaws in the subarctic (Jorgenson et al. 2013). Greater precipitation in more concentrated events is exposing floodplain forests—typically some of the most productive in a region—to more frequent flooding, with longer periods of inundation and more erosive damage than in the recent past. Many fungal forest pathogens, such as *Dothistroma* needle blight in British Columbia (Woods et al. 2005) and *Heterobasidion* root rot in Finland (Venäläinen et al. 2020), seem to be expanding their range or acting with greater virulence than in the past. Threats from invasive species—ranging from plant and animal diseases, herbivorous insects, and vascular plants to ecosystem-disrupting vertebrates—also continue to increase (Hulme 2009; Seebens et al. 2017). Individually, similar forest disturbances have occurred in the past, but they are accelerating as the climate changes, and it now behooves us to address the associated risks and cumulative impacts.

8.3 Planning for wildfire

Based on a global compilation of satellite observations, it is estimated that fires burned an average of 4.393 million km^2/yr from 1982 to 2000, and 4.788 million km^2/yr from 2001 to 2018, with the greatest fire activity regularly occurring in Africa (Bowman et al. 2011; Otón et al. 2021). Another estimate reports a mean of 3.406 million km^2/yr burned from 2003 to 2012, but with only 666,330 km^2/yr being in forests (van Lierop et al. 2015), the rest presumably in grassland, scrub, tundra, and savanna. It is clear that fire—whether planned or unplanned, and in closed-canopy forest or in more open habitats—is one of the most prevalent forces that modifies our planet's vegetation, and forests in particular, over a short period of time (Bond and Keeley 2005). Wildfire impacts include direct mortality of trees and other organisms, carbon and nitrogen emissions, habitat degraded for wildlife associated with mature and closed-canopy forests, soils exposed to erosion, and human life and property threatened. It is widely accepted that forests should be managed to avoid, resist, minimize, and recover from such negative effects of forest fires, while still acknowledging their role in nutrient cycling, pest reduction, forest regeneration, and habitat diversification (Pausas and Keeley 2019).

Most wildfire resilience strategies are quite understandably about promoting wildfire avoidance and resistance: human life, property, and infrastructure can be threatened by severe fire, which often spreads from forests to towns and homes. Indeed, catastrophic wildfires have been the incentive for creating government forest services around the world, often organized like a paramilitary defense force with the primary objective of fighting forest fires (Pyne 1982: 401–3). Early wildfire detection historically depended on staffed lookout towers and reports from the public, but increasingly uses automated electronic lightning-strike mapping and smoke detection, supplemented by aerial inspection and satellite monitoring (Alkhatib 2014). Prompt detection facilitates rapid containment before a wildfire gets too big to control. Paradoxically, the success of such fire-suppression efforts has resulted in fuel buildups reflecting this "fire deficit" (Marlon et al. 2012). Insect infestations after long periods free of fire leave forests more flammable than they would be under natural disturbance regimes characterized by frequent low-severity or mixed-severity fires (Keane et al. 2002; Hessburg et al. 2019). It is only in recent decades that most forest managers have conceded that we need a more sophisticated and nuanced approach to living with fire in protected and multiple-use forests on public lands, while still recognizing the need to protect private land and plantation investments.

The primary tools available for engendering resistance to fire include forest thinning, tree pruning, surface fuel removal or modification treatments, tree species selection, the creation and maintenance of fuel breaks, and strategic deployment of those options within landscapes. These treatments (described in further detail below) are premised on the assumption that recurrent fires are part of the natural (historical or preindustrial) disturbance regime of a region, and that wildfires will inevitably occur on the landscape again. That is not to say that such practices will substitute for fire-prevention and fire-suppression efforts, but rather can be considered a supplement to such activities, which must necessarily prioritize vulnerable forest values and human investments. Wildfire prevention programs and suppression actions are still needed in peopled forest landscapes, supported by strategic deployment of firefighting personnel and resources, and increasingly informed by near real-time satellite monitoring of wildfire occurrence and spread (Thangavel et al. 2023). Prevention and suppression efforts can focus on the protection of sociocultural, economic, and sensitive ecological assets. Mapped values and vulnerabilities, coupled with an understanding of forest ecosystem ecology, can guide forest managers to let some wildfires burn in some places at some times, and to employ purposeful (controlled, prescribed) fire selectively to reduce fuel loads and promote habitat change. Collectively, the minimization of "bad fire" and the acceptance and use of "good fire" promote overall forest landscape health and resilience (Kaufmann et al. 2005; Doerr and Santín 2016).

Stand-level treatments to enhance forest resistance to wildfire generally have the goal of preventing crown fires and reducing canopy tree mortality (Agee and Skinner 2005). Techniques employed include increasing the height to the base of live crowns by pruning, removing "ladder fuels" (low branches, conifer saplings, leaning deadwood) between the forest floor and tree crowns, and decreasing the density of overstory trees by thinning while selectively retaining big trees of fire-resistant species (see Table 8.2). As with most risk-reduction treatments, however, these widely practiced techniques are likely to be ineffective in all forest types or under severe fire weather conditions. Reducing surface fuel biomass (in large dead wood, fallen branches, dense shrubs) can be done by hand with help from small tractors/yarders to remove logs and bundles of cut brush and branches. Alternatively, mechanical mulching/mastication treatments may be feasible where trees are widely spaced, and if plant community integrity and soil health can be protected. Prescribed burning under cool, controlled conditions is also an option, but (depending on the forest type) is most effective in conjunction with thinning and the removal of ladder fuels. Prescribed burning with multiple objectives (often referred to as "cultural burning") has been the traditional tool employed for millennia by Indigenous peoples in fire-prone landscapes around the world (Pyne 1982; Bowman et al. 2011; Hoffman et al. 2021; see Section 11.5).

Pruning and the strategic removal of saplings or pole-sized trees with low crown bases serve to increase the distance between the forest floor and the live tree crowns making up the bulk of the forest canopy. Such work almost always requires manual labor. The resulting forest structure is one of a much more open understory with forest regeneration (seedlings and saplings) greatly reduced in density or largely restricted to clumps in canopy gaps. That open understory may also enhance berry production and forage for ungulates, with open conditions being better for wildlife viewing and hunting. Stand density is typically reduced using a "thin-from-below" approach, which removes sub-dominant trees in the canopy and the long-crowned small trees that may provide ladder fuels, thereby reducing overall canopy bulk density. Those trees may be cut and yarded by a combination of manual and mechanical means but may be too small to sell for lumber and so are often used for firewood, pulp, or bioenergy. Further reductions in crown

Table 8.2 Rationale for stand-level treatment options employed to reduce risk of crown fire and mature tree mortality from wildfire

Principle	Risk reduction mechanisms	Treatments
Reduce surface fuels	• reduces surface fire intensity and potential flame length • reduces likelihood of torching • makes control easier	• manual removal followed by chipping or pile burning • mechanical mulching or mastication • prescribed burning (underburns)
Increase height to live crown	• ladder fuels are removed • requires greater surface fire intensity or flame length to transition to crown fire • reduces likelihood of torching	• removal of low-crowned trees (e.g. saplings and sub-canopy shade-tolerant conifers), except in open gaps • pruning of lower branches • flammable material is removed from tree bases, chipped, or safely burned
Decrease crown density	• reduces crown fire potential and intensity • reduces potential for an independent or running crown fire	• low thinning, with material removed, chipped, or burned • commercial (uniform) thinning with careful yarding so as not to damage remaining trees
Retain big trees of fire-resistant or fire-resilient trees	• species and individuals with thick bark and high bases to live crown have greater probability of surviving surface fire and resisting crown fire initiation • depending on fire severity and species, may be able to resprout from bole or base and continue growing	• selective thinning, identifying "crop" or "valued" trees to leave • species selection in plantation establishment • crop tree focused brushing and spacing to accelerate growth of preferred individuals

Based on Agee and Skinner (2005).

density may extend to the removal of co-dominant and dominant trees to reduce canopy bulk density and potential wildfire intensity where the threat of an independent crown fire needs to be seriously reduced (for example, outside a community located uphill from prevailing winds) or if the fuel-reduction treatments need to be financed by the sale of logs to sawmills (see Section 11.2 for an example).

Attention is paid not just to which trees to remove, but to which trees to retain; thinning treatments for fire-risk reduction often employ a "mark-to-leave" approach to assure reliable implementation of the fuel-reduction prescription. Retained trees should be the larger, healthy ones that have good fire resistance (thick bark, high branches) or fire recovery (resprouting) potential (see Figure 8.3). Those trees are intended to serve as future crop trees or anchors ("mother trees") for the provision of desired forest values and the future recovery, adjustment, reconfiguration, or transformation of the forest in the event of wildfire. An ecological template for the target forest structure might employ the natural range of variability (NRV) in tree sizes, stand densities, and spatial distributions of trees (for example, clumped or uniform) characteristic of that particular

forest type, obtained from historical surveys, photographs, or models reflecting a preindustrial or natural disturbance regime (Swetnam et al. 1999). A forward-looking resilience plan might look for such templates in climate-analog forests instead. The retention of large trees and clusters of recruitment in gaps implies an uneven-aged or multi-aged forest structure that is characteristic of many dry forests that are also prone to frequent low- or mixed-severity wildfires. Such stand structures can also be found in wet old-growth forests characterized by gap dynamics where stand-replacing disturbances are rare; but, if fire risks are low, fuel-reduction treatments would not normally be undertaken. Modified application of these principles can also promote fire-resistance treatments in some even-aged stand types, while treating other forest types will depend more on landscape-level fuel breaks and natural barriers (water bodies, utility corridors, roads) to reduce the spread of wildfires.

In contrast to rather intensive stand-level fuel-reduction treatments, forest- or landscape-level treatments to promote wildfire resistance employ spatial land cover configurations strategically to constrain the growth and spread of wildfires.

Figure 8.3 Examples of fire-resilient trees that are prime candidates for retention during fuel reduction and fire-smart forest-thinning operations: (a) Thick-barked, fire resistant species with no low branches, such as *Pinus ponderosa* in western North America (bare trunk in the right of the photo); and (b) species that reliably resprout after all but the most severe wildfires, such as *Eucalyptus obliqua* in Australia, here shown a few months after the 2019/2020 Badja Forest bushfire (photo by Maralyn Callaghan).

Landscape approaches are especially appropriate in moist, high-elevation, and boreal forests where fires may be infrequent, but tend to be severe, stand-replacing events when they occur. Understory burns, low thinning, and other fuel modification efforts within stands can be relatively ineffective in such forests (Weir et al. 1995; Taylor et al. 2020). Treatments to prevent the development and spread of crown fires in closed-canopy, even-aged conifer forests need to consist of wider fuel breaks and much more aggressive stand thinning (for example, retaining as little as 10 percent of original basal area, or 10 m spacing between residual tree crowns) than in forests and climates that experience frequent surface fires but infrequent crown fires (Mooney 2010). More important, however, is to promote a mosaic of variable resistance to fire, thereby limiting the likely extent of any individual wildfire.

Proactive wildfire risk-reduction planning should consider the location of low-fuel terrain (rivers, lakes, wetlands, rock outcrops, cliffs, grasslands), low-fuel land uses (roads, utility rights of way, pastureland, agricultural fields, open-pit mines and quarries), and vegetation with lower flammability (for example, broadleaf trees and shrubs, compared to needle-leaved or sclerophyllous vegetation). A combination of existing natural and man-made features can then anchor a network of strategically located stand treatments to fill gaps in the existing network (see Box 8.2). Natural anchors may include bodies of water, moist shaded hillslopes, broadleaf forest stands, or old-growth forest that evidently avoided burning several times before and is likely to do so again (Wei et al. 2008; Alcasena et al. 2018; Yemshanov et al. 2021). Risk-reduction treatments are then conducted at selected fire-control locations, occupying a much smaller footprint than would be required if implemented across the entire forest.

Barriers to reduce the chance of wildfire spread may consist of full fuel breaks—"fire guards" bulldozed down to mineral soil, or at least with all woody material removed. Other fire-barrier treatments may consist of overstory removal or heavy crown thinning, prescribed burning under safe conditions, or any of the fuel-reduction treatments listed in Table 8.2. Whether implemented in strips designed to limit fire spread, throughout targeted forest stands, or in wildland–urban interface (WUI) zones, silvicultural practices such as selective thinning and planting can promote dominance by less-flammable broadleaf species in conifer-dominated landscapes (Cumming 2001; Wang et al. 2021). In some parts of the world, options exist to promote tree species with higher foliar moisture content and lower concentrations of flammable oils (Popović et al. 2021). In Australia, for example, one might promote more fire-resistant species of *Acacia*, *Banksia*, or *Casuarina* at selected locations in *Eucalyptus* woodlands (Gill and Zylstra 2005). In production and multiple-use forests, the strategic deployment of wide fuel breaks (often having an alternative land use), thinning and/or prescribed burning for fuel reduction, and pre-planting site preparation (in the form of plowing or disc trenching) can be expected to reduce fire spread. Timber plantations in a primarily agricultural landscape may be at lower risk of wildfire spread, but owners and managers may want to limit the size or contiguity of individual stands in order to minimize losses should fire erupt.

All the above efforts to improve fire resistance can only increase the chances of avoiding severe crown wildfires and the associated degradation of ecosystem services and values. There inevitably will be unwanted wildfires in forests and woodlands, for which emphasis may then shift to promoting recovery. Natural regeneration and the healing powers of ecological succession will generally be preferable after all but the most severe burns. Natural recovery can also be counted on in most protected areas and multipurpose forests having fire-adapted vegetation (for example, forests dominated by species with serotinous or semi-serotinous cones that essentially constitute "aerial seed banks," or those with reliable resprouting abilities). There is ongoing debate about whether active restoration or reforestation is warranted in burnt multiple-use forests (e.g. Shatford et al. 2007; Ouzts et al. 2015), depending on burn severity, the natural disturbance regime, land-use objectives, and climatic trends (Chen et al. 2014; White and Long 2019). Where catchments serve as municipal or residential water sources, or in steep terrain subject to high levels of precipitation, post-fire rehabilitation is often focused on grass seeding for erosion control, especially for the exposed mineral soils of severely burned areas and fuel breaks

---

Box 8.2 Strategic landscape-level wildfire-containment planning

The concept of proactive wildfire spread control has been around for a long time, with narrow "fire guards" of fuel-free strips of land often created to provide some level of protection for residences, ranches, villages, and recreational camps and lodges embedded in forest environments. Depending on the wildfire climate and vegetation of a particular region, there are differences in the recommended width and composition (bare ground, grass or shrubs allowed, fully open or shaded) of such fuel breaks. Questions remain about how effective such fuel breaks are, especially under conditions of severe fire behavior (Mooney 2010; Syphard et al. 2011; Shinneman et al. 2019).

In recent years, efforts to optimize the location of fuel breaks and fuel-reduction treatments in wildland landscapes has been assisted by the simulation modelling of fire spread behavior (e.g. Wei et al. 2008; Rytwinski and Crowe 2010; Alcasena et al. 2018). Large areas of forest with a history

of fire suppression often require fuel-reduction treatments in order to reduce wildfire risk and restore forest health, but available resources are rarely sufficient to address this need over large areas in a timely matter, so it is important to prioritize locations for treatment. Professional expertise and local knowledge can be supplemented with simulations or statistical modelling to identify landscape features—such as lakes, rivers, rock outcrops, ridge tops, roads, cropland, or even transitions in vegetation (from conifer to broadleaf or from forest to grassland)—that are likely to stop or slow a wildfire (O'Connor et al. 2017). With those features mapped, planners are then able to supplement them by targeting fuel treatments and containment lines at strategic locations to fill gaps.

One application of such strategic wildfire-containment planning at the landscape level has been the development of "potential wildfire operational delineations" or PODs (e.g.

Box 8.2 *Continued*

Figure 8.4). PODs are spatial units of analysis that integrate fire risk and hazard mapping with the identification of locations where wildfire control is more likely to be successful. Analyzing POD boundaries along with the evolving conditions of a given wildfire incident can help wildfire managers decide where, how, and when to engage the fire at

predetermined potential control locations (O'Connor et al. 2016; Yemshanov et al. 2021). As such, PODs support both proactive fire-risk reduction through the identification of the most efficacious fuel-treatment locations, and reactive fire suppression through the designation of predetermined containment locations.

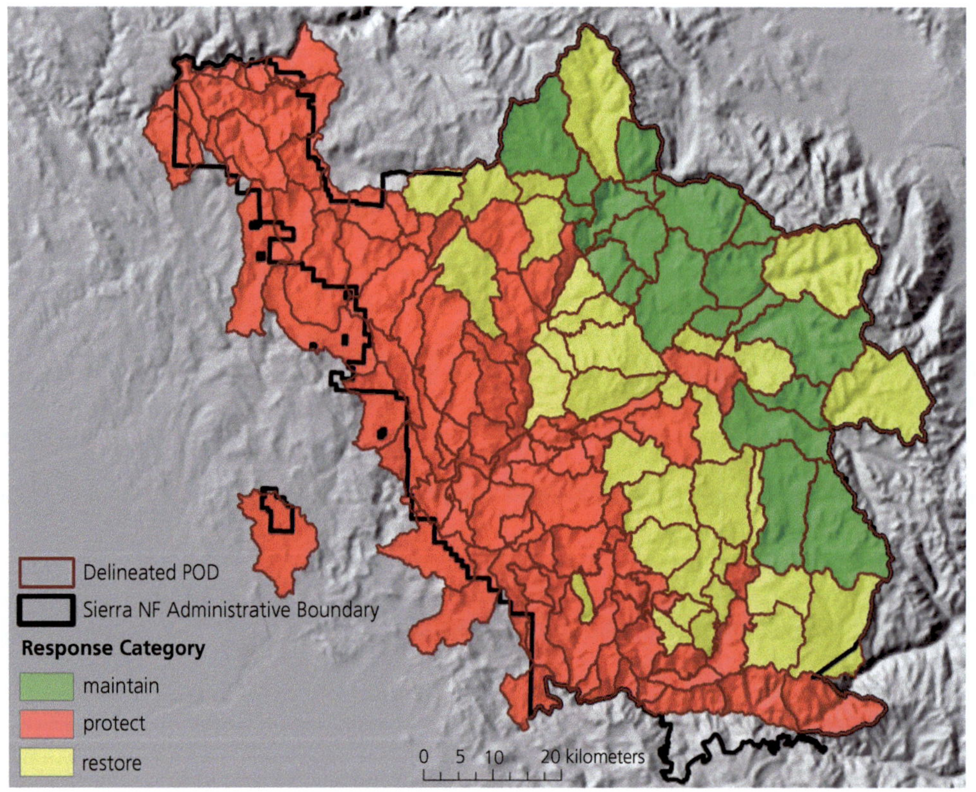

Figure 8.4 Potential operational delineations (PODs) for wildfire management mapped in Sierra National Forest (California, US) to identify locations at which fire spread can most likely be contained. Based on valued assets and resources, fuel conditions, and fire suppression difficulty, PODs have been zoned according to three different wildfire preparation and response categories: widespread mechanical fuel treatments and full wildfire suppression (protect); strategic fuel modification using prescribed fire or mechanical treatment and selective wildfire suppression (restore); and untreated fuels where wildfires can generally burn freely where environmental benefits outweigh likely impacts (maintain) (reproduced from Thompson et al. (2016) under Creative Commons, CC-BY-4.0 DEED, http://creativecommons.org/licenses/by/4.0/).

made by heavy equipment. In all but dedicated production forests, such efforts should sow only native species and rigorously exclude the introduction and

spread of exotic plant materials. Grassed fire guards might also be sown to dense sod-forming grasses in order to inhibit tree establishment and thereby maintain a low-vegetation fuel break to facilitate future fire control, a decision that has to be balanced against a reduction to the timber-producing and carbon-sequestering land base. Efforts to accelerate tree establishment in burnt forest by planting or seeding may be limited by safety concerns regarding the stability of burned tree boles and roots, so artificial regeneration activities sometimes occur only after timber-salvage operations (see Section 8.6).

When fire resistance is futile, and the active promotion of post-fire recovery is contemplated, it is often preferable to promote forest reconfiguration or transformation rather than simple recovery or historical restoration (which, in a changing climate, may no longer be effective; Harris et al. 2006). Disturbances such as fire typically provide opportunities that otherwise might not arise for adjusting forest composition and structure, and our expectation of forest values. McWethy et al. (2019) provide a framework for anticipating the circumstances under which wildfire resistance and recovery (which they term "basic resilience") should give way to adaptive or transformative resilience. Appropriate actions depend on where a landscape falls with respect to three considerations: (1) the exposure of human communities and ecosystem values to wildfire; (2) wildfire severity and impacts; and (3) fire novelty, in terms of the degree of change from historical patterns. A resistance strategy is generally appropriate where the novelty of fire frequency and behavior is low, and where the exposure of valued socio-ecological elements (human communities, rare species, and so on) is moderate. An adaptive strategy (adjustment and reconfiguration) is warranted under conditions of moderate fire novelty, moderate to high fire severity, and moderate to high exposure. A transformative strategy, however, is called for under any conditions of high fire novelty, where that landscape has not historically experienced such a pronounced shift in the disturbance regime (McWethy et al. 2019). In many respects, use of such a framework can be considered analogous to and an extension of a vulnerability assessment (see Section 6.4), as applied in the specific context of the fire regime under a changing climate.

8.4 Planning for windstorms

Extreme weather, including heavy precipitation, strong winds, and prolonged droughts, is estimated to affect about 384,000 km^2/yr of forest around the world, mostly in temperate zones (van Lierop et al. 2015). Like wildfires, storms with sufficient energy to kill mature trees (primarily through stem breakage and uprooting) are episodic events that occur over a relatively short period of time. Also like wildfires, there are less severe events, in which defoliation and branch loss may occur but trees survive and overall habitat values are only lightly affected, making it important to recognize thresholds for the transition from "thinning and pruning" events to very destructive events that upset the overarching physical structure of the forest ecosystem. The legacies of all storms, however, are patches of more open conditions and accumulations of litter and fallen woody material that are important to some organisms and ecological processes.

The damage track of major storms can be relatively narrow (a few hundreds of meters or less in the case of tornados, thunderstorms, and downbursts), a few kilometers or more in the case of hurricanes or derechos), but can be very long (hundreds of kilometers), surrounded by bands of lesser damage crossing entire landscapes or regions. Geographic features such as hills can result in impacts being intermittent and heterogeneous within those tracks as well. The potential extent and ecological impact of wind events is exemplified by cyclonic storms *Lothar* and *Martin*, which damaged more than 240 million m^3 of timber across fifteen countries in Europe in December 1999. From 1950 to 2010, it is estimated that wind was directly responsible for 51 percent of recorded damage in European forests (Schuck and Schelhaas 2013).

Unusual storms with extreme wind speeds are only one cause of wind damage in forests. Perhaps more common is the windthrow associated with the creation of new stand edges when adjacent forest is clearcut logged (Burton 2002; Mason and Valinger 2013). Conversely, forest stands that have matured with permanent high-contrast edges (as is characteristic of many forest plantations in agriculture-dominated landscapes) typically have very windfirm edges, so damage from high winds tends to occur in stand interiors (Peltola et al. 2013). Trees growing under conditions of intense competition for light (as in fully stocked, even-aged stands of a single tree species) allocate resources to height growth at the expense of potential stem diameter growth or root growth. Such low-taper stem profiles are desirable for maximum lumber recovery from a sawmilling perspective, and make excellent power poles, house logs, and other specialty products, but they make trees susceptible to being toppled by winds when the buffering effect of neighboring trees is removed. Other traits generally associated with susceptibility to windthrow include trees that are taller and fast-growing, with low wood density, and with shallow rooting (Turton and Alamgir 2015).

Depending on the direction of prevailing winds and topographic exposure, trees at forest edges adjacent to recent clearcut boundaries are often susceptible to being snapped or uprooted by wind speeds that are predictably reached every year or two (Mitchell et al. 2001; Zeng et al. 2004). That degree of topographic exposure or shelter can be measured using a topographic map or mapped digital elevation model (DEM) to derive a "topex" score for specified locations, calculated as the sum of angular elevation deviations from the horizontal exhibited by terrain in eight directions (MacKenzie 1974). Such scores can be weighted by the length of fetch and according to the prevalence of wind directions. If the management response is simply to salvage the damaged timber (see Section 8.6) and establish a new cutblock boundary, foresters may find themselves "chasing windthrow" in a futile effort to curtail ongoing forest damage. Instead, it is more prudent to "windfirm" cutblock boundaries where topographic and directional exposure indicates that windthrow is likely to occur. Suitable treatments may involve top-pruning, stand edge serration or feathering, and selective thinning for one to two tree heights into standing forest (Rowan et al. 2003; Mason and Valinger 2013). Ideally, the more vulnerable trees are removed or topped in a wedge-shaped pattern that absorbs or disperses wind energy, leaving residual trees to provide progressively more protection over a distance of one to two tree heights, so that trees in the bulk of the uncut forest remain standing.

As described in Chapter 7 and in Section 8.3, uniform forest thinning throughout a stand might be undertaken to minimize drought impacts, reduce the risk of crown fire, and help limit wildfire spread. But, if undertaken in closed-canopy stands, those same operations can make trees more vulnerable to windthrow when the wind-buffering protection offered by neighboring trees is removed (Gardiner et al. 1997; Hanewinkel et al. 2013). To prevent unnecessary losses, such thinning operations should remove trees with lower stem taper, or a high ratio of tree height (Ht) to stem diameter at breast height (DBH). It is those trees with a comparatively high slenderness coefficient (defined as Ht/DBH) that are more susceptible to being broken or uprooted by wind (Munishi and Chamshana 1994; Ribeiro et al. 2016; Skrzyszewski and Pach 2020; see Figure 8.5). Trees with very high slenderness coefficients are also more likely to be broken or toppled by heavy loads of wet snow. Adjusting a forest to be more windfirm and snowfirm can be more efficiently achieved if stands are established (through planting or early pre-commercial thinning) at low stocking densities, so that trees experience minimal shading from each other, retain full crowns, and develop a more pronounced taper and stronger root system than trees grown in close proximity with neighbors (Hanewinkel et al. 2013, Peltola et al. 2013). Such a practice provides all the benefits of stronger drought, fire, and wind resistance, though at the cost of lower carbon sequestration and timber yields per unit land area, and a greater ratio of knots to clear wood in the resulting solid wood products.

An alternative to treating clearcut boundaries to enhance windfirmness is to avoid creating long fetches of open ground in the first place—namely, to reduce the use of clearcut logging (and particularly large cutblocks) in wind-prone terrain. In forest stands—especially coniferous ones—established under high densities, it can be a challenge to undertake commercial thinning or other partial harvesting practices without leaving the remaining trees at risk of wind damage when their immediate neighbors are removed, as described above. With windthrow occurring in proportion to wind fetch up to about five tree heights in width (Gardiner et al. 1997; Mason and Valinger 2013), it is useful

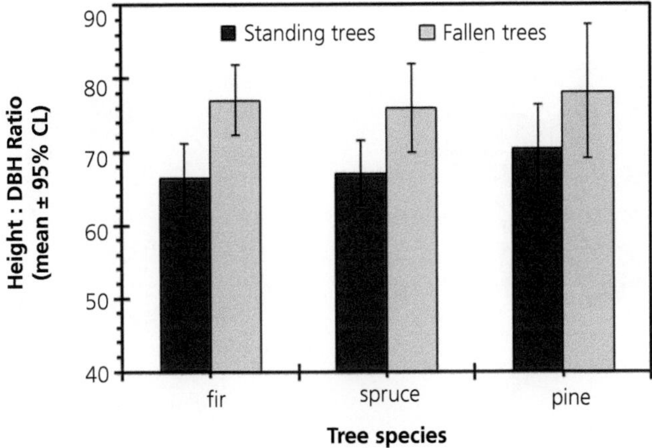

Figure 8.5 Contrasting height-to-diameter (DBH, diameter at breast height) ratios of standing mature trees and those that fell after adjacent sub-boreal forest in central British Columbia (Canada) was clearcut logged six to twenty years earlier. This ratio, also known as a slenderness coefficient, is indicative of windthrow susceptibility. Fir is *Abies lasiocarpa*, spruce denotes natural hybrids of *Picea englemannii* x *glauca*, and pine is *Pinus contorta* var. *latifolia* (reproduced from Burton (2002)).

to think in terms of making harvest openings as small as is operationally and silviculturally feasible in order to minimize that effect. Continuous cover forestry, irregular stand structure, and maintaining a diverse array of tree sizes within stands can reduce wind damage in some forest types on some sites (Mason 2002; Brang et al. 2014; Pukkala et al. 2016). However, partial cuts should not be placed directly beside recent clearcuts (Ruel 2020). For the same volume of wood harvested, partial cutting methods also can entail an increase in road lengths over clearcutting approaches, with increased wind exposure and the likelihood of landslides and other disturbances.

Variable retention harvesting and a greater use of green-tree retention is being widely advocated and adopted as another alternative to clearcut logging in many forests around the world (Gustafsson et al. 2012; Martínez Pastur et al. 2020). These practices have the benefit of providing some wildlife cover and mature forest structure and propagules in the secondary forest that develops around them, potentially serving a "lifeboat" role for some species that otherwise might be lost. A common objection to the use of retention-based logging and silviculture is the perception that "all those remaining trees will just blow down." Notwithstanding the benefits of downed woody material in ecosystems (Harmon

et al. 2004; Jia-Bing et al. 2005), surveys indicate high levels of treefall in very small clusters of retained conifers (Burton 2001; Beese et al. 2019), but that windthrow in large (>1 ha) retention patches does not differ from that in uncut forest (Urgenson et al. 2013; Coates et al. 2020). Windthrow in retention patches can also be minimized with careful planning. The clustering of residual trees (rather than leaving individual boles uniformly dispersed) provides some protection from winds. If the long narrow axis of a retention patch is orientated in the direction of prevailing winds, the leading edge may suffer some damage but will provide protection to the rest of the retention patch (see Figure 8.6).

Forest management practices to promote resilience to wind in parks and protected areas are usually quite minimal, mostly limited to removing trees at risk of falling on roads, trails, campgrounds, cottages, and other human infrastructure. There can also be important disturbance interactions that make the prevention or mitigation of storm damage and windthrow more important, notably where dead and damaged trees may support intense wildfire or a buildup in herbivorous or xylophagous insects (see Box 8.1). As with fire, most of the practices promoting forest integrity in the face of potential wind damage focus on resistance strategies—namely, the minimization of damage to

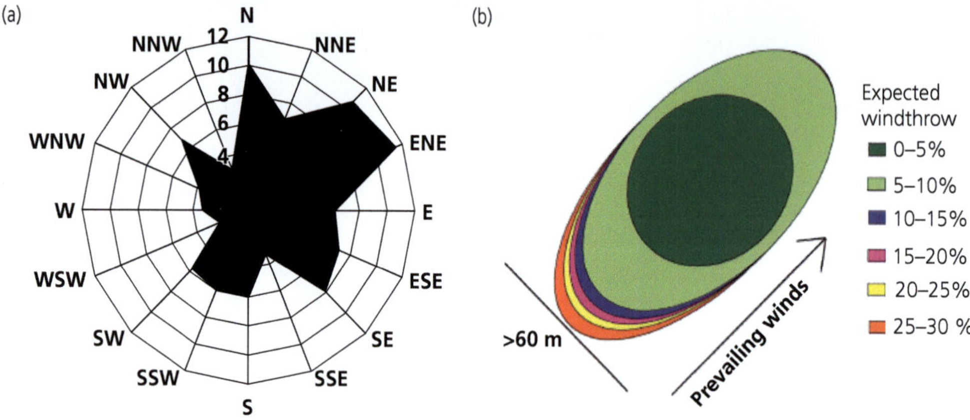

Figure 8.6 Analysis of the direction of high winds at nearby weather stations or (a) the direction of observed treefall (% of 368 fallen stems) in forest interiors informs (b) the recommended configuration and orientation of green tree retention patches in order to minimize windthrow. Empirical analysis of windthrow distribution in retention patches indicates how partial treefall in the leading edge of patches buffers the bulk of a patch from wind damage. Color-coded values associated with (b) are expected values of windthrow, based on transects surveyed through eleven retention patches in the central interior of British Columbia, Canada (reproduced from Burton (2001)).

existing forests. Selective thinning to promote high-taper, more windfirm, tree species or individuals can result in a gradual adjustment of the forest to being more wind resistant, though newly exposed trees may be more vulnerable to windthrow for a few years until they become acclimated (Duperat et al. 2022).

In boreal, hemi-boreal, and subalpine regions, spruces (*Picea* spp.) are generally more vulnerable than associated tree species owing to relatively shallow, plate-like root structures. The gradual reduction in European land managed for *Picea abies* in favor of more windfirm species such as *Abies alba* and broadleaf species such as *Fagus sylvatica* and *Quercus* spp. can, in part, be seen as a response to the greater likelihood of windthrow in spruce stands (Schütz et al. 2006; Mason and Valinger 2013). In the US southeast, live oak (*Quercus virginiana*) is known to be more resistant to wind damage and other hurricane-induced stresses than most other bottomland tree species, with windthrow of bald-cypress (*Taxodium distichum*) being rare (Stanturf et al. 2007). In the temperate rainforests of western North America, western redcedar (*Thuja plicata*) has greater stem taper and greater windfirmness than most associated species. Mixed-species stands in general are more resistant to wind damage (Schütz et al. 2006; Griess and Knoke 2011). Adjustment and reconfiguration (rather than resistance) approaches

promote such shifts in stand composition, which may be accompanied by clearcut and even-aged management systems being replaced by those characterized by small forest openings, mixed species, and multiple tree cohorts within stands (Potterf et al. 2024).

8.5 Planning for insect outbreaks

Large populations of herbivorous insects are supported by the primary productivity of trees and forests around the world. Native populations of those insect species usually coexist with their host trees at relatively low and stable "endemic" population densities. But, occasionally, various biotic and abiotic constraints on insect population growth may be relaxed, especially in this era of warming winters and the global transfer of insects to environs free of their natural predators and parasites. These conditions can result in a population explosion of insects to the degree that large areas of trees are killed or repeatedly stressed (Berryman 1988; Nair 2007; Wermelinger 2021). Although we may tend to think of wildfires as being the most important agent of forest disturbance and renewal, in fact the global forest area annually affected by outbreaking insects is greater, estimated at 855,000 km^2/yr, mostly in temperate regions (van Lierop et al. 2015).

Similar to fires, large mortality-causing herbivory events are not necessarily novel; there is a long history of major insect outbreaks and subsequent widespread tree mortality in natural systems (Brunelle et al. 2008). Recurrent insect outbreaks can be a characteristic agent of the disturbance regime, forest regeneration, and habitat diversification in many landscapes (Boulanger et al. 2012; Burton and Boulanger 2018). Insect damage is typically specific to certain species or sizes of trees, reflecting a long history of coevolution between specialized insect species and their host tree species, which often have herbivore-specific phytochemical defenses (Steinberg 2012). The intercontinental transfer of insect species is a common cause of explosive forest insect outbreaks as the new arrivals encounter naive host trees and associated animal communities with no history of coevolved interactions (Brockerhoff et al. 2006; Ramsfield et al. 2016). An average of twenty-five new insect species per decade have become established in the continental US, of which 14 percent typically cause notable damage to trees (Aukema et al. 2010). Vigilant phytosanitary measures (focused on the inspection of imported plant and soil materials, including wooden crates and dunnage) and prompt eradication of any incipient naturalization of invasive exotic insects and pathogens are urged to protect the integrity of native ecosystems and the economic value of commercial forests (Brockerhoff et al. 2010).

Many jurisdictions conduct annual surveys, typically by aircraft, to detect and map the extent of forest insect outbreaks in wild and extensively managed natural forests. Satellite-based detection methods are challenged by widespread false positives associated with other causes of tree stress such as drought, although discriminatory ability continues to improve (Senf et al. 2017). Indicators such as foliage loss and discoloration then prompt ground-based inspections to confirm the causal stress, pathogen, or insect. Insect-outbreak assessments in industrial plantations, particularly in the tropics and southern hemisphere, are primarily assessed on the ground in terms of insect density per tree (Wylie and Speight 2012), making comparisons with aerial overview surveys difficult. The most important forest insect pests include defoliators (notably the larvae of many moth species, order Lepidoptera) and sap-sucking insects such as aphids and adelgids (order Hemiptera, superfamily Aphidoidea), which may drain tree resources over several years after which trees often die. Bark beetles (order Coleoptera, family Curculionidae, subfamily Scolytinae) typically employ chemical pheromones to coordinate mass attacks of colonizing females that lay galleries of eggs that develop into larvae that feed on the phloem and cambium, especially in conifer species. A wide range of other insects feed on the vegetative and reproductive buds, seeds, and roots of forest trees (Wylie and Speight 2012; Wermelinger 2021). Insect populations are typically constrained by a combination of food (host tree) availability, predators (including vertebrates as well as other insects), parasites and diseases, and weather conditions during key life phases such as dispersal, mating, and overwintering (Raffa et al. 2008; Kneeshaw et al. 2015). The complexity of insect food web interactions, weather sensitivities, and response to forest composition makes their population dynamics very difficult to understand and control (Eveleigh et al. 2007; Kneeshaw et al. 2021).

As insect outbreaks often take several years to grow and spread, their early detection can sometimes inform direct or indirect methods of insect population control. Defoliators are now often treated with biological control agents such as aerial spraying with formulations containing spores of the bacterium *Bacillus thuringiensis* var. *kurstaki* (commonly known as Btk), the pathogenic effects of which are specific to lepidopteran caterpillars. Although widely replacing the use of broad spectrum and persistent pesticides (organophosphates such as malathion, carbamates such as carbaryl, and the widely banned DDT), Btk also causes collateral mortality of non-pest lepidopteran species, making its use inadvisable in woodland habitats with rich or rare butterfly and moth fauna. Another option for controlling localized populations of some insect pests is to use a combination of repellant and attractant semio-chemicals (insect pheromones or pheromone analogs) in a "push–pull strategy" to concentrate insects into a few "trap trees" that can then be felled and disposed of safely (Lindgren and Borden 1993; Witzgall et al.

2010). Another approach, more appropriate in parks and the urban forests for protection of individual high-value trees, is the use of injectable systemic insecticides, with azadirachtin (derived from the neem tree, *Azadirachta indica*) and emamectin benzoate showing noteworthy promise (McCullough 2020).

Indirect control methods were employed for many decades before the advent of chemical insecticides and pheromone-based approaches. Sanitation harvesting, or targeted pre-emptive salvage logging, is the most widely used approach in multi-value and production forests. The intent is for infested trees and those immediately around them to be extracted and processed before insect larvae on and in the harvested trees can successfully mature and reproduce. There is a danger, however, that log transport at specific times in the insect life cycle can result in some insects (notably bark beetles) being dispersed from infestation centers to other locations. Where poor access or high costs do not warrant directing logs to mills, small bark-beetle infestations may be controlled by felling the trees and then stripping their bark or burning them on site. Ideally, the use of chemical and biological control methods can be minimized by improved cultural practices in a program of integrated pest management (Wainhouse 2005). Still, history has many examples of insect population growth outstripping the ability of forest managers to harvest infested trees and control insect spread (see Table 8.3). The growing extent of forest insect outbreaks is recognized as a symptom of climate change, prompting an increasing interest in designing forests to be more insect resilient in a proactive rather than reactive manner (Klapwijk and Björkman 2018; Kneeshaw et al. 2021).

The premise that forest composition and structure can be managed to reduce the impact of herbivorous insect outbreaks is sometimes referred to as "the silvicultural hypothesis" of forest pest management (Miller and Rusnock 1993). In general, stand management methods to reduce insect herbivore damage have much in common with general tactics to enhance forest resilience, as described in Chapter 5. Managing for a variety of tree or stand ages, and a diversity of tree species—especially of different taxonomic genera or families—is a

promising defense. Many outbreaking insects (depending on species) have preferences for younger or older trees as hosts, and are finely coevolved to breed successfully on host trees of specific genera. There is growing evidence that maintaining a high proportion of broadleaf trees in stands managed primarily for conifer production provides "associational resistance" to herbivory by specialist insects, but not generalists (see Figure 8.7). Promoting two or more tree genera in managed stands will reduce overall losses to such specialist herbivores, often owing to reduced insect-dispersal success (Griess and Knoke 2011; Bauhus et al. 2017; Jactel et al. 2017). Examples of associational resistance are provided by population densities of the European pine sawfly (*Neodiprion sertifer*) and the eastern spruce budworm (*Choristoneura fumiferana*) being lower on suitable host trees in mixed forests than in conifer monocultures (Campbell et al. 2008; Klapwijk and Björkman 2018; Zhang et al. 2018, 2020). It has also been found that a *Populus tremuloides* overstory can hide the leaders of *Picea glauca* from attack by the *Pissodes strobi* terminal weevil (Taylor et al. 1996). These are just a few examples of how forest diversity can confer resistance to insect herbivory, with novel cases and mechanisms being increasingly discovered (see Figure 8.7).

Silvicultural manipulations have had mixed success in containing forest insect damage. Stand thinning can sometimes improve net resistance to defoliation by eastern spruce budworm, but this varies with site moisture regime and with tree species (Bauce and Fuentealba 2013). Conversely, populations of the balsam fir sawfly (*Neodiprion abietis*) and levels of associated conifer damage have been found to be greater in stands subject to pre-commercial thinning (Ostaff et al. 2006). For small localized bark-beetle populations (that is, pre-outbreak conditions), stand thinning in susceptible stands can enhance the vigor of the remaining conifer crop trees and their ability to generate resin, which can physically and chemically deter beetles attempting to burrow under the bark. Stand thinning to promote "beetle proofing" of conifer stands has been widely tested in western North America, but has not been practiced in Europe (Fettig et al. 2007; Marini et al. 2021). Once again, promoting compositional diversity at the stand level may be the most

Table 8.3 Selected examples of some large and notable forest insect outbreaks documented from around the world

Insect species	Principal host tree species	Outbreak location and period	Area[*] affected	References
Nun Moth (***Lymantria monacha***) **and following Spruce Bark Beetle** (***Ips typographus***)	*Picea abies*	Prussia (modern-day Poland) and Russia, 1853–63	403,000 km^2	Bejer (1988)
European Spruce Bark Beetle (***Ips typographus***)	*Picea abies*	Southern Norway, late 1970s–early 1980s	140,000 km^2	Bakke (1991), in Kautz et al. (2017)
Mountain Pine Beetle (*Dendroctonus ponderosae***)**	*Pinus* spp., especially *P. contorta* var. *latifolia*	British Columbia (Canada), 1999–2015 Western US, 2008–10	180,000 km^2 97,000 km^2	Canadian Forest Service (2021); USDA Forest Service (2016), in Kautz et al. (2017)
Spongy Moth[**] (*Lymantria dispar***)**	*Quercus* spp., other broadleaf species	NE US, 1981 SE Europe, 1997–99	52,000 km^2 20,000 km^2	USDA Forest Service (2010), in Kautz et al. (2017); Allard et al. (2003)
Spruce Beetle (*Dendroctonus rufipennis***)**	*Picea glauca, P. sitchensis*	Southern Alaska (US) and SW Yukon (Canada), 1990–2000	12,000 km^2	Werner et al. (2006), in Kautz et al. (2017)
Eastern Spruce Budworm (*Choristoneura fumiferana***)**	*Abies balsamea, Picea* spp.	Eastern Canada, 1967–93 Quebec (Canada), 2006–present	519,000 km^2 96,000 km^2 (as of 2019)	Canadian Forest Service (2016), in Kautz et al. (2017); Canadian Forest Service (2022)
Bronze Bug[**] (***Thaumastocoris peregrinus***)	*Eucalyptus* spp.	Brazil, 2008–18	5,000 km^2	Junqueira et al. (2018)
Ormiscodes amphimone	Especially *Nothofagus pumilio*	Southern Chile, 2000–15	1,640 km^2	Estay et al. (2019)
Gumleaf skeletonizer (***Uraba lugens***)	*Eucalyptus marginata, Corymbia calophylla*	Western Australia, 1982–88 2009–11	3,000 km^2 2,500 km^2	Farr (2002), in Estay et al. (2019); Farr and Wills (2012)
Hemlock woolly Adelgid [**] (*Adelges tsugae***)**	*Tsuga canadensis*	Northeastern US, 1988–present	>10,000 km^2	Orwig et al. (2002); Ellison et al. (2018)
Forest Tent caterpillar (***Malacosoma disstria***)	*Populus* spp., *Quercus* spp., *Acer saccharum*	Northern Ontario and Quebec (Canada), 1990–2001 NY and VT (US), 2002–5	143,000 km^2 3,530 km^2	Canadian Forest Service (2016), in Kautz et al (2017); https://en.wikipediaorg/wiki/Forest_tent_caterpillar_moth
Siberian Silk Moth (***Dendrolimus sibiricus***)	Conifers	Central Siberia (Russia), 1994–6 Sakha Republic (Russia), 2000	7,000 km^2 59,000 km^2	Khark et al. (2007), in Kautz et al. (2017); Gninenko and Orlinskii (2002), in Kautz et al. (2017)
Pine Caterpillars (***Dendrolimus* spp.**)	*Pinus* spp.	Shandong (China) 1992 Vietnam, 1987	1,250 km^2 565 km^2	Bao et al. (2019); Billings (1991)
Pine shoot beetle (*Tomicus piniperda***)**	*Pinus yunnanensis*	Yunnan (China), late 1970s–early 1990s	15,000 km^2	Hui and Xue-Song (1991), in Kautz et al. (2017)

* Variously includes area affected, not just moderate or severe defoliation and mortality.
** Species exotic to the outbreak region.

effective strategy for resilient forest management, including the retention of non-crop and understory species that can serve as diversionary food sources and as habitat for the predators and parasites of insect pests (Jactel and Brockerhoff 2007; Jactel et al. 2009). Site selection, site preparation techniques,

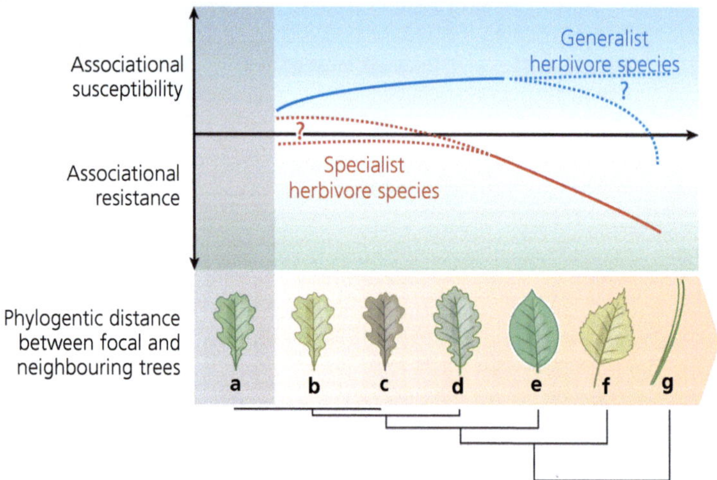

Figure 8.7 Phylogenetic diversity may deter population growth and spread of specialist insect herbivores, described as associational resistance. Such diversity has little effect (except possibly at the greatest levels of phylogenetic difference) on generalist insect herbivores, which can spill over from one species to another, known as associational susceptibility. The focal tree is represented on the left (e.g. an oak), with the dendrogram representing the genetic distances between tree genotypes (a–c), species (d), genera or families (e, f), and divisions/growthforms (g) (reproduced from Jactel et al. (2021); used with permission).

and species selection are important, after which natural stand development free of silvicultural interventions may be the more promising approach to minimize the impacts of herbivorous forest insects.

Uneven-aged management or continuous-cover forestry has the potential to reduce the proportion of trees susceptible to attack by some insect species, notably European spruce beetle and large pine weevil (*Hylobius abietis*; Nevalainen 2017). Klapwijk et al. (2016) evaluated three alternatives to standard European forest management practices in terms of their expected effects on defoliators, bark beetles, and insects that damage regeneration. They concluded that mixed-species stands and short-rotation management (with no thinning) can increase the abundance of natural enemies (predators, parasites), with improved long-term continuity of enemy abundance likely in mixed stands and under continuous-cover forestry. However, in other ecosystem contexts, an uneven-aged stand structure makes regenerating trees more susceptible to some insect species, such as the western spruce budworm, *Choristoneura occidentalis* in interior *Pseudotsuga menziesii* var. *glauca* stands (Alfaro and Maclauchlan 1992).

At the landscape level, maintaining a variety of tree species and age classes is advisable, reducing connectivity between similar forest stands as much as possible. Nonetheless, the efficacy of all practices depends on the biology of particular insect threats, the status of regionally synchronized insect populations, and the structure and composition of the forest landscape. High levels of forest fragmentation and associated edge effects constrain parasitoid populations that limit damage by forest tent caterpillar (*Malacosoma disstria*), resulting in high levels and duration of defoliation on *Populus tremuloides* in Ontario, Canada (Roland and Taylor 1997). The warmer microclimate associated with clearcut edges promotes local population increases in European spruce beetle (*Ips typographus*; Kautz et al. 2013), whereas forest fragmentation and interspersion with conifer plantations reduces insect dispersal effectiveness and damage by winter moth (*Operophtera brumata*) on multiple broadleaf tree species in Poland (Wesołowski and Rowiński 2006). A combination of stand-level and landscape-level diversity, along with monitoring and early intervention (especially to control invasive exotic insects), is widely advised (see Box 8.3).

Box 8.3 Guidance to minimize forest damage by biotic disturbance agents

A good understanding of local pest biology, tree-species silvics, and their interaction with site and climate is clearly needed in order to adopt an appropriate strategy for promoting resilience under the threat of biotic disturbances. Several authors (Liebhold 2012; Alfaro and Langor 2016; Ramsfield et al. 2016) offer some general advice for planning and implementing a holistic or ecosystem-based approach to forest pest management in a changing world:

1. Recognize that it is practically impossible to suppress insect outbreaks over large geographic areas, and that such outbreaks are part of a healthy forest landscape.
2. Silvicultural practices (including site-preparation options, careful tree-species choice, and spacing decisions in the context of site or stand vulnerability and perceived insect pressures) can sometimes be effective at reducing forest susceptibility to outbreaks.
3. Mixed-species stands and diverse landscapes are typically more resistant to pest outbreaks (see Figure 8.8), and should be promoted.
4. Minimizing the invasion of non-native pest species is essential, and eradication efforts must be undertaken early in the invasion process in order to be effective.
5. Monitoring and research into insect population dynamics should take place during endemic (quiescent) population levels, not just during outbreak conditions.
6. Insect outbreaks are often facilitated by drought and other stresses, indicating the need for managers to monitor insect populations and cumulative effects on susceptible sites and during drought events.
7. Classical biological control (Kenis et al. 2017) and the selection of species and genotypes for host resistance (Wagner et al. 2007) are important approaches.
8. Plantations of exotic tree species, especially where native congenerics are present in the landscape, are prone to catastrophic damage.

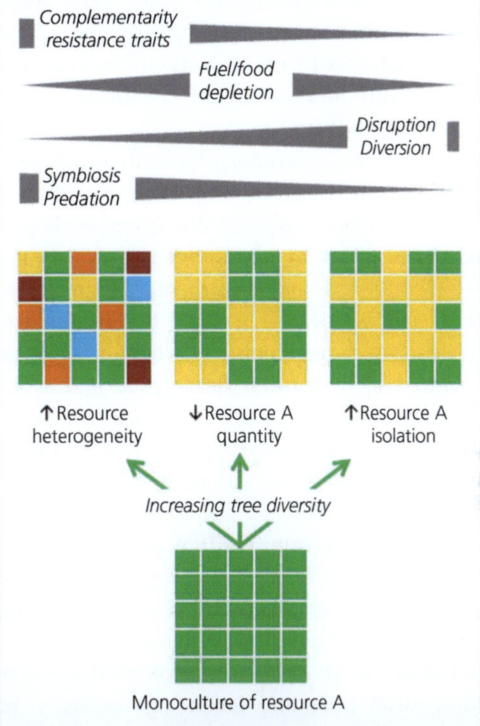

Figure 8.8 Potential mechanisms by which forest stand diversity generates associational resistance to biotic disturbances (e.g. from eruptive insects and fungal pathogens) (reproduced from Jactel et al. (2017); used with permission).

Most of these same principles largely apply to the management of forest pathogens (viral, bacterial, and fungal diseases of trees) as well (Liebhold 2012; Sturrock 2012; Ramsfield et al. 2016).

8.6 A place for salvage logging

The most common forest management response to natural disturbances is to undertake salvage logging in the affected areas. Motives are primarily to recover as much economic value from the damaged timber as possible, but also to reduce fuel loads and the risk of wildfire (in the case of insect outbreaks and wind damage), to prevent the spread of bark beetles and saproxylic insects, to remove potentially dangerous trees in public use areas, and to prepare sites for artificial regeneration. Those are legitimate objectives in many situations, particularly in production forests, but salvage logging is not always effective or necessary in achieving such objectives; conversely, there may be no economic disadvantage in refraining from salvaging damaged timber (Knoke et al. 2021). Indeed, salvage operations can often do more harm than good when viewed through all aspects of sustainable forest management (Lindenmayer et al. 2008; Thorn et al. 2018). Consequently, salvage-logging operations after natural disturbances should be deployed very selectively and strategically, with an aim to minimize their adverse impacts.

Because salvage operations typically occur soon after natural disturbances, they often contribute a "double hit" to forest habitat features and ecological processes, resulting in cumulative impacts greater than either the original disturbance or green-tree logging on their own (see Table 8.1). If delayed for a year or more (common when large areas are affected by a natural disturbance and when new access roads must first be built), the naturally recovering vegetation and newly germinated or released tree seedlings can be damaged by logging machinery. This has frequently been found to be the case during wildfire salvage (Fraser et al. 2004; Donato et al. 2006; Saint-Germain and Greene 2009), but not so much in other situations (for example, after windthrow; Peterson and Leach 2008; Kramer et al. 2014). The ecological value of standing dead trees and fallen logs needs to be appreciated, especially where a history of forest

conversion to agriculture, extensive firewood gathering, and the prompt removal of dead trees makes them a rare feature in the landscape (Siitonen 2001; Swanson et al. 2011). Dead trees in various stages of decay are important to woodpeckers, raptors, cavity-nesting birds and mammals, and many insect species (such as long-horn beetles, Cerambycidae) that specialize on colonizing recently burned trees (Thorn et al. 2020). Large woody material derived from fallen trees is a vital component of healthy streams, providing structure that moderates flow rates and creates habitat for fish and the prey on which they depend (Harmon et al. 2004).

A global meta-analysis of biodiversity in naturally disturbed forest stands that have been salvage logged or not reveals that most groups of organisms are richer *without* further disturbance by logging (Thorn et al. 2018). Dead standing trees and undisturbed understory vegetation (such as prevails after windthrow events and forest insect outbreaks) also provide some level of hydrological regulation and erosion control that is compromised when forests are further disturbed by large-scale clearcut logging. With clearcut harvesting prevailing in most commercial salvage-logging operations, it is usual for mature living trees to be harvested along with the dead ones, with large amounts of wood waste

Figure 8.9 Piles of waste wood (small trees, tops, and logs culled because of damage) remaining from clearcut salvage logging of beetle-killed lodgepole pine (*Pinus contorta* var. *latifolia*) in central British Columbia, Canada. Alternatively, forest cover can be maintained if careful logging retains high levels of green trees (visible in the background). Wood waste can be reduced further if material unsuitable for sawmilling is directed to pulp or biofuel use.

accruing from immature trees and damaged stems (see Figure 8.9). Despite the widespread perception that concentrations of dead standing or fallen trees constitute a fire danger or a nucleus for further insect outbreaks, evidence indicates that weather plays a more important role, and salvage logging cannot be expected to prevent future disturbances (Leverkus et al. 2021). Finally, it should be noted that the surge in regional fiber supplies associated with widespread salvage logging after large-scale disturbances can result in a glut of timber on the market, prompting a decline in prices (Prestemon et al. 2006) and a reduced availability of mature timber for several years in the future (Burton 2010).

With natural disturbances expected to increase as the global climate warms, and management actions unable to prevent their occurrence, what role (if any) can timber salvage play in promoting resilient forests? As with all stewardship options, the appropriate response to natural disturbances depends on forest type, landscape context, and the prioritized values designated for the affected tract of forest. Forest disturbances need not undermine the general principles, plans, and practices for sustainable forest management, so long as their inevitable occurrence is incorporated (as a welcome agent of forest renewal and habitat diversification) into forest plans. This means that salvage logging is to be avoided in national parks, ecological reserves, and other protected areas where it is intended that

natural processes dominate, although there can be localized exceptions (see Box 8.4).

In public forests managed extensively for multiple values, there may be opportunities to redirect predetermined levels of sustainable timber harvesting from undisturbed stands to those damaged by wildfire, insects, or wind. With the frequency and extent of natural disturbances on the increase (Seidl et al. 2017), salvage logging can be expected to make up an increasing proportion of the total wood volume harvested. The decision to do so, however, might first consider whether affected stands can be accessed by existing roads, and whether additional logging at this time would compromise the hydrological functioning of a catchment basin. Another important question is whether the biodiversity habitat value of naturally disturbed forest is greater than that of the mature or old-growth stands that would otherwise be harvested (that is, which is rarer in the landscape: old-growth forest or stands of dead trees and naturally regenerating forest?). If naturally disturbed young forests are rare in the landscape, they should be left to develop without interference as a unique and important habitat type; conversely, it will often be preferable to undertake post-disturbance salvage logging where old-growth forest is rare and otherwise would be logged. It is imperative, however, that salvaged timber volumes fall within predetermined levels of allowable harvesting, contrary to widespread policies

---

Box 8.4 Post-disturbance management in protected forests

When natural disturbances occur in parks and other protected areas (as defined under IUCN categories Ia, Ib, II, and sometimes III or IV; Dudley 2008), the default management direction is to let nature take its course. Salvage logging and related management activities (such as tree planting, or selective brushing and spacing) may be undertaken to promote habitat values for targeted rare and endangered species, or to assure public safety from falling trees and branches in heavily used front-country locations.

Natural disturbances in parks and protected areas are often important for scientific research and public education, because stands of naturally dead and dying trees—and their associated stand dynamics and biodiversity—can be rather

rare in many human-dominated parts of the world. Leaving such stands unsalvaged and free to recover naturally often provides a stark contrast and valuable insights when compared with actively managed forests nearby (Thorn et al. 2017). Users of backcountry trails and wilderness areas expect a higher level of adventure and risk, so all dead trees need not be removed near trails, though may be felled and bucked into firewood in the vicinity of campgrounds. Signage can alert park users to the dangers of undertaking recreational activities in disturbed forests (see Figure 8.10(a)), and can serve an important role in demonstrating the dynamics of disturbance and recovery in natural forest landscapes that the public tends to think of as static (see Figure 8.10(b)).

Figure 8.10 Promoting acceptance of natural disturbances in protected areas: (a) Signage in Jasper National Park, Alberta, Canada, warning backcountry trail users of the potential for falling trees in forest damaged by fire; (b) signage in Nationalpark Bayerischer Wald, Bavaria, Germany, inviting visitors to interpret the dead trees left by a bark-beetle outbreak as dramatic evidence of a dynamic and self-renewing forest.

and incentives to accelerate post-disturbance logging beyond sustainable levels (Saint-Germain and Greene 2009; Burton 2010).

Additional considerations—especially in the case of insect outbreaks or storm damage—include whether there is sufficient advance regeneration or secondary structure (living trees that have survived the disturbance) for the forest to recover in a timely manner, or whether forest renewal could be accelerated by logging and prompt planting (Stanturf et al. 2007; Burton et al. 2016; see Figure 8.11). Unnecessary amounts of waste wood can be avoided and some mature forest habitat values can be maintained if salvage logging is done carefully and selectively, with high concentrations of live and dead trees retained wherever possible. Salvage operations can sometimes be justified to reduce wildfire risk near communities and other infrastructure, or to prevent population buildup in eruptive insects

such as the spruce beetles (*Ips typographus* in Europe, *Dendroctonus rufipennis* in North America) in dead and weakened trees. Even in such situations however, small infested patches (e.g. <20 individual trees or <5 m^2 of basal area) do not warrant sanitation or salvage logging (Kausrud et al. 2012). In general, accelerated levels of timber harvesting, new road building, and clearcut logging primarily compromise, rather than improve, the resilience of forests to natural disturbances.

The decision to salvage damaged timber from production forests, where economic costs and benefits are paramount, is usually more straightforward. Nonetheless, basic expectations for environmental protection and sustainability suggest the need to proceed cautiously, not to accelerate logging unnecessarily, and not to compromise soil and watershed health as a result of poorly planned or excessive logging. In some cases, land-use priorities

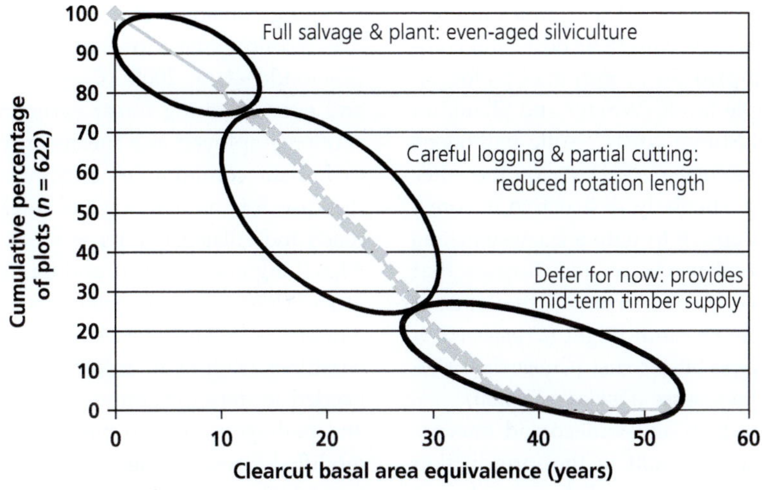

Figure 8.11 Recommendations for post-disturbance stand management devised by comparing the current basal area of surviving trees in sample plots with the time projected (using an accepted growth-and-yield model) to grow an equivalent basal area if damaged stands were to be clearcut-logged and promptly planted to full stocking (courtesy of K. D. Coates, based on Coates et al. (2006)).

or values may have changed, such that a plantation of exotic tree species, for example, might now be nurturing the recovery of a more valuable native forest plant community, which would be damaged if salvage logging were to proceed. But, in general, the redirection of a sustainable level of timber harvesting from undisturbed forest to dead and damaged trees is the most prudent response. Understanding the "shelf life" (period of time over which the wood quality of dead trees remains acceptable for specific uses) of local tree species under the prevailing climate can help prioritize and schedule salvage operations (Lewis and Hartley 2006; Barrette et al. 2015; Vaughan et al. 2018). The high levels of "waste" wood typically associated with salvage logging can sometimes be offset by putting that material to good use to fuel boilers at processing facilities, for municipal heating or power generation, or processed into pellets for commercial sale. While salvage operations should be undertaken promptly after a disturbance and stands regenerated quickly thereafter, any harvest over and above sustainable levels will have unfavorable consequences down the road. If markets are not able to absorb a surge in fiber supplies without a concomitant drop in prices, deliveries can be drawn out by maintaining inventories of salvaged logs (by storage in water, under snow, or in dryland sort yards)

or as finished wood products over several years. But, if sustainable harvest levels are exceeded, or if they were set without due consideration for the likelihood of losses to natural disturbances, disease, climatic stress, and other disruptions, then future harvest levels (and revenues and employment) will inevitably have to be curtailed (Burton 2010).

8.7 Expect the unexpected

Wildfires, windstorms, and insect outbreaks are the most widespread forms of natural disturbance in the world's forests, but there are several other disturbance agents (for example, landslides, debris flows, snow avalanches, flooding) that can be locally important. Tree pathogens (particularly fungal parasites but bacterial and viral diseases as well) can behave as localized, landscape, or regional disturbances when virulent strains, vulnerable hosts, and favorable environmental conditions coincide, with all three axes of this disease triangle being in a continuous state of flux (Sturrock et al. 2011). Endemic levels of native tree pathogens, such as the conifer root rots (for example, *Armillaria* and *Tomentosus* species) contribute important diversity to natural forests through the creation of canopy gaps and the promotion of tree species diversity (Van der

Kamp 1991; see Figure 4.6(b)). The complex interactions of forest pathogens with their hosts and competitors have prompted the evolution of biological diversity at multiple levels (Winder and Shamoun 2006; Comita and Stump 2020). Exotic pathogens introduced to naive congeneric hosts, on the other hand, have been particularly destructive to forest resources and disruptive to native ecosystems, as witnessed by several waves of tree diseases that have driven native North American tree species nearly to extinction: chestnut blight (*Cryphonectria parasitica*), white pine blister rust (*Cronartium ribicola*), and Dutch elm disease (*Ophiostoma ulmi*).

Global trade practices and widespread international travel inevitably result in the introduction of exotic species that can establish, spread, and seriously compromise the ecological processes of forests in novel habitats. Examples include efficient predators (such as European stoats, *Mustela erminea*, in New Zealand and rats, *Rattus* spp., accidentally introduced to many of the world's islands) that decimate native bird populations, vertebrate herbivores (such as introduced rabbits eating tree seedlings and ringbarking small trees in Australia) and insect herbivores such as the spongy moth (*Lymantria dispar dispar*) and emerald ash borer (*Agrilus planipennis*) now widespread in North America. Approximately six new insect pest species per year have been detected in Europe since 2000 (Roques et al. 2020), indicative of the constant threat associated with invasive species. Invasive plant species can also threaten forest regeneration (for example, Scotch broom, *Cytisus scoparius*, in the US Pacific Northwest; Isaacson 2000) or forest growth and yield (for example, kudzu, *Pueraria montana*, in the US southeast; Forseth and Innis 2004). The ranges of these and many other invasive exotic species are expected to expand under climate change, along with a growing risk from abiotic disturbances such as wildfire (see Section 8.3), extreme hurricanes, and soil collapse owing to permafrost thaw.

A broadly framed risk and vulnerability analysis, as outlined for climate change in Section 6.4, will consider the historical disturbance regime of landscapes and the response of those disturbance agents to expected changes in regional climate. For geophysical disturbances such as volcanic eruptions, floods, and mass movements, risk and hazard mapping considers signs of past events and utilizes terrain analysis to identify vulnerable locations (Remondo et al. 2008). Conserving, maintaining, and even restoring forest cover in such locations is often employed as a means of reducing impacts to human communities and infrastructure, as is the case for maintaining tall forest upslope from roads and villages to intercept rock fall in the Swiss Alps (Lingua et al. 2020), and for maintaining mangroves (*Rhizophora*, *Avicennia*, and *Aegiceras* spp.) to dampen tsunami impacts along low-lying tropical coastlines (Koh et al. 2018). Reliable monitoring is needed to detect eruptions of native forest insects and pathogens, and to limit their spread through modified forest management practices as much as possible (Logan et al. 2003). Similarly, vigilant interception and eradication programs are needed to prevent the establishment of invasive species new to a continent or island, and to minimize populations of those pests already there.

Worldwide research and monitoring efforts can alert forest managers to potential threats and promising methods for mitigating them. For example, it has been suggested that conifer seedlings can be inoculated with selected bacterial and fungal endophytes that can reduce herbivory by insect defoliators (Vacher et al. 2021). Research also indicates that the resident competitors, herbivores, and overall diversity of intact ecosystems can help resist invasive organisms (Levine et al. 2004; Byun et al. 2018), so minimizing soil disturbance and species loss are important in the fight against invasive exotics. There is widespread interest among forest professionals in mitigating the impacts of natural disturbances, but there is insufficient research supporting the most promising strategies of increasing the abundance of broadleaf tree species, increasing structural heterogeneity, and decreasing monoculture management (Nikinmaa et al. 2024).

All disturbances, disruptions, and stresses to forest ecosystems and forest management cannot be anticipated, although preparation for the specific threats identified in a vulnerability analysis can help avoid surprises. Planning for general resilience (see Chapter 5)—facilitating recovery, adjustment, reconfiguration, and transformation—is an important backup for targeted disturbance-resistant strategies. While planning for an asteroid collision or "zombie apocalypse" is not warranted, it is still worthwhile to have contingencies in place for avoiding or responding to low-probability but high-impact disruptions. Depending on the degree to which a forest management plan depends on labor, fossil fuels, social license, public subsidies, or market access, it may be wise to incorporate contingency plans to maintain resilience in the face of labor strife, a global pandemic, petroleum shortages, public protests and blockades, changes in government priorities, and the imposition of forest product tariffs. Those general resilience strategies of fostering flexibility and adaptive capacity, diversification at all levels, maintaining intact networks, and exercising the precautionary principle (in both biophysical and socioeconomic realms) within a framework of collaborative adaptive management (see Chapter 10) will serve forest managers well in an uncertain future (Millar et al. 2007; Seidl and Lexer 2013; Mina et al. 2022).

Box 8.5 Key points

- Natural disturbances, from the scale of single-tree gaps to massive wildfires and insect outbreaks, are characteristic of the ecology of all forests; they are important agents of forest renewal and create conditions on which many species depend, thereby supporting forest productivity and biodiversity.
- Global change (in the form of a warming climate, intercontinental species introductions, and homogenized landscapes) is making forests more susceptible to various disturbances, which then quickly threaten their ability to deliver many valuable ecosystem services.
- Common strategies to minimize the negative impacts of disturbances include geographically scattered holdings, compositional and structural diversity at stand and landscape levels, and rigorous monitoring, early disturbance detection, and control.
- Not all types of disturbance are equally likely in all forests, so vulnerability analysis can better prepare forest managers for specific disruptions such as wildfire, windstorms, and insect outbreaks.
- Although some silvicultural interventions (for example, stand thinning) can increase resistance to some disturbances under low or moderate levels of threat, they are generally ineffective under severe conditions,

and are rarely feasible on a large scale or have other undesirable impacts.
- Post-disturbance management in the form of salvage logging often compromises the inherent resilience of forest ecosystems to recover, adjust, and reconfigure themselves after natural disturbances, so must be done carefully, selectively, and primarily where harvest levels are diverted from rarer (for example, old-growth) forest types.
- Disturbance effects and contingency response plans need to be incorporated in forest plans, but not all disturbances and disruptions can be anticipated; general principles of resilient forest management are an important backup to approaches intended to promote resistance to specific threats.

Things will work out—maybe just not the way you plan.
Rick Riordan (2012)

References cited

Agee, J. K., and Skinner, C. N. (2005). "Basic Principles of Forest Fuel Reduction Treatments," *Forest Ecology and Management*, 211/1–2: 83–96.

Alcasena, F. J., Ager, A. A., Salis, M., Day, M. A., and Vega-Garcia, C. (2018). "Optimizing Prescribed Fire Allocation for Managing Fire Risk in Central Catalonia," *Science of the Total Environment*, 621: 872–85.

Alfaro, R. I., and Langor, D. (2016). "Changing Paradigms in the Management of Forest Insect Disturbances," *Canadian Entomologist*, 148/S1: S7–S18.

Alfaro, R. I., and Maclauchlan, L. E. (1992). "A Method to Calculate the Losses Caused by Western Spruce Budworm in Uneven-Aged Douglas Fir Forests of British Columbia," *Forest Ecology and Management*, 55/1–4: 295–313.

Alkhatib, A. A. (2014). "A Review on Forest Fire Detection Techniques," *International Journal of Distributed Sensor Networks*, 10/3: 597368.

Allard, G. B., Fortuna, S., See, L. S., Novotny, J., Baldini, A., and Courtinho, T. (2003). "Global Information on Outbreaks and Impact of Major Forest Insect Pests and Diseases." Paper presented at the XII World Forestry Congress, Quebec City, Canada. Available online at https://www.fao.org/3/xii/1019-b3.htm (accessed February 14, 2024).

Angelstam, P. K. (1998). "Maintaining and Restoring Biodiversity in European Boreal Forests by Developing Natural Disturbance Regimes," *Journal of Vegetation Science*, 9/4: 593–602.

Aukema, J. E., McCullough, D. G., Von Holle, B., Liebhold, A. M., Britton, K., and Frankel, S. J. (2010). "Historical Accumulation of Nonindigenous Forest Pests in the Continental United States," *BioScience*, 60/11: 886–97.

Bao, Y., Wang, F., Tong, S., et al. (2019). "Effect of Drought on Outbreaks of Major Forest Pests, Pine Caterpillars (*Dendrolimus* spp.), in Shandong Province, China," *Forests*, 10/3: 264.

Barrette, J., Thiffault, E., Saint-Pierre, F., Wetzel, S., Duchesne, I., and Krigstin, S. (2015). "Dynamics of Dead Tree Degradation and Shelf-Life Following Natural Disturbances: Can Salvaged Trees from Boreal Forests 'Fuel' the Forestry and Bioenergy Sectors?" *Forestry*, 88/3: 275–90.

Bauce, É., and Fuentealba, A. (2013). "Interactions between Stand Thinning, Site Quality and Host Tree Species on Spruce Budworm Biological Performance and Host Tree Resistance over a 6 Year Period after Thinning," *Forest Ecology and Management*, 304: 212–23.

Bauhus, J., Forrester, D. I., Gardiner, B., Jactel, H., Vallejo, R., and Pretzsch, H. (2017). "Ecological Stability of Mixed-Species Forests," in H. Pretzsh, D. I. Forrester, and J. Bauhus (eds), *Mixed-Species Forests: Ecology and Management*. Berlin: Springer, 337–82.

Beese, W. J., Rollerson, T. P., and Peters, C. M. (2019). "Quantifying Wind Damage Associated with Variable Retention Harvesting in Coastal British Columbia," *Forest Ecology and Management*, 443: 117–31.

Bejer, B. (1988). "The Nun Moth in European Spruce Forests," in A. A. Berryman (ed.), *Dynamics of Forest Insect Populations*. New York: Plenum, 211–31.

Bengtsson, J., Angelstam, P., Elmqvist, T., et al. (2003). "Reserves, Resilience and Dynamic Landscapes," *Ambio*, 32/6: 389–96.

Berryman, A. A. (ed.). (1988). *Dynamics of Forest Insect Populations*. New York: Plenum. 602 pp.

Billings, R. F. (1991). "The Pine Caterpillar *Dendrolimus punctatus* in Viet Nam; Recommendations for Integrated Pest Management," *Forest Ecology and Management*, 39: 97–106.

Bond, W. J., and Keeley, J. E. (2005). "Fire as a Global 'Herbivore': The Ecology and Evolution of Flammable Ecosystems," *Trends in Ecology & Evolution*, 20/7: 387–94.

Boulanger, Y., Arseneault, D., Morin, H., Jardon, Y., Bertrand, P., and Dagneau, C. (2012). "Dendrochronological Reconstruction of Spruce Budworm (*Choristoneura fumiferana*) Outbreaks in Southern Quebec for the Last 400 Years," *Canadian Journal of Forest Research*, 42/7: 1264–76.

Bowman, D. M., Balch, J., Artaxo, P., et al. (2011). "The Human Dimension of Fire Regimes on Earth," *Journal of Biogeography*, 38/12: 2223–36.

Brang, P., Spathelf, P., Larsen, J. B., et al. (2014). "Suitability of Close-to-Nature Silviculture for Adapting Temperate European Forests to Climate Change," *Forestry*, 87/4: 492–503.

Brice, M.-H., Vissault, S., Vieira, W., Gravel, D., Legendre, P., and Fortin, M.-J. (2020). "Moderate Disturbances Accelerate Forest Transition Dynamics under Climate Change in the Temperate–Boreal Ecotone of Eastern North America," *Global Change Biology*, 26/8: 4418–35.

Brockerhoff, E. G., Liebhold, A. M., and Jactel, H. (2006). "The Ecology of Forest Insect Invasions and Advances in their Management," *Canadian Journal of Forest Research*, 36: 263–8.

Brockerhoff, E. G., Liebhold, A. M., Richardson, B., and Suckling, D. M. (2010). "Eradication of Invasive Forest Insects: Concept, Methods, Costs and Benefits," *New Zealand Journal of Forest Science*, 40: S117–S135.

Brunelle, A., Rehfeldt, G. E., Bentz, B., and Munson, A. S. (2008). "Holocene Records of *Dendroctonus* Bark Beetles in High Elevation Pine Forests of Idaho and Montana, USA," *Forest Ecology and Management*, 255/3–4: 836–46.

Buma, B. (2015). "Disturbance Interactions: Characterization, Prediction, and the Potential for Cascading Effects," *Ecosphere*, 6/4: 1–15.

Buma, B., and Schultz, C. (2020). "Disturbances as Opportunities: Learning from Disturbance-Response Parallels in Social and Ecological Systems to Better Adapt to Climate Change," *Journal of Applied Ecology*, 57/6: 1113–23.

Buma, B., Weiss, S., Hayes, K., and Lucash, M. (2020). "Wildland Fire Reburning Trends across the US West Suggest Only Short-term Negative Feedback and Differing Climatic Effects," *Environmental Research Letters*, 15(3): 034026.

Burton, P. J. (2001). "Windthrow Patterns on Cutblock Edges and in Retention Patches in the SBSmc," in S. J. Mitchell and J. Rodney (compilers), *Windthrow Assessment and Management in British Columbia: Proceedings of the Windthrow Researchers Workshop held January 31–February 1, 2001, in Richmond, British Columbia*. Vancouver: BC Forestry Continuing Studies Network, 19–31.

Burton, P. J. (2002). "Effects of Clearcut Edges on Trees in the Sub-Boreal Spruce Zone of Northwest-Central British Columbia," *Silva Fennica*, 36: 329–52.

Burton, P. J. (2010). "Striving for Sustainability and Resilience in the Face of Unprecedented Change: The Case of the Mountain Pine Beetle Outbreak in British Columbia," *Sustainability*, 2: 2403–23.

Burton, P. J., and Boulanger, Y. (2018). "Characterizing the Combined Fire and Insect Outbreak Disturbance Regimes of British Columbia, Canada," *Landscape Ecology*, 33/11: 1997–2011.

Burton, P.J., Jentsch, A., and Walker, L. R. (2020). "The Ecology of Disturbance Interactions," *BioScience*, 70/10: 854–70.

Burton, P. J., Parisien, M.-A., Hicke, J. A., Hall, R. J., and Freeburn, J. T. (2008). "Large Fires as Agents of Ecological Diversity in the North American Boreal Forest," *International Journal of Wildland Fire*, 17/6: 754–67.

Burton, P. J., Svoboda, M., Kneeshaw, D., and Gottschalk, K. W. (2016). "Options for Promoting the Recovery and Rehabilitation of Forests Affected by Severe Insect Outbreaks," in J. A. Stanturf (ed.), *Restoration of Boreal and Temperate Forests*. 2nd edn. Boca Raton, FL: CRC Press, 495–517.

Byun, C., de Blois, S., and Brisson, J. (2018). "Management of Invasive Plants through Ecological Resistance," *Biological Invasions*, 20: 13–27.

Campbell, E. M., MacLean, D. A., and Bergeron, Y. (2008). "The Severity of Budworm-Caused Growth Reductions in Balsam Fir/Spruce Stands with the Hardwood Content of Surrounding Forest Landscapes," *Forest Science*, 54: 195–205.

Canadian Forest Service (2021). *Mountain Pine Beetle (Factsheet)*. Available online at https://www.nrcan.gc.ca/forests/fire-insects-disturbances/top-insects/13397 (accessed February 14, 2024).

Canadian Forest Service (2022). *Spruce Budworm*. Available online at https://natural-resources.canada.ca/our-natural-resources/forests/wildland-fires-insects-disturbances/top-forest-insects-and-diseases-canada/spruce-budworm/13383 (accessed February 14, 2024).

Chen, W., Moriya, K., Sakai, T., Koyama, L., and Cao, C. (2014). "Post-Fire Forest Regeneration under Different Restoration Treatments in the Greater Hinggan Mountain Area of China," *Ecological Engineering*, 70: 304–11.

Coates, K. D., DeLong, C., Burton, P. J., and Sachs, D. L. (2006). *Abundance of Secondary Structure in Lodgepole Pine Stands Affected by Mountain Pine Beetle: Report for the Chief Forester*. Smithers, BC: Bulkley Valley Research Centre. 17 pp. Available online at https://www.for.gov.bc.ca/hfd/library/FIA/2007/FSP_Y072184b.pdf (accessed February 14, 2024).

Coates, K. D., Lilles, E. B., Dhar, A., and Hall, E. C. (2020). "Wind Damage over 21 Years across Different Levels of Tree Removal in Natural-Origin Mixed Forests of Northwestern British Columbia," *Canadian Journal of Forest Research*, 50/9: 946–52.

Comita, L. S., and Stump, S. M. (2020). "Natural Enemies and the Maintenance of Tropical Tree Diversity: Recent Insights and Implications for the Future of Biodiversity in a Changing World," *Annals of the Missouri Botanical Garden*, 105/3: 377–92.

Coop, J. D., Parks, S. A., Stevens-Rumann, C. S., et al. (2020). "Wildfire-Driven Forest Conversion in Western North American Landscapes," *BioScience*, 70/8: 659–73.

Cumming, S. G. (2001). "Forest Type and Wildfire in the Alberta Boreal Mixedwood: What Do Fires Burn?" *Ecological Applications*, 11/1: 97–110.

Dale, V. H., Joyce, L. A., McNulty, S., et al. (2001). "Climate Change and Forest Disturbances," *BioScience*, 51/9: 723–34.

Doerr, S. H., and Santín, C. (2016). "Global Trends in Wildfire and its Impacts: Perceptions versus Realities in a Changing World," *Philosophical Transactions of the Royal Society B*, 371: 20150345.

Donato, D. C., Fontaine, J. B., Campbell, J. L., Robinson, W. D., Kauffman, J. B., and Law, B. E. (2006). "Post-Wildfire Logging Hinders Regeneration and Increases Fire Risk," *Science*, 311/5759: 352.

Dudley, N. (ed.). (2008). *Guidelines for Applying Protected Area Management Categories*. Best Practice Protected Area Guidelines Series, No. 21. Gland, Switzerland: International Union for Conservation of Nature. 86 pp. Available online at https://portals.iucn.org/library/sites/library/files/documents/PAG-021.pdf (accessed February 19, 2024).

Duperat, M., Gardiner, B., and Ruel, J.-C. (2022). "Effects of a Selective Thinning on Wind Loading in a Naturally Regenerated Balsam Fir Stand," *Forest Ecology and Management*, 505: 1198878.

Ellison, A. M., Orwig, D. A., Fitzpatrick, M. C., and Preisser, E. L. (2018). "The Past, Present, and Future of the Hemlock Woolly Adelgid (*Adelges tsugae*) and Its Ecological Interactions with Eastern Hemlock (*Tsuga canadensis*) Forests," *Insects*, 9(4): 172.

Estay, S. A., Chávez, R. O., Rocco, R., and Gutiérrez, A. G. (2019). "Quantifying Massive Outbreaks of the Defoliator Moth *Ormiscodes amphimone* in Deciduous *Nothofagus*-Dominated Southern Forests Using Remote Sensing Time Series Analysis." *Journal of Applied Entomology*, 143(7): 787–96.

Eveleigh, E. S., McCann, K. S., McCarthy, P. C., et al. (2007). "Fluctuations in Density of an Outbreak Species Drive Diversity Cascades in Food Webs," *Proceedings of the National Academy of Sciences*, 104/43: 16976–81.

Falk, D. A., van Mantgem, P. J., Keeley, J. E., et al. (2022). "Mechanisms of Forest Resilience," *Forest Ecology and Management*, 512: 120129.

Farr, J. D. (2002). "Biology of the Gumleaf Skeletoniser, *Uraba lugens* Walker (Lepidoptera: Noctuidae), in the Southern Jarrah Forest of Western Australia." *Australian Journal of Entomology*, 41/1: 60–69.

Farr, J. D., and Wills, A. J. (2012). "Field Testing Desire® Delta trap and GLS Sex Pheromone Lure System for Measuring Outbreak and Basal Populations of

Uraba lugens Walker (Lepidoptera: Nolidae)," *Australian Forestry*, 75: 175–9.

Fettig, C. J., Klepzig, K. D., Billings, R. F., et al. (2007). "The Effectiveness of Vegetation Management Practices for Prevention and Control of Bark Beetle Infestations in Coniferous Forests of the Western and Southern United States," *Forest Ecology and Management*, 238/1–3: 24–53.

Finney, D. L., Doherty, R. M., Wild, O., Stevenson, D. S., MacKenzie, I. A., and Blyth, A. M. (2018). "A Projected Decrease in Lightning under Climate Change," *Nature Climate Change*, 8/3: 210–13.

Fleming, R. A., Candau, J. N., and McAlpine, R. S. (2002). "Landscape-Scale Analysis of Interactions between Insect Defoliation and Forest Fire In Central Canada," *Climatic Change*, 55/1: 251–72.

Forrest, J. R. (2016). "Complex Responses of Insect Phenology to Climate Change," *Current Opinion in Insect Science*, 17: 49–54.

Forseth, I. N., and Innis, A. F. (2004). "Kudzu (*Pueraria montana*): History, Physiology, and Ecology Combine to Make a Major Ecosystem Threat," *Critical Reviews in Plant Sciences*, 23/5: 401–13.

Franklin, J. F., Lindenmayer, D., MacMahon, J. A., et al. (2000). "Threads of Continuity," *Conservation Biology in Practice*, 1/1: 8–16.

Fraser, E., Landhäusser, S., and Lieffers, V. (2004). "The Effect of Fire Severity and Salvage Logging Traffic on Regeneration and Early Growth of Aspen Suckers in North-Central Alberta," *Forestry Chronicle*, 80/2: 251–6.

Gardiner, B. A., Stacey, G. R., Belcher, R. E., and Wood, C. J. (1997). "Field and Wind Tunnel Assessments of the Implications of Respacing and Thinning for Tree Stability," *Forestry*, 70/3: 233–52.

Gill, A. M., and Zylstra, P. (2005). "Flammability of Australian Forests," *Australian Forestry*, 68/2: 87–93.

Girard, F., Payette, S., and Gagnon, R. (2009). "Origin of the Lichen–Spruce Woodland in the Closed-Crown Forest Zone of Eastern Canada," *Global Ecology and Biogeography*, 18/3: 291–303.

Griess, V. C., and Knoke, T. (2011). "Growth Performance, Windthrow, and Insects: Meta-Analyses of Parameters Influencing Performance of Mixed-Species Stands in Boreal and Northern Temperate Biomes," *Canadian Journal of Forest Research*, 41/6: 1141–59.

Gustafsson, L., Baker, S. C., Bauhus, J., et al. (2012). "Retention Forestry to Maintain Multifunctional Forests: A World Perspective," *BioScience*, 62/7: 633–45.

Hall, J., Muscarella, R., Quebbeman, A., et al. (2020). "Hurricane-Induced Rainfall Is a Stronger Predictor of Tropical Forest Damage in Puerto Rico than Maximum Wind Speeds," *Scientific Reports*, 10/1: 4318.

Hansen, A. J., Spies, T. A., Swanson, F. J., and Ohmann, J. L. (1991). "Conserving Biodiversity in Managed Forests," *BioScience*, 41/6: 382–92.

Hanewinkel, M., Albrecht, A., and Schmidt, M. (2013). "Influence of Stand Characteristics and Landscape Structure on Wind Damage," in B. Gardiner, A. R. T. Schuck, M. J. Schelhaas, C. Orazio, K. Blennow, and B. Nicoll (eds), *Living with Storm Damage to Forests*. Joensuu, Finland: European Forest Institute, 39–45.

Harmon, M. E., Franklin, J. F., Swanson, F. J., et al. (2004). "Ecology of Coarse Woody Debris in Temperate Ecosystems," *Advances in Ecological Research*, 34: 59–234.

Harris, J. A., Hobbs, R., Higgs, E., and Aronson, J. (2006). "Ecological Restoration and Global Climate Change," *Restoration Ecology*, 14/2: 170–6.

Harvey, B. J., Donato, D. C., and Turner, M. G. (2016). "Burn Me Twice, Shame on Who? Interactions between Successive Forest Fires across a Temperate Mountain Region," *Ecology*, 97/9: 2272–82.

Harvey, B. J., Hart, S. J., and Cansler, C. A. (2021). "The Disturbance Regime Concept," in R. A. Francis, J. D. A. Millington, G. L. W. Perry, and E. S. Minor (eds), *The Routledge Handbook of Landscape Ecology*. New York: Routledge, 159–74.

Haughian, S. R., Burton, P. J., Taylor, S. W., and Curry, C. L. (2012). "Expected Effects of Climate Change on Forest Disturbance Regimes in British Columbia," *Journal of Ecosystems and Management*, 13/1: 1–24.

Hawley, R. C. (1937). *Forest Protection*. New York: John Wiley & Sons. 262 pp.

Herms, D. A., and Mattson, W. J. (1992). "The Dilemma of Plants: To Grow or to Defend," *Quarterly Review of Biology*, 67: 283–335.

Hessburg, P. F., Miller, C. L., Parks, S. A., et al. (2019). "Climate, Environment, and Disturbance History Govern Resilience of Western North American Forests," *Frontiers in Ecology and Evolution*, 7: 239.

Hoffman, K. M., Davis, E. L., Wickham, S. B., et al. (2021). "Conservation of Earth's Biodiversity Is Embedded in Indigenous Fire Stewardship," *Proceedings of the National Academy of Sciences*, 118/32: e2105073118.

Huggel, C., Allen, S., Deline, P., Fischer, L., Noetzli, J., and Ravanel, L. (2012). "Ice Thawing, Mountains Falling—Are Alpine Rock Slope Failures Increasing?" *Geology Today*, 28/3: 98–104.

Hulme, P. E. (2009). "Trade, Transport and Trouble: Managing Invasive Species Pathways in an Era of Globalization," *Journal of Applied Ecology*, 46/1: 10–18.

Isaacson, D. L. (2000). "Impacts of Broom (*Cytisus scoparius*) in Western North America," *Plant Protection Quarterly*, 15/4: 145–7.

Jactel, H., Bauhus, J., Boberg, J., et al. (2017). "Tree Diversity Drives Forest Stand Resistance to Natural Disturbances," *Current Forestry Reports*, 3/3: 223–43.

Jactel, H., and Brockerhoff, E. (2007). "Tree Diversity Reduces Herbivory by Forest Insects," *Ecology Letters*, 10/9: 835–48.

Jactel, H., Moreira, X., and Castagneyrol, B. (2021). "Tree Diversity and Forest Resistance to Insect Pests: Patterns, Mechanisms, and Prospects," *Annual Review of Entomology*, 66: 277–96.

Jactel, H., Nicoll, B. C., Branco, M., et al. (2009). "The Influences of Forest Stand Management on Biotic and Abiotic Risks of Damage," *Annals of Forest Science*, 66/7: 701.

Jentsch, A., Kreyling, J., and Beierkuhnlein, C. (2007). "A New Generation of Climate Change Experiments: Events not Trends," *Frontiers in Ecology and the Environment*, 6: 315–24.

Jentsch, A., and von Heßberg, A. (2019). "Die störungsregime und klimaextreme der Vegetationszonen der Erde," in T. Wohlgemuth, A. Jentsch, and R. Seidl (eds), *Störungsökologie*. Bern, Switzerland: UBT/Haupt Verlag, 45–74.

Jentsch, A., and White, P. S. (2019). "A Theory of Pulse Dynamics and Disturbance in Ecology," *Ecology*, 100/7: e02734.

Jia-bing, W., De-xin, G., Shi-jie, H., Mi, Z., and Chang-Jie, J. (2005). "Ecological Functions of Coarse Woody Debris in Forest Ecosystem," *Journal of Forestry Research*, 16/3: 247–52.

Johnson, C. J., Ehlers, L. P., and Seip, D. R. (2015). "Witnessing Extinction: Cumulative Impacts across Landscapes and the Future Loss of an Evolutionarily Significant Unit of Woodland Caribou in Canada," *Biological Conservation*, 186: 176–86.

Johnstone, J. F., Allen, C. D., Franklin, J. F., et al. (2016). "Changing Disturbance Regimes, Ecological Memory, and Forest Resilience," *Frontiers in Ecology and the Environment*, 14/7: 369–78.

Johnstone, J. F., Hollingsworth, T. N., Chapin, F. S., and Mack, M. C. (2010). "Changes in Fire Regime Break the Legacy Lock on Successional Trajectories in Alaskan Boreal Forest," *Global Change Biology*, 16: 1281–95.

Jorgenson, M. T., Harden, J., Kanevskiy, M., et al. (2013). "Reorganization of Vegetation, Hydrology and Soil Carbon after Permafrost Degradation across Heterogeneous Boreal Landscapes," *Environmental Research Letters*, 8/3: 035017.

Junqueira, L. R., Barbosa, L. R., and Wilcken, C. F. (2018). "Quantification of Damages by *Thaumastocoris peregrinus* (Hemiptera: Thaumastocoridae) in Eucalypt," in *Improving Forest Health on Commercial Plantations: Book of Abstracts*. IUFRO Working Party 7.02.13 meeting, Punta del Este, Uruguay, 21–23 March 2018, 38. Available online at https://www.iufro.org/download/file/28424/6631/70213-puntadeleste18-abstracts_pdf/ (accessed February 14, 2024).

Kausrud, K., Økland, B., Skarpaas, O., Grégoire, J. C., Erbilgin, N., and Stenseth, N. C. (2012). "Population Dynamics in Changing Environments: The Case of an Eruptive Forest Pest Species," *Biological Reviews*, 87/1: 34–51.

Kaufmann, M. R., Shlisky, A., and Marchand, P. (2005). *Good Fire, Bad Fire: How to Think about Forest Land Management and Ecological Processes*. Fort Collins, CO: USDA Forest Service. 16 pp. Available online at https://www.fs.usda.gov/research/treesearch/20296 (accessed February 14, 2024).

Kautz, M., Meddens, A., Hall, R. J., and Arneth, A. (2017). "Biotic Disturbances in Northern Hemisphere Forests: A Synthesis of Recent Data, Uncertainties and Implications for Forest Monitoring and Modelling," *Global Ecology and Biogeography*, 26/5: 533–52.

Kautz, M., Schopf, R., and Ohser, J. (2013). "The 'Sun-Effect': Microclimatic Alterations Predispose Forest Edges to Bark Beetle Infestations," *European Journal of Forest Research*, 132/3: 453–65.

Keane, R. E., Ryan, K. C., Veblen, T. T., Allen, C. D., Logan, J. A., and Hawkes, B. (2002). "The Cascading Effects of Fire Exclusion in Rocky Mountain Ecosystems," in J. S. Baron. (ed.), *Rocky Mountain Futures: An Ecological Perspective*. Washington: Island Press, 133–52.

Kenis, M., Hurley, B. P., Hajek, A. E., and Cock, M. J. (2017). "Classical Biological Control of Insect Pests of Trees: Facts and Figures," *Biological Invasions*, 19: 3401–17.

Klapwijk, M. J., and Björkman, C. (2018). "Mixed Forests to Mitigate Risk of Insect Outbreaks," *Scandinavian Journal of Forest Research*, 33/8: 772–80.

Klapwijk, M. J., Bylund, H., Schroeder, M., and Björkman, C. (2016). "Forest Management and Natural Biocontrol of Insect Pests," *Forestry*, 89/3: 253–62.

Kneeshaw, D., and Bergeron, Y. (2016). "Applying Knowledge of Natural Disturbance Regimes to Develop an Ecosystem Management Approach in Forestry," in G. R. Larocque (ed.), *Ecological Forest Management Handbook*. Boca Raton, FL: CRC Press, 3–31.

Kneeshaw, D., Sturtevant, B. R., Cooke, B., Work, T., Pureswaran, D., DeGrandpre, L., and MacLean, D. A. (2015). "Insect Disturbances in Forest Ecosystems." in Peh, K. S.-H., Corlett, R. T., and Bergeron, Y. (eds), *Routledge Handbok of Forest Ecology*. Abington, UK: Routledge, 93–113.

Kneeshaw, D., Sturtevant, B., DeGrandpé, L., et al. (2021). "The Vision of Managing for Pest-Resistant Landscapes: Realistic or Utopic?" *Current Forestry Reports*, 7: 97–113.

Knoke, T., Gosling, E., Thom, D., Chreptun, C., Rammig, A., and Seidl, R. (2021). "Economic Losses from Natural Disturbances in Norway Spruce Forests: A Quantification Using Monte-Carlo Simulations," *Ecological Economics*, 185: 107046.

Koh, H. L., Teh, S. Y., Kh'Ng, X. Y., and Raja Barizan, R. S. (2018). "Mangrove Forests: Protection against and

Resilience to Coastal Disturbances," *Journal of Tropical Forest Science*, 30/5: 446–60.

Kramer, K., Brang, P., Bachofen, H., Bugmann, H., and Wohlgemuth, T. (2014). "Site Factors Are More Important than Salvage Logging for Tree Regeneration after Wind Disturbance in Central European Forests," *Forest Ecology and Management*, 331: 116–28.

Kreyling, J. (2010). "Winter Climate Change: A Critical Factor for Temperate Vegetation Performance," *Ecology*, 91: 1939–48.

Kulakowski, D., Veblen, T. T., and Bebi, P. (2003). "Effects of Fire and Spruce Beetle Outbreak Legacies on the Disturbance Regime of a Subalpine Forest in Colorado," *Journal of Biogeography*, 30/9: 1445–56.

Kuosmanen, N., Čada, V., Halsall, K., et al. (2020). "Integration of Dendrochronological and Palaeoecological Disturbance Reconstructions in Temperate Mountain Forests," *Forest Ecology and Management*, 475: 118413.

Kuuluvainen, T., and Grenfell, R. (2012). "Natural Disturbance Emulation in Boreal Forest Ecosystem Management: Theories, Strategies, and a Comparison with Conventional Even-Aged Management," *Canadian Journal of Forest Research*, 42/7: 1185–203.

Lehmann, P., Ammunét, T., Barton, M., et al. (2020). "Complex Responses of Global Insect Pests to Climate Warming," *Frontiers in Ecology and the Environment*, 18/3: 141–50.

Leverkus, A. B., Buma, B., Wagenbrenner, J., et al. (2021). "Tamm Review: Does Salvage Logging Mitigate Subsequent Forest Disturbances?" *Forest Ecology and Management*, 481: 118721.

Levine, J. M., Adler, P. B., and Yelenik, S. G. (2004). "A Meta-Analysis of Biotic Resistance to Exotic Plant Invasions," *Ecology Letters*, 7/10: 975–89.

Lewis, K. J., and Hartley, I. D. (2006). "Rate of Deterioration, Degrade, and Fall of Trees Killed by Mountain Pine Beetle," *Journal of Ecosystems and Management*, 7/2: 11–19.

Liebhold, A. M. (2012). "Forest Pest Management in a Changing World," *International Journal of Pest Management*, 58/3: 289–95.

Lindenmayer, D. B., Burton, P. J., and Franklin, J. F. (2008). *Salvage Logging and its Ecological Consequences*. Washington: Island Press. 227 pp.

Lindenmayer, D. B., Hunter, M. L., Burton, P. J., and Gibbons, P. (2009). "Effects of Logging on Fire Regimes in Moist Forests," *Conservation Letters*, 2/6: 271–7.

Lindgren, B. S., and Borden, J. H. (1993). "Displacement and Aggregation of Mountain Pine Beetles, *Dendroctonus ponderosae* (Coleoptera: Scolytidae), in Response to their Antiaggregation and Aggregation Pheromones," *Canadian Journal of Forest Research*, 23/2: 286–90.

Lindner, M., and Ramukainen, M. (2013) "Climate Change and Storm Damage Risk in European Forests," in B. Gardiner, A. R. T. Schuck, M. J. Schelhaas, C. Orazio, K. Blennow, and B. Nicoll (eds), *Living with Storm Damage to Forests*. Joensuu, Finland: European Forest Institute, 109–15.

Lingua, E., Bettella, F., Pividori, M., et al. (2020). "The Protective Role of Forests to Reduce Rockfall Risks and Impacts in the Alps under a Climate Change Perspective," in W. L. Filho, G. J. Nagy, M. Borga, P. D. Chávez Muñoz, and A. Magnuszewski (eds), *Climate Change, Hazards and Adaptation Options*. Cham, Switzerland: Springer, 333–47.

Logan, J. A., Régnière, J., and Powell, J. A. (2003). "Assessing the Impacts of Global Warming on Forest Pest Dynamics," *Frontiers in Ecology and the Environment*, 1/3: 130–7.

McCullough, D. G. (2020). "Challenges, Tactics and Integrated Management of Emerald Ash Borer in North America," *Forestry*, 93/2: 197–211.

McInnes, K. L., Erwin, T. A., and Bathols, J. M. (2011). "Global Climate Model Projected Changes in 10m Wind Speed and Direction due to Anthropogenic Climate Change," *Atmospheric Science Letters*, 12/4: 325–33.

MacKenzie, R. F. (1974). "Some Factors Influencing the Stability of Sitka in Northern Ireland," *Irish Forestry*, 31/2: 110–29.

McWethy, D. B., Schoennagel, T., Higuera, P. E., et al. (2019). "Rethinking Resilience to Wildfire," *Nature Sustainability*, 2/9: 797–804.

Marini, L., Ayres, M. P., and Jactel, H. (2021). "Impact of Stand and Landscape Management on Forest Pest Damage," *Annual Review of Entomology*, 67: 181–99.

Marlon, J. R., Bartlein, P. J., Gavin, D. G., et al. (2012). "Long-Term Perspective on Wildfires in the Western USA," *Proceedings of the National Academy of Sciences*, 109/9: E535–E543.

Martínez Pastur, G. J., Vanha-Majamaa, I., and Franklin, J. F. (2020). "Ecological Perspectives on Variable Retention Forestry," *Ecological Processes*, 9/1: 1–6.

Mason, B., and Valinger, E. (2013). "Managing Forests to Reduce Storm Damage," in B. Gardiner, A. R. T. Schuck, M. J. Schelhaas, C. Orazio, K. Blennow, and B. Nicoll (eds), *Living with Storm Damage to Forests*. Joensuu, Finland: European Forest Institute, 87–96.

Mason, W. L. (2002). "Are Irregular Stands More Windfirm?" *Forestry*, 75/4: 347–55.

Maxwell, J. T., Ortegren, J. T., Knapp, P. A., and Soulé, P. T. (2013). "Tropical Cyclones and Drought Amelioration in the Gulf and Southeastern Coastal United States," *Journal of Climate*, 26/21: 8440–52."

Means, J. E., and Winjun, J. K. (1983). Road to Recovery after Eruption of Mt St. Helens," in J. Hayes (ed.),

The Yearbook of Agriculture 1983: Using our Natural Resources. Washington: United States Department of Agriculture, 204–5.

Millar, C. I., Stephenson, N. L., and Stephens, S. L. (2007). "Climate Change and Forests of the Future: Managing in the Face of Uncertainty," *Ecological Applications*, 17/8: 2145–51.

Miller, A., and Rusnock, P. (1993). "The Rise and Fall of the Silvicultural Hypothesis in Spruce Budworm (*Choristoneura fumiferana*) Management in Eastern Canada," *Forest Ecology and Management*, 61: 171–89.

Mina, M., Messier, C., Duveneck, M. J., Fortin, M.-J., and Aquilué, N. (2022). "Managing for the Unexpected: Building Resilient Forest Landscapes to Cope with Global Change," *Global Change Biology*, 28/14: 4323–41.

Mitchell, S. J., Hailemariam, T., and Kulis, Y. (2001). "Empirical Modelling of Cutblock Edge Windthrow Risk on Vancouver Island, Canada, Using Stand Level Information," *Forest Ecology and Management*, 154/1–2: 117–30.

Mooney, C. (2010). *Fuelbreak Effectiveness in Canada's Boreal Forests: A Synthesis of Current Knowledge.* Vancouver: FP Innovations. 53 pp. Available online at https://wildfire.fpinnovations.ca/74/FuelbreakEffectivenessFinalReport.pdf (accessed February 14, 2024).

Moritz, M. A., Hessburg, P. F., and Povak, N. A. (2010). "Native Fire Regimes and Landscape Resilience," in D. McKenzie, C. Miller, and D. A. Falk (eds), *The Landscape Ecology of Fire*. Dordrecht: Springer, 51–86.

Munishi, P. K. T., and Chamshama, S. A. O. (1994). "A Study of Wind Damage on *Pinus patula* Stands in Southern Tanzania," *Forest Ecology and Management*, 63/1: 13–21.

Nair, K. S. S. (2007). *Tropical Forest Insect Pests: Ecology, Impact, and Management*. Cambridge: Cambridge University Press. 422 pp.

Narayanaraj, G., and Wimberly, M. C. (2012). "Influences of Forest Roads on the Spatial Patterns of Human- and Lightning-Caused Wildfire Ignitions," *Applied Geography*, 32/2: 878–88.

Netherer, S., and Schopf, A. (2010). "Potential Effects of Climate Change on Insect Herbivores in European Forests: General Aspects and the Pine Processionary Moth as Specific Example," *Forest Ecology and Management*, 259/4: 831–8.

Nevalainen, S. (2017). "Comparison of Damage Risks in Even-and Uneven-Aged Forestry in Finland," *Silva Fennica*, 51/3: 1741.

Nikinmaa, L., de Koning, J. H., Derks, J., et al. (2024). "The Priorities in Managing Forest Disturbances to Enhance Forest Resilience: A Comparison of a Literature Analysis and Perceptions of Forest Professionals," *Forest Policy and Economics*, 158: 103119.

O'Connor, C. D., Calkin, D. E., and Thompson, M. P. (2017). "An Empirical Machine Learning Method for Predicting Potential Fire Control Locations for Pre-Fire Planning and Operational Fire Management," *International Journal of Wildland Fire*, 26/7: 587–97.

O'Connor, C. D., Thompson, M. P., and Rodríguez Y Silva, F. (2016). "Getting Ahead of the Wildfire Problem: Quantifying and Mapping Management Challenges and Opportunities," *Geosciences*, 6/3: 35.

Orwig, D. A., Foster, D. R., and Mausel, D. L. (2002). "Landscape Patterns of Hemlock Decline in New England Due to the Introduced Hemlock Woolly Adelgid," *Journal of Biogeography*, 29(10–11), 1475–87.

Ostaff, D. P., Piene, H., Quiring, D. T., Moreau, G., Farrell, J. C., and Scarr, T. (2006). "Influence of Pre-Commercial Thinning of Balsam Fir on Defoliation by the Balsam Fir Sawfly," *Forest Ecology and Management*, 223/1–3: 342–8.

Otón, G., Lizundia-Loiola, J., Pettinari, M. L., and Chuvieco, E. (2021). "Development of a Consistent Global Long-Term Burned Area Product (1982–2018) Based on AVHRR-LTDR Data," *International Journal of Applied Earth Observation and Geoinformation*, 103: 102473.

Ouzts, J., Kolb, T., Huffman, D., and Meador, A. S. (2015). "Post-Fire Ponderosa Pine Regeneration with and without Planting in Arizona and New Mexico," *Forest Ecology and Management*, 354: 281–90.

Palik, B. J., D'Amato, A. W., Franklin, J. F., and Johnson, K. N. (2021). *Ecological Silviculture: Foundations and Applications*. Long Grove, IL: Waveland Press. 343 pp.

Parise, M., and Cannon, S. H. (2012). "Wildfire Impacts on the Processes that Generate Debris Flows in Burned Watersheds," *Natural Hazards*, 61: 217–27.

Parks, S. A., Parisien, M.-A., Miller, C., Holsinger, L. M., and Baggett, L. S. (2018). "Fine-Scale Spatial Climate Variation and Drought Mediate the Likelihood of Reburning," *Ecological Applications*, 28/2: 573–86.

Parry, M. L., Canziani, O. F., Palutikof, J. P., et al. (2007). "Technical Summary," in M. L. Parry, O. F. Canziani, J. P. Palutikof, P. J. van der Linden, and C. E. Hanson (eds), *Climate Change 2007: Impacts, Adaptation and Vulnerability. Contribution of Working Group II to the Fourth Assessment Report of the Intergovernmental Panel on Climate Change*. Cambridge: Cambridge University Press, 23–78. Available online at https://www.ipcc.ch/report/ar4/wg2/ (accessed February 14, 2024).

Parthiban, K. T., Suganthy, M., and Krishnakumar, N. (2019). *Forest Protection: Principles and Applications*. New Delphi, India: Jain Brothers. 274 pp.

Patacca, M., Lindner, M., Lucas-Borja, M. E., et al. (2023). "Significant Increase in Natural Disturbance Impacts on European Forests since 1950," *Global Change Biology*, 29/5: 1359–76.

Patton, A. I., Rathburn, S. L., and Capps, D. M. (2019). "Landslide Response to Climate Change in Permafrost Regions," *Geomorphology*, 340: 116–28.

Pausas, J. G., and Keeley, J. E. (2019). "Wildfires as an Ecosystem Service," *Frontiers in Ecology and the Environment*, 17/5: 289–95.

Peltola, H., Gardiner, B., and Nicoll, B. (2013). "Mechanics of Wind Damage," in B. Gardiner, A. R. T. Schuck, M. J. Schelhaas, C. Orazio, K. Blennow, and B. Nicoll (eds), *Living with Storm Damage to Forests*. Joensuu, Finland: European Forest Institute, 31–7.

Perera, A. H., Buse, L. J., and Weber, M. G. (eds). (2004). *Emulating Natural Forest Landscape Disturbances: Concepts and Applications*. New York: Columbia University Press. 315 pp.

Peters, D. P. C., Lugo, A. E., Chapin, F. S., et al. (2011). "Cross-System Comparisons Elucidate Disturbance Complexities and Generalities," *Ecosphere*, 2/7: 81.

Peterson, C. J., and Leach, A. D. (2008). "Limited Salvage Logging Effects on Forest Regeneration after Moderate-Severity Windthrow," *Ecological Applications*, 18/2: 407–20.

Pickett, S. T. A., and White P. S. (eds). (1985). *The Ecology of Natural Disturbance and Patch Dynamics*. San Diego, CA: Academic Press. 472 pp.

Popović, Z., Bojović, S., Marković, M., and Cerdà, A. (2021). "Tree Species Flammability Based on Plant Traits: A Synthesis," *Science of the Total Environment*, 800: 149625.

Potterf, M., Eyvindson, K., Blattert, C., et al. (2024). "Diversification of Forest Management Can Mitigate Wind Damage Risk and Maintain Biodiversity," *European Journal of Forest Research*, 143: 419–36.

Prestemon, J. P., Wear, D. N., Stewart, F. J., and Holmes, T. P. (2006). "Wildfire, Timber Salvage, and the Economics of Expediency," *Forest Policy and Economics*, 8/3: 312–22.

Pukkala, T., Laiho, O., and Lähde, E. (2016). "Continuous Cover Management Reduces Wind Damage," *Forest Ecology and Management*, 372: 120–7.

Pureswaran, D. S., Roques, A., and Battisti, A. (2018). "Forest Insects and Climate Change," *Current Forestry Reports*, 4/2: 35–50.

Pyne, S. J. (1982). *Fire in America: A Cultural History of Wildland and Rural Fire*. Princeton: Princeton University Press. 654 pp.

Raffa, K. F., Aukema, B. H., Bentz, B. J., Carroll, A. L., Hicke, J. A., and Kolb, T. E. (2015). "Responses of Tree-Killing Bark Beetles to a Changing Climate," in J. Björkman and P. Niemelä (eds), *Climate Change and Insect Pests*. Wallingford, UK: CAB International, 173–201.

Raffa, K. F., Aukema, B. H., Bentz, B. J., Carroll, A. L., Hicke, J. A., Turner, M. G., and Romme, W. H. (2008). "Cross-scale Drivers of Natural Disturbances Prone to Anthropogenic Amplification: The Dynamics of Bark Beetle Eruptions," *BioScience*, 58(6), 501–17.

Ramsfield, T. D., Bentz, B. J., Faccoli, M., Jactel, H., and Brockerhoff, E. G. (2016). "Forest Health in a Changing World: Effects of Globalization and Climate Change on Forest Insect and Pathogen Impacts," *Forestry*, 89: 245–52.

Remondo, J., Bonachea, J., and Cendrero, A. (2008). "Quantitative Landslide Risk Assessment and Mapping on the Basis of Recent Occurrences," *Geomorphology*, 94/3–4: 496–507.

Reynolds, K. M., and Holsten, E. H. (1994). "Relative Importance of Risk Factors for Spruce Beetle Outbreaks," *Canadian Journal of Forest Research*, 24/10: 2089–95.

Ribeiro, G. H. P. D. M., Chambers, J. Q., Peterson, C. J., et al. (2016). "Mechanical Vulnerability and Resistance to Snapping and Uprooting for Central Amazon Tree Species," *Forest Ecology and Management*, 380: 1–10.

Riordan, R. (2012). *The Lost Hero*. London: Puffin Books. 554 pp.

Robinson, W. A. (2021). "Climate Change and Extreme Weather: A Review Focusing on the Continental United States," *Journal of the Air & Waste Management Association*, 71/10: 1186–209.

Roland, J., and Taylor, P. D. (1997). "Insect Parasitoid Species Respond to Forest Structure at Different Spatial Scales," *Nature*, 386/6626: 710–13.

Roques, A., Shi, J., Auger-Rozenberg, M. A., Ren, L., Augustin, S., and Luo, Y. Q. (2020). "Are Invasive Patterns of Non-Native Insects Related to Woody Plants Differing between Europe and China?" *Frontiers in Forests and Global Change*, 2: 91.

Rowan, C. A., Mitchell, S. J., and Temesgen, H. (2003). "Effectiveness of Clearcut Edge Windfirming Treatments in Coastal British Columbia: Short-Term Results," *Forestry*, 76/1: 55–65.

Ruel, J.-C. (2020). "Ecosystem Management of Eastern Canadian Boreal Forests: Potential Impacts on Wind Damage," *Forests*, 11: 578.

Rytwinski, A., and Crowe, K. A. (2010). "A Simulation–Optimization Model for Selecting the Location of Fuel-Breaks to Minimize Expected Losses from Forest Fires," *Forest Ecology and Management*, 260/1: 1–11.

Saint-Germain, M., and Greene, D. F. (2009). "Salvage Logging in the Boreal and Cordilleran Forests of Canada: Integrating Industrial and Ecological Concerns in Management Plans," *Forestry Chronicle*, 85/1: 120–34.

Schneider, R. R., Stelfox, J. B., Boutin, S., and Wasel, S. (2003). "Managing the Cumulative Impacts of Land Uses in the Western Canadian Sedimentary Basin: A Modelling Approach," *Conservation Ecology*, 7/1: 8.

Schuck, A., and Schelhaas, M. J. (2013). "Storm Damage in Europe: An Overview," in B. Gardiner, A. R. T. Schuck, M. J. Schelhaas, C. Orazio, K. Blennow, and B. Nicoll (eds), *Living with Storm Damage to Forests*. Joensuu, Finland: European Forest Institute, 15–23.

Schütz, J. P., Götz, M., Schmid, W., and Mandallaz, D. (2006). "Vulnerability of Spruce (*Picea abies*) and Beech (*Fagus sylvatica*) Forest Stands to Storms and Consequences for Silviculture," *European Journal of Forest Research*, 125: 291–302.

Seebens, H., Blackburn, T. M., Dyer, E. E., et al. (2017). "No Saturation in the Accumulation of Alien Species Worldwide," *Nature Communications*, 8/1: 1–9.

Seidl, R., and Lexer, M. J. (2013). "Forest Management under Climatic and Social Uncertainty: Trade-offs between Reducing Climate Change Impacts and Fostering Adaptive Capacity," *Journal of Environmental Management*, 114: 461–9.

Seidl, R., Thom, D., Kautz, M., et al. (2017). "Forest Disturbances under Climate Change," *Nature Climate Change*, 7/6: 395–402.

Senf, C., Seidl, R., and Hostert, P. (2017). "Remote Sensing of Forest Insect Disturbances: Current State and Future Directions," *International Journal of Applied Earth Observation and Geoinformation*, 60: 49–60.

Shatford, J. P. A., Hibbs, D. E., and Puettmann, K. J. (2007). "Conifer Regeneration after Forest Fire in the Klamath-Siskiyous: How Much, How Soon?" *Journal of Forestry*, 105/3: 139–46.

Shinneman, D. J., Germino, M. J., Pilliod, D. S., Aldridge, C. L., Vaillant, N. M., and Coates, P. S. (2019). "The Ecological Uncertainty of Wildfire Fuel Breaks: Examples from the Sagebrush Steppe," *Frontiers in Ecology and the Environment*, 17/5: 279–88.

Siitonen, J. (2001). "Forest Management, Coarse Woody Debris and Saproxylic Organisms: Fennoscandian Boreal Forests as an Example," *Ecological Bulletins*, 49: 11–41.

Skrzyszewski, J., and Pach, M. (2020). "The Use of the Slenderness Coefficient in Diagnosing Wind Damage Risks," *Acta Silvestria*, 57: 7–24.

Stanturf, J. A., Goodrick, S. L., and Outcalt, K. W. (2007). "Disturbance and Coastal Forests: A Strategic Approach to Forest Management in Hurricane Impact Zones," *Forest Ecology and Management*, 250: 119–35.

Steinberg, C. E. W. (2012). "Arms Race between Plants and Animals: Biotransformation System," in C. E. W. Steinberg, *Stress Ecology: Environmental Stress as Ecological Driving Force and Key Player in Evolution*. Dordrecht: Springer, 61–106.

Sturrock, R. N. (2012). "Climate Change and Forest Diseases: Using Today's Knowledge to Address Future Challenges," *Forest Systems*, 21/2: 329–36.

Sturrock, R. N., Frankel, S. J., Brown, A. V., et al. (2011). "Climate Change and Forest Diseases," *Plant Pathology*, 60/1: 133–49.

Sturtevant, B. R., and Fortin, M.-J. (2021). "Understanding and Modelling Forest Disturbance Interactions at the Landscape Level," *Frontiers in Ecology and Evolution*, 9: 653647.

Swanson, M. E., Franklin, J. F., Beschta, R. L., et al. (2011). "The Forgotten Stage of Forest Succession: Early-Successional Ecosystems on Forest Sites," *Frontiers in Ecology and the Environment*, 9/2: 117–25.

Swetnam, T. W., Allen, C. D., and Betancourt, J. L. (1999). "Applied Historical Ecology: Using the Past to Manage for the Future," *Ecological Applications*, 9/4: 1189–206.

Syphard, A. D., Keeley, J. E., and Brennan, T. J. (2011). "Comparing the Role of Fuel Breaks across Southern California National Forests," *Forest Ecology and Management*, 261/11: 2038–48.

Taylor, C., Blanchard, W., and Lindenmayer, D. B. (2020). "Does Forest Thinning Reduce Fire Severity in Australian Eucalypt Forests?" *Conservation Letters*, 14: e12766.

Taylor, S. P., Alfaro, R. I., DeLong, C, and Rankin, L. (1996). "The Effects of Overstory Shading on White Pine Weevil Damage to White Spruce and its Effects on Spruce Growth Rates," *Canadian Journal of Forest Research*, 26/2: 306–12.

Thangavel, K., Spiller, D., Sabatini, R., Marzocca, P., and Esposito, M. (2023). "Near Real-Time Wildfire Management Using Distributed Satellite System," *IEEE Geoscience and Remote Sensing Letters*, 20: 5500705.

Thom, D., Rammer, W., and Seidl, R. (2017). "Disturbances Catalyze the Adaptation of Forest Ecosystems to Changing Climate Conditions," *Global Change Biology*, 23: 269–82.

Thorn, S., Bässler, C., Brandl, R., et al. (2018). "Impacts of Salvage Logging on Biodiversity: A Meta-Analysis," *Journal of Applied Ecology*, 55/1: 279–89.

Thorn, S., Bässler, C., Svoboda, M., and Müller, J. (2017). "Effects of Natural Disturbances and Salvage Logging on Biodiversity: Lessons from the Bohemian Forest," *Forest Ecology and Management*, 388: 113–19.

Thorn, S., Seibold, S., Leverkus, A. B., et al. (2020). "The Living Dead: Acknowledging Life after Tree Death to Stop Forest Degradation," *Frontiers in Ecology and the Environment*, 18/9: 505–12.

Thompson, M. P., Bowden, P., Brough, A., et al. (2016). "Application of Wildfire Risk Assessment Results to Wildfire Response Planning in the Southern Sierra Nevada, California, USA," *Forests*, 7/3: 64.

Tradowsky, J. S., Philip, S. Y., Kreienkamp, F., et al. (2023). "Attribution of the Heavy Rainfall Events Leading to Severe Flooding in Western Europe during July 2021," *Climatic Change*, 176/7: 90.

Turner, M. G. (2010). "Disturbance and Landscape Dynamics in a Changing World," *Ecology*, 91/10: 2833–49.

Turton, S. M., and Alamgir, M. (2015). "Ecological Effects of Strong Winds on Forests," in K. S.-H. Peh, R. T. Corlett, and Y. Bergeron (eds), *Routledge Handbook of Forest Ecology*. New York: Routledge, 127–40.

Urgenson, L. S., Halpern, C. B., and Anderson, P. D. (2013). "Level and Pattern of Overstory Retention Influence Rates and Forms of Tree Mortality in Mature, Coniferous Forests of the Pacific Northwest, USA," *Forest Ecology and Management*, 308: 116–27.

Vacher, C., Castagneyrol, B., Jousselin, E., and Schimann, H. (2021). "Trees and Insects Have Microbiomes: Consequences for Forest Health and Management," *Current Forestry Reports*, 7: 81–96.

Van der Kamp, B. J. (1991). "Pathogens as Agents of Diversity in Forested Landscapes," *Forestry Chronicle*, 67/4: 353–4.

van Lierop, P., Lindquist, E., Sathyapala, S., and Franceschini, G. (2015). "Global Forest Area Disturbance from Fire, Insect Pests, Diseases and Severe Weather Events," *Forest Ecology and Management*, 352: 78–88.

Vaughan, D., Mackes, K., and Webb, J. B. (2018). "Time-since-Death and its Effect on Wood from Beetle-Killed Engelmann Spruce in Southwest Colorado," *Forest Science*, 64/3: 316–23.

Venäläinen, A., Lehtonen, I., Laapas, M., et al. (2020). "Climate Change Induces Multiple Risks to Boreal Forests and Forestry in Finland: A Literature Review," *Global Change Biology*, 26/8: 4178–96.

Venier, L. A., Walton, R., and Brandt, J. P. (2021). "Scientific Considerations and Challenges for Addressing Cumulative Effects in Forest Landscapes in Canada," *Environmental Reviews*, 29/1: 1–22.

Wainhouse, D. (2005). *Ecological Methods in Forest Pest Management*. Oxford: Oxford University Press. 227 pp.

Wagner, M. R., Clancy, K. M., Lieutier, F., and Paine, T. D. (eds). (2007). *Mechanisms and Deployment of Resistance in Trees to Insects*. New York: Kluwer Academic Publishers. 332 pp.

Walker, L. R., and Wardle, D. A. (2014). "Plant Succession as an Integrator of Contrasting Ecological Time Scales," *Trends in Ecology and Evolution*, 29: 504–10.

Wang, H. H., Finney, M. A., Song, Z. L., Wang, Z. S., and Li, X. C. (2021). "Ecological Techniques for Wildfire Mitigation: Two Distinct Fuelbreak Approaches and their Fusion," *Forest Ecology and Management*, 495: 119376.

Wei, Y., Rideout, D., and Kirsch, A. (2008). "An Optimization Model for Locating Fuel Treatments across a Landscape to Reduce Expected Fire Losses," *Canadian Journal of Forest Research*, 38/4: 868–77.

Weir, J. M. H., Chapman, K. J., and Johnson, E. A. (1995). "Wildland Fire Management and the Fire Regime in the Southern Canadian Rockies," in J. K. Brown, R. W. Mutch, C. W Spoon,.and R. H. Wakimoto (technical coordinators), *Proceedings: Symposium on Fire in Wilderness and Park Management, 1993, March 30–April 1, 1993, Missoula, MT*. General Technical Report INT-GTR-320. Ogden, UT: USDA Forest Service, 275–80.

Wermelinger, B. (2004). "Ecology and Management of the Spruce Bark Beetle *Ips typographus*: A Review of Recent Research," *Forest Ecology and Management*, 202/1–3: 67–82.

Wermelinger, B. (2021). *Forest Insects in Europe: Diversity, Functions and Importance*. Boca Raton, FL: CRC Press. 365 pp.

Wesołowski, T., and Rowiński, P. (2006). "Tree Defoliation by Winter Moth *Operophtera brumata* L. during an Outbreak Affected by Structure of Forest Landscape," *Forest Ecology and Management*, 221/1–3: 299–305.

White, A. M., and Long, J. W. (2019). "Understanding Ecological Contexts for Active Reforestation Following Wildfires," *New Forests*, 50/1: 41–56.

White, P. S. (1979). "Pattern, Process, and Natural Disturbance in Vegetation," *Botanical Review*, 45/3: 229–99.

White, R. H., Anderson, S., Booth, J. F., et al. (2023). "The Unprecedented Pacific Northwest Heatwave of June 2021," *Nature Communications*, 14/1: 727.

Williams, E. R. (2005). "Lightning and Climate: A Review," *Atmospheric Research*, 76/1–4: 272–87.

Williams, J. D., Ponder, H. G., and Gilliam, C. H. (1987). "Reducing Moisture Stress in *Cornus florida*," *Journal of Environmental Horticulture*, 5: 131–3.

Witzgall, P., Kirsch, P., and Cork, A. (2010). "Sex Pheromones and their Impact on Pest Management," *Journal of Chemical Ecology*, 36/1: 80–100.

Winder, R. S., and Shamoun, S. F. (2006). "Forest Pathogens: Friend or Foe to Biodiversity?" *Canadian Journal of Plant Pathology*, 28/S1: S221–S227.

Wohlgemuth, T., Jentsch, A., and Seidl, R. (eds). (2019). *Störungsökologie*. Bern, Switzerland: Haupt Verlag, and Stuttgart, Germany: UTB GmbH. 396 pp.

Woods, A., Coates, K. D., and Hamann, A. (2005). "Is an Unprecedented *Dothistroma* Needle Blight Epidemic Related to Climate Change?" *BioScience*, 55/9: 761–9.

Wylie, F. R., and Speight, M. R. (2012). *Insect Pests in Tropical Forestry*. 2nd edn. Wallingford, UK: CAB International. 365 pp.

Yemshanov, D., Liu, N., Thompson, D. K., et al. (2021). "Detecting Critical Nodes in Forest Landscape Networks to Reduce Wildfire Spread," *PloS One*, 16/10: e0258060.

Zeng, H., Peltola, H., Talkkari, A., et al. (2004). "Influence of Clear-Cutting on the Risk of Wind Damage at Forest Edges," *Forest Ecology and Management*, 203/1–3: 77–88.

Zhang, B., MacLean, D. A., Johns, R. C., and Eveleigh, E. S. (2018). "Effects of Hardwood Content on Balsam Fir Defoliation during the Building Phase of a Spruce Budworm Outbreak," *Forests*, 9: 104–18.

Zhang, B., MacLean, D. A., Johns, R. C., Eveleigh, E. S., and Edwards, S. (2020). "Hardwood-Softwood Composition Influences Early-Instar Larval Dispersal Mortality during a Spruce Budworm Outbreak," *Forest Ecology and Management*, 463: 118035.

PART III

Synthesis and Direction

The application of resilient forest management is a place-based exercise, informed by local conditions and vulnerabilities, and by stakeholder priorities. The world's forests differ tremendously in their composition, structure, and inherent abilities to absorb a variety of stresses and disruptions, and in their capacities to adjust or transform themselves in response. Similarly, forest enterprises and communities differ in the financial and human resources on which they can draw, with resulting differences in their flexibility and adaptive capacity. Furthermore, forest owners, managers, and stakeholders can differ considerably from place to place in their objectives, priorities, expectations, and tolerances to risk. Managing socio-ecological systems for resilience often entails processes of diversification, scenario analysis, and contingency planning with the goal of maintaining a healthy portfolio of options for persistence in the long term. Managing for resilience does not usually optimize any one forest ecosystem service in the near term, but should improve prospects for long-term vitality of forests, forest enterprises, and forest communities.

Many forests around the world have a history of exploitation and degradation, for which resilient forest management must start out as a process of ecological restoration. Forest regeneration or afforestation is often a first step, but reintroducing compositional and structural complexity in forests previously managed like agricultural crops is also desirable. Forest restoration in a changing climate can still be informed by our understanding of natural processes that confer resilience. Monitoring, evaluation, and situational awareness are needed to guide selection of promising management policies and practices at any given location and point in time. Potential tipping points and stakeholder tolerance thresholds may inform switching between different modes of resilient forest management. Progressive approaches adopted here and there around the world illustrate the feasibility of resilient forest management and offer templates for implementation. A paradigm shift to forest stewardship informed by ecological principles and socio-cultural understanding has the best chance of promoting the resilience we all desire.

CHAPTER 9

Restoring Forests for Resilience

The work of restoration cannot begin until a problem is fully faced.

Dan B. Allender (2014)

9.1 Forest loss and degradation

The preceding chapters have outlined potential approaches to prevent the loss of forest values, and to adjust forest plans and practices in the face of foreseeable and unforeseeable change. Large areas of the earth have already lost forest cover to agricultural, residential, commercial, and industrial use. It is estimated that forests and woodlands covered 61.2 million km^2 prior to the Neolithic agricultural revolution that started about 10,000 years ago (Matthews 1983), while today it is estimated that wooded land covers only 40.6 million km^2 (FAO 2020), a drop of 34 percent. Many forested landscapes were gradually cleared to grow cultivated crops and to support pasturage and fodder production for domesticated livestock. With a growing human population, forest loss to urbanization and transportation infrastructure (including rail lines, highways, pipeline, and electrical transmission corridors) has increased greatly since the mid-twentieth century. Although these land-use changes seem permanent (or certainly long term within a human lifetime), there are numerous examples where human depopulation owing to plague, war, or cultural collapse has resulted in the natural reforestation of agricultural lands (see Chapter 1). Over recent decades or centuries (depending on the country), we have also seen improvements in food-production efficiency, replacement of fuelwood by fossil fuels, and growing urbanization, which collectively have resulted in a "forest transition" characterized by a net shift from deforestation to one of forest expansion (Barbier et al. 2010).

Empirical evidence from places such as western Europe, Central America, and the eastern US indicate that human land use can be considered temporary in the long run, and that forests can return with enough time or appropriate interventions (see Figure 9.1).

Forest loss has not been the only consequential impact of human utilization of forest ecosystems. Broad areas of the planet that are nominally dominated by trees are nonetheless suffering from ecological degradation, as indicated by a loss of the defining attributes of primary forest, such as very old trees, disturbance-sensitive fauna, complex vertical and horizontal structure, and abundant deadwood (Putz and Redford 2010). While appearing structurally intact, degraded forests may be suffering from depauperate floras and faunas, or conversely an influx of invasive exotic species, each of which can have serious knock-on effects (Ellison et al. 2005). Forest understory plant communities can reflect a history of logging and forest management for decades or centuries (Duffy and Meier 1992; Bratton 1994; Schmiedinger et al. 2012). Much nominal "forest" cover (when mapped by satellite imagery or reported by government agencies) around the world today consists of single species of *Pinus*, *Populus*, *Eucalyptus*, or other tree crops such as oil palm (*Elaeis guineensis*), olive (*Olea europaea*), fruit and nut trees (for example, *Malus*, *Prunus*, *Citrus* spp.), or coconut (*Cocos nucifera*), often supporting minimal plant or animal diversity. Such production lands might better be considered an agricultural land use, even though they provide

Resilient Forest Management. Philip J. Burton, Oxford University Press. © Philip J. Burton (2025). DOI: 10.1093/oso/9780198832997.003.0009

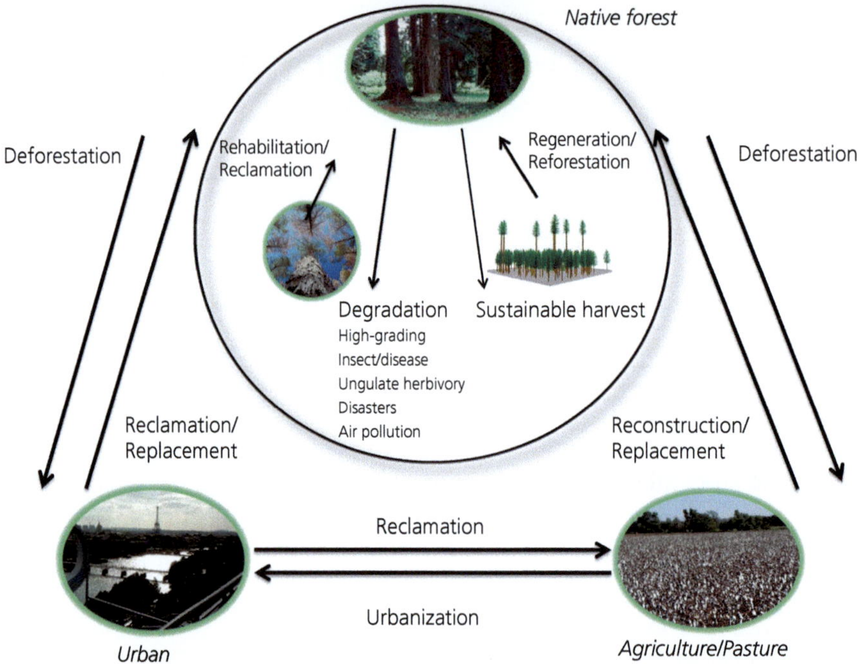

Figure 9.1 Lands subject to deforestation, degradation, or harvesting can be restored to mature forest by reclamation, afforestation, rehabilitation, and reforestation (reproduced from Stanturf (2016) with permission of CRC Press).

some forest ecosystem services (Demestihas et al. 2017). Many wild and otherwise natural forests are fragmented and damaged by edge effects, such that species that depend on interior forest conditions are threatened (Harper et al. 2005). More than half of the world's forest cover—including much currently within protected areas—is simplified or degraded in the manners described above. It has been suggested that 25 million km^2 of the world's current forest land will require a long time to recover ecological integrity, or would benefit from ecological restoration to accelerate the process (Grantham et al. 2020).

As a result of forest loss, forest degradation, and the compromised status of one or more forest values, there is widespread and growing interest in forest restoration (Burton and Macdonald 2011; Budiharta et al. 2014; Stanturf 2016). Forest restoration is seen as a necessary and "win–win" response to the global crises of biodiversity loss and climate change. International support for such initiatives is reflected in a number of indicators and initiatives:

- growing international membership in the Society for Ecological Restoration (www.ser.org);
- Aichi Biodiversity Target 15 that called for restoration of 15 percent of the world's degraded ecosystems (https://www.cbd.int/sp/targets/);
- the United Nations declaration that 2021–30 shall serve as the Decade on Ecosystem Restoration (www.decadeonrestoration.org); and
- the 2011 Bonn Challenge to restore 350 million hectares of degraded and deforested landscapes by 2030 (www.bonnchallenge.org).

The United Nations Framework Convention on Climate Change (UNFCCC) also includes a climate mitigation mechanism for Reducing Emissions from Deforestation and Forest Degradation in developing countries (REDD+), facilitating conservation, restoration, and tree-planting programs through transfer payments from the industrialized world (Alexander et al. 2011; see https://redd.unfccc.int/). It is widely accepted that the planet needs more trees. Trees actively extract CO_2 from the atmosphere and store carbon for long periods of time while providing multiple ecosystem services, including erosion control, hydrological regulation, and habitat for a broad array of animals, plants, fungi, and microbes. But reforestation and afforestation initiatives require more than the planting of trees if they are to be successful at engendering resilience in the world's current and future forests. Such work needs to proceed holistically and with caution so as not to compromise biodiversity any further in the process (Gardner et al. 2012). Practitioners will want to apply the popular Hippocratic Oath invocation "to abstain from doing harm" in order to avoid unintended consequences associated with good intentions to fix things.

Ensuring a halt to degradation, over-harvesting, and the simplification of forest composition and structure is the most important (but sometimes ignored) step in ecological restoration. Only after such processes of degradation are stopped can their effects gradually be reversed, using a variety of targeted interventions (see Table 9.1) to accelerate forest recovery, adjustment, reconfiguration, and overall resilience (Stanturf et al. 2014b; Lamb 2015). Successful forest restoration effectively boosts or

Table 9.1 Common forms of ecological degradation and corresponding restorative actions

Category	Examples	Responses
Over-harvesting	• of targeted spp. (trees, medicinal plants) • lack of deadwood (snags, logs) • human disturbances unparalleled in evolutionary history, threatening population viability of some species • timber overexploitation, unsustainable age-class structures • absence/paucity of old forest	• reduced, regulated, and monitored harvesting • introduction of deadwood • emulation of natural disturbance patterns, frequency, severity, and event sizes • logging deferrals, thinning programs to fill some size-class gaps, extended rotations • accelerate development of old-growth features in second-growth stands
Too many trees	• lack of fire (e.g. oak woodlands, US inland west, dry pine forests); see Altered fire regime • drained peatlands (e.g. Finland, Indonesia); see Altered hydrology	• variable density thinning, prescribed burning • block drainage ditches
Biotic imbalance	• overgrazing (understory depauperation, poor tree regeneration) by deer, cattle, goats, pigs • exotic tree spp.; e.g. Douglas-fir and Sitka spruce in Europe, pines and eucalypts in South America and Africa • invasive species (plants, pathogens, vertebrates, insects, earthworms)	• reduced stocking, fencing, regeneration protection, promotion of hunting • stand diversification and replacement with native tree species • species-specific removal or control strategies, employing careful biocontrol and chemical use as necessary
Altered fire regime	• more frequent or more intense fire than biota and ecological processes are adapted to, especially in wetter forests • fire too infrequent to maintain habitats to which biota and ecological processes are adapted, especially in drier forests; see Too many trees	• early wildfire detection and control; careful management of residual biomass associated with logging • reintroduce prescribed and cultural burning; may need to be supported by preparatory thinning and pruning
Altered hydrology	• forest roads and lost tree cover resulting in rapid runoff, poor water retention on hillslopes; see Landcape fragmentation • drained floodplains and swamps no longer contain flood waters or support flood-adapted species	• reforest hillslopes and decommission roads as rapidly as possible • remove dikes and levees that prevent floodplain flooding, block drainage ditches

continued

Table 9.1 *Continued*

Category	Examples	Responses
Homogenization	• homogenous substrate, microtopography • single and offsite tree species, e.g. spruce plantations • even-aged and single-layer stands; identical species, stand structures across the landscape	• mechanical preparation of berms and mounds • stand diversification and conversion to site-adapted native species • creation of canopy gaps, some left to shrubs and herbs, some regenerated to diverse tree species
Landscape fragmentation	• watershed deterioration • predominance of edge effects, lack of interior forest • lack of forest connectivity • roads and utility corridors that facilitate poacher access and the spread of exotic species	• control erosion and sediment generation from roads and landslides, prioritize riparian restoration • protect remaining old, mature, primary forest, especially in contiguous blocks and networks • control forest access with barriers and gates; decommission and restore roadbeds • decommission unnecessary roads so they are self-maintaining; minimize open corridors and reforest them where possible

accelerates processes of natural forest dynamics and the inherent natural mechanisms of resilience (Fischer et al. 2016).

9.2 Resilient reforestation, afforestation, and forest regeneration

Once the processes of deforestation and degradation are halted, the most fundamental step in forest restoration is to support the recovery mode of resilience (see Section 4.3). Tree species have evolved a wide range of pollination, seed dispersal, germination regulation, and vegetative reproduction mechanisms that facilitate species persistence and recovery after disturbances. However, if disturbances are too frequent, too severe, or too widespread, forest regeneration can be much delayed or too sparse to provide important ecosystem services such as hydrological regulation, wildlife habitat, and desired levels of carbon sequestration or fiber production. Basic sustained-yield forestry requires that fellings must be followed by successful forest regeneration (or released growth of remaining trees) to generate sufficient growth increment that will eventually replace the volume of wood that has been harvested. That regeneration often occurs naturally through the growth of seedlings and saplings that had already existed at the time of harvest, through seed germination, or by the sprouting of cut stumps or belowground buds. Although natural regeneration is an expression of healthy ecosystem processes and may be expedient, dependence on natural

regeneration will typically result in highly variable tree densities (some locations overstocked, with tree growth suppressed, while other spots have no trees), and may have long delays. Depending on land-use priorities (for example, wildlife habitat, watershed protection, or timber production), those traits may be desirable or undesirable.

Where natural regeneration is hindered, too slow, or too sparse, forest managers typically employ artificial methods of regeneration, which may consist of direct seeding (for example, of *Eucalyptus* spp. in Tasmania; Baker 2013), the planting of container-grown seedlings (for example, of *Pinus* and *Picea* species in boreal Canada), or the planting of unrooted cuttings (for example, of hybrid *Populus* clones in southern Europe; Oliveira et al. 2015). There is a wide variety of forest regeneration, plant propagation, and nursery technologies available to facilitate the establishment of new trees and accelerate forest recovery, some of which are more suitable for some tree species, site conditions, and management goals than others (see Table 9.2). There is a long history of practical experience in "regeneration silviculture" (e.g. Stoeckeler and Jones 1957; Cleary et al. 1978; Lavender et al. 1990), with ongoing research published in journals such as *New Forests* (https://link.springer.com/journal/11056).

A number of silvicultural techniques have been developed over the years to ensure the establishment of tree seedlings and to enhance their growth. Aerial or other broadcast seeding methods typically result in most tree seeds being wasted (unless sown on freshly burned or tilled ground), often falling on unsuitable substrates or being eaten by rodents, but

Table 9.2 Strengths and weaknesses of broad categories of forest regeneration

Regeneration method	Advantages	Disadvantages
NATURAL REGENERATION	Inexpensive, no nursery infrastructure needed	Limited ability to shift species composition
Vegetative sprouting	Fast-growing, supported by mature root systems	Typically requires thinning
Advance regeneration release	Several years of growth already achieved	Extra care must be taken to avoid damage during logging; rarely an option for light-demanding species
Seed bank release	Germination is prompted by open conditions	Few tree species have long-lived seeds, may not be the desired species
Seed dispersal/inseeding	Local species and genes perpetuated, but with sexual recombination	Requires some overstory retention or proximity; seed production may be unreliable
ARTIFICIAL REGENERATION	Strong control of species and genetic composition	Usually needs advance planning, storage, and/or nursery facilities, deployment logistics
Direct seeding	Rapid deployment (if seeds are in stock)	Inefficient use of seed if many seeds fall on unsuitable microsites; seed predation by small mammals and birds
Wildling transplants	Little advance planning needed; may have several years of growth already achieved	Inefficient use of labor; roots often damaged in transplanting; sensitive to drought in first season
Live-stake planting	Can be integrated in wattle fences and brush layers to stabilize slopes, restore riparian zones.	Not suitable for all desired species; must usually be collected in dormant season; may dry out before roots establish
Planting of nursery-grown stock	Strong potential for quality assurance	Requires nursery infrastructure and labor; mechanical planting largely limited to agricultural soils
Rooted cuttings	Superior genotypes propagated by cloning	Care must be taken to maintain genetic diversity in stands
Bare-root seedlings	Good root form, promotes stable trees; stock can be quite large	Planting can be labor intensive, requiring large planting holes and backfilling
Container-grown seedlings	Efficient production and planting	Can have poor root form, resulting in unstable trees
Somatic embryogenesis	Superior genotypes propagated by cloning	requires high-technology, multi-stage infrastructure
Genetically modified planting stock	Potential for very selective stock improvement of traits	widespread public opposition limits deployment on public lands; recommended primarily for lands owned by industry

directed or "spot" seeding can be effective. New technologies such as the use of remotely piloted aerial drones are also being developed to provide directed tree-seeding services (Mohan et al. 2021). Site-preparation techniques prior to sowing or planting range from broadcast burning or surface scarification (to reduce the forest floor thickness and expose mineral soil) to more disruptive methods of plowing, disc trenching, and mounding that also seek to create warmer, moister, or better-drained microsites (Löf et al. 2016). Where sites are especially severe, soil amendments can be applied at the level of individual trees. On nutrient-poor sites, packets of slow-release fertilizer can be inserted in each planting hole near the base of a tree seedling (e.g. Walker 2003). Where growing-season drought is anticipated, hydrogels (superabsorbent polymers) can be deposited during planting to absorb soil water during moist conditions, then gradually release that water during drought (Crous 2017). Inoculum-containing spores of rhizosphere bacteria and mycorrhizal fungi can boost soil health, resource acquisition, and the resilience of tree seedlings (Tedersoo 2015; see Section 5.6). Simple precautions such as avoiding outplanting of tender nursery-grown seedlings under conditions of high vapor pressure deficit (high evaporative demand, reflecting some combination of high air temperatures, low relative humidity, or windy conditions) can also improve tree establishment success.

On productive sites, competition from a few fast-growing herb or shrub species can constrain the establishment and growth of tree seedlings, while also limiting the diversity of understory plant and animal species (Walstad and Kuch 1987; Burton 1996; Balandier et al. 2006). Consequently, vegetation control using chemical herbicides, manual cutting, or controlled grazing is often undertaken on such sites to facilitate full stocking and optimal growth by desired tree species. Boreal and temperate forests that have not undergone a period of deforestation, cultivation, or sustained grazing tend to exhibit substantial resilience in the recovery of native plant communities after logging, site preparation, and vegetation control (Sullivan and Sullivan 2003; Haeussler et al. 2021). On the other hand, a minimally interventionist approach is usually effective on sites characterized by moderate environmental and productivity conditions (Prach and Hobbs 2008; Holl and Aide 2011), where natural recovery or supplementation by the planting of large healthy tree seedlings is all that is needed to facilitate forest restoration. A sound understanding of local disturbance severity, tree species silvics (autecology), and overall ecosystem resilience to silvicultural disturbances thus informs regeneration and restoration planning (Trevenen et al. 2017). If multiple-use lands and their associated stakeholders call for a minimal disruption to soils and the natural plant community, while still promoting the survival and eventual dominance of trees, then a greater tolerance of vegetation competing with young trees is in order. Conversely, in production forests where tree growth is designed to be optimized, more aggressive intervention to control competing vegetation is appropriate (see Figure 9.2). Such interventions should nonetheless be "results based" in terms of demonstrable tree performance, rather than being implemented on an arbitrary or regulatory basis of perceived competition (Burton 1993).

Tree plantations—if suitable for the local climate and soils—can usually be relied upon to provide the ecosystem services of timber production, erosion control, and hydrological regulation. But it is widely understood that simply planting trees does not constitute forest ecosystem restoration (Chazdon 2008; Putz and Redford 2010). Nonetheless, if the right trees are planted at sufficient densities in the right places, they will provide the distinctive superstructure upon which most other forest ecosystem attributes and values depend. Following the simple restoration mantra of "build it and they will come," there is evidence that, once trees are well established, a wide variety of native forest bird, mammal, and understory plant species can be found in plantations (Parrotta et al. 1997; Lindenmayer and Hobbs 2004; Irwin et al. 2014). Characteristic forest invertebrate and fungal

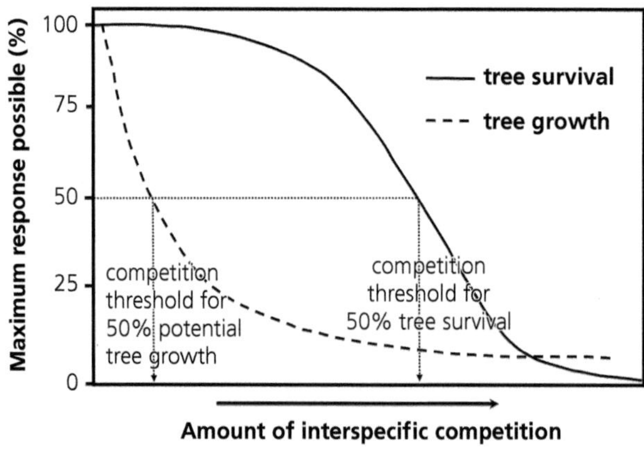

Figure 9.2 Lower thresholds for actively reducing the amount of vegetation competing with young trees may be adopted for production forests, where loss of potential productivity is to be avoided, than for multiple-value forests, where minimal disruption of native plant communities and natural ecological processes is desired, as long as most trees survive and can be expected to dominate eventually (redrawn and modified from Wagner et al. (1989)).

communities also can be found in plantations, although trends vary among taxonomic groups (Ferris et al. 2000; Irwin et al. 2014) and diversity and abundance rarely match those of diverse native forests (Cifuentes-Croquevielle et al. 2020). The ability of planted forests to support biodiversity restoration and resilience (Brockerhoff et al. 2008) is promoted when plantations consist of multiple species, are adjacent to intact native forest, and are permitted to reach over-mature ages—that is, older than rotation lengths optimal for timber production (Parrotta et al. 1997; Ferris et al. 2000; Lindenmayer and Hobbs 2004; Dejene et al. 2017).

A general consensus has now emerged in the ecological restoration community that restoration should be informed by historical ecosystem attributes, but should not blindly attempt to reconstruct past ecosystems under conditions that no longer prevail (Harris et al. 2006; Higgs et al. 2014). This approach makes forest restoration, particularly at the regeneration phase, a powerful and necessary component of forest adjustment, reconfiguration, or transformation in adapting to climate change. In particular, options to include tree species and provenances from warmer or more drought-prone habitats are being increasingly considered in all forest regeneration and forest restoration planning (see Chapter 7). Through such actions of assisted migration and climate-smart silviculture (e.g. Nagel et al. 2017; see Figure 7.11), forest restoration is shifting from its backward-looking "historical-ecology" roots (Higgs et al. 2014; Stanturf et al. 2014b) to a forward-looking model of restoring naturalness and diversity for enhanced current and future resilience (see Section 9.5).

Another application of forest restoration techniques is in the afforestation of lands that had been used for agricultural, industrial, or urban/residential purposes for many years, although it is debatable whether the term "restoration" is applicable if the land had not supported forests in the past. The return, or even the expansion, of tree cover to provide forest connectivity at key locations is an important tool in forest and landscape restoration (see Section 9.4), especially to ensure the mobility and migration of many species as they adapt to a changing climate (Krosby

et al. 2010; see Chapter 7). Afforestation has the added benefit of climate-change mitigation, by promoting additional carbon sequestration in long-lived cellulose and lignin on lands that had not been providing that service. Although some analysts have identified a huge global capacity for forest expansion through afforestation (Bastin et al. 2019), care must be taken not to replace grasslands, shrublands, savannas, peatlands, and other habitats that are equally as important as native forests in providing habitat for their own distinctive forms of biodiversity and (especially in the case of peatlands) carbon retention. Instead, it would be more productive to direct afforestation efforts to closed mines and quarries and other industrial sites, and to abandoned or marginal farmlands. For example, large areas of the Amazon basin and other tropical domains have been (and continue to be) deforested in support of agricultural expansion. But the lateritic soils are rapidly exhausted of nutrients by cultivated crops or have become dominated by invasive shrubs (for example, *Lantana camara*) or exotic grasses (for example, *Brachiaria decumbens, Paspalum virgatum, Imperata cylindrica*) that are maintained by recurrent fires (Veldman and Putz 2011; Silvério et al. 2013). Such lands—originally very biodiverse—today are prime candidates to be restored as closed-canopy forest, open woodland, or (at least) savanna ecosystems. The success of Costa Rica in returning pastureland to tropical forest since the 1980s (Klemens 2020; Tafoya et al. 2020; Tucker et al. 2023) provides an inspiring example of what can be done to rebuild the world's forests when the will, expertise, and supportive policies are there.

The regeneration phase of forest dynamics represents a window of opportunity to direct and adjust the next several decades of forest ecosystem properties. In deciding what plant materials to introduce or suppress, at what densities, and in what spatial configurations, the forest manager has the option of designing a wide range of future forests. Options exist to reconstitute the same kind of forest that was most recently harvested or degraded, to configure the kind of forest found on similar sites nearby or in the past, or to build a forest that is expected to be better suited for the future climate and future stakeholder priorities. Those options are represented in the resistance-resilience-transformation framework of experimental treatments being tested in the Adaptive Silviculture for Climate Change (ASCC) network in North America (Nagel et al. 2017).

9.3 Redirecting forest stands for resilience

As noted above, forest restoration is not just about establishing or regenerating trees. Many treed lands that are nominally considered to be managed woodlands and forests are simplified ecosystems that would benefit from restoration treatments to promote the resilience described in previous chapters. Decades or even centuries of exploitative or well-intended forest management conducted to optimize a single function—typically sawlog, fuelwood, or fiber production—have often resulted in degraded forests that are vulnerable to pests and climate extremes, often in landscapes that can no longer sustain indigenous biodiversity (Stanturf 2016). Much of the world's wood supply is now produced on former agricultural lands with a history of cultivation and/or grazing that has removed most forest legacies, making it especially difficult to restore indigenous forest ecosystems. Forests tend to be degraded and simplified in a variety of recurrent and multiple ways, each of which requires its own site-specific remedies.

An often-unappreciated form of forest ecosystem degradation is biotic imbalance, owing to a combination of missing and exotic species resulting from forest exploitation and globalization. Large carnivores such as wolves, bears, and tigers have been hunted to local extinction (extirpation) from large portions of their natural ranges, while human introductions of rats, rabbits, foxes, cats, and mustelids have wreaked havoc on forest biodiversity in other parts of the world (Clout and Russell 2008). Selective harvesting of high-value tree species has resulted in the endangerment of wild populations of species such as big-leaf mahogany (*Swietenia macrophylla*) and Guatemalan fir (*Abies guatemalensis*), Brazilian rosewood (*Dalbergia nigra*), African ebony (*Diospyros crassiflora*), and Serbian spruce (*Picea omorika*), for example (Oldfield et al. 1998). Understory plant communities have likewise been ravaged by unregulated harvesting of

plants used for herbal remedies, such as American ginseng (*Panax quinquefolius*) and goldenseal (*Hydrastis canadensis*) in eastern North America (Duke 1997).

Invasive plant and fungal species have also changed the composition and dynamics of many forests. Although many ruderal and invasive plant species have followed human activities around the world, most do not persist in closed-canopy forests. However, shade-tolerant species, such as garlic mustard (*Alliaria petiolata*), Amur honeysuckle (*Lonicera maackii*), and English ivy (*Hedera helix*) in North America, and black cherry (*Prunus serotina*) in Europe, are especially challenging to control. Several exotic fungal and insect pests have been notably effective at decimating important tree populations in eastern North America: chestnut blight (*Cryphonectria parasitica*), Dutch elm disease (*Ophiostoma ulmi*), and emerald ash borer (*Agrilus planipennis*) having removed or reduced the abundance and ecological contributions of *Castanea dentata*, *Ulmus americana*, and *Fraxinus* spp., respectively. Ameliorating the impacts of invasive species can depend on valiant community volunteer efforts at manual removal, targeted hunting or trapping and fencing programs, pesticide applications, and the screening and introduction of species-specific biological control agents or natural enemies (Simberloff et al. 2005; Clout and Williams 2009). Other invasive species, such as European earthworms in North America and New Zealand, have well-documented impacts (Bohlen et al. 2004), but with uncertain prospects for control or ecosystem restoration (Boyer et al. 2016).

Unprecedented populations of native species can also arise if climatic or predation constraints are relaxed—for example, bark-beetle outbreaks in western and boreal North America (Bentz et al. 2010), and growing densities of white-tailed deer (*Odocoileus virginianus*) in the broadleaf forests and mixed landscapes of midwestern and eastern USA (Rawinski 2008). Causing widespread mortality of adult trees or tree seedlings and understory plants, these animals certainly degrade forest ecosystems, but if and how to ameliorate an imbalance of native species and restore affected forests is widely debated (Burton 2006; Royo et al. 2010). Where grazing by domestic ungulates (cattle, sheep, goats, swine) is also an accepted use of a multipurpose forest landscape, there can be a fine line between the beneficial (for example, control of exotic weeds, fine fuel reduction) and degrading effects of that grazing pressure. Integrating ungulate herbivory into forest management and restoration often requires careful monitoring (often informed by comparing vegetation inside and outside fenced animal exclosures), extensive or movable fencing in order to concentrate grazing where and when it is most appropriate, and cooperation among land managers and users (Rooney et al. 2016).

Restorative actions for threatened plant and animal populations typically require species-specific recovery plans, with associated propagation, facilitated or selective breeding, and reintroduction programs supported by legal protection (Falk et al. 1996; Foin et al. 1998). Ecosystem restoration may sometimes depend on restoring trophic cascades through the reintroduction of extirpated indigenous animal species such as beavers (*Castor* spp.; Law et al. 2017) or wolves (*Canis lupus*; Ripple and Beschta 2004), which can improve landscape hydrology or keep ungulate browsing pressure in check, respectively. The protection, management, and recovery of rare, threatened, and endangered species constitutes a large subfield of the discipline of conservation biology, often requiring site-specific research into the biology and ecology of the populations being restored. To the extent that threatened species usually depend on native vegetation, ecosystem restoration can often be supportive of species recovery plans. Yet, if the recovery of endangered species is the priority, more of a "functional restoration" approach (Stanturf

et al. 2014b) may be required, which could include greater latitude in accepting some useful exotic species and blended (novel) ecosystems as an endpoint (Casazza et al. 2016).

Some of the most common deficiencies in the world's exploited and managed forests include the absence or rarity of features associated with old forests: large quantities of deadwood (standing snags and fallen logs), large old trees, trees with hollows and broken tops or branches, variable tree heights and foliage layers, and multiple canopy gaps supporting dense vegetation or regenerating trees. Consequently, many forest restoration efforts around the world consist of introducing old-growth elements into younger, more homogenous forests (Bauhus et al. 2009; Gower et al. 2016). Efforts to diversify stand structure—for the benefit of understory vegetation and multiple animal species—are undertaken in homogenous second-growth forests around the world through the introduction of gaps, snags, and woody material, and the use of variable density thinning (see, e.g., Box 9.1). It is not always the presence of old trees per se that supports high levels of biodiversity in old forests, but rather the history of undisrupted continuity of forest cover and the complete suite of variable horizontal, vertical, age class, and compositional attributes that characterize old-growth and climax forests. This means that restoration treatments in forest stands and landscapes that have a long history of human exploitation and homogenization can set the stage for old-growth recovery, but ultimately a long period of time is still needed (Crouzeilles et al. 2016).

Box 9.1 Initiating old-growth structure in homogenous second-growth forests

The primary forests of the Kitimat Valley of northwestern British Columbia, Canada, consisted of uneven-aged stands of western hemlock (*Tsuga heterophylla*), western redcedar (*Thuja plicata*), Pacific silver fir (*Abies amabilis*), and Sitka spruce (*Picea sitchensis*). Having a gentle terrain and good road access, most of the natural forest was clearcut logged in continuous swaths in the 1960s and 1970s, with the specific intent of converting the wild forest into a landscape of even-aged managed stands. Although planting took place,

primarily of the favored Sitka spruce, most of the logged areas regenerated naturally with full stocking by the prolific and easily dispersed western hemlock and the very shade tolerant Pacific silver fir. Tree growth has been outstanding on the deep soils and under the favorable temperate rainforest climate of this location, such that both commercial thinning and clearcut harvesting are being undertaken in fifty- and sixty-year-old stands, well ahead of the expected eighty-year rotation (see Figure 9.3).

Box 9.1 Continued

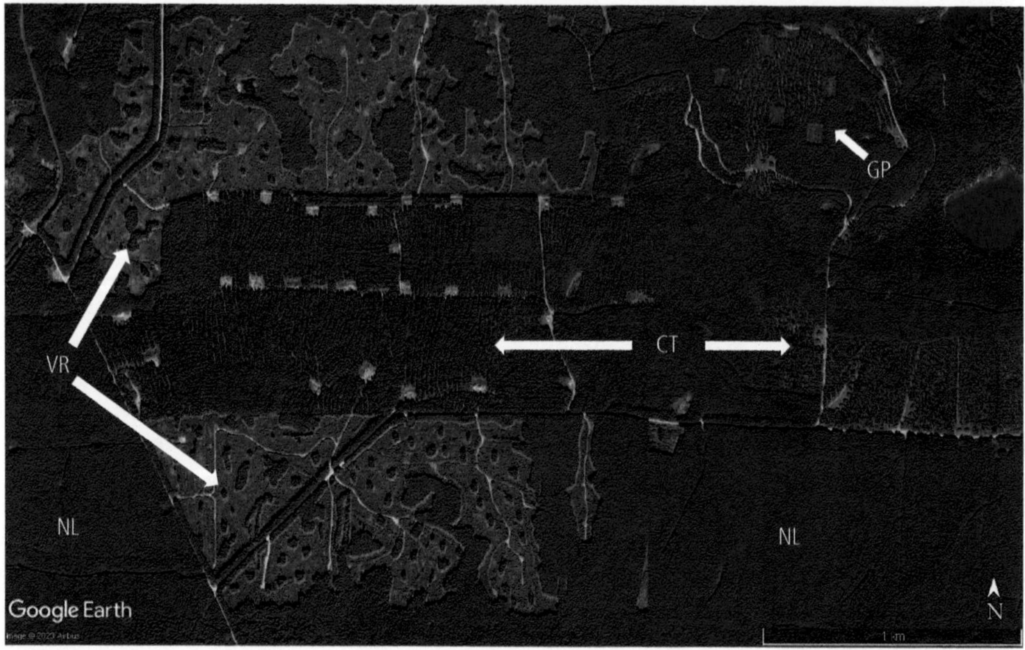

Figure 9.3 Even-aged second-growth stands in the Kitimat Valley of northwestern British Columbia, Canada

Note: Aerial photographs show recent harvesting operations designed to enhance structural diversity: commercial thinning (CT), retention of roadside and wetland buffers and variably sized retention patches in clearcuts (VR), and 0.45 ha biodiversity gap-based "pods" (GP) with some girdled trees remaining; see Figure 9.4). For comparison, non-logged (NL) second–growth areas are shown on lower left and lower right, and are interspersed in some of the CT areas. (reproduced from Google Earth[TM], image © Airbus, downloaded November 29, 2023).

Most current harvesting practices in British Columbia differ from the wall-to-wall clearcut logging that prevailed in the mid- to late twentieth century, with greater use of riparian buffers, wildlife tree patches, and other forms of variable retention, such as the VR areas in Figure 9.3 (Gustafsson et al. 2012; Beese et al. 2019). While such efforts will contribute much-needed structure and spatial diversity to third-growth forests, they do not address the lack of such diversity in the current expanse of homogenous second-growth forest. To remedy that immediate deficit of old-growth habitat for many bird, mammal, and berry-producing species, and to offset a nearby land-use conversion, local researchers, managers, and stakeholders embarked on a program of introducing old-growth attributes within the remaining matrix of young forest. This initiative built on an earlier pilot project (described in Gower et al. 2016) and research in similar forests on "gap-based silviculture" (Coates and Burton 1997). Interventions consisting of the following manipulations were combined in gap-based "pods" (GP in Figure 9.3; see also Figure 9.4) at a few locations in the Kitimat Valley and are now being implemented elsewhere in the region:

- most trees within selected gaps or small patch cuts are felled or stubbed (leaving 2–4 m of the tree base standing) using a small feller-buncher, allowing shrubs important for browse, berry production, and traditional medicinal use to thrive under higher light conditions;
- within each gap, several of the largest trees are girdled so they can be retained as standing snags; their habitat value can be further accelerated by inoculation with wood-rotting fungi or by drilling artificial cavities;

Box 9.1 *Continued*

- some large felled logs may be left on the ground individually, but most logs are yarded and sold, or are stacked in bundles (visible in Figure 9.4, with another example in Figure 9.5(b)), in lieu of the very large woody material no longer available; and
- small and medium-sized piles of woody debris and logging slash are built to provide cover for small mammals and furbearers, and potential denning sites for bears.

Those general treatments are supplemented with bat houses and nest platforms for raptors in some cases. Monitoring results within three years of implementation indicate greater abundance and diversity of birds and bats than found in untreated locations.

Additional diversification treatments are being undertaken by the Terrace Community Forest in its Kitimat Valley tenure area. Commercial thinning is being conducted in forty–fifty-year-old stands that previously had been subject to pre-commercial thinning some twenty years earlier. Those operations (CT areas in Figure 9.3) open up the treated stands with benefits to understory and wildlife diversity, while also stimulating enhanced growth in the remaining

trees. Throughout these operational treatments, priority was given to removing trees with poor form (particularly spruce trees that had been attacked by a terminal weevil, *Pissodes strobi*, which resulted in many forked stems), while retaining the rarer tree species, including western redcedar, which is highly valued by local Indigenous people. Planting programs (in the variable retention cutblocks) now include all of the original primary forest species (not just spruce) and anticipate climate change by testing an array of species (not presently native to the area) brought in from the south. These introduced species include Douglas-fir (*Pseudotsuga menziesii*), ponderosa pine (*Pinus ponderosa*), and western larch (*Larix occidentalis*). Analogous "restoration silviculture" treatments are being undertaken in uniform managed stands elsewhere in western North America. Commercial thinning, variable retention harvesting, enrichment planting, and variable-density thinning are being increasingly practiced to promote in-stand heterogeneity, the growth of selected large or rare trees, and overall biodiversity value (Franklin et al. 2006; O'Hara et al. 2010; Brodie and Harrington 2020).

Figure 9.4 A representative pod of treatments newly implemented for enhancing old-growth attributes in a uniform second-growth stand, showing reduced overstory density and cover, girdled trees to create snags, and bundles of logs left on the ground

Snags can be created by girdling live trees (see Figure 9.4), while greater densities of downed wood can be generated by felling or mechanical uprooting (Lilja-Rothsten et al. 2008; Fischer et al. 2016). Artificial cavities (tree hollows) for use by nesting birds or mammals can be created using portable drills (Hooper and McAdie 1996; Rueegger 2017). In order to promote more rapid fungal colonization and wood decomposition, trees can be purposely topped (using a chainsaw or explosives) to leave a jagged edge at considerable height, or may be inoculated with fungal spores (Franklin et al. 2006; Gower et al. 2016). Large pieces of deadwood may need to be found elsewhere and hauled onto the site, and can be "planted" upright to simulate naturally occurring dead trees (see Figure 9.5(a)). Tree boles with intact root wads also can be planted in an inverted fashion to provide ready-made platforms for raptor nests. The trees felled to create gaps may be yarded off the site and sold to finance restoration operations, or some may be retained on site or bundled to simulate large logs no longer found in degraded forests (see Figure 9.5(b)).

The conversion of uniform tree plantations, especially of exotic or off-site species, to native forest is a widespread goal of conservation biologists and resilience-minded foresters around the world. Monotypic forests may lend themselves to industrial management and harvesting, but they tend to be more vulnerable to insect pests, fungal diseases, and windstorm damage (Jactel et al. 2017). Societal preferences around much of the world increasingly call for the protection and promotion of social amenity values (aesthetics, recreation) and non-consumptive ecosystem services in forests over the commercial benefits of wood production and harvesting (Gundersen and Frivold 2008; Kijazi and Kant 2010). Consequently, there is increasing interest and experimentation in promoting multi-species, multilayered stands of native tree species. If plantation history is short (that is, with legacies such as viable stump or root buds, or propagules of indigenous flora and fauna, still found on site), or if the surrounding landscape has those components and is relatively free of invasive species, then the gradual recovery of a native forest ecosystem can occur passively (see Figure 9.6). On the other hand, if there is a long history of cultivation prior to plantation establishment, or if a more rapid transition to native forest is desired, then a series of restorative interventions need to be undertaken. Whether spontaneous or highly managed, ecological restoration can be considered a process of directed or accelerated succession (Luken 1990; Walker and del Moral 2009). The probability of succession reaching climax or target conditions increases with latitude and decreases with the degree of anthropogenic landscape transformation (Prach and Walker 2020).

Figure 9.5 Large woody material reintroduced into homogenous managed forests: (a) Large pieces of deadwood installed on a former log landing in a *Pseudotsuga menziesii* plantation (Galiano Island, British Columbia, Canada). (b) Piled logs six years after variable-density thinning in second-growth mixed conifer forest to provide substrate for mosses and lichens, serve as nurse logs for regenerating trees, and provide protection and foraging habitat for amphibians, mammals, and birds (Olympic Peninsula, Washington, US). (reproduced from Brodie and Harrington (2020), courtesy of US Department of Agriculture)

Figure 9.6 Examples of natural forest recovery in the understory of planted exotic tree crops, indicative of the potential for passive forest restoration: (a) Native tree ferns (*Cyathea* and *Dicksonia* spp.) and hardwoods established in a plantation of exotic *Pinus radiata*, North Island, New Zealand. (b) Diverse native trees, lianas, shrubs, herbs, and bird life thriving in a planted stand of big-leaf mahogany (*Swietenia macrophylla*), part of a forest reserve now subject to limited harvesting owing to its designation as a park and protected watershed for the city of Suva, Viti Luvu, Fiji.

Similarly, successful regeneration of native tree species in monospecific even-aged plantations has been greatest at higher elevations, proximate to native seed sources, under relatively open canopies with sparse ground vegetation and high litter cover (Kremer and Bauhus 2020).

Some of the best examples of naturalizing monospecific plantations can be seen in efforts across Europe to convert spruce (*Picea abies* or *P. sitchensis*) plantations to more compositionally and structurally diverse stands containing native broadleaf species (see Box 9.2). In other situations, particularly in the tropics, the long-term restoration of native forests can be promoted in plantations of fast-growing broadleaf trees such as *Acacia* spp. (Norisada et al. 2005; Chazdon 2008) and *Albizia* spp. (Parrotta 1992), which function as nurse crops while slow-growing, shade-tolerant native species become well established. Other common plantation species (*Eucalyptus* spp., *Pinus* spp., *Tectona grandis*) may or may not be suitable as nurse crops,

depending on site conditions and stand densities (Harrington 1999; Feyera et al. 2002; Healey and Gara 2003). Although widely promoted as a cost-effective restoration strategy (e.g. Lugo et al. 1993; Keenan et al. 1997; Feyera et al. 2002; Holl 2017), the efficacy of using exotic nurse crops to facilitate the restoration of native tropical ecosystems remains to be fully demonstrated. To be successful, such an approach typically requires careful harvesting of planted trees (that is, in order to recover restoration costs from timber sales) while protecting the native understory, and agreement that the land will be repurposed for non-timber uses.

Conversion of pastures and grass-dominated old fields to forest can be especially challenging because of the degree to which well-established grasses inhibit tree seed germination and seedling growth (Burton and Bazzaz 1991, 1995). Overcoming this resistance to tree establishment often requires multiple interventions over long periods of time. One approach is first to apply a broad-spectrum

Box 9.2 Protocols for naturalizing spruce plantations in Europe

European reforestation and afforestation programs from the eighteenth through the twentieth centuries emphasized the establishment of spruce plantations throughout much of the continent to alleviate widespread timber supply shortfalls (Johann et al. 2004). Most widely planted was Norway spruce (*Picea abies*), native to northern and mountain regions of Europe but introduced to many areas further south and at lower elevations. Those lands may have been cultivated for generations or had been previously protected or managed as forests of mixed broadleaf species, including oaks (*Quercus* spp.), beech (*Fagus sylvatica*), ash (*Fraxinus excelsior*), maple (*Acer* spp.), and lime (*Tilia* spp.). Some 5.7 million to 7.3 million ha of pure Norway spruce are found outside its natural range, most of it on sites naturally dominated by broadleaf tree species (Spiecker et al. 2004a). In recent years, susceptibility to storm and snow damage, sensitivity to drought and bark beetles, combined with social pressure to provide more natural habitat for biodiversity, have prompted many land managers to undertake programs of gradually transforming monotypic spruce forests into multi-species forests dominated by native broadleaf tree species (von Lüpke et al. 2004: Hansen and Spiecker 2016). Researchers and practitioners are now actively exploring several alternative approaches to converting pure spruce plantations to more diverse and more spatially complex forests (see Figure 9.7).

Summarizing a European Forest Institute program examining the conversion of Norway spruce stands, Spiecker et al. (2004(b), 261) point out that "the choice of an appropriate conversion strategy has to integrate stand history, site conditions, stand and landscape characteristics, current as well as expected perceptions and needs expressed by owners, other stakeholders and the public." Climate change, including expected increases in the frequency and severity of drought, has become an important consideration in evaluating potential species mixtures (Pretzsch et al. 2020). Different approaches and timelines are followed depending on what combination of tree-species enrichment, degree of replacement with broadleaf species, or conversion to an uneven-aged structure is desired. A complete replacement of conifers with broadleaf trees may not be desirable, as some combination of those growth forms can support elevated yields, richer biodiversity, and reduced risk of loss (Knoke et al. 2008; Pretzsch 2022). The promotion of broadleaf tree species is now a common response to disturbance events such as storm damage or bark-beetle outbreaks.

If dead and damaged spruce trees are harvested from a concentrated area, the resulting high-light environments can support the successful establishment and growth of shade-intolerant species such as *Quercus robur*. Proactive and more gradual conversion strategies may consist of selective thinning (both commercial and non-commercial), shelterwood harvests, or single-tree selection to favor naturally established or planted seedlings of the desired broadleaf species, which vary in their shade tolerance and optimal overstory cover (Löf et al. 2007; Yousefpour and Hanewinkel 2014). In addition to the widespread underplanting of shade-tolerant *Fagus sylvatica*, direct seeding of that species and the shade-tolerant *Abies alba* is also employed as a stand-enrichment technique (Baumhauer et al. 2005:

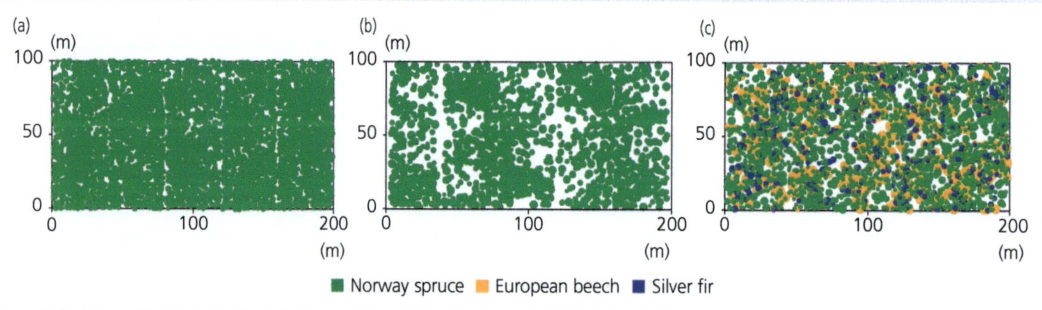

Figure 9.7 Projected stand development upon conversion of an initial stand (a) of pure Norway spruce (*Picea abies*), using a combined shelterwood-femel (group selection) system to open the stand, (b), thereby reducing competition sufficiently to allow European beech (*Fagus sylvatica*) and silver fir (*Abies alba*) to establish and thrive in an uneven-aged mixed-species stand, (c). This and other transition strategies have been explored using spatially explicit stand simulation models such as SILVA. (reproduced from Pretzsch (2020) with permission of Burleigh Dodds Publishing).

Box 9.2 *Continued*

Huth et al. 2017). Gap-based openings or strip cuts, in combination with target-diameter harvesting, are predicted to balance timber, carbon sequestration, biodiversity, and economic considerations effectively (Hilmers et al. 2020: Reventlow et al. 2021). Full conversion to a more resilient forest with the diverse composition and multilayered structures characteristic of natural European forests will typically take repeated silvicultural interventions over several decades (von Lüpke et al. 2004).

---

herbicide (for example, one containing glyphosate as the active ingredient) to kill the existing vegetation across the area to be treated, or in the immediate neighborhood of spots designated for tree planting. Alternatively, dense clusters of native trees can be planted such that they will more quickly form a closed canopy and shade out the grasses. Such "nucleation centers" can become floristically enriched by attracting additional seeds dispersed by birds and mammals, and will eventually expand outward by showering adjacent areas with seedfall (Holl et al. 2020). Harnessing the site amelioration potential of cover crops and nucleation sites can be described as successional restoration and is especially important where simple tree planting often fails, owing to severe abiotic or competitive constraints. Suppression of highly competitive grasses (for example, many African grasses introduced to South America) is the greatest challenge in restoring pastureland to forest, so it is better to concentrate limited planting material in high-density clusters rather than spreading tree seedlings or cuttings out over a larger area (Stanturf et al. 2014a). Nurse crop and cluster planting approaches to forest restoration can be strategically implemented where the probability of native ecosystem recovery is greatest—namely, adjacent to undisturbed forest, where animal-dispersed native tree species are found nearby, or where shade-tolerant indigenous tree species can be easily propagated and planted.

9.4 Forest landscape restoration

Scaling up forest restoration from stand-level and in-stand treatments to address widespread and cumulative ecological degradation across entire catchment basins, landscapes, and regions is obviously very challenging (Mansourian et al. 2017; Noulèkoun et al. 2021). Limited resources may first be focused on critical elements such as riparian zones (Apostol and Berg 2006; Bentrup et al. 2012), which can be restored to protect stream and wetland water quality and aquatic resources while supporting particularly rich (in terms of both productivity and diversity) habitats that can also provide connectivity corridors across the landscape. In the long run, however, many conservation biologists call for widespread "rewilding" of vast areas, not just connective corridors such as riparian zones. There is growing evidence that some 30–50 percent of landscapes is needed to maintain robust populations of native wildlife and resilient biodiversity (Dinerstein et al. 2017; Dotson and Pereira 2022). In addition to thinking on continent-wide scales and calling for the protection or restoration of broad, interconnected networks of natural habitat, rewilding also calls for the reintroduction of keystone species such as large herbivores and top carnivores, or simply a hands-off approach to allow natural succession and other ecological processes to prevail over large areas of land (Stanturf et al. 2023).

Forest landscape restoration (FLR), or, more correctly, "forest and landscape restoration," has emerged as a distinct paradigm for land management, representing more than large-scale tree planting and technical (biophysical) solutions for reversing deforestation and forest degradation (Mansourian and Vallauri 2005; Stanturf et al. 2012). One definition of FLR is "a process that aims to regain ecological integrity and enhance human well-being in deforested or degraded forest landscapes" (Maginnis and Jackson 2007: 10). Recognizing that ecological degradation has typically stemmed from historic and ongoing human exploitation of a region, FLR seeks to balance the promotion of ecological integrity with ongoing human use, including the integration and amelioration of agricultural, residential, and infrastructure land uses into landscape rehabilitation. As such, it has much in common with integrative and utilitarian traditions of resource conservation, sustainable development, and sustainable forest management (see Chapter 2). However, it explicitly depends on broad stakeholder involvement and requires careful attention to the governance structure behind decision-making and the prioritization of the many actions needed to reverse—not just halt—ecological and socioeconomic impacts over large areas of once-forested land (Lamb et al. 2012; Mansourian et al. 2017).

A key feature of forest landscape restoration is that it is a participatory process based on adaptive management (see Chapter 10) in a manner that is responsive to social, economic, and environmental priorities and changes. It seeks to restore ecological integrity with a focus on functionality rather than composition, with the added criterion that it must also enhance human well-being. This "frameshift" from small plot-based research studies aimed at restoring pre-disturbance ecological composition to integrating forest restoration in a mosaic of land uses driven by community needs improves the prospects of successful implementation (Holl 2017). As the term implies, FLR must be implemented at a landscape level, or rather that site-level actions are to be prioritized in the context of landscape-wide cost efficiency and likely efficacy (Maginnis and Jackson 2007; Latawiec et al. 2015). From a conservation planning perspective, FLR can include combinations of "land sharing" (multiple use) and "land sparing" (protected area restoration) initiatives. Part of the solution may be adoption of some variation of the triad system (see Box 2.3), including intensifying production functions on a smaller portion of the land base. Actions may promote agroforestry, silvopastoral systems, and urban trees (see Section 11.7), enhancement of old forest attributes in secondary forest (see Section 9.3), natural regeneration and tree planting on degraded lands and marginal agricultural sites (see Section 9.2), and selected efforts at full ecosystem restoration.

Mansuy et al. (2020) suggest that FLR objectives often correspond with those of other existing provincial or national policies and programs. It is therefore worthwhile for FLR projects to be selected, planned, and coordinated in such a manner that they qualify for funding support from programs promoting reforestation/afforestation, carbon sequestration, endangered species recovery, soil conservation, watershed restoration, agricultural diversification, or rural employment and economic development. For example, many restoration undertakings in developing tropical countries qualify for transfer payments from the industrialized world under REDD+ criteria, while forest restoration in the critically endangered Atlantic Forest of Brazil might find support through the Atlantic Forest Restoration Pact (Pinto et al. 2014). Initiatives for FLR are also generally supportive of social equity, poverty reduction, clean water, and several other United Nations Sustainable Development Goals (see Box 1.2). Undertaking FLR helps countries respond to the Bonn Challenge and meet treaty obligations under the Aichi targets of the Convention on Biological Diversity and the Land Degradation Neutrality targets of the United Nations Convention to Combat Desertification (Stanturf and Mansourian 2020).

9.5 Forest restoration for the future

Ecological restoration has traditionally focused on site- or stand-level treatments to reconstruct or reconfigure the composition of vegetation to resemble some pre-disturbance template or reference ecosystem. Such efforts were often implemented with volunteer labor and limited resources, motivated by naturalists and conservation biologists seeking to repair or rebuild plant communities and habitats that had become rare in a region (Egan and Howell 2001). But the need for forest restoration on a massive scale has now taken on a degree of urgency in this era of massive human domination of the planet, habitat destruction, species extirpations and extinctions, accelerating climate change, and the need to sequester as much carbon as possible in long-lived biomass and soil reservoirs (Perring et al. 2018). Forest management will often require forward-looking but nature-informed restorative efforts to recover, adjust, reconfigure, and transform our existing forests for them to be more resilient going forward.

There will always be a need for expertise and manuals to guide the restoration of specific ecosystem types to conditions prevailing prior to their degradation by human activities. For example, Allison and Murphy (2017) have individual chapters dedicated to the restoration of boreal forests, temperate broadleaf forests, temperate savannas and woodlands, Mediterranean-type shrublands and woodlands, mangroves, tropical savannas, and tropical forests. Apostol and Sinclair (2006) provide guidance on restoring the oak woodlands and savannas, old-growth conifer forests, riparian woodlands, ponderosa pine forests, and various non-wooded ecosystems specific to the US Pacific Northwest. Elliot et al. (2013) provide practical advice for tropical forest restoration, with a focus on Thailand.

If we step back for a broader perspective, ecological restoration can simply be viewed as giving nature a helping hand, facilitating or accelerating natural succession and other healthy ecological processes to generate ecosystem composition and structure that can be expected to persist under current and future conditions (Jackson and Hobbs 2009; Burton and Macdonald 2011). That sort of motivation for restoration reinforces the principles of managing ecosystems for resilience, embracing diversity and flexibility as key components of any restoration plan. The discipline of restoration ecology has evolved to recognize that it can be guided by historical relationships but not constrained by them, even though we are understandably nostalgic about the way the world used to be. Historic ecosystem attributes in many places around the world no longer match current realities where human activities, exotic species, climate, and disturbance regimes are very different from those that prevailed a century or two ago (Harris et al. 2006; Jackson and Hobbs 2009). Some illustrations of the potential application of this principle of expanded flexibility in the practice of forest restoration are provided in Box 9.3.

So how can one repair and restore nature—features of the Earth system that dominated for the vast majority of evolutionary history—in a changing world where the human species and our impacts are preeminent? The answer with respect to

---

Box 9.3 Flexibility and adaptability in forest restoration

The desirability of considering the natural range of variability (NRV) in the undisturbed ecosystem rather than trying to match a single reference ecosystem has been advocated before (White and Walker 1997; Burton 2005). But we might also consider the landscape or ecoregion level in considering what that "natural ecosystem" is or can be, embracing site-constrained and successional variants as perfectly legitimate and acceptable targets in a world of increasingly unpredictable disturbance and extreme weather events. So, with reference to boreal forest restoration (see Figure 9.8), it is perfectly reasonable to accept or direct recovery of a logged white spruce (*Picea glauca*) stand to domination by trembling aspen (*Populus tremuloides*) or some mixture of white spruce, black spruce (*Picea mariana*), jack pine (*Pinus banksiana*), and balsam fir (*Abies balsamea*). Some differences in dominant species emerge in response to site differences (for example, in soil moisture and nutrient regimes, as affected by slope position and parent materials), while others change during the course of succession. Especially in a changing climate, where moisture availability and nutrient cycling at any given site are being altered (Campbell et al. 2009), it is prudent to recognize a wide range of boreal compositions and structures as acceptable endpoints for restoration.

Further direction to incorporate climate-change adaptation explicitly into forest restoration can be informed by bioclimatic envelope modelling, which may inform the facilitated migration of provenances (reflecting genetic differences within species) or of species (Pedlar et al. 2012; Prober et al. 2015), as discussed in Chapter 7. Furthermore, by reference to current combinations of forest composition and structure supported elsewhere by climates expected to prevail at a site in the future (that is, forests found under analogs to future climates), managers can decide to accept, nudge, or replace naturally recovering vegetation (see Figure 9.9).

Box 9.3 *Continued*

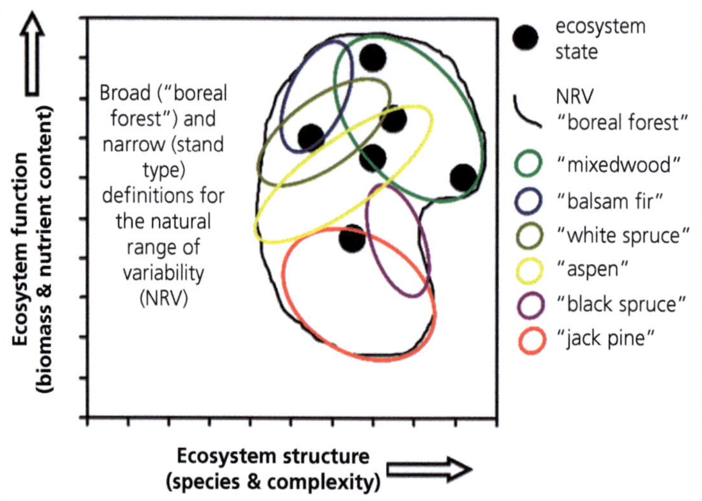

Figure 9.8 Target restoration options portrayed by natural range of variability (NRV) envelopes defined by ecosystem structure and function, as introduced in Figure 4.3(a), illustrating hypothetical narrow and broad definitions of North American boreal forest stand types

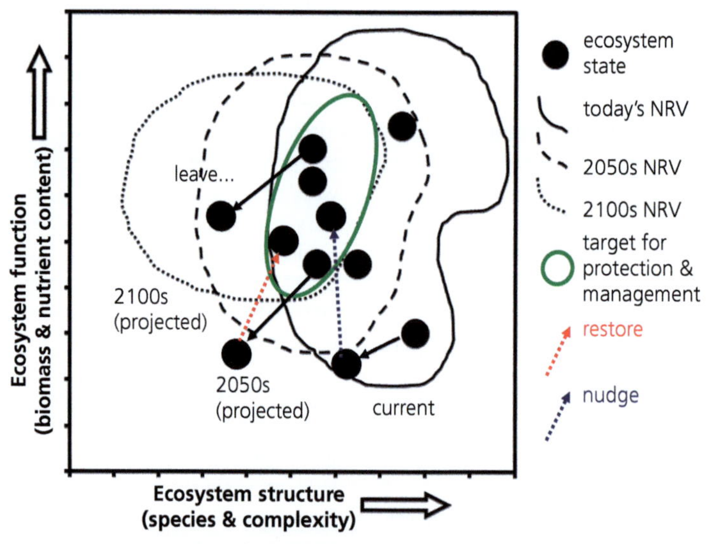

Figure 9.9 Hypothetical NRV envelopes for forests found under analogs of projected future climates. Disturbance or degradation (solid black arrows) may generate ecosystem states outside the current NRV, but potentially within a projected future NRV, where they can be left to develop naturally. Restoration (dotted red arrow) should seek to match ecosystem structure and function combinations suitable under both current and future conditions. A "nudge" (dotted blue arrow) to adjust or reconfigure currently acceptable vegetation to match future conditions better is also advisable.

Box 9.3 *Continued*

This proposed use of bioclimatic envelope modelling adapts the stability domain concept (as introduced in Figure 4.3) to recognize that equilibrial conditions (which can no longer be considered to confer stability) are shifting in a changing climate. The solid black polygon in Figure 9.9 designates the multidimensional natural range of variability (NRV) for a recognized ecosystem type under current conditions. The black dashed and dotted polygons in Figure 9.9 represent expected ecosystem attributes under mid- and long-range climate change. As species must survive current as well as future conditions, the green polygon describes the "no-regrets" or "win–win" target condition common to all three black polygons. Forest stands currently characterized

by conditions defined by the green polygon are already preadapted to future climates and can be left unmanaged. Disturbances or processes of degradation (solid black arrows) that shift stands outside both current and future envelopes should be restored (dotted red arrow) to green polygon conditions. Disturbances that shift stands outside the NRV for current conditions but within the expected NRV for a future climate can probably be left, but could also be restored. Disturbed stands that remain within the current NRV might also be tweaked or nudged (blue dotted arrow) to match the future climate better, if desired.

---

the practice of forest management may be more akin to gardening and applied research (or "intelligent tinkering," as Aldo Leopold (1949) would say) than to what have been the historical practices of forestry or ecosystem restoration. In order to keep repaired forests on track to a resilient future, regular assessments of current conditions and trends are required (Burton 2014; see Chapter 10) to assure appropriate

trajectories for ecosystem recovery or development, and for the eventual provision of desired ecosystem services. Efforts are being made to articulate principles that are sufficiently general and flexible to guide widespread expansion and recovery of forests and woodlands while supporting biodiversity and carbon sequestration (see Box 9.4).

Box 9.4 Guidelines and standards for forest restoration.

Several authors and organizations offer guidelines for training and best practices in the implementation of forest restoration activities, often with a focus on particular regions or values. Di Sacco et al. (2021) propose the following ten "golden rules" for reforestation to optimize carbon sequestration, biodiversity recovery, and livelihood benefits:

(1) protect existing forest first;
(2) work together, involving all stakeholders;
(3) aim to maximize biodiversity recovery to meet multiple goals;
(4) select appropriate areas for restoration;
(5) use natural regeneration wherever possible;
(6) select species to maximize biodiversity;
(7) use resilient plant material, with appropriate genetic variability and provenance;
(8) plan ahead for infrastructure, capacity, and seed supply;

(9) learn by doing, using an adaptive management approach; and
(10) ensure the economic sustainability of any project.

The Global Biodiversity Standard (BGCI 2023) takes those rules further, proposing a system of site-based assessment and certification of tree planting, habitat restoration, and agroforestry practices that prioritizes biodiversity protection, enhancement, and restoration. In particular, appropriate areas for reforestation must not displace natural biodiversity, operations should refrain from planting invasive alien species, should use native species, and should incorporate threatened species in planting wherever possible; robust monitoring, evaluation, and adaptive management should explicitly include biodiversity.

Similarly, the United Nations Decade on Ecosystem Restoration (https://www.decadeonrestoration.org/) is

Box 9.4 *Continued*

supported by ten guiding principles (UNEP 2021)— namely, that restoration programs should:

(1) contribute to the United Nations Sustainable Development Goals (see Box 1.2; Katila et al. 2020) and the goals of the Rio Conventions (Saint-Laurent 2005; Antofie et al. 2010);

(2) promote inclusive and participatory governance, social fairness, and equity;

(3) include a continuum of activities (see Figure 9.9);

(4) aim to achieve the highest level of recovery for biodiversity, ecosystem health and integrity, and human well-being;

(5) address the direct and indirect causes of ecosystem degradation;

(6) incorporate all types of knowledge and promote their exchange and integration;

(7) be based on well-defined short-, medium-, and long-term ecological, cultural, and socioeconomic goals and objectives;

(8) be tailored to local ecological, cultural, and socio-economic contexts while considering the larger landscape;

(9) include monitoring, evaluation, and adaptive management throughout and beyond the lifetime of an individual project or program; and

(10) be enabled by policies and measures that promote their long-term progress, fostering replication and scaling-up.

There is noticeable overlap in those rules, criteria, and principles, which are further bolstered by a set of eight international principles (Gann et al. 2019), that should underpin the full continuum of ecological restoration options (Figure 9.10) as endorsed by the Society for Ecological Restoration (https://www.ser.org/). Those principles are then elaborated upon and supported by an evolving collection of regional and national standards, such as those for Australia (SERA 2021), which provide a hierarchical set of expectations for planning and design, for project implementation, and for monitoring, documentation, evaluation, and reporting. Key to the long-term success of any such guidelines and standards are the monitoring, assessment, and willingness to adjust course inherent to adaptive management (see Chapter 10). Although it is tempting to fall back on standards as professionally defensible goals, they should never constrain creativity and experimentation.

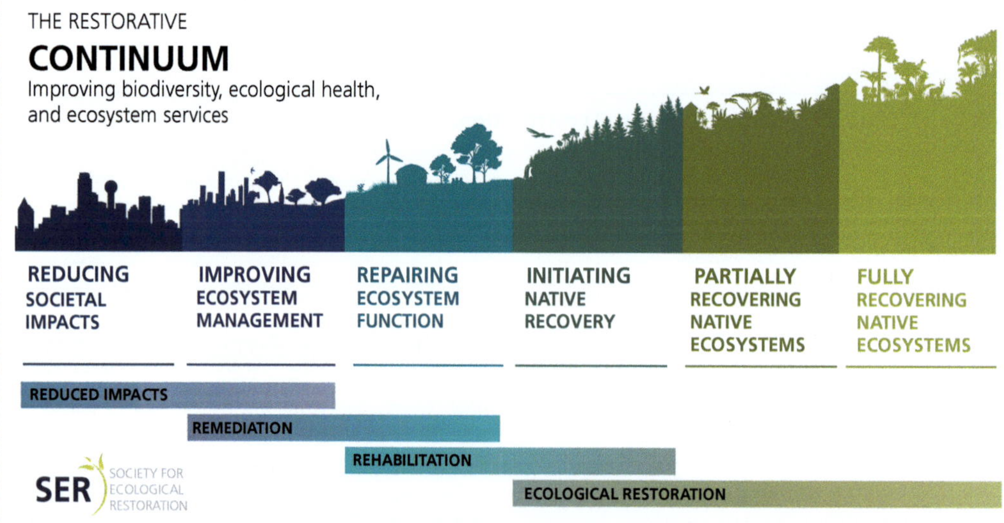

Figure 9.10 The broad spectrum of efforts to improve the health, integrity, and naturalness of ecosystems, as embraced by the Society for Ecological Restoration (SER) in its International Principles and Standards for the Practice of Ecological Restoration (https://www.ser.org/page/SERstandards). This graphic portrays the continuum of restorative actions, from reducing human impacts to fully reconstituted native ecosystems ("ecosystem restoration"), although use and interpretation of the terms "remediation" and "rehabilitation" differs among practitioners; reproduced from Gann et al. (2019) under terms of the Creative Commons Attribution–Noncommercial License).

In conclusion, it is worth recognizing the earliest roots and many widespread practices of forestry (see Chapter 1) as forms of ecological restoration. Reforestation, silviculture, and selective harvesting practices aimed at promoting ecosystem health and resistance to disturbance are important aspects of forest restoration, especially when working with native tree species and informed by natural disturbance regimes (Burton 1999). Initiatives to promote "continuous cover forestry" (Pommerening and Murphy 2004), "closer to nature" forest management (Larsen et al. 2022), and the "emulation of natural disturbance" in forestry (Kuuluvainen et al. 2021) can be described as efforts to restore naturalness in forests, and to reform the practice of forestry to follow more closely the principles of ecological restoration rather than those of agriculture (Bauhus et al. 2009; Puettmann et al. 2009). As such, forest restoration cannot be undertaken as a project, but rather as a mindset or a paradigm, with work requiring multiple and ongoing interventions.

Much like recent interest in "regenerative agriculture" (Newton et al. 2020), and "regenerative tourism" (Dredge 2022), forest restoration through regenerative silviculture strives to undo and repair past exploitative damage. Resilience concepts (see Chapter 4) inform restoration planning by identifying the need for a variety of management interventions, and by promoting attributes that will promote development and persistence of the desired future forest under changing and uncertain conditions (Trevenen et al. 2017). If one considers the restoration continuum (as portrayed in Figure 9.10), it is clear that restoration forestry has a role to play in all major categories of the forest land use triad (see Box 2.3). Restoring key features supporting productivity, disturbance resistance, and climate adaptability obviously make sense for production forests. Facilitating full recovery of future-adapted native ecosystems is a logical priority for protected areas, while various balances of climate-smart restorative approaches are expected in multipurpose forests. These approaches and the ongoing evolution of forest management to promote resilient forests are incorporated in a renewed emphasis on ecologically based forest stewardship (Larocque 2016; Franklin et al. 2018; see Chapter 12).

Box 9.5 Key points

- Large areas of the world have suffered from deforestation and forest degradation (biotic simplification and compromised ecosystem service delivery), and would benefit from ecological restoration efforts.
- The first and most important step is to halt forest loss and reduce society's impacts and processes of degradation, after which a wide range of techniques can be employed to regenerate forests, enrich them with native species, promote complexity, and restore ecological processes.
- Forest regeneration efforts can draw upon a long history of silvicultural research and practical experience; practices often need to be customized to local species and site properties while working within the constraints of local resources and community values.
- Many forests having a history of exploitation or timber management can support more biodiversity if they are modified to match the composition and structure of local primary forests more closely; this work often entails the reintroduction of deadwood (both standing and fallen), canopy gaps, and late-successional species.
- Interventions to repair and naturalize forests provide the opportunity to make landscapes better adapted to the future, informed by but not mimicking historical forest templates.
- Ongoing monitoring and assessment are needed to assure that stand development is on track to meet objectives; maintenance and repeated treatments are usually needed in any program of ecological restoration.

The job, for the rest of this century, is repairing the damage we did over the past two centuries of industrialization . . .

Gwynne Dyer (2008)

References cited

Allender, D. B. (2014). *The Wounded Heart*. Colorado Springs, CO: NavPress. 304 pp.

Alexander, S., Nelson, C. R., Aronson, J., et al. (2011). "Opportunities and Challenges for Ecological Restoration within REDD+," *Restoration Ecology*, 19/6: 683–9.

Antofie, M. M., Pop, O. G., and Gruia, R. (2010). "Resilience of Ecosystems: A Target for Assessing the Implementation of the Rio Conventions Synergy," *Environmental Engineering & Management Journal*, 9/12: 1629–35.

Apostol, D., and Berg, D. R. (2006). "Riparian Woodlands," in D. Apostol and M. Sinclair (eds), *Restoring the Pacific Northwest: The Art and Science of Ecological Restoration in Cascadia*. Washington: Island Press, 122–49.

Apostol, D., and Sinclair, M. (eds). (2006). *Restoring the Pacific Northwest: The Art and Science of Ecological Restoration in Cascadia*. Washington: Island Press. 475 pp.

Allison, S. K., and Murphy, S. D. (eds). (2017). *Routledge Handbook of Ecological and Environmental Restoration*. London: Taylor & Francis. 620 pp.

Baker, S. (2013). "Management of Tasmanian Eucalypt Forests as Complex Adaptive Systems," in C. Messier, K. J. Puettmann, and K. D. Coates, (eds), *Managing Forests as Complex Adaptive Systems: Building Resilience to the Challenge of Global Change*. New York: Routledge, 269–98.

Balandier, P., Collet, C., Miller, J. H., Reynolds, P. E., and Zedaker, S. M. (2006). "Designing Forest Vegetation Management Strategies Based on the Mechanisms and Dynamics of Crop Tree Competition by Neighbouring Vegetation," *Forestry*, 79/1: 3–27.

Barbier, E. B., Burgess, J. C., and Grainger, A. (2010). "The Forest Transition: Towards A More Comprehensive Framework," *Land Use Policy*, 27: 98–107.

Bastin, J. F., Finegold, Y., Garcia, C., et al. (2019). "The Global Tree Restoration Potential," *Science*, 365/6448: 76–9.

Bauhus, J., Puettmann, K., and Messier, C. (2009). "Silviculture for Old-Growth Attributes," *Forest Ecology and Management*, 258/4: 525–37.

Baumhauer, H., Madsen, P., and Stanturf, J. (2005). "Regeneration by Direct Seeding: A Way to Reduce Costs of Conversion," *Restoration of Boreal and Temperate Forests*. Boca Raton, FL: CRC Press, 349–54.

Beese, W. J., Deal, J., Dunsworth, B. G., Mitchell, S. J., and Philpott, T. J. (2019). "Two Decades of Variable Retention in British Columbia: A Review of its Implementation and Effectiveness for Biodiversity Conservation," *Ecological Processes*, 8: 33.

Bentrup, G., Dosskey, M., Wells, G., and Schoeneberger, M. (2012). "Connecting Landscape Fragments through Riparian Zones," in A. Stanturf, D. Lamb, and P. Madsen (eds), *Forest Landscape Restoration: Integrating Natural and Social Sciences*. New York: Springer, 93–109.

Bentz, B. J., Régnière, J., Fettig, C. J., et al. (2010). "Climate Change and Bark Beetles of the Western United States and Canada: Direct and Indirect Effects," *BioScience*, 60/8: 602–13.

BGCI (2023). *The Global Biodiversity Standard*. London: Botanic Gardens Conservation International. Available online at https://www.biodiversitystandard.org (accessed February 15, 2024).

Bohlen, P. J., Groffman, P. M., Fahey, T. J., et al. (2004). "Ecosystem Consequences of Exotic Earthworm Invasion of North Temperate Forests," *Ecosystems*, 7: 1–12.

Boyer, S., Kim, Y. N., Bowie, M. H., Lefort, M. C., and Dickinson, N. M. (2016). "Response of Endemic and Exotic Earthworm Communities to Ecological Restoration," *Restoration Ecology*, 24/6: 717–21.

Bratton, S. P. (1994). "Logging and Fragmentation of Broadleaved Deciduous Forests: Are We Asking the Right Ecological Questions?" *Conservation Biology*, 8/1: 295–7.

Brockerhoff, E. G., Jactel, H., Parrotta, J. A., Quine, C. P., and Sayer, J. (2008). "Plantation Forests and Biodiversity: Oxymoron or Opportunity?" *Biodiversity and Conservation*, 17/5: 925–51.

Brodie, L. C., and Harrington, C. A. (2020). *Guide to Variable-Density Thinning Using Skips and Gaps*. General Technical Report PNW-GTR-989. Portland, OR: USDA Forest Service. 37 pp.

Budiharta, S., Meijaard, E., Erskine, P. D., Rondinini, C., Pacifici, M., and Wilson, K. A. (2014). "Restoring Degraded Tropical Forests for Carbon and Biodiversity," *Environmental Research Letters*, 9/11: 114020.

Burton, P. J. (1993). "Some Limitations Inherent to Static Indices of Plant Competition," *Canadian Journal of Forest Research*, 23/10: 2141–52.

Burton, P. J. (1996). "When Is Vegetation Control Necessary?" in P. G. Comeau, G. L. Harper, Blache, M. E., Boateng, J.O., and Gilkeson, L.A. (eds), *Integrated Forest Vegetation Management: Options and Applications*. FRDA Report 251. Victoria, BC: Canadian Forest Service and British Columbia Ministry of Forests, 11–16. Available online at https://www.for.gov.bc.ca/hfd/pubs/docs/frr/Frr251-1.pdf (accessed February 15, 2024).

Burton, P. J. (1999). "An Assessment of Silvicultural Practices and Forest Policy in British Columbia from the Perspective of Restoration Ecology," in B. Egan. (ed.), *Helping the Land Heal: Ecological Restoration in British Columbia Conference Proceedings*. Vancouver: British Columbia Environmental Network Educational Foundation, 173–8.

Burton, P. J. (2005). "Ecosystem Management and Conservation Biology," in S. B. Watts and L. Tolland (eds), *Forestry Handbook for British Columbia*. 5th edn. Vancouver: Faculty of Forestry, University of British Columbia, 307–22.

Burton, P. J. (2006). "Restoration of Forests Attacked by Mountain Pine Beetle: Misnomer, Misdirected, or Must-Do?" *Journal of Ecosystems and Management*, 7/2: 1–10.

Burton, P. J. (2014). "Considerations for Monitoring and Evaluating Forest Restoration," *Journal of Sustainable Forestry*, 33/S1: S149–S160.

Burton, P. J., and Bazzaz, F. A. (1991). "Tree Seedling Emergence on Interactive Temperature and Moisture Gradients and in Patches of Old-Field Vegetation," *American Journal of Botany*, 78/1: 131–49.

Burton, P. J., and Bazzaz, F. A. (1995). "Ecophysiological Responses of Tree Seedlings Invading Different Patches of Old-Field Vegetation," *Journal of Ecology*, 83/1: 99–112.

Burton, P. J., and Macdonald, S. E. (2011). "The Restorative Imperative: Assessing Objectives, Approaches and Challenges to Restoring Naturalness in Forests," *Silva Fennica*, 45/5: 843–63.

Campbell, J. L., Rustad, L. E., Boyer, E. W., et al. (2009). "Consequences of Climate Change for Biogeochemical Cycling in Forests of Northeastern North America," *Canadian Journal of Forest Research*, 39/2: 264–84.

Casazza, M. L., Overton, C. T., Bui, T. V. D., et al. (2016). "Endangered Species Management and Ecosystem Restoration: Finding the Common Ground," *Ecology and Society*, 21/1: 19.

Chazdon, R. L. (2008). "Beyond Deforestation: Restoring Forests and Ecosystem Services on Degraded Lands," *Science*, 320: 1458–60.

Cifuentes-Croquevielle, C., Stanton, D. E., and Armesto, J. J. (2020). "Soil Invertebrate Diversity Loss and Functional Changes in Temperate Forest Soils Replaced by Exotic Pine Plantations," *Scientific Reports*, 10/1: 7762.

Cleary, B. D., Greaves, R. D., and Hermann, R. K. (eds). (1978). *Regenerating Oregon's Forests: A Guide for the Regeneration Forester*. Corvallis, OR: Oregon State University Extension Service. 286 pp.

Clout, M. N., and Russell, J. C. (2008). "The Invasion Ecology of Mammals: A Global Perspective," *Wildlife Research*, 35/3: 180–4.

Clout, M. N., and Williams, P. A. (eds). (2009). *Invasive Species Management: A Handbook of Principles and Techniques*. Oxford: Oxford University Press. 320 pp.

Coates, K. D., and Burton, P. J. (1997). "A Gap-Based Approach for Development of Silvicultural Systems to Address Ecosystem Management Objectives," *Forest Ecology and Management*, 99/3: 337–54.

Crous, J. W. (2017). "Use of Hydrogels in the Planting of Industrial Wood Plantations," *Southern Forests: A Journal of Forest Science*, 79/3: 197–213.

Crouzeilles, R., Curran, M., Ferreira, M. S., Lindenmayer, D. B., Grelle, C. E., and Rey Benayas, J. M. (2016). "A Global Meta-Analysis on the Ecological Drivers of Forest Restoration Success," *Nature Communications*, 7/1: 11666.

Dejene, T., Oria-de-rueda, J. A., and Martín-Pinto, P. (2017). "Fungal Diversity and Succession Following Stand Development in *Pinus patula* Schiede ex Schltdl. & Cham. Plantations in Ethiopia," *Forest Ecology and Management*, 395: 9–18.

Demestihas, C., Plénet, D., Génard, M., Raynal, C., and Lescourret, F. (2017). "Ecosystem Services in Orchards: A Review," *Agronomy for Sustainable Development*, 37/2: 12.

Dinerstein, E., Olson, D., Joshi, A., et al. (2017). "An Ecoregion-Based Approach to Protecting Half the Terrestrial Realm," *BioScience*, 67/6: 534–45.

Di Sacco, A., Hardwick, K. A., Blakesley, D., et al. (2021). "Ten Golden Rules for Reforestation to Optimize Carbon Sequestration, Biodiversity Recovery and Livelihood Benefits," *Global Change Biology*, 27/7: 1328–48.

Dotson, T., and Pereira, H. M. (2022). "From Antagonistic Conservation to Biodiversity Democracy in Rewilding," *One Earth*, 5/5: 466–9.

Dredge, D. (2022). "Regenerative Tourism: Transforming Mindsets, Systems and Practices," *Journal of Tourism Futures*, 8/3: 269–81.

Duffy, D. C., and Meier, A. J. (1992). "Do Appalachian Herbaceous Understories Ever Recover from Clearcutting?" *Conservation Biology*, 6: 196–201.

Duke, J. A. (1997). "Phytomedicinal Forest Harvest in the United States," in G. Bodeken, K. K. S. Bhat, J. Burley, and P. Vantomme (eds), *Medicinal Plants for Forest Conservation and Health Care*. Non-Wood Forest Products 11. Rome: FAO, 147–58.

Dyer, G. (2008). *Climate Wars*. Toronto: Random House. 264 pp.

Egan, D. and Howell, E. A. (eds). (2001). *The Historical Ecology Handbook: A Restorationist's Guide to Reference Ecosystems*. Washington: Island Press. 457 pp.

Elliot, S. D., Blakesley, D., and Hardwick, K. (2013). *Restoring Tropical Forests: A Practical Guide*. London: Kew Publishing, Royal Botanic Gardens. 361 pp.

Ellison, A. M., Bank, M. S., Clinton, B. D., et al. (2005). "Loss of Foundation Species: Consequences for the Structure and Dynamics of Forested Ecosystems," *Frontiers in Ecology and the Environment*, 3/9: 479–86.

Falk, D. A., Millar, C. I., and Olwell, M. (eds). (1996). *Restoring Diversity: Strategies for Reintroduction of Endangered Plants*. Washington: Island Press. 528 pp.

FAO (2020). *Global Forest Resources Assessment 2020: Main Report*. Rome: Food and Agriculture Organization of the United Nations. 164 pp. Available online at https://doi.org/10.4060/ca9825en (accessed February 15, 2024).

Ferris, R., Peace, A. J., and Newton, A. C. (2000). "Macrofungal Communities of Lowland Scots Pine (*Pinus sylvestris* L.) and Norway Spruce (*Picea abies* (L.) Karsten.) Plantations in England: Relationships with Site Factors and Stand Structure," *Forest Ecology and Management*, 131/1–3: 255–67.

Feyera, S., Beck, E., and Lüttge, U. (2002). "Exotic Trees as Nurse-Trees for the Regeneration of Natural Tropical Forests," *Trees*, 16/4: 245–9.

Fischer, H., Huth, F., Hagermann, U., and Wagner, S. (2016). "Developing Restoration Strategies for Temperate Forests Using Natural Regeneration Processes," in J. A. Stanturf (ed.), *Restoration of Boreal and Temperate Forests*. 2nd edn. Boca Raton, FL: CRC Press, 103–64.

Foin, T. C., Pawley, A. L., Ayres, D. R., Carlsen, T. M., Hodum, P. J., and Switzer, P. V. (1998). "Improving Recovery Planning for Threatened and Endangered Species," *BioScience*, 48/3: 177–84.

Franklin, J. F., Berg, D. R., Carey, A. B., and Hardt, R. A. (2006). "Old-Growth Conifer Forests," in D. Apostol and M. Sinclair (eds), *Restoring the Pacific Northwest: The Art and Science of Ecological Restoration in Cascadia*. Washington: Island Press, 97–121.

Franklin, J. F., Johnson, K. N., and Johnson, D. L. (2018) (eds). *Ecological Forest Management*. Long Grove, IL: Waveland Press. 646 pp.

Gann, G. D., McDonald, T., Walder, B., et al. (2019). "International Principles and Standards for the Practice of Ecological Restoration," *Restoration Ecology*, 27: S1–S46.

Gardner, T. A., Burgess, N. D., Aguilar-Amuchastegui, N., et al. (2012). "A Framework for Integrating Biodiversity Concerns into National REDD+ Programmes," *Biological Conservation*, 154: 61–71.

Gower, T. L., Burton, P. J., and Fenger, M. (2016). "Integrating Forest Restoration into Mainstream Land Management in British Columbia, Canada," in J. A. Stanturf (ed.), *Restoration of Boreal and Temperate Forests*. 2nd edn. Boca Raton, FL: CRC Press, 271–97.

Grantham, H. S., Duncan, A., Evans, T. D., et al. (2020). "Anthropogenic Modification of Forests Means Only 40% of Remaining Forests Have High Ecosystem Integrity," *Nature Communications*, 11: 5978.

Gundersen, V. S., and Frivold, L. H. (2008). "Public Preferences for Forest Structures: A Review of Quantitative Surveys from Finland, Norway and Sweden," *Urban Forestry & Urban Greening*, 7/4: 241–58.

Gustafsson, L., Baker, S. C., Bauhus, J., et al. (2012). "Retention Forestry to Maintain Multifunctional Forests: A World Perspective," *BioScience*, 62/7: 633–45.

Haeussler, S., Kabzems, R., McClarnon, J., and Bedford, L. (2021). "Successional Change, Restoration Success, and Resilience in Boreal Mixedwood Vegetation Communities over Three Decades," *Canadian Journal of Forest Research*, 51/6: 766–80.

Hansen, J., and Spiecker, H. (2016). "Conversion of Norway Spruce (*Picea abies* [L.] Karst.) Forests in Europe," in J. A. Stanturf (ed.), *Restoration of Boreal and Temperate Forests*. 2nd edn. Boca Raton, FL: CRC Press, 355–64.

Harris, J. A., Hobbs, R. J., Higgs, E., and Aronson, J. (2006). "Ecological Restoration and Global Climate Change," *Restoration Ecology*, 14/2: 170–6.

Harper, K. A., Macdonald, S. E., Burton, P. J., et al. (2005). "Edge Influence on Forest Structure and Composition in Fragmented Landscapes," *Conservation Biology*, 19/3: 768–82.

Harrington, C. A. (1999). "Forests Planted for Ecosystem Restoration or Conservation," *New Forests*, 17/1: 175–90.

Healey, S. P., and Gara, R. I. (2003). "The Effect of a Teak (*Tectona grandis*) Plantation on the Establishment of Native Species in an Abandoned Pasture in Costa Rica," *Forest Ecology and Management*, 176/1–3: 497–507.

Higgs, E., Falk, D. A., Guerrini, A., et al. (2014). "The Changing Role of History in Restoration Ecology," *Frontiers in Ecology and the Environment*, 12/9: 499–506.

Hilmers, T., Biber, P., Knoke, T., and Pretzsch, H. (2020). "Assessing Transformation Scenarios from Pure Norway Spruce to Mixed Uneven-Aged Forests in Mountain Areas," *European Journal of Forest Research*, 139/4: 567–84.

Holl, K. D. (2017). "Restoring Tropical Forests from the Bottom up," *Science*, 355/6324: 455–6.

Holl, K. D., and Aide, T. M. (2011). "When and Where to Actively Restore Ecosystems?" *Forest Ecology and Management*, 261/10: 1558–63.

Holl, K. D., Reid, J. L., Cole, R. J., Oviedo-Brenes, F., Rosales, J. A., and Zahawi, R. A. (2020). "Applied Nucleation Facilitates Tropical Forest Recovery: Lessons Learned from a 15-Year Study," *Journal of Applied Ecology*, 57/12; 2316–28.

Hooper, R. G., and McAdie, C. J. (1996). "Hurricanes and the Long-Term Management of the Red-Cockaded Woodpecker," in J. L. Haymond, D. D. Hook, and W. R. Harms (eds), *Hurricane Hugo: South Carolina Forest Land Research and Management Related to the Storm*. General Technical Report SRS-GTR-5. Asheville, NC: USDA Forest Service, 417–36.

Huth, F., Wehnert, A., Tiebel, K., and Wagner, S. (2017). "Direct Seeding of Silver Fir (*Abies alba* Mill.) to Convert Norway Spruce (*Picea abies* L.) Forests in Europe: A Review," *Forest Ecology and Management*, 403: 61–78.

Irwin, S., Pedley, S. M., Coote, L., et al. (2014). "The Value of Plantation Forests for Plant, Invertebrate and Bird Diversity and the Potential for Cross-Taxon Surrogacy," *Biodiversity and Conservation*, 23/3: 697–714.

Jackson, S. T., and Hobbs, R. J. (2009). "Ecological Restoration in the Light of Ecological History," *Science*, 325/5940: 567–9.

Jactel, H., Bauhus, J., Boberg, J., et al. (2017). "Tree Diversity Drives Forest Stand Resistance to Natural Disturbances," *Current Forestry Reports*, 3/3: 223–43.

Johann, E., Agnoletti, M., Axelsson, A. L., et al. (2004). "History of Secondary Norway Spruce Forests in Europe," in H. Spiecker, J. Hansen, E. Klimo, J. P. Skovsgaard, H. Sterba, and K. von Teuffel (eds), *Norway Spruce Conversion: Options and Consequences*. European Forest Institute Research Report 18. Leiden: The Netherlands: Brill, 25–62.

Katila, P., Pierce Colfer, C. J., de Jong, W., Galloway, G., Pacheco, P., and Winkel, G. (eds). (2020). *Sustainable Development Goals: Their Impacts on Forests and People*. New York: Cambridge University Press. 617 pp.

Keenan, R., Lamb, D., Woldring, O., Irvine, T., and Jensen, R. (1997). "Restoration of Plant Biodiversity beneath Tropical Tree Plantations in Northern Australia," *Forest Ecology and Management*, 99/1-2: 117–31.

Kijazi, M. H., and Kant, S. (2010). "Forest Stakeholders' Value Preferences in Mount Kilimanjaro, Tanzania," *Forest Policy and Economics*, 12/5: 357–69.

Klemens, J. A. (2020). "Insights from Five Decades of Biodevelopment and Long-Term Research at Area de Conservación Guanacaste, Costa Rica," *Biotropica*, 52/6: 1014–16.

Knoke, T., Ammer, C., Stimm, B., and Mosandl, R. (2008). "Admixing Broadleaved to Coniferous Tree Species: A Review on Yield, Ecological Stability and Economics," *European Journal of Forest Research*, 127/2: 89–101.

Kremer, K. N., and Bauhus, J. (2020). "Drivers of Native Species Regeneration in the Process of Restoring Natural Forests from Mono-Specific, Even-Aged Tree Plantations: A Quantitative Review," *Restoration Ecology*, 28/5: 1074–86.

Krosby, M., Tewksbury, J., Haddad, N. M., and Hoekstra, J. (2010). "Ecological Connectivity for a Changing Climate," *Conservation Biology*, 24/6: 1686–9.

Kuuluvainen, T., Angelstam, P., Frelich, L. E., et al. (2021). "Natural Disturbance-Based Forest Management: Moving beyond Retention and Continuous-Cover Forestry," *Frontiers in Forests and Global Change*, 4: 629020.

Lamb, D. (2015). "Restoration of Forest Ecosystems," in K.S.-H. Peh, R. T. Corlett, and Y. Bergeron (eds), *Routledge Handbook of Forest Ecology*. London: Taylor & Francis, 397–410.

Lamb, D., Stanturf, J., and Madsen, P. (2012). "What Is Forest Landscape Restoration?" in A. Stanturf, D. Lamb, and P. Madsen (eds), *Forest Landscape Restoration: Integrating Natural and Social Sciences*. New York: Springer, 3–23.

Latawiec, A. E., Strassburg, B. B., Brancalion, P. H., Rodrigues, R. R., and Gardner, T. (2015). "Creating Space for Large-Scale Restoration in Tropical Agricultural Landscapes," *Frontiers in Ecology and the Environment*, 13/4: 211–18.

Larocque, G. R. (ed.). (2016). *Ecological Forest Management Handbook*. Boca Raton, FL: CRC Press. 604 pp.

Larsen, J. B., Angelstam, P., Bauhus, J., et al. (2022). *Closer-to-Nature Forest Management*. From Science to Policy 12. Joensuu, Finland: European Forest Institute. 53 pp. Available online at https://doi.org/10.36333/fs12 (accessed February 15, 2024).

Lavender, D. P., Parish, R., Johnson, C. M., et al. (eds). (1990). *Regenerating British Columbia's Forests*. Vancouver: University of British Columbia Press. 372 pp.

Law, A., Gaywood, M. J., Jones, K. C., Ramsay, P., and Willby, N.J. (2017). "Using Ecosystem Engineers as Tools in Habitat Restoration and Rewilding: Beaver and Wetlands," *Science of the Total Environment*, 605: 1021–30.

Leopold, A. (1949). *A Sand County Almanac, and Sketches Here and There*. New York: Oxford University Press. 226 pp.

Lilja-Rothsten, S., de Chantal, M., Peterson, C., Kuuluvainen, T., Vanha-Majamaa, I., and Puttonen, P. (2008). "Microsites before and after Restoration in Managed *Picea abies* Stands in Southern Finland: Effects of Fire and Partial Cutting with Dead Wood Creation," *Silva Fennica*, 42/2: 250.

Lindenmayer, D. B., and Hobbs, R. J. (2004). "Fauna Conservation in Australian Plantation Forests: A Review," *Biological Conservation*, 119/2: 151–68.

Löf, M., Ersson, B. T., Hjältén, J., Nordfjell, T., Oliet, J. A., and Willoughby, I. (2016). "Site Preparation Techniques for Forest Restoration," in J. A. Stanturf. (ed.), *Restoration of Boreal and Temperate Forests*. 2nd edn. Boca Raton, FL: CRC Press, 85–102.

Löf, M., Karlsson, M., Sonesson, K., Welander, T. N., and Collet, C. (2007). "Growth and Mortality in Underplanted Tree Seedlings in Response to Variations in Canopy Closure of Norway Spruce Stands," *Forestry*, 80/4: 371–83.

Lugo, A. E., Parrotta, J. A., and Brown, S. (1993). "Loss in Species Caused by Tropical Deforestation and their Recovery through Management," *Ambio*, 22/2–3: 106–9.

Luken, J. O. (1990). *Directing Ecological Succession*. London: Chapman and Hall. 251 pp.

Maginnis, S., and Jackson, W. (2007). "What is FLR and how Does it Differ from Current Approaches?" in J. Rietbergen-McCracken, S. Maginnis, and A. Sarre (eds), *The Forest Restoration Handbook*. London: Earthscan, 2–20.

Mansourian, S., Dudley, N., and Vallauri, D. (2017). "Forest Landscape Restoration: Progress in the Last

Decade and Remaining Challenges," *Ecological Restoration*, 35/4: 281–8.

Mansourian, S., and Vallauri, D. (eds). (2005). *Forest Restoration in Landscapes: Beyond Planting Trees*. New York: Springer. 438 pp.

Mansuy, N., Burton, P. J., Stanturf, J., et al. (2020). "Scaling up Forest Landscape Restoration in Canada in an Era of Cumulative Effects and Climate Change," *Forest Policy and Economics*, 116: 102177.

Matthews, E. (1983). "Global Vegetation and Land Use: New High-Resolution Data Bases for Climate Studies," *Journal of Climate and Applied Meteorology*, 22: 474–87.

Mohan, M., Richardson, G., Gopan, G., et al. (2021). "UAV-Supported Forest Regeneration: Current Trends, Challenges and Implications," *Remote Sensing*, 13: 2596.

Nagel, L. M., Palik, B. J., Battaglia, M. A., et al. (2017). "Adaptive Silviculture for Climate Change: A National Experiment in Manager–Scientist Partnerships to Apply an Adaptation Framework," *Journal of Forestry*, 115/3: 167–78.

Newton, P., Civita, N., Frankel-Goldwater, L., Bartel, K., and Johns, C. (2020). "What Is Regenerative Agriculture? A Review of Scholar and Practitioner Definitions Based on Processes and Outcomes," *Frontiers in Sustainable Food Systems*, 4: 194.

Norisada, M., Hitsuma, G., Kuroda, K., et al. (2005). "*Acacia mangium*, a Nurse Tree Candidate for Reforestation on Degraded Sandy Soils in the Malay Peninsula," *Forest Science*, 51/5: 498–510.

Noulèkoun, F., Mensah, S., Birhane, E., Son, Y., and Khamzina, A. (2021). "Forest Landscape Restoration under Global Environmental Change: Challenges and a Future Roadmap," *Forests*, 12/3: 276.

O'Hara, K. L., Nesmith, J. C., Leonard, L., and Porter, D. J. (2010). "Restoration of Old Forest Features in Coast Redwood Forests Using Early-Stage Variable-Density Thinning," *Restoration Ecology*, 18: 125–35.

Oldfield, S., Lusty, C., and MacKinven, A. (eds). (1998). *The World List of Threatened Trees*. Cambridge: World Conservation Monitoring Centre; Gland, Switzerland: International Union for the Conservation of Nature IUCN. 650 pp.

Oliveira, N., Sixto, H., Cañellas, I., Rodríguez-Soalleiro, R., and Pérez-Cruzado, C. (2015). "Productivity Model and Reference Diagram for Short Rotation Biomass Crops of Poplar Grown in Mediterranean Environments," *Biomass and Bioenergy*, 72: 309–20.

Parrotta, J. A. (1992). "The Role of Plantation Forests in Rehabilitating Degraded Tropical Ecosystems," *Agriculture, Ecosystems & Environment*, 41/2: 115–33.

Parrotta, J. A., Knowles, O. H., and Wunderle, J. M. (1997). "Development of Floristic Diversity in 10-Year-Old Restoration Forests on a Bauxite Mined Site in Amazonia," *Forest Ecology and Management*, 99/1–2: 21–42.

Pedlar, J. H., McKenney, D. W., Aubin, I., et al. (2012). "Placing Forestry in the Assisted Migration Debate," *BioScience*, 62/9: 835–42.

Perring, M. P., Erickson, T. E., and Brancalion, P. H. (2018). "Rocketing Restoration: Enabling the Upscaling of Ecological Restoration in the Anthropocene," *Restoration Ecology*, 26/6: 1017–23.

Pinto, S. R., Melo, F., Tabarelli, M., et al. (2014). "Governing and Delivering a Biome-Wide Restoration Initiative: The Case of Atlantic Forest Restoration Pact in Brazil," *Forests*, 5/9: 2212–29.

Pommerening, A., and Murphy, S. T. (2004). "A Review of the History, Definitions and Methods of Continuous Cover Forestry with Special Attention to Afforestation and Restocking," *Forestry*, 77/1: 27–44.

Prach, K., and Hobbs, R. J. (2008). "Spontaneous Succession versus Technical Reclamation in the Restoration of Disturbed Sites," *Restoration Ecology*, 16: 363–6.

Prach, K., and Walker, L. R. (2020). *Comparative Plant Succession among Terrestrial Biomes of the World*. Cambridge: Cambridge University Press. 399 pp.

Pretzsch, H. (2020). "Transitioning Monocultures to Complex Forest Stands in Central Europe: Principles and Practice," in J. A. Stanturf (ed.), *Achieving Sustainable Management of Boreal and Temperate Forests*. Cambridge: Burleigh Dodds, 355–96.

Pretzsch, H. (2022). "Mixing Degree, Stand Density, and Water Supply Can Increase the Overyielding of Mixed versus Monospecific Stands in Central Europe," *Forest Ecology and Management*, 503: 119741.

Pretzsch, H., Grams, T., Häberle, K. H., Pritsch, K., Bauerle, T., and Rötzer, T. (2020). "Growth and Mortality of Norway Spruce and European Beech in Monospecific and Mixed-Species Stands under Natural Episodic and Experimentally Extended Drought: Results of the KROOF Throughfall Exclusion Experiment," *Trees*, 34: 957–70.

Prober, S. M., Byrne, M., Mclean, E. H., Et Al. (2015). "Climate-Adjusted Provenancing: A Strategy for Climate-Resilient Ecological Restoration," *Frontiers in Ecology and Evolution*, 3: 65.

Puettmann, K. J., Coates, K. D., and Messier, C. (2009). *A Critique of Silviculture: Managing for Complexity*. Washington: Island Press. 189 pp.

Putz, F. E., and Redford, K. H. (2010). "The Importance of Defining 'Forest': Tropical Forest Degradation, Deforestation, Long-Term Phase Shifts, and Further Transitions," *Biotropica*, 42/1: 10–20.

Rawinski, T. J. (2008). *Impacts of White-Tailed Deer Overabundance in Forest Ecosystems: An Overview*. Newton Square, PA: USDA Forest Service. 8 pp.

Available online at https://longpointbiosphere.com/download/mammals/Impacts-of-white-tailed-deer-overabundance-in-forest-Rawinski-2008.pdf (accessed February 15, 2024).

Reventlow, D. O. J., Nord-Larsen, T., Biber, P., Hilmers, T., and Pretzsch, H. (2021). "Simulating Conversion of Even-Aged Norway Spruce into Uneven-Aged Mixed Forest: Effects of Different Scenarios on Production, Economy and Heterogeneity," *European Journal of Forest Research*, 140/4: 1005–27.

Ripple, W. J., and Beschta, R. L. (2004). "Wolves and the Ecology of Fear: Can Predation Risk Structure Ecosystems?" *BioScience*, 54/8: 755–66.

Rooney, T. P., Buttenshon, R., Madsen, P., Olesen, C. R., Royo, A. A., and Stout, S. L. (2016). "Integrating Ungulate Herbivory into Forest Landscape Restoration," in J. A. Stanturf (ed.), *Restoration of Boreal and Temperate Forests*. 2nd edn. Boca Raton, FL: CRC Press, 69–83.

Royo, A. A., Stout, S. L., deCalesta, D. S., and Pierson, T. G. (2010). "Restoring Forest Herb Communities through Landscape-Level Deer Herd Reductions: Is Recovery Limited by Legacy Effects?" *Biological Conservation*, 143/11: 2425–34.

Rueegger, N. (2017). "Artificial Tree Hollow Creation for Cavity-Using Wildlife: Trialling an Alternative Method to that of Nest Boxes," *Forest Ecology and Management*, 405: 404–12.

Saint-Laurent, C. (2005). "Optimizing Synergies on Forest Landscape Restoration between the Rio Conventions and the UN Forum on Forests to Deliver Good Value for Implementers," *Review of European Community and International Law*, 14: 39–49.

Schmiedinger, A., Kreyling, J., Steinbauer, M. J., Macdonald, S. E., Jentsch, A., and Beierkuhnlein, C. (2012). "A Continental Comparison Indicates Long-Term Effects of Forest Management on Understory Diversity in Coniferous Forests," *Canadian Journal of Forest Research*, 42/7: 1239–52.

SERA (2021). *National Standards for the Practice of Ecological Restoration in Australia, Edition 2.2.* Society for Ecological Restoration Australasia. 49 pp. Available online at www.seraustralasia.org (accessed February 15, 2024).

Silvério, D. V., Brando, P. M., Balch, J. K., et al. (2013). "Testing the Amazon Savannization Hypothesis: Fire Effects on Invasion of a Neotropical Forest by Native Cerrado and Exotic Pasture Grasses," *Philosophical Transactions of the Royal Society B: Biological Sciences*, 368/1619: 20120427.

Simberloff, D., Parker, I. M., and Windle, P. N. (2005). "Introduced Species Policy, Management, and Future Research Needs," *Frontiers in Ecology and the Environment*, 3/1: 12–20.

Spiecker, H., Hansen, J., Klimo, E., Skovsgaard, J. P., Sterba, H., and von Teuffel, K. (2004a). "Summarizing Discussion," in H. Spiecker, J. Hansen, E. Klimo, J. P. Skovsgaard, H. Sterba, and K. von Teuffel (eds), *Norway Spruce Conversion—Options and Consequences*. European Forest Institute Research Report 18. Leiden, Netherlands: Brill, 253–60.

Spiecker, H., Hansen, J., Klimo, E., Skovsgaard, J. P., Sterba, H., and von Teuffel, K. (2004b). "Conclusions," in H. Spiecker, J. Hansen, E. Klimo, J. P. Skovsgaard, H. Sterba, and K. von Teuffel (eds). *Norway Spruce Conversion: Options and Consequences*. European Forest Institute Research Report 18. Leiden, Netherlands: Brill, 261–4.

Stanturf, J. A. (2016). "What Is Forest Restoration?" in J. A. Stanturf (ed.), *Restoration of Boreal and Temperate Forests*. 2nd edn. Boca Raton, FL: CRC Press, 1–16.

Stanturf, J. A., Harvey, W. W. J., Petrokofsky, G., et al. (2023). *Forest Related Nature-Based Solutions: Review of Terms and Concepts—From Afforestation to Forest Landscape Restoration*. IUFRO Occasional Paper, No. 36. Vienna, Austria: International Union of Forest Research Organizations. 54 pp. Available online at https://www.iufro.org/uploads/media/op36.pdf (accessed February 14, 2024).

Stanturf, J., Lamb, D., and Madsen, P. (eds). (2012). *Forest Landscape Restoration: Integrating Natural and Social Sciences*. Dordrecht: Springer. 330 pp.

Stanturf, J. A., and Mansourian, S. (2020). "Forest Landscape Restoration: State of Play," *Royal Society Open Science*, 7: 201218.

Stanturf, J. A., Palik, B. J., and Dumroese, R. K. (2014a). "Contemporary Forest Restoration: A Review Emphasizing Function," *Forest Ecology and Management*, 331: 292–323.

Stanturf, J. A., Palik, B. J., Williams, M. I., Dumroese, R. K., and Madsen, P. (2014b) "Forest Restoration Paradigms," *Journal of Sustainable Forestry*, 33: S161–S194.

Stoeckeler, J. H., and Jones, G. W. (1957). *Forest Nursery Practice in the Lake States*. Agriculture Handbook 110. Washington: USDA Forest Service. 124 pp.

Sullivan, T. P., and Sullivan, D. S. (2003). "Vegetation Management and Ecosystem Disturbance: Impact of Glyphosate Herbicide on Plant and Animal Diversity in Terrestrial Systems," *Environmental Reviews*, 11/1: 37–59.

Tafoya, K. A., Brondizio, E. S., Johnson, C. E., et al. (2020). "Effectiveness of Costa Rica's Conservation Portfolio to Lower Deforestation, Protect Primates, and Increase Community Participation," *Frontiers in Environmental Science*, 8: 580724.

Tedersoo, L. (2015). "Mycorrhizal Symbiosis in Forest Ecosystems," in K. S.-H. Peh, R. T. Corlett, and Y.

Bergeron (eds), *Routledge Handbook of Forest Ecology*. London: Taylor & Francis, 309–24.

Trevenen, E., Standish, R., Price, C., and Hobbs, R. (2017). "Restoration and Resilience," in S. K. Allison and S. D. Murchy (eds), *Routledge Handbook of Ecological and Environmental Restoration*. London: Taylor & Francis, 509–21.

Tucker, N. I., Elliott, S., Holl, K. D., and Zahawi, R. A. (2023). "Restoring Tropical Forests: Lessons Learned from Case Studies on Three Continents," in S. Florentine, P. Gibson-Roy, K. W. Dixon, and L. Broadhurst (eds), *Ecological Restoration: Moving Forward Using Lessons Learned*. Cham, Switzerland: Springer Nature, 63–101.

UNEP (2021). *Principles for Ecosystem Restoration to Guide the United Nations Decade 2021–2030*. Nairobi: United Nations Environment Programme. 20 pp. Available online at https://www.decadeonrestoration.org/publications/principles-ecosystem-restoration-guide-united-nations-decade-2021-2030 (accessed February 14, 2024).

Veldman, J. W., and Putz, F. E. (2011). "Grass-Dominated Vegetation, not Species-Diverse Natural Savanna, Replaces Degraded Tropical Forests on the Southern Edge of the Amazon Basin," *Biological Conservation*, 144: 1419–29.

von Lüpke, B., Ammer, C., Bruciamacchie, M., et al. (2004). "Silvicultural Strategies for Conversion," in

H. Spiecker, J. Hansen, E. Klimo, J. P. Skovsgaard, H. Sterba, and K. von Teuffel (eds), *Norway Spruce Conversion: Options and Consequences*. European Forest Institute Research Report 18. Leiden, The Netherlands: Brill, 121–64.

Wagner, R. G., Petersen, T. D., Ross, D. W., and Radosevich, S. R. (1989). "Competition Thresholds for the Survival and Growth of Ponderosa Pine Seedlings Associated with Woody and Herbaceous Vegetation," *New Forests*, 3/2: 151–70.

Walker, L. R., and Del Moral, R. (2009). "Lessons from Primary Succession for Restoration of Severely Damaged Habitats," *Applied Vegetation Science*, 12/1: 55–67.

Walker, R. F. (2003). "Comparison of Organic and Chemical Soil Amendments Used in the Reforestation of a Harsh Sierra Nevada Site," *Restoration Ecology*, 11/4: 466–74.

Walstad, J. D., and Kuch, P. J. (eds). (1987). *Vegetation Management for Conifer Production*. New York: John Wiley & Sons. 542 pp.

White, P. S., and Walker, J. L. (1997). "Approximating Nature's Variation: Selecting and Using Reference Information in Restoration Ecology," *Restoration Ecology*, 5/4: 338–49.

Yousefpour, R., and Hanewinkel, M. (2014). "Balancing Decisions for Adaptive and Multipurpose Conversion of Norway Spruce (*Picea abies* L. Karst) Monocultures in the Black Forest Area of Germany," *Forest Science*, 60/1: 73–84.

CHAPTER 10

Resilience through Adaptive Management

Stay calm, be brave, wait for the signs.
sign-off of the *Dead Dog Café Comedy Hour*, written by Thomas King (1997)

10.1 An overview of adaptive management

"Adaptive management" is a formal system for continuous improvement in any set of operations, but especially where some degree of system uncertainty exists. Promoted in natural-resource sectors since the late 1970s and early 1980s (Holling 1978; Walters 1986), it saw early applications in fisheries management, where fish stocks and their population dynamics are notably difficult to assess. Adaptive management is even more relevant today as we wrestle with the uncertainties of global change and its effects. It could be asserted that all resilience promotion is or incorporates adaptive management of necessity. At its essence, adaptive management assumes that better decisions and improved management are needed (or at least are possible) and can be achieved by treating plans or management actions as hypotheses that are then implemented, monitored, assessed, and subsequently improved at the next opportunity. By its very nature, adaptive management has the goal of aiding adaptation to the sorts of environmental changes and socioeconomic challenges discussed throughout this book (Allen et al. 2011). Adaptive management might be considered an intentional version of the complex adaptive cycle (see Section 4.5) that characterizes the inherent process of change in both ecological and social systems. It is noteworthy that the ground work for resilience theory, understanding complex adaptive systems, and their application in adaptive management of renewable resources was undertaken by the same team of C. S. Holling, C. Walters, and their colleagues at the International Institute for Applied Systems Analysis (Laxenburg, Austria) and at the University of British Columbia (Vancouver, Canada; Allen and Garmestani 2015).

On the one hand, some resource practitioners may consider adaptive management as unnecessarily academic or a threat to established procedures. On the other hand, it can be interpreted as simply an explicit description of how our species has learned by trial and error for millennia, and how any aspect of human culture evolves. Many natural resource management agencies claim that they practice adaptive management, but without explicit hypotheses and purposely directed monitoring to inform management alternatives. Consequently, examples of true adaptive monitoring are rare (Westgate et al. 2013), and can be especially challenging in forestry (Failing and Gregory 2003; Bormann et al. 2007). Common barriers have included: poor leadership, collaboration, or funding; a lack of clear objectives; an absence of intermediaries between researchers and practitioners; poor identification of system scale; and unsupportive institutional, policy, or social environments (Garmestani and Allen 2015). Nonetheless, some system of monitoring, evaluation, and redirection is needed if system resilience is to be achieved under uncertain social and environmental

Resilient Forest Management. Philip J. Burton, Oxford University Press. © Philip J. Burton (2025). DOI: 10.1093/oso/9780198832997.003.0010

stresses, constraints, and shifts that are likely to arise within the planning horizon of a forest management plan. Those uncertainties are accentuated by our fundamental ignorance of many ecological processes. In other words, resilient forest management must be embedded in an adaptive management process.

There are several versions of the adaptive management cycle, ranging from a simple five- or six-point cycle (see Figure 10.1(a)) to various customized versions, depending on the procedures and opportunities associated with a given management system. One version especially relevant to forest management, for which a complete management cycle (for example, often a timber crop rotation) typically lasts decades or more than a century, is the opportunity for adjustment or realignment before the cycle is complete (see Figure 10.1(b)). Where there is only one "system" (for example, corporation, fishery, forest estate, or political unit) subject to the decision-making jurisdiction of a manager, adaptive management follows a passive or recursive approach. Alternatively, where the opportunity exists to compare two hypotheses simultaneously (e.g. on similar stands in a forest landscape), an active or parallel adaptive management approach (see Figure 10.1(c)) is possible. Active approaches are preferred for questions of natural-resources management, where feasible, in order to avoid the complications of changes in weather, climate, commodity prices, and social values over time. Active adaptive management is more analogous to scientific experimentation, although often without the random application or replication of the treatments being compared.

While not meeting the formal requirement of adaptive management to learn from purposeful but full-scale operational decisions, options also exist for comparative lessons to be drawn from almost every set of forest management options. It is worthwhile to leave some untreated "control" areas in the application of a forestry practice (for example, stand thinning; see Figure 9.3), even if full-scale implementation of two or more parallel approaches is not feasible. It is surprising that even "research forests" managed by academic institutions or government agencies do not typically follow this simple practice of leaving side-by-side treated/untreated areas to facilitate comparisons of the effect of every intervention. The power of providing visual comparisons of the impacts of each management decision—as applied in a specific forest at a given time—reinforces the concept that forests can be sustainably managed in many different ways, that all forest management is experimental, and that our understanding of management effects is imperfect. Demonstrating treatment effects that are visually evident can often have a greater chance of modifying managerial behavior than highly technical reports with statistical tests showing significant treatment differences

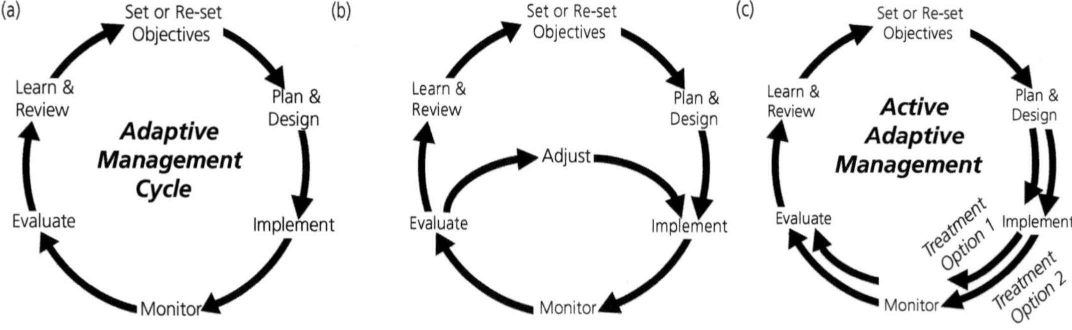

Figure 10.1 Alternative approaches to adaptive management: (a) The basic (passive) adaptive management cycle; (b) a modified version of the passive adaptive management cycle, in which implementation adjustments can be made without changing the overall approach; (c) an active adaptive management cycle, comparing two or more treatment options

on the basis of "p values." Statistical significance (of trends or treatment differences) rarely convinces policymakers or resource managers to change course unless also accompanied by demonstrable levels of ecological, economic, or social significance.

Duinker and Trevisen (2003), Fernandez-Gimenez et al. (2008), and Allan and Stankey (2009) provide several examples of moderately effective adaptive management in multiple-use forests and in forests managed primarily for conservation. Challenges abound, however, with social, political, and value-based issues tending to prevail over biophysical information gaps or technical barriers to understanding the system and agreeing on preferred management options or how current approaches might benefit from change. Support is always needed for people to engage, acquire knowledge, and negotiate solutions in the place-based context of each project's biophysical, cultural, and political environment. With enough support and enough time, adaptive management can be successful in mapping a way forward under uncertainty, if there is a sincere commitment to improve policies and practices on the part of managers and stakeholders. Other criteria for effective and successful adaptive management include a robust institutional framework, with adequate support for funding and evaluation over long periods of time.

10.2 Monitoring, standards, and certification

As outlined briefly in Chapter 3, having a reliable understanding of the forest resource is a requirement for sustained yield forestry and sustainable forest management (SFM). Forest inventory is a fundamental step in forest planning, and is increasingly being modified to incorporate values such as watershed protection, wildlife habitat, wild foods and medicinal plants, in addition to timber stocks (Abderhalden and Coch 2003). Using a combination of remote sensing, ground-based sampling, and database updates to account for harvesting, natural disturbances, and estimated tree growth, it is tempting to conduct a comprehensive forest inventory only once or as infrequently

as possible. When redone and compared with a former inventory—as is increasing feasible using remotely sensed imagery—recurrent forest inventories or "state of the forest" reports (e.g. Commissioner for Environmental Sustainability 2019; OMNDMNRF 2021) can be considered a broad form of ecosystem monitoring. Regularly updated information on the extent of forest cover by dominant tree species and age class in a given management unit can be used to assess forest sustainability and inform future management decisions. Such reports often highlight achievement of selected criteria and indicators of SFM (see Box 10.2), which are also used in support of forest products certification, thus serving to promote corporate or government agendas. But forest inventories and status reports, whether done once or recurrently, do not in themselves provide information with which to evaluate forest practices or policies in a manner to guide managers in promoting forest resilience. Rather, they have to be part of an adaptive management system in which specific questions are asked.

Environmental monitoring programs, broadly speaking, tend to be undertaken for one of three reasons: (1) passive monitoring, out of curiosity and in the hope that something interesting might turn up; (2) monitoring mandated by government, political, or certification requirements, usually to assure regulatory compliance but potentially to identify trends; and (3) question-driven or scenario monitoring, guided by a conceptual model and a rigorous design (Lindenmayer and Likens 2010). Passive monitoring programs tend to lack consistency in their methods and implementation, and rarely generate useful information. Mandated monitoring often covers wide geographic or disciplinary areas but cannot usually address ecological mechanisms or management options. This contrasts with question-driven monitoring, which conversely (in attempting to limit variability) is more constrained in its ability to extrapolate to broader spatial scales or ecosystem types. Whether mandated or not, a monitoring program needs to incorporate a number of features—including clear questions and a receptive audience—in order to be useful in an adaptive management context (see Table 10.1). Continuity and sustained support over

Table 10.1 Some features of effective monitoring programs

1. Design the program around clear and compelling management questions.

   - There is no point to initiating a monitoring program without articulating its intended applications in advance; be explicit in the questions it can answer or the management options it can inform.

2. Ensure that a robust experimental and sampling design underpins the monitoring program.

   - There must be sufficient statistical power to detect significant treatment differences and trends over time, without bias in sample selection and with adequate controls and replication.

3. Include review, feedback, and adaptation in the design.

   - There must be balance between continuity over time and flexibility to remain relevant; any change in methods should include a period of overlap to allow calibration and continuity.

4. Choose and conduct measurements carefully, and with the future in mind.

   - Not everything can or should be measured, so select basic measures of system function, indicators of change, or variables of particular management interest.
   - Measurements should be rigorous, repeatable, well documented, affordable, and employ accepted methods.

5. Maintain the quality, consistency, and accessibility of the data.

   - Include provisions for curation, long-term archiving, and accessibility.
   - Raw data, metadata, and descriptions of procedures should be stored in multiple locations.

6. Continually examine, interpret, and present the monitoring data.

   - Repeatedly review, analyze, and use the data, reporting trends and treatment differences to the public, and to provide value to the corporation, institution, or agency funding the program.

7. Include monitoring within an integrated management program, especially when new policies or practices are being introduced.

   - Monitoring is integral to adaptive management and any program of continuous improvement and the evaluation of options for resilience (see Section 10.4).

Source: Based on Lovett et al. (2007) and Lindenmayer et al. (2022).

long periods of time are also critical, and can be difficult to secure without broadly based community and agency buy-in (McKay and Johnson 2017). Lindenmayer and Likens (2018) provide further guidance and tips on how to avoid common weaknesses in ecological monitoring programs while implementing robust and relevant frameworks for applied scientific research and adaptive resource management.

Within the mandated monitoring category, at least two kinds of monitoring are used to provide relevant information in various management systems, including forest management: these can broadly be described as compliance monitoring and effectiveness monitoring. Compliance monitoring or implementation monitoring evaluates the degree to which prescribed activities are being carried out as planned, whether regulatory or contractual obligations are being satisfied, and how specific management objectives are being met. That is, compliance monitoring has a normative underpinning (that is, an explicit or implicit assumption of desired conditions), in which actions or their effects are evaluated against accepted standards. Those standards might be specified in a forest practices code (e.g. Hoberg 2016; Gale and Baker 2017), as professional best management practices (BMPs— e.g. Hammond 1997; Cristan et al. 2016), or as site- or project-specific objectives. For example, has a tree-planting contractor achieved the prescribed goal of establishing a specific density of correctly planted healthy trees of the desired species, or have loggers left the required width of uncut buffer strips adjacent to streams? Such assessments should not be confused with non-normative monitoring for resource change, nor recurrent inventories or selected resource updates, as described in the previous paragraph. The mere existence of standards against which performance is judged presupposes that those standards are adequate, appropriate, and are supportive of forest policy and societal goals— assumptions that need to be tested from time to time.

Consequently, it is just as important to undertake effectiveness monitoring, in which the efficacy of management decisions and practices at achieving stated goals is assessed (e.g. Mulder et al. 1999). Note the distinction here in evaluating progress toward landowner and societal goals, which may include consideration of various elements of forest policy, not simply whether more narrowly defined action-based objectives are being achieved. Effectiveness monitoring can be considered a partial implementation of the adaptive management cycle, ending at the evaluation stage, though implicitly providing information for future improvements in management. Unfortunately, effectiveness monitoring tends to be practiced sporadically rather than continuously, typically taking the form of a procedural audit, commission, or external review. Such reviews are often undertaken in response to a change in leadership or under (public, stakeholder, or shareholder) pressure to change direction. Examples of authorized forest policy effectiveness evaluation (not "monitoring" per se, being episodic rather than ongoing) include establishment of India's National Forest Commission in 2003 (Joshi et al. 2011), and an assessment of the 1994 Northwest Forest Plan in the United States as required by the 2012 Planning Rule of the US Forest Service (Spies et al. 2019). Limited assessments of forest management decisions, such as determining contemporary constraints on the allowable level of timber harvesting, may be undertaken regularly, as in the timber-supply reviews mandated to take place for each forest management unit on a ten-year cycle in British Columbia, Canada (BCMOF 2021), although larger policy directions are not evaluated.

The need to demonstrate and certify forest products as responsibly sourced prompted the development of various criteria and indicators of SFM in the 1990s (see Box 10.1). Forest policies and management plans are judged effective (sustainable) if they successfully meet a set of broad socio-ecological criteria, which are generally assessed by independent third parties. Success at meeting those criteria is in turn evidenced by the status of numerous measurable biophysical and socioeconomic indicators. Certification systems are essentially normative in nature, a form of voluntary compliance monitoring intended to assure market access or price premiums for commercial products sourced from a given forest management entity. Although large price differentials have not always materialized or been sustained (Chen et al. 2010), some sort of forest product certification has become essential in many first-world markets, and there is a gradual increase in the willingness to pay a premium for certified products (Cai and Aguilar 2013). The effectiveness of forest certification systems as an incentive to reduce deforestation and forest degradation has been mixed and rather disappointing (Wolff and Schweinle 2022).

Box 10.1 Criteria and indicators of sustainable forest management

Because sustainable forest management (SFM) requires the successful stewardship of multiple forest values in addition to sustained timber yield, it is clear that a system of multiple resource accounting and reporting is required (Adamowicz and Burton 2003; Bosela et al. 2016). Several processes were developed from the late 1980s through the early 2000s to develop SFM monitoring and inspection systems carried out by independent third parties. It could be argued that this widespread movement was incentivized by widespread public distrust of government forest policies and oversight, which inevitably involved a conflict of interest where public revenues and short-term economic activity benefit from rapid forest exploitation. Variously referred to as principles, protocols, standards, processes, or certification systems (Kadam et al. 2021), more or less parallel (but probably mutually informed) characterizations of SFM requirements were articulated by:

- the International Tropical Timber Organization (ITTO) in 1992 (Innes 2017);
- the Rio Forest Principles, adopted at the 1992 United Nations Conference on Environment and Development (Grubb et al. 1993; Innes 2017);
- the Helsinki Protocol in 1993, with a focus on European forests (Kadam et al. 2021);
- the American Forest and Paper Association (AFPA) in 1994, with a focus on the US, later expanded to include Canada (Vogt et al. 1997);
- the Montreal Process and its 1995 endorsement in the Santiago Declaration, with a focus on boreal and temperate forests (Siry et al. 2018; Gilani and Innes 2020); and
- advocates of ecoforestry in 1997, with an emphasis on protecting native biodiversity and natural ecosystem processes (Hammond 1997).

Box 10.1 *Continued*

Several other processes for guiding and documenting the improvement of forest practices have arisen at different times and places, some with roots back to the 1970s. These include initiatives of the African Timber Organization and the Tarapoto Process (Bosela et al. 2016).

Those principles and criteria were subsequently revised, expanded, and elaborated to include various descriptive and measurable standards and indicators. The AFPA principles evolved into the Sustainable Forestry Initiative (SFI), the Montreal Process into several national systems of SFM evaluation such as the Canadian Standards Association (CSA), while the Helsinki Protocol gave birth to the Pan-European Forest Certification (PEFC) system (later rebranded as the Programme for the Endorsement of Forest Certification), which subsequently endorsed the SFI and CSA systems. Those certification systems can be described as requiring that management- or process-based standards are met, in contrast to performance-based standards of the Forest Stewardship Council (FSC) that evolved from eco-forestry perspectives (Hammond 1997; Duinker and Trevisan 2003). Nonetheless, there have been convergence and some degree of harmonization among the various SFM certification systems while different strengths and weaknesses for evaluating forest management plans and for assuring the chain of custody of sustainably sourced wood products have been retained (Bosela et al. 2016; Kadam et al. 2021). Shared principles or criteria among these systems include:

- adherence to governmental laws and regulations;
- legitimate access to resources expressed through defined tenure or use rights;
- protection and conservation of soil and water resources;
- protection and conservation of biological diversity;
- promotion of healthy and productive forests;
- demonstrably sustained yield of timber and any other harvested forest products; and
- maintenance and enhancement of multiple socioeconomic benefits.

Notably absent in many certification systems is an explicit requirement to respect Indigenous rights, presumably because the legal status of such rights is unresolved or varies considerably among national and subnational jurisdictions.

Each of those criteria, and the additional ones that characterize each particular certification system, is typically elaborated in terms of subcriteria articulated for specific forest types or countries. Those criteria or subcriteria in turn are supported by several indicators that can regularly be evaluated objectively by quantitative or qualitative means. For example, the sustainability of wood production in the Czech Republic version of PEFC certification is indicated by metrics reported at the regional level (mean annual biomass and carbon increments per hectare, annual harvest levels, and average rotation length) and by indicators reported by individual forest owners (meeting regulatory requirements for felling, for forest regeneration, and with appropriate stock used for afforestation; PEFC Czech Republic 2016). Analogously, the FSC standard for implementation in the continental US supports the maintenance of ecological functions and values by required reporting on five landscape-level indicators (land area by successional stages, rare ecological communities, old growth, animal species and habitat diversity, and riparian management zones) and six stand- or site-scale indicators (plant species diversity, local seed sources, range of tree and deadwood sizes, even-aged retention, invasive species control, and fuels management; FSC-US 2010). Some forest certification systems incorporate both top-down direction and local priorities. For example, the CSA system is informed by criteria set by the Canadian Council of Forest Ministers, while also requiring public advisory groups to devise a local set of indicators for each individual applicant. The criteria and indicators of SFM continue to evolve, reflecting both national priorities (for example, reconciliation with Indigenous peoples) and local priorities (for example, enhanced protection for particular endangered species; McDermott et al. 2023).

As of the end of 2020, some 430 million hectares of the world's forests were certified as being sustainably managed under the PEFC or FSC umbrella systems (Fernholz et al. 2021). Although ITTO has not extended its SFM guidelines into a formal system of certification, it is estimated that 53 million hectares of tropical forest (including plantations, natural, and semi-natural forest) meet the ITTO criteria for sustainable management (Ehrenberg-Azcárate and Peña-Claros 2020). Collectively, this represents about 12% of global forest area, or 24% of the area covered by long-term forest management plans (FAO and UNEP 2020).

Certification, its supportive criteria and indicators, and the various auditing and reporting systems associated with their regular evaluation, are intended to prompt improvements in forest management policies and practices where accepted standards are not yet being met. Yet these systems typically do not require evidence of continuous improvement, evaluation of the suitability of indicators, nor whether standards for sustainability are sufficient to confer resilience in an uncertain future. The degree to which policymakers and managers embrace adaptive management and utilize forest updates, sustainability audits, or any other form of monitoring can be strongly influenced by personal leadership and corporate or agency culture (Lindenmayer and Likens 2010; Greig et al. 2013). All too often, such information ends up being filed away, or reported openly but not acted on. The potential for monitoring and reporting systems to inform resilient management depends on them being linked to receptivity ("receptive capacity" and willingness for improvement) within the management hierarchy—that is, embedded within a system (or at least a spirit) of adaptive management (Failing and Gregory 2003; Klenk and Wyatt 2015).

Newsom et al. (2006) reported that all eighty US forestry operations they evaluated that were aspiring to Forest Stewardship Council (FSC) certification were in the process of undertaking substantial improvements in ecological, social, and procedural practices in order to qualify. Lier et al. (2021) reported international progress on thirteen of twenty-four Pan-European criteria for SFM. It is reasonable to conclude that the incentive of forest certification is generating substantive and measurable improvements in forest management where it is undertaken. However, those incentives (market access, attracting investment, or social license advantages) relative to the investments and trade-offs required are insufficient to motivate many companies to pursue certification, particularly in the tropics (Gullison 2003; Chen et al. 2010). Many companies nonetheless appreciate the benefits of improved management processes, regular and smoother communication with stakeholders, and enhanced public image that come with certification requirements (Araujo et al. 2009). Suitable government, governance, and economic structures are still required for SFM, which can greatly limit the adoption of SFM and forest certification in developing countries (Ebeling and Yasué 2009).

10.3 Indicators, thresholds, and bellwethers

The monitoring and assessment of indicators provide a reasonable solution to the challenge of keeping track of a complex and multidimensional system such as a national economy or a forested landscape. The selection of indicators is a critical aspect of tracking and improving forest management, requiring users to respect their strengths and limitations (Gauthier et al. 2014). An indicator is a quantitative or qualitative attribute that can provide simplified information on the current state of a system, or, if measured periodically, can detect whether changes are occurring (FAO 2011). The following traits are commonly required for environmental indicators as well as those intended for performance evaluation (Gudmundsson 2003); they:

- condense large amounts of information;
- are sensitive to signals of change in the identified element or system;
- describe states, flows, or changes within systems;
- can be descriptive or normative (evaluating performance relative to standards or goals); and
- may ascribe "agency" (that is, cause and effect).

There is a danger, however, in focusing too much on indicators at the expense of the larger picture and a holistic understanding of whole-system status. Peter Drucker's mantra that "what gets measured is what gets managed," often promoted in business schools, should be considered a cautionary warning rather than sage advice when selecting indicators of complex systems. Capmourteres and Anand (2016) outline some basic indicators for assessing forest complexity and a forest's condition as a self-organizing, complex adaptive system. We can generally evaluate the status of most forest ecosystem services (Bell et al. 2016), but many of the sociocultural values associated with forests do not lend themselves to measurement or objective scoring (Wilson 2021). Selected indicators must also be sufficiently practical, simple, and inexpensive to be consistently measured by different monitoring teams over long periods of time, in order to establish reliable temporal trends and comparisons.

Many forest monitoring programs regularly assess a common set of key features, paying particular attention to how they relate to different management approaches and trends over time. Some multipurpose attributes to monitor at the stand level might include:

- density of trees by size class and species (stems/ha);
- density of large trees (living, dead, and fallen) (stems/ha);
  - where "large" is system specific (e.g. > 30m tall, or >1.0 m^3 bole volume, or simply in the 95th percentile of sizes for the system being assessed);
- dead woody material by size class and decay class (m^3/ha or kg/ha);
- biomass or carbon stocks, preferably of all layers (above and below ground) (kg/ha);
- timber/fiber productivity (m^3·ha^{-1}·yr^{-1}) and carbon sequestration rates (kg·ha^{-1}·yr^{-1});
- abundance and productivity of locally valued non-timber forest products;
  - e.g. wild berries, mushrooms, medicinal plants that are actively collected;
- ground cover by vegetation and litter or forest floor;
- frequency, density, or proportion of trees and other organisms observed to be dead, dying, or in poor vigor;
- species richness and evenness of easy-to-sample organism groups;
  - e.g. vascular plants, ground beetles, birds;
- presence and signs of habitat use by any rare, endangered, threatened, or protected species (see Box 10.2); and
- abundance and status of rare or declining habitat features (Noss 1999);
  - e.g. trees with natural hollows.

That is by no means a complete or necessary list, as indicators must be customized for each forest, community context, and set of managerial questions or options. For forest-level reporting, those same indicators can be rolled up to describe the average and range of levels found across a landscape, catchment basin, or forest estate (defined forest management area). In addition, the following features (among others) might be regularly updated and assessed:

- nominal age-class or structure-class distribution of stands (ha or %);
- size-class distribution of mappable stands, by stand type;
  - as defined by composition and age or structure;
- fragmentation or balance of edge to interior habitat;
  - e.g. ratios of artificial and natural stand edges to stand areas (km/km^2);
- road density (km/km^2);
- integrity of watershed, stream, and aquatic systems;
  - e.g. uniformity of seasonal streamflow, water quality, benthic invertebrate community structure, fish populations;
- abundance and status of rare or declining habitats (Noss 1999);
  - e.g. ancient forest stands, springs and caves;
- health of wide-ranging wildlife populations, including large herbivores and predators.

In addition to biophysical indicators, the ability of SFM to persist or the need for it to adjust to changing times requires monitoring of socioeconomic indicators as well. To that end, the following combination of quantitative and qualitative indicators might be monitored, generally at the forest estate level:

- market value of logs, lumber, or other products (e.g. mushrooms) generated by the forest;
- number of people employed in forest management,
  - including planning, timber and non-timber harvesting, silviculture and restoration, fish and wildlife management, fire management, and providing recreational, educational, or access services to forest users;
- subsistence or non-market forest use;
  - e.g. fuelwood harvesting, gathering of berries, mushrooms, and medicinal plants;
- recreational forest use;
  - e.g. hunting, fishing, bird-watching, hiking/tramping/skiing, camping, mountain biking, off-road motorized recreation;

Box 10.2 Species as indicators

Assessments of the status of focal species—in terms of species presence, population sizes, reproductive rates, and the health and vigor of individuals—are frequently conducted as an effective means of monitoring forest integrity. Focal species often provide the impetus for ecosystem protection and special management, and their populations may warrant monitoring because they serve one or several of the following roles (Simberloff 1998; Primack 2014):

- "flagship species" capture public attention, have symbolic value, or may be crucial to ecotourism; they are sufficiently iconic to attract public campaigns and political support for protection (for example, "charismatic megafauna" such as the giant panda (*Ailuropoda melanoleuca*), Siberian tiger (*Panthera tigris altaica*), or the Kermode variant of the American black bear (*Ursus americanus*));
- "umbrella species" such as brown bears (*Ursus arctos*) have large home ranges and diverse habitat requirements during the course of their life cycles, such that healthy populations are presumed to indicate that most other species and ecological processes in their range are doing well too;
- "endangered species," such as the northern spotted owl (*Strix occidentalis caurina*) in North America, are at risk of extinction or extirpation; they have legal protection in some jurisdictions, and species recovery plans typically require regular assessments and monitoring, which may extend to "threatened species" (those at risk of becoming endangered) as well;
- "keystone species" serve ecological roles in driving or regulating ecosystem composition, structure, or function to a degree greater than would be expected from their abundance or biomass alone; such species may be ecosystem engineers such as beavers (*Castor* sp.) or burrowing rodents, or carnivores such as the grey wolf (*Canis lupus*) or African lion (*Panthera leo*) at the top of trophic cascades;
- a "bellwether" or "early-warning" species is known to be sensitive to some aspect of habitat degradation, exhibiting changes before other species or ecosystem processes detectably respond to stressors; for example, lichens such as *Lobaria pulmonaria* are particularly sensitive to air pollution (Ravera et al. 2023), while others such as *Alectoria sarmentosa* are sensitive to the loss of interior forest conditions (Esseen and Renhorn 1998).

Simberloff (1998) advocates for the monitoring of keystone species as being particularly useful, as they integrate species population health with ecosystem processes. On the other hand, not all ecosystems may have keystone species, our understanding of their role and population dynamics is frequently imperfect, and each socio-ecological landscape may need to prioritize different sorts of indicator species. Bell et al. (2016) argue that biodiversity indicators (including structural and functional diversity as well as compositional diversity) need to be monitored within a framework that evaluates pressures on species and informs potential policy and management responses that could address worrying trends.

In monitoring the status of any focal species or group of species, it is understood that they are serving as surrogates for a wide variety of other species, ecosystem integrity, or biodiversity in general. We might point to the presence or abundance of a single species as strongly indicative of old growth, clean air, or particular soil conditions. But the very nature of evolution by natural selection and competitive exclusion means that each species is unique in its habitat preferences and tolerances; there is no proverbial "canary in a coal mine" whose status can portray overall ecosystem integrity. While not always calibrated or tested, indicator species monitoring is most effective when applied to a set of species having different but complementary requirements. Selecting that set of species (or other indicators) presupposes identification of the limiting factors or drivers of change in the forest of interest (Mulder et al. 1999; Gauthier et al. 2014; Sesser et al. 2019). For example, a suite of species selected to monitor forest integrity might include populations significantly associated with:

- specific, rare habitat features such as fruiting trees, springs;
- standing and fallen deadwood;
- large, old living trees, or old-growth forest;
- interior forest conditions, free of edge effects;
- natural disturbances such as fire, insect outbreaks, or windthrow;
- a diversity of habitats, differing in stand composition, age or openness; or
- geographic range limits, where conditions are marginal and may be expected to change.

Box 10.2 *Continued*

Where the goal is to monitor and maintain the biodiversity found in primary forests, Lindenmayer et al. (2000) suggest that it is more prudent to monitor general indicators of forest structure, as listed above for stand and landscape levels. Particular attention needs to be paid to structural complexity and plant-species composition at the stand level, and to forest connectivity and heterogeneity at the landscape level. Not all species indicators need to be associated with intact primary forests. It can be just as important to monitor invasive species (Poland et al. 2021), and the damage and population status of insect and fungal pests that can cause extensive mortality in trees or other species (Stone and Coops 2004; Kautz et al. 2017). Comprehensive species presence or abundance data, or stand structural indicators, can be further rolled up using multivariate ordination methods, with axis scores (of sampling sites in species space) then correlated with independent measures of ecosystem integrity. With trends empirically derived for a particular forest type or region, the extension of such methods to facilitate rapid assessment of key species (for example, within vascular plant, ground beetle, benthic invertebrate, or bird communities) can provide useful measures of forest integrity and forest change (e.g. Keddy and Drummond 1996; Burke et al. 2016). There is a suite of well-developed statistical tools dedicated to indicator-species identification and analysis (Dufrêne and Legendre 1997). There is also an entire journal (*Ecological Indicators*, https://www.sciencedirect.com/journal/ecological-indicators) addressing ecological and environmental indicators more broadly. Yet these techniques and the growing body of empirically calibrated indicator knowledge are not yet well integrated into adaptive forest management.

- relative dependence of nearby communities on forest cover and forest-related employment;
- frequency and extent of public engagement or complaints/protests about forest management activities or decisions.

By definition, all indicators are incomplete in giving a full picture of a system; that would require one to measure and track everything of interest, which is impractical. Consequently, a number of complex or multidimensional indices have been developed, such as a consumer price index to track the cost of living, the Dow Jones Industrial Average to track stock-market activity, or a forest health index to report on the ability to deliver valued ecosystem services (Arnott et al. 2015). Individual indicators are also imperfect in that most exhibit some variance or measurement error among the data collected (for example, of tree densities and volumes), and various strengths of correlation between the values of interest (for example, timber volume increment) and measurements undertaken (for example, trees sampled for height, girth, or annual rings). Several categories of indicators can be recognized, based not only on the domain (physical, ecological, social) being portrayed, but also on the nature of the functional relationship between what is being monitored

and what is being interpreted or valued. Those functional relationships might serve as proxy portrayals of current conditions, predictors of future conditions, or measures of precariousness and the potential for imminent shifts in the value of interest (see Figure 10.2). Whether linear or non-linear, it is important for the statistical relationship of indicators to values to be well documented, including consideration of uncertainty and variations in sensitivity.

Some of the most useful indicators—whether of climate change, ecological integrity, sustainability, human well-being, or system resilience—are those that exhibit a tight statistical relationship between what is being measured and what is being interpreted and valued. Those relationships do not, however, need to exhibit a unidirectional linear or log-linear relationship like that shown in Figure 10.2(a). Distinct breaks or thresholds in relationships (e.g. Figure 10.2(b)) are often associated with state changes or "system flips" from one stability domain to another (see Figure 4.3). Such breakpoints or functional thresholds can be detected using techniques such as segmented or piecewise linear regression (Toms and Villard 2015). Many other relationships are not unidirectional, such that all increasing values should not be interpreted as

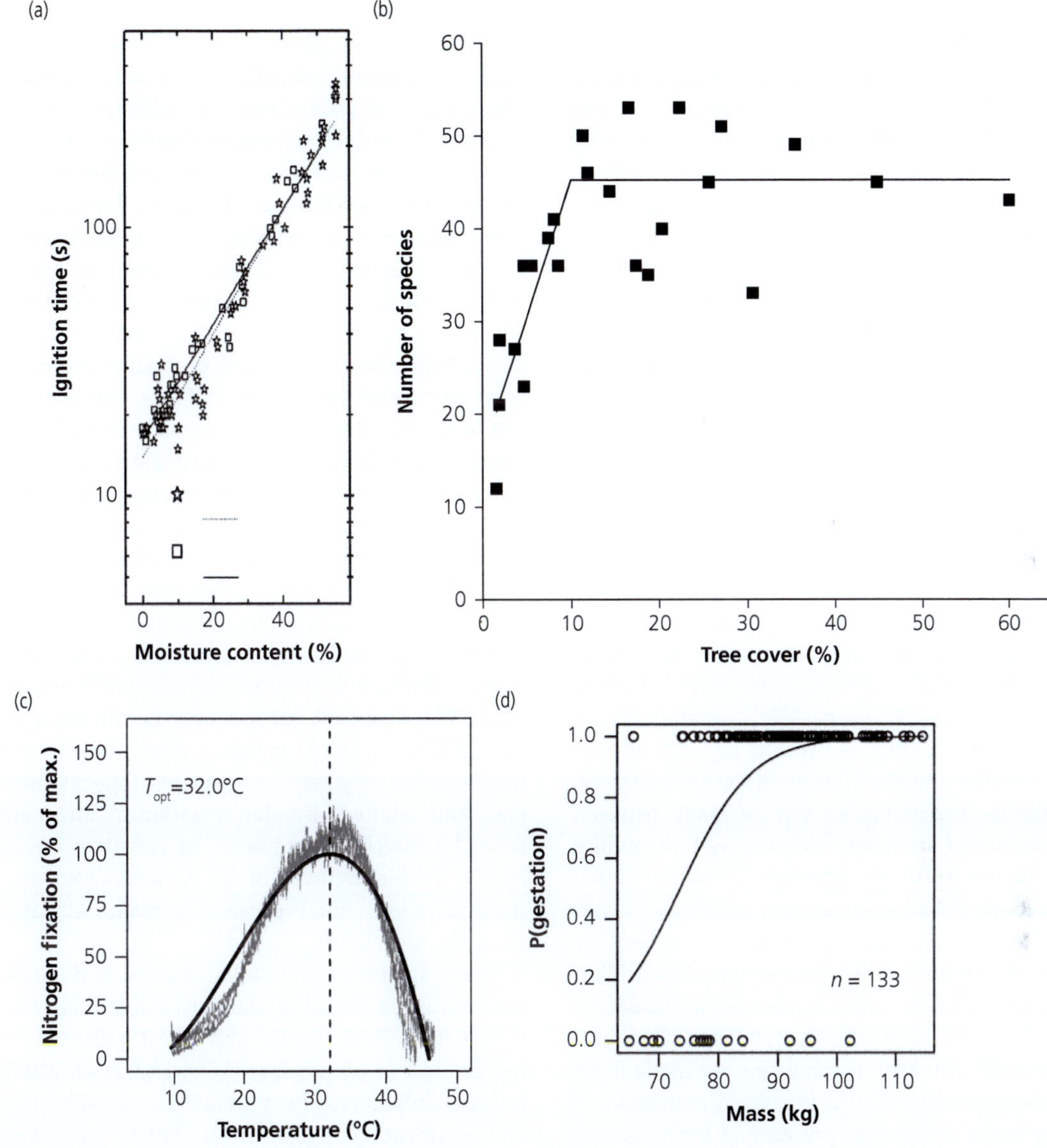

Figure 10.2 Examples of some alternative relationships between values of interest and predictor or indicator attributes that can be more readily monitored: (a) A continuous, unidirectional predictor: time to ignition in *Pinus halepensis* needles with different moisture content and in different seasons (here made linear on a logarithmic scale (Terrah et al. (2020), reproduced with permission); (b) a sharp ecological threshold: bird species loss below 10% forest cover in southeastern Australia (Radford et al. (2005), reproduced with permission); (c) a parabolic response: nitrogen fixation in the leguminous tree, *Robinia pseudoacacia*, over a range of temperatures (Bytnerowicz et al. (2022), reproduced with permission); and (d) a binary or presence/absence response probability described by a logistic equation: reproductive success of female caribou (*Rangifer tarandus*) in eastern Canada as a function of body weight (Pachkowski et al. (2013), reproduced with permission).

indicative of an ongoing positive response in the value of interest. In the parabolic example shown in Figure 4.2(c), note that a broad central range of temperature provides more or less optimal levels of nitrogen fixation (the value of interest), with a rapidly rising response at lower temperatures and a rapidly falling response at higher temperatures. This is in contrast to many relationships that can be described by logistic equations (see, e.g., Figure 4.2(d)), which are relatively unresponsive at low and high values of the indicator, but can be quite responsive to slight changes at intermediate levels, thereby denoting sensitivity to a state change.

In the complex world of socio-ecological systems, many driving or influential factors can be at play in affecting a value of interest, as indicated by the level of variance explained in the relationship of a response variable to the factor being tested, typically expressed as the R^2 statistic, the coefficient of variation. As nicely articulated by Welden and Slauson (1986), competition (for example) can be important—as indicated by a high R^2 in a relationship of performance plotted against the abundance of competitors (as in Figure 9.2)—but not intense where there is little slope to the fitted regression line. The converse can also hold true in a different situation, where competition may be intense (where performance shows a steep responsivity to slight differences in competitive pressure), but of little importance (that is, exhibits low R^2) relative to many other untested factors in the environment. Analogous interpretations can be made between any measured indicator and the system quality or value it purports to represent. Many functional interactions between ecosystem attributes would be better described as impositions of "ceiling" or "floor" constraints, rather than serving as a driving factor described by central tendencies in the data—an underused area of ecological statistics (Cade and Noon 2003). All such relationships are made more complex and difficult to use in guiding management owing to the widespread presence of the hysteresis phenomenon, in which the response to decreasing levels of a stressor or other predictor is not the same as its response to increasing levels. Despite these and other weaknesses (Lindenmayer et al. 2000; Failing and Gregory 2003; Niemi and McDonald 2004), it is fairly straightforward to use indicators to inform and update our understanding of system condition. That understanding can then be used in management models and to report patterns over space or trends over time to the public, government agencies, or forest certification bodies.

At the landscape level, the hydrological, geomorphological, and biological attributes of streams can serve as efficient integrators of forest condition. Well-vegetated watersheds with effective erosion control exhibit seasonal streamflows (as portrayed in hydrographs) that are relatively drawn out and stable, rather than having extremes of flooding and intermittent flow. Watersheds characterized by high levels of logging and wildfire can result in highly unstable stream channels, bank erosion, and floodplain rearrangement as a result of flooding (Chang 2013). High sediment content in streams and other receiving waterbodies (ponds, lakes, estuaries) is not only indicative of excessive watershed disturbance but has direct and indirect negative effects on aquatic life. Consequently, the relative abundance of different benthic invertebrates, for example, has emerged as a useful integrator of catchment basin integrity (Kerans and Karr 1994; Karr 1998). Nutrient export detectable in streamwater sampling as a result of vegetation loss or suppression is also recognized as indicative of forest degradation (Bormann et al. 1974; Rust et al. 2018). More recently, environmental DNA (eDNA) collected from streamwater facilitates the very rapid assessment of species presence and relative abundance within a catchment basin by matching fragments of genetic material against online archives of DNA sequences documented for individual species of animals and plants (Ruppert et al. 2019).

Many attributes of wildlife habitat and biodiversity can be linked to standard forest inventory information (tree species and sizes) collected on the ground or by satellite (Löfstrand et al. 2003), but such inferences always require local calibration and verification (Noss 1999). Other important information, such as the presence and use of non-timber forest products (NTFPs) gathered for food, medicine, and crafting, requires more comprehensive biophysical inventories and social analysis. Often missing in the monitoring of forest condition is the compilation of community-based socioeconomic information, ranging from the perception of forest health, utility, and aesthetics in the eyes of forest dwellers to the levels of employment, economic activity, and subsistence dependence associated with nearby forests (Ford et al. 2017;

Nerfa et al. 2020). Engaging community members as forest monitors is increasingly feasible as mobile phone applications powered by image analysis and artificial intelligence become available and are increasingly refined. For example, iNaturalist (https://www.inaturalist.org/) allows users to photograph and identify plants, insects, and fungi and registers those observations in a geographical database. The Merlin app (https://merlin.allaboutbirds.org/) empowers novice users to identify birds by sound recordings, greatly expanding the potential to conduct rapid point-count surveys. Often described as "citizen scientists," people who regularly visit the forest and contribute to its description not only provide an expanded set of eyes in the forest to detect important changes, but also tend to become invested in its sustainable management (McKinley et al. 2017).

Indicators of shifting conditions or impending change are perhaps most useful, in that they alert managers to the need to review assumptions and management direction. The likelihood of change is often greatest where conditions are approaching the limits of geographic ranges, of the historical range of variability, or of the feasibility/efficacy of management practices, suggesting that monitoring intensity should be increased in such situations (Johnson and Ray 2021). As an example, signs of significant changes from the previously stable climate started to become apparent in the 1990s, prompting the documentation of trends that continue to this day (see Box 10.3). Consistent deviations of abiotic conditions from previously documented norms (for example, for peak summer temperatures, the timing of autumn freeze-up and spring break-up, or annual maximum and minimum streamflows) are indicative that regional climate or landscape disturbance has shifted into a new realm to which species and stewardship practices may not be well adapted. The new levels of those same indicators (and often their specific combination) can serve as search parameters to find bioclimatic analogs elsewhere in space or in history, which can then provide guidance for more suitable species and management practices (MacKenzie and Mahony 2021; Mette et al. 2021; Esperon-Rodriguez et al. 2022). Where such indicators and their trends approach conditions characterized by system boundaries or transitions, managers should be alerted to the possibility of impending state change, an even more obvious and important shift in previous conditions. Such changes at species or ecosystem range limits (for example, forest–grassland transition zones; Michaelian et al. 2011) or at the margins of successful management systems (for example, maple syrup production; Matthews and Iverson 2017) warrant focused monitoring and attention, even though they may not necessarily meet the definition of bellwethers.

Simply because we cannot monitor or measure every one of the innumerable properties of a forest ecosystem, the use of indicators is bound to continue despite their limitations. Research to improve the selection and calibration of indicators for many

---

Box 10.3 Indicators of climate change in Canadian forests

Knowledge of changes in, or impending threats to, the forest's ability to provide goods and services in a reliable and predictable manner are required before adaptation measures (see Chapter 7) can be undertaken. Such changes, and potential indicators of change, have been seen in climate, forest, and human systems in recent years.

Climate Drivers

Most areas of Canada have experienced increases in mean annual temperature in recent decades, especially pronounced in terms of less extreme minimum temperatures in winter. Many areas are experiencing longer growing seasons, with a greater number of days between late spring frosts and early fall frosts. More precipitation is falling as rain rather than snow, resulting in reduced snowpacks. Extreme events such as summer drought and windstorms—but of unexpected heat or cold too—are becoming more frequent. Integrative indices of drought or dryness such as the climate moisture index (CMI; Hogg et al. 2013) and fire weather index (FWI; Wotton 2009) are particularly relevant to forests.

Box 10.3 *Continued*

Forest System

Several drivers of forest composition, structure, and function are sensitive to a changing climate. Landforms are changing as a result of thawing permafrost in the north, and mass movements are triggered by glacial debutressing, alpine permafrost thaw, and soil saturation during high rainfall events. Glacial melt, rapid spring thaws, and high rainfall events are resulting in short-term flood events. Other natural disturbances such as wildfires and some insect outbreaks are also affecting larger areas than in the past, and with longer seasons of activity. Bud-break and flowering in trees and other plants are occurring earlier in the spring, insect larval development is proceeding more rapidly, and vertebrate migration and hibernation times have changed. Latitudinal and elevational shifts in species ranges or their productivity have been detected, and forests are failing to regenerate after some disturbances. Such shifts are highly species specific, with some species benefiting and other species suffering at any location, often varying by site type. With decomposition rates being sensitive to both temperature and moisture, ecosystem processes such as mineralization rates and edaphic traits such as organic matter content are changing as well, but with high variability among regions and among site types within regions.

Human System

Social and engineered systems are also showing evidence of impacts from climate change, and some show evidence of gradual adaptation. Although climate-induced human migration is not pronounced in Canada, temporary and localized displacements owing to storms, flooding, and wildfire have caused significant disruption, stress, and expense. Damage to roads and stream crossings by flood events are prompting the installation of larger culverts and reinforced bridges. Human health is being compromised by smoke inhalation from long and persistent wildfires, and from expanded ranges of vector-borne diseases such as West Nile virus and Lyme disease. Institutions are responding with revised risk assessments, updated emergency management plans, and policies promoting assisted migration for tree provenances and species that will better match the observed changes and projected trends in climatic conditions.

Source: Gauthier et al. (2014).

---

values of interest is clearly needed (Noss 1999; Gao et al. 2015). Nonetheless, there is broad agreement on the need to evaluate aspects of ecosystem composition, structure, and function, along with some measure of landscape structure. To that end, we continue to see the development of composite indicators and the use of multivariate statistics to assess conditions and trends in forest health and integrity. Where individual measurements are numerous and complex, and especially when dealing with public lands, it is important to provide highly summarized and visual "status and trends" reports that can be readily interpreted by decision-makers and the public (see Box 10.4). Although assessments are often categorically summarized (for example, as "good," "fair," or "poor"), the ability to identify such distinctions objectively is limited, because functional thresholds in ecological systems are poorly understood and may be relatively infrequent (Lindenmayer and Luck 2005; Groffman et al. 2006; Johnson 2013). Thresholds for intervention, for the selection of a management approach, or for changing course (see Section 10.4) will consider the status, precariousness, or trends exhibited by biophysical indicators, but will inevitably reflect the articulated tolerances for risk or loss on the part of managers and stakeholders too (Johnson and Ray 2021).

Box 10.4 Communicating forest status and trends

Given the complex and multidimensional nature of forest stands, landscapes, and their many components, it can be difficult to describe patterns or trends in ecological health and integrity. At the same time, forest managers, policymakers, and the general public need "actionable information," presented with enough clarity that land management decisions can be made. A common approach is to digest monitoring data into simple color-coded "stoplight" symbology, with green denoting "good," yellow "fair," and red "poor" conditions, or "improving," "unchanged," or "deteriorating" trends.

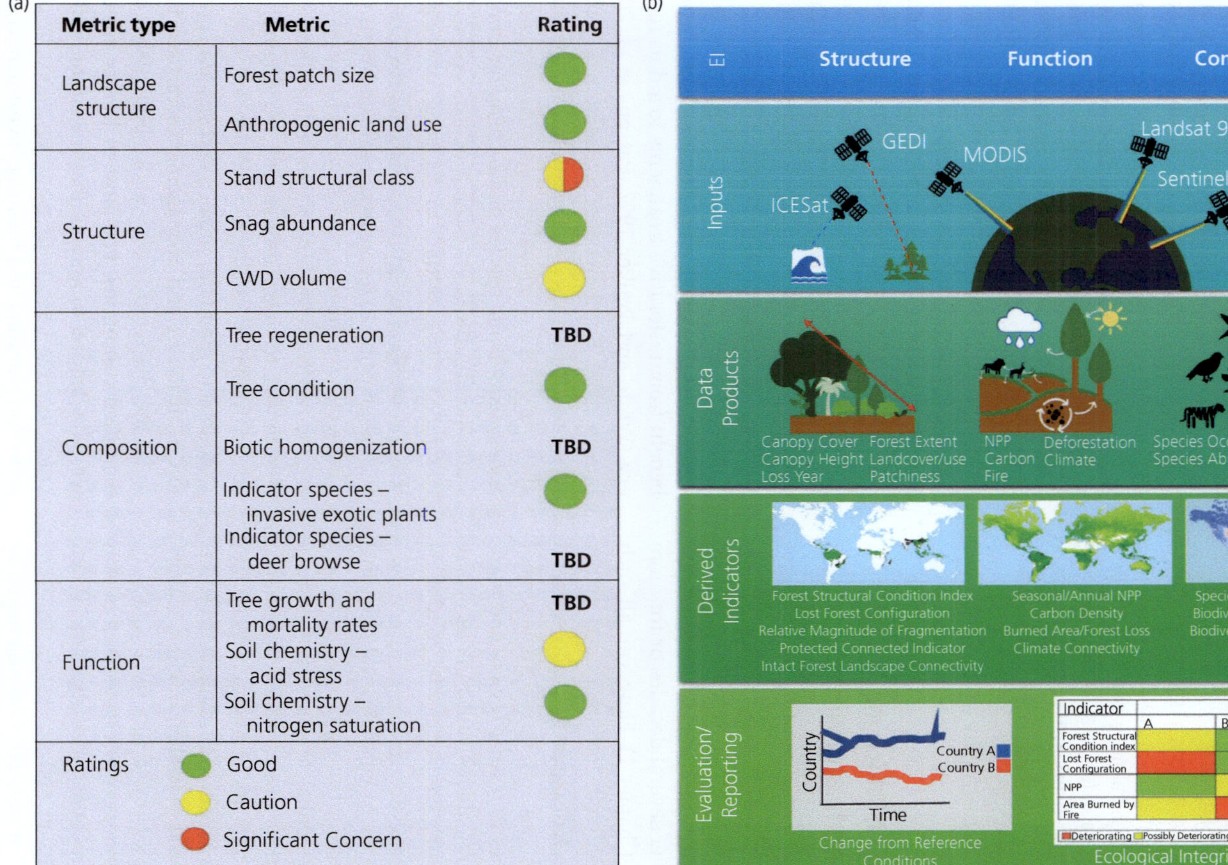

Figure 10.3 Communicating ecological integrity trends to the public and to forest managers: (a) An example of "stoplight" forest integrity reporting in Acadia National Park, Maine, US, where attributes with insufficient data are openly identified with the notation TBD, "to be determined" (Tierney et al. (2009), reproduced with permission); (b) an infographic portraying the multiple sources of data and their value for the interpretation of ecological integrity for use in a monitoring program (reproduced under Creative Commons CC BY 3.0 from Hansen et al. (2021)). Both examples emphasize collecting multiple data products (remotely and in situ) to characterize ecosystem structure, function, and composition.

Such summaries can be provided for different values of interest, for different ecosystem types in a jurisdiction, or for different attributes of ecosystem composition, structure, and function, as in Figure 10.3(a). Each such icon must be transparently backed up by tabular data summaries and clearly described sampling protocols and interpretative criteria, and ideally would link to digital archives of the original data, especially where collected on public lands. Burton (2014) suggests the use of polar ordination to condense multivariate data along a gradient between the most degraded and most intact condition, without necessarily assigning categorical interpretations. Carefully designed graphical abstracts (see, e.g., Figure 10.3(b)) can be very effective in communicating the complex nature of data acquisition and interpretation that goes into the reporting of forest status and trends.

10.4 Shifting gears: Which mode of resilience to promote and when?

Forestry is notable in the realm of human endeavors, not only for its historical focus on the technical determination of requirements for sustainability, but also for the temporal scope of management plans that extend over several decades or for more than a century (see Chapter 1). However, all such plans must be considered "rolling plans" rather than strict blueprints. If designed and followed with sufficient flexibility, resilient forest management plans lend themselves to regular review and readjustment in order to remain relevant and responsive to current and trending environmental conditions and socioeconomic values.

A formal system of monitoring and adaptive management is a standard component of SFM and can be considered essential to sound forest stewardship. These practices take on added importance when managing forests for resilience. Ecosystem status and trends are critical in guiding tactical decisions to support aspects of resilience with a reasonable chance of success. How can one tell if a particular forest and forest management system is robust in the face of current and anticipated stressors? When does forest regeneration and ecosystem recovery need to be promoted or assisted? Can a damaged or altered forest continue to deliver desired ecosystem services, or do some trade-offs and compromises

need to be evaluated? Under what circumstances should forest structure, composition, and associated silvicultural systems be adjusted? At what stage is a wholesale transformation of the forest ecosystem warranted?

In the past, a healthy ecosystem and its normal cycles of disturbance and recovery were considered resilient if their attributes remained within their natural range of variability (NRV) (Morgan et al. 1994; Landres et al. 1999). Although inevitably limited by incomplete data, that NRV is typically informed by historical information (and may be referred to as the historical range of variability (HRV)) or by spatial variation across a relatively homogenous climatic region and range of site types. Under stationary conditions, evidence of habitat loss might simply prompt enhanced monitoring, followed by modified practices or restricted activities once a cautionary threshold has been crossed, or a complete cessation of industrial activities once a critical threshold has been crossed (Johnson and Ray 2021). Where a disturbance or the cumulative effects of multiple stressors or disruptions—especially if induced by human actions—pushes a system beyond its NRV, then it can be considered degraded and would benefit from restorative treatments to aid its recovery (White and Walker 1997; Parsons et al. 2000; Burton 2005). But the goal of returning a degraded ecosystem to a previous condition may be unrealistic in an era of global change characterized by widespread exotic species and a changing climate. Consequently, the science and practice of ecological restoration have shifted to focusing on ecosystem processes and greater "naturalness" rather than attempting to reconstruct past conditions (see Chapter 9). A similar approach is required in assessing the relative feasibility of promoting absorption (resistance), recovery, adjustment, reconfiguration, or transformation as adaptive resilience options.

Actions to maintain, protect, and sustain current forest attributes and management practices tend to be the preferred option under both stable and disrupted conditions. For most of human history, we have tended to assume that the future will be much like the past, with many decisions in conservation planning and business investment counting on a more or less stable environment. So it is understandable that the first aspect of resilience thinking that is applied in the face of impending threats or stressors is to protect existing resources and investments, and to defend previous decisions. This means that tactics to promote system resistance or recovery as a means of absorbing change and outside threats make up the first set of tools for resilient forest management. Where sustainability is threatened, the decision process may then proceed along the following lines (as illustrated in the left portion of Figure 10.4):

(1) recognize that valued ecosystem services or indicators are outside their NRV or HRV, or are failing to meet (or are approaching) limits or target levels set by forest stakeholders;
   • this indicates the need for some sort of intervention or change in management approach;

(2) determine whether large levels of key resources, species, or ecosystem elements remain, and effective methods are available to bolster their vigor and survival;
   • this suggests that tactics to enhance resistance (for example, stand thinning to address rising drought, wildfire, or bark beetle threats) may be viable, and that general application of the precautionary principle (for example, avoiding or reducing resource harvesting) is needed;

(3) evaluate whether the ecosystem is recovering naturally or can be expected to do so in a reasonable period of time;
   • if not, some degree of artificial regeneration or ecological restoration is called for;

(4) consider whether current stressors and disruptions are unusual and infrequent, or part of a larger trend by which they can be expected to be more frequent, severe, and important.

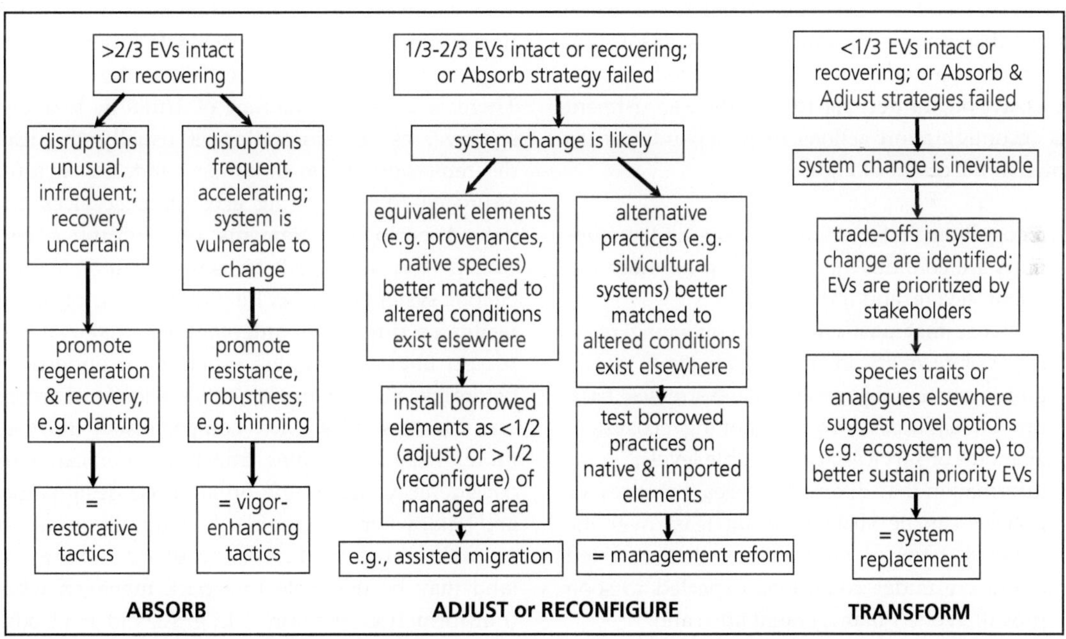

Figure 10.4 A decision support aid to guide in the selection of which approach to resilient forest management should be promoted, based on monitored status and trends and on stakeholder priorities. Motivation to manage for resilience typically starts with recognition of disruption or stress that threatens sustainability of ecosystem elements, services, or their indicators (ecosystem values (EVs)), pushing the system outside its historical range of variability or threatening to do so. Thresholds for acceptable levels of change or disruption before shifting tactics (here portrayed as arbitrary levels of one-third and two-thirds, for convenience), and what proportion of the managed area to which a resilience tactic should be applied, should be determined by landowner/manager/stakeholder consensus. The distinction between adjustment or reconfiguration is largely a matter of whether a minority or a majority of the system is altered to match changing conditions better.

- if the latter, then the absorption strategy may be less appropriate than moving on to consider substantial adjustments to forest composition, structure, or management.

The portfolio of tactics for promoting forest resilience is broad, and can exhibit considerable overlap among the nominal categories outlined in Figure 10.4 and in previous chapters. Absorption approaches presume that current system condition can be sustained with appropriate interventions to promote resistance and/or recovery. Shifting into adjustment mode involves the inherent acceptance of signs that the world is changing, that the forest and forest management approaches cannot remain the same as in the past. Where the climate has already undergone a discernible shift, exotic species are pervasive, or there has been a huge swing in public opinion, it is no longer appropriate or possible to restore an ecosystem after a single disturbance or disruption; nor is it feasible simply to regenerate a forest that has been harvested or subject to severe natural disturbance. Picking up where absorption options of enhanced resistance and assisted regeneration leave off, the decision to undertake adjustment and reconfiguration actions then depends on (as shown in the center of Figure 10.4):

(1) recognition that current stressors and disruptions are confidently expected to be more frequent, severe, and important;
(2) evidence that past efforts at enhancing resistance or recovery in this system (or in analogous systems subject to the same stresses) have proven unsuccessful at sustaining ecosystem elements or services at acceptable levels;
(3) information that different species, practices, or policies can be sustainable (at least over the medium term of decades), as demonstrated elsewhere under conditions expected to soon prevail in one's management area; and
(4) knowledge of individual component (for example, species, silvicultural system) traits that are successful under a wide range of conditions if faced with a "no-analog future" or great uncertainty.

Reducing anthropogenic stress in all its forms (pollution, over-harvesting, unregulated recreational use, and so on) is once again called for when implementing a strategy of adjustment or reconfiguration, at least until its effectiveness has been demonstrated. The avoidance and reduction of cumulative effects from multiple stressors must, of necessity, be part of the response strategy to indications of current or impending ecological deterioration. Those cumulative effects are often non-linear and synergistic, making assessments of the impacts of a single disturbance, stressor, or industrial proposal inadequate in portraying the full picture (Gillingham et al. 2016; Blakley and Franks 2021). Rather, the monitoring, assessment, and mitigation of cumulative effects must be wide in scope and strongly place-based in their approach (Millar and Stephenson 2015; Miller and McGill 2019; França et al. 2020).

Undertaking a transformative approach to resilient forest management may seem like "giving up," a strategy to be undertaken only if absorption and adjustment approaches fail or indicators suggest they will be ineffective in the long run. There is a certain amount of truth in that perspective, as transformation is usually the least desirable option. On the other hand, a structured approach to prioritizing ecosystem services and valued ecological elements can inform a controlled and managed transition, rather than a chaotic system collapse. Intentional transformation facilitates directional interventions that can still protect important values, especially if undertaken before key ecosystem elements are lost and critical ecosystem services are compromised. Furthermore, what constitutes transformation can vary considerably, depending on land-use designations and landowner/stakeholder/community goals. For example, transformation of a forest to a native grassland may be desirable to a park manager, while transformation of a forest to grassland is at odds with the goals of an industrial timberland manager. Conversion of a conifer forest to a broadleaf forest may be considered a wholesale transformation by some managers and members of the public, while it can simultaneously be considered a modest adjustment to maintain forest cover by others. Transformation can variously humanize or naturalize a landscape (for example, long-term land-use changes; see Figure 9.1), depending on the particular alignment of management vision with biophysical changes. Consequently, guidance to undertake a strategy of transformation (see the right portion of Figure 10.4) tends to be more vague, more general, and more dependent on sociocultural endorsement than that offered for promoting resilience through absorption, adjustment, or reconfiguration. System transformation may be the most appropriate option where:

(1) it is recognized that conditions have changed (or will soon change) to such an extent that all valued ecosystem components cannot be retained nor ecosystem services provided in the future, such that trade-offs will have to be made, with some attributes saved or promoted at the expense of others;

(2) past efforts at enhancing stress absorption or promoting adaptive adjustment (for example, through component substitution) in this system, or in analogous ones subject to the same stresses, have proven unsuccessful at sustaining all desired ecosystem elements or services at acceptable levels;

(3) some ecosystem services (for example, carbon storage, erosion control, wildlife habitat) can still be provided, even if there is a wholesale conversion of tree cover, vegetation, or land use; and

(4) there is widespread stakeholder consultation, community engagement, and broad-based support for significant alteration of the ecosystem or its management system.

Figure 10.4 and the numbered lists above provide some conceptual criteria for employing alternative approaches to promoting forest resilience. But to incorporate these considerations fully in an adaptive management framework, a more formal use of indicator monitoring with comparison of indicator levels and trends to target conditions or risk tolerance levels is required (Failing and Gregory 2003; Borchers 2005). For example,

managers and stakeholders in one situation might prefer that an absorption strategy be followed so long as observed or projected reductions in some valued indicator are less than 30 percent; those in another situation might set the threshold to shift to an adjustment strategy as being a 20 percent or 50 percent change. Other situations will focus on indicators approaching particular thresholds that represent potential tipping points in key system attributes. For example, Figure 10.5 provides a graphic portrayal of using 50 percent and 100 percent levels of persistent change as criteria for switching between adjustment and reconfiguration approaches, and between reconfiguration and transformation approaches, in contrast to the 33 percent and 67 percent thresholds in Figure 10.4.

In all likelihood, indicators will show conflicting trends and responses to management alternatives (for example, organism group responses to salvage logging; Thorn et al. 2018), forcing stakeholders and managers to prioritize ecosystem services and values in decision-making, even if drastic transformation is not being considered. With conflicting values and trade-offs in ecosystem services widespread (Rodríguez et al. 2006; Wang and Fu 2013; see Chapter 2), but potentially shifting over time, there may be a need to revisit triad or other land-use and resource-emphasis zoning decisions. The collation of sufficient information to calibrate and exercise statistical or simulation models that can be used to explore the implications of alternative management options (e.g. Seidl et al. 2009; Caves et al. 2013) is also important. Spatially-explicit forest estate models and landscape simulation models are becoming widely used to explore and communicate forest development alternatives (see Box 10.5). Model projections can explore alternative socio-ecological scenarios and forest management decisions, which not only serve as an aid to decision-making under uncertainty, but also help managers and stakeholders to prepare mentally for multiple futures. Making such explicit projections and testing them against observations are often considered central aspects of adaptive resource management (Walters 1986; Duinker and

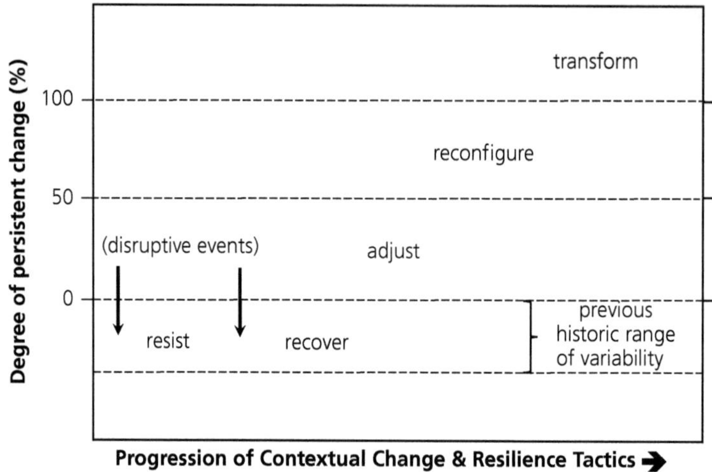

Figure 10.5 An alternative portrayal of thresholds for adjusting resilience tactics. Challenges are to assess the extent of socio-environmental change, and for managers and stakeholders to prioritize values denoting system functionality, and to agree on criteria, indicators, and thresholds for switching tactics.

Trevisan 2003). Ideally, resilient management incorporates contingency planning and the flexibility to respond to a robust and comprehensive system of signals from the forest system and its socioeconomic context.

Box 10.5 Spatial forest landscape modelling and visualization

Management goals at the level of an entire forest estate, landscape, or catchment are ultimately more important than individual stand-level objectives and treatments. Coupled with the sensitivity of many forest values to features such as fragmentation, edge effects, and connectivity (see Box 5.3), it is important for forest planners to extrapolate the implications of management options through space and over time. That is the purpose of manual approaches to full-rotation planning and landscape design such as those expounded by Diaz and Apostol (1992), which can be considered akin to those of a landscape architect.

Exploring the spatiotemporal dynamics of forest management options has become much more feasible and sophisticated with the development of a wide variety of computer simulation modelling tools. Those spatially-explicit forest simulation models have evolved from two different origins, yet have converged in their abilities and outputs. The first category developed as evermore sophisticated timber harvest scheduling models, built on optimization algorithms for accessing mature timber, but now can address multiple objectives and can include outputs for a range of non-timber values too (Gustafson et al. 2006, Baskent et al. 2024). Software in that category includes FPS-ATLAS (e.g., Seely et al. 2004, Griess et al. 2019), Patchworks™ (e.g., Moore and Tink 2008, https://spatial.ca/), and Woodstock™ (e.g., Corrigan and Nieuwenhuis 2016, https://remsoft.com/). The second category of models have developed from forest landscape dynamics simulators such as SELES (Fall and Fall 2001) or LANDIS (He and Mladenoff 1999, Gustafson et al. 2000), which focus on stochastic natural disturbances and succession but have evolved to include realistic patterns of forest road development and logging as a disturbance as well.

Many of these computerized models draw upon yield tables or output from other forest stand development models, and can pass output to additional decision-support tools that evaluate economic, hydrological, or carbon budget implications. All such models can project the status of key value indicators over time under alternative management scenarios, including planimetric maps and oblique-view visualizations of what the forest management unit will look like many decades into the future. Such visualizations (e.g. Figure 10.6) are especially important in conveying (or achieving) a vision of the desired future forest in collaboration with stakeholders and the general public (Sheppard and Meitner 2005).

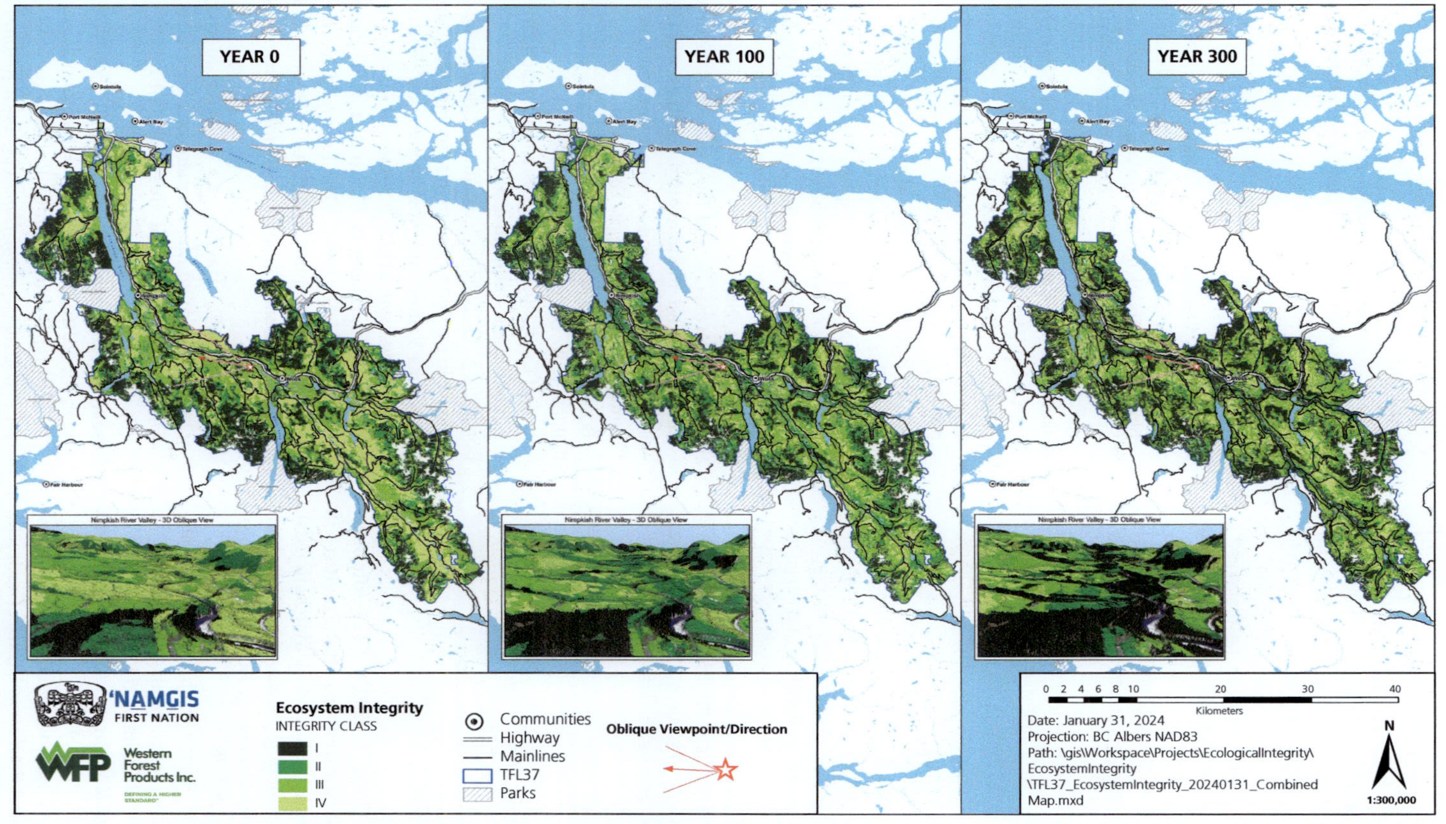

Figure 10.6 Spatially explicit projections of future forest integrity (largely reflecting forest age) for a watershed on northern Vancouver Island, British Columbia, Canada. Other projected outputs and indicators include forest age class structure, forest habitat types, timber volume available for harvest, total road network length, large woody debris input to stream channels, numbers of redcedar (*Thuja plicata*) trees available for traditional bark harvesting, and carbon balance. Projections represent a concensus direction between Western Forest Products Inc. and the 'Nimgis First Nation, as simulated using the Patchworks™ forest planning model (imagery courtesy of Western Forest Products).

Box 10.6   Key points

- Adaptive management is the structured embodiment of managing for resilience, in which the monitoring of system conditions and the socio-ecological milieu are regularly evaluated to guide management decisions in a cycle of continuous improvement and adjustment.
- Not everything that is important can be monitored, so selected indicators are tracked instead. Indicators should show strong links to forest values of interest, but may show complex response functions that need to be understood before status and trends are used to guide management.
- Indicators can be descriptive or normative, with international criteria considered indicative of sustainable forest management being widespread in certifying forest management plans and wood products around the world.
- Monitoring of system conditions and their context informs whether existing forest values and management approaches can absorb socio-ecological changes, disturbances, and stressors, or whether strategies of adjustment, reconfiguration, or transformation need to be adopted.
- When and how to respond to socio-ecological changes and stressors depend not only on indicators of such change and scientific interpretations of likely impacts, but also on manager and stakeholder values and risk tolerance.

It is a bad plan that admits of no modification.
Publilius Syrus (85–43 BCE)

References cited

Abderhalden, W., and Coch, T. (2003). "Key-Attributes for the Monitoring of Non-Timber Forest Resources in Europe," in P. Corona, M. Köhl, and M. Marchetti (eds), *Advances in Forest Inventory for Sustainable Forest Management and Biodiversity Monitoring*. Dordrecht: Kluwer Academic Publishers, 267–78.

Adamowicz, W. L., and Burton, P. J. (2003). "Sustainability and Sustainable Forest Management," in P. J. Burton, C. Messier, D. W. Smith, and W. L. Adamowicz (eds), *Towards Sustainable Management of the Boreal Forest*. Ottawa: NRC Research Press, 41–64.

Allan, C., and Stankey, G. H. (eds). (2009). *Adaptive Environmental Management: A Practitioner's Guide*. Dordrecht: Springer; Melbourne: CSIRO Publishing. 367 pp.

Allen, C. R., Fontaine, J. J., Pope, K. L., and Garmestani, A. S. (2011). "Adaptive Management for a Turbulent Future," *Journal of Environmental Management*, 92: 1339–45.

Allen, C. R., and Garmestani, A. S. (eds). (2015). *Adaptive Management of Social-Ecological Systems*. Dordrecht: Springer. 264 pp.

Araujo, M., Kant, S., and Couto, L. (2009). "Why Brazilian Companies Are Certifying their Forests?" *Forest Policy and Economics*, 11/8: 579–85.

Arnott, J. C., Osenga, E. C., Cundiff, J. L., and Katzenberger, J. W. (2015). "Engaging Stakeholders on Forest Health: A Model for Integrating Climatic, Ecological, and Societal Indicators at the Watershed Scale," *Journal of Forestry*, 113/5: 447–53.

Baskent, E. Z., Borges, J. G., and Kašpar, J. (2024). "An Updated Review of Spatial Forest Planning: Approaches, Techniques, Challenges, and Future Directions," *Current Forestry Reports*, 10: 299–321.

BCMOF (British Columbia Ministry of Forests. (2021). *Timber Supply Review Backgrounder*). Victoria, BC: British Columbia Ministry of Forests. 5 pp. Available online at https://www2.gov.bc.ca/gov/content/industry/forestry/managing-our-forest-resources/timber-supply-review-and-allowable-annual-cut (accessed February 24, 2024).

Bell, F. W., Dacosta, J., and Larocque, G. R. (2016). "Considering Forest Biodiversity Indicators within a Pressure, State, Benefit, and Response Framework," in G. R. Larocque (ed.), *Ecological Forest Management Handbook*. Boca Raton, FL: CRC Press, 337–59.

Blakley, J. A. E., and Franks, D. M. (eds). (2021). *Handbook of Cumulative Impact Assessment*. Cheltenham, UK: Edward Elgar Publishing. 385 pp.

Borchers, J. G. (2005). "Accepting Uncertainty, Assessing Risk: Decision Quality in Managing Wildfire, Forest Resource Values, and New Technology," *Forest Ecology and Management*, 211/1–2: 36–46.

Bormann, F. H., Likens, G. E., Siccama, T. G., Pierce, R. S., and Eaton, J. S. (1974). "The Export of Nutrients and Recovery of Stable Conditions Following Deforestation at Hubbard Brook," *Ecological Monographs*, 44/3: 255–77.

Bormann, B. T., Haynes, R. W., and Martin, J. R. (2007). "Adaptive Management of Forest Ecosystems: Did Some Rubber Hit the Road?" *BioScience*, 57/2: 186–91.

Bosela, M., Larocque, G. R., Baycheva, T., Valbuena, R., and Lier, M. (2016). "Criteria and Indicators of Sustainable Forest Management," in G. R. Larocque (ed.), *Ecological Forest Management Handbook*. Boca Raton, FL: CRC Press, 381–413.

Burke, D. J., Knisely, C., Watson, M. L., Carrino-Kyker, S. R., and Mauk, R. L. (2016). "The Effects of Agricultural History on Forest Ecological Integrity as Determined by a Rapid Forest Assessment Method," *Forest Ecology and Management*, 378: 1–13.

Burton, P. J. (2005). "Ecosystem Management and Conservation Biology," in S. B. Watts and L. Tolland (eds), *Forestry Handbook for British Columbia*. 5th edn. Vancouver: Faculty of Forestry, University of British Columbia, 307–22.

Burton, P. (2014). "Using Polar Ordination to Evaluate Ecosystem Recovery and Restoration Progress." Presented at a conference on *Forest Landscape Mosaics: Disturbance, Restoration and Management at Times of Global Change*, August 10–14, Tartu, Estonia.

Bytnerowicz, T. A., Akana, P. R., Griffin, K. L., and Menge, D. N. (2022). "Temperature Sensitivity of Woody Nitrogen Fixation across Species and Growing Temperatures," *Nature Plants*, 8/3: 209–16.

Cade, B. S., and Noon, B. R. (2003). "A Gentle Introduction to Quantile Regression for Ecologists," *Frontiers in Ecology and the Environment*, 1/8: 412–20.

Cai, Z., and Aguilar, F. X. (2013). "Meta-Analysis of Consumer's Willingness-to-Pay Premiums for Certified Wood Products," *Journal of Forest Economics*, 19/1: 15–31.

Capmourteres, V., and Anand, M. (2016). "Assessing Abundance, Biomass, and Complexity in the Context of Ecological Forest Management," in G. R Larocque. (ed.), *Ecological Forest Management Handbook*. Boca Raton, FL: CRC Press, 305–35.

Caves, J. K., Bodner, G. S., Simms, K., Fisher, L. A., and Robertson, T. (2013). "Integrating Collaboration, Adaptive Management, and Scenario-Planning: Experiences at Las Cienegas National Conservation Area," *Ecology and Society*, 18/3: 43.

Chang, M. (2013). *Forest Hydrology: An Introduction to Water and Forests*. 3rd edn. Boca Raton, FL: CRC Press. 598 pp.

Chen, J., Innes, J. L., and Tikina, A. (2010). "Private Cost-Benefits of Voluntary Forest Product Certification," *International Forestry Review*, 12/1: 1–12.

Commissioner for Environmental Sustainability (2019). *Victoria's State of the Forests Report* 2018. Victoria, Australia: Commissioner for Environmental Sustainability. 242 pp. Available online at https://www.ces.vic.gov.au/publications-library/state-forests-2018-report (accessed February 16, 2024).

Corrigan, E., and Nieuwenhuis, M. (2016). "A Linear Programming Model to Biophysically Assess some Ecosystem Service Synergies and Trade-offs in Two Irish Landscapes," *Forests*, 7(7), 128.

Cristan, R., Aust, W. M., Bolding, M. C., Barrett, S. M., Munsell, J. F., and Schilling, E. (2016). "Effectiveness of Forestry Best Management Practices in the United States: Literature Review," *Forest Ecology and Management*, 360: 133–51.

Diaz, N., and Apostol, D. (1992). Forest Landscape Analysis and Design: A Process for Developing and Implementing Land Management Objectives for Landscape Patterns. R6 ECO-TP-043-92. USDA Forest Service, Portland, OR. 118 pp. Available on-line at https://www.fs.usda.gov/treesearch/pubs/6268 (accessed June 17, 2024).

Dufrêne, M., and Legendre, P. (1997). "Species Assemblages and Indicator Species: The Need for a Flexible Asymmetrical Approach," *Ecological Monographs*, 67/3: 345–66.

Duinker, P. N., and Trevisan, L. M. (2003). "Adaptive Management: Progress and Prospects for Canadian Forests," in P. J. Burton, C. Messier, D. W. Smith, and W. L. Adamowicz (eds), *Towards Sustainable Management of the Boreal Forest*. Ottawa: NRC Research Press, 857–92.

Ebeling, J., and Yasué, M. (2009). "The Effectiveness of Market-Based Conservation in the Tropics: Forest Certification in Ecuador and Bolivia," *Journal of Environmental Management*, 90/2: 1145–53.

Ehrenberg-Azcárate, F., and Peña-Claros, M. (2020). "Twenty Years of Forest Management Certification in the Tropics: Major Trends through Time and among Continents," *Forest Policy and Economics*, 111: 102050.

Esperon-Rodriguez, M., Ordoñez, C., van Doorn, N. S., Hirons, A., and Messier, C. (2022). "Using Climate Analogues and Vulnerability Metrics to Inform Urban Tree Species Selection in a Changing Climate: The Case for Canadian Cities," *Landscape and Urban Planning*, 228: 104578.

Esseen, P. A., and Renhorn, K. E. (1998). "Edge Effects on an Epiphytic Lichen in Fragmented Forests," *Conservation Biology*, 12/6. 1307–17.

Failing, L., and Gregory, R. (2003). "Ten Common Mistakes in Designing Biodiversity Indicators for Forest Policy," *Journal of Environmental Management*, 68/2 121–32.

FAO (2011). *Framework for Assessing and Monitoring Forest Governance*. Rome: Food and Agriculture Organization of the United Nations. 36 pp. Available online at https://www.fao.org/3/i2227e/i2227e.pdf (accessed February 25, 2024).

FAO and UNEP (2020). *The State of the World's Forests 2020: Forests, Biodiversity and People*. Rome: Food and Agriculture Organization of the United Nations and United Nations Environment Programme. 188 pp. Available on-line at https://www.fao.org/3/ca8642en/ca8642en.pdf (accessed February 16, 2024).

Fernandez-Gimenez, M. E., Ballard, H. L., and Sturtevant, V. E. (2008). "Adaptive Management and Social

Learning in Collaborative and Community-Based Monitoring: A Study of Five Community-Based Forestry Organizations in the Western USA," *Ecology and Society*, 13/2: 4.

Fernholz, K., Bowyer, J., Erickson, G., et al. (2021). *Forest Certification Update 2021: The Pace of Change*. Minneapolis, MN: Dovetail Partners. 14 pp. Available online at https://dovetailinc.org/upload/tmp/1611160123.pdf (accessed February 16, 2024).

Ford, R. M., Anderson, N. M., Nitschke, C., Bennett, L. T., and Williams, K. J. (2017). "Psychological Values and Cues as a Basis for Developing Socially Relevant Criteria and Indicators for Forest Management," *Forest Policy and Economics*, 78: 141–50.

França, F. M., Benkwitt, C. E., Peralta, G., et al. (2020). "Climatic and Local Stressor Interactions Threaten Tropical Forests and Coral Reefs," *Philosophical Transactions of the Royal Society B*, 375/1794: 20190116.

FSC-US (2010). *FSC-US Forest Management Standard (v1.0) (Complete with FF Indicators and Guidance)*. Minneapolis, MN: Forest Stewardship Council. 122 pp. Available online at https://us.fsc.org/download.fsc-us-forest-management-standard-with-family-forest-indicators.96.htm (accessed February 16, 2024).

Gale, F., and Baker, T. (2017). "Conceptualising 'Code Complexes': A Case Study of Harvesting-Related Codes Applying to Forest Operations in Tasmania, Australia," *Forest Policy and Economics*, 81: 57–64.

Gao, T., Nielsen, A. B., and Hedblom, M. (2015). "Reviewing the Strength of Evidence of Biodiversity Indicators for Forest Ecosystems in Europe," *Ecological Indicators*, 57: 420–34.

Garmestani, A. S., and Allen, C. R. (2015). "Adaptive Management of Social-Ecological Systems: The Path Forward," in C. R. Allen and A. S. Garmestani (eds), *Adaptive Management of Social-Ecological Systems*. Dordrecht: Springer, 255–62.

Gauthier, S., Lorente, M., Kremsater, L., et al. (2014). *Tracking Climate Change Effects: Potential Indicators for Canada's Forests and Forest Sector*. Natural Resources Canada, Ottawa: Canadian Forest Service. 86 pp. Available online at https://cfs.nrcan.gc.ca/pubwarehouse/pdfs/35231.pdf (accessed February 16, 2024).

Gilani, H. R., and Innes, J. L. (2020). "The State of Canada's Forests: A Global Comparison of the Performance on Montréal Process Criteria and Indicators," *Forest Policy and Economics*, 118: 102234.

Gillingham, M. P., Halseth, G. R., Johnson, C. J., and Parkes, M. W. (eds). (2016). *The Integration Imperative: Cumulative Environmental, Community and Health Effects of Multiple Natural Resource Developments*. Cham, Switzerland: Springer. 256 pp.

Greig, L. A., Marmorek, D. R., Murray, C., and Robinson, D. C. (2013). "Insight into Enabling Adaptive Management," *Ecology and Society*, 18/3: 24. Available online at http://dx.doi.org/10.5751/ES-05686-180324 (accessed February 16, 2024).

Groffman, P. M., Baron, J. S., Blett, T., et al. (2006). "Ecological Thresholds: The Key to Successful Environmental Management or an Important Concept with No Practical Application?" *Ecosystems*, 9: 1–13.

Grubb, M., Koch, M., Thomson, K., Sullivan, F., and Munson, A. (1993). *The "Earth Summit" Agreements: A Guide and Assessment—An Analysis of the Rio '92 UN Conference on Environment and Development*. London: Routledge. 202 pp.

Gudmundsson, H. (2003). "The Policy Use of Environmental Indicators: Learning from Evaluation Research," *Journal of Transdisciplinary Environmental Studies*, 2: 1–12.

Gullison, R. E. (2003). "Does Forest Certification Conserve Biodiversity?" *Oryx*, 37/2: 153–65.

Gustafson, E. J., Roberts, L. J., & Leefers, L. A. (2006). "Linking Linear Programming and Spatial Simulation Models to Predict Landscape Effects of Forest Management Alternatives," *Journal of Environmental Management*, 81/4: 339–50.

Gustafson, E. J., Shifley, S. R., Mladenoff, D. J., Nimerfro, K. K., and He, H. S. (2000). "Spatial Simulation of Forest Succession and Timber Harvesting using LANDIS," *Canadian Journal of Forest Research*, 30: 32–43.

Hammond, H. (1997). "Standards for Ecologically Responsible Forest Use," in A. Drengson and D. Taylor (eds), *Ecoforestry: The Art and Science of Sustainable Forest Use*. Gabriola, BC: New Society Publishers, 204–10.

Hansen, A. J., Noble, B. P., Veneros, J., et al. (2021). "Toward Monitoring Forest Ecosystem Integrity within the Post-2020 Global Biodiversity Framework," *Conservation Letters*, 14/4: e12822.

He, H. S., and Mladenoff, D. J. (1999). "Spatially Explicit and Stochastic Simulation of Forest Landscape Fire Disturbance and Succession," *Ecology*, 80: 81–99.

Hoberg, G. (2016). "The British Columbia Forest Practices Code: Formalization and its Effects," in M. Howlett (ed.), *Canadian Forest Policy: Adapting to Change*. Toronto: University of Toronto Press, 348–77.

Hogg, E. H., Barr, A. G., and Black, T. A. (2013). "A Simple Soil Moisture Index for Representing Multi-Year Drought Impacts on Aspen Productivity in the Western Canadian Interior," *Agricultural and Forest Meteorology*, 178: 173–82.

Holling, C. S. (ed.). (1978). *Adaptive Environmental Assessment and Management*. New York: John Wiley & Sons. 377 pp.

Innes, J. L. (2017). "Criteria and Indicators of Sustainable Forest Management," in J. L. Innes and A. V. Tikina (eds), *Sustainable Forest Management: From Concept to Practice*. New York: Routledge, 33–43.

Johnson, C. J. (2013). "Identifying Ecological Thresholds for Regulating Human Activity: Effective Conservation or Wishful Thinking?" *Biological Conservation*, 168: 57–65.

Johnson, C. J., and Ray, J. C. (2021). "The Challenge and Opportunity of Applying Ecological Thresholds to Environmental Assessment Decision Making," in J. A. E. Blakley and D. M. Franks (eds), *Handbook of Cumulative Impact Assessment*. Cheltenham, UK: Edward Elgar Publishing, 140–57.

Joshi, A. K., Pant, P., Kumar, P., Giriraj, A., and Joshi, P. K. (2011). "National Forest Policy in India: Critique of Targets and Implementation," *Small-Scale Forestry*, 10/1: 83–96.

Kadam, P., Dwivedi, P., and Karnatz, C. (2021). "Mapping Convergence of Sustainable Forest Management Systems: Comparing Three Protocols and Two Certification Schemes for Ascertaining the Trends in Global Forest Governance," *Forest Policy and Economics*, 133: 102614.

Karr, J. R. (1998). "Rivers as Sentinels: Using the Biology of Rivers to Guide Landscape Management," in R. J. Naiman and R. E. Bilby (eds), *River Ecology and Management: Lessons from the Pacific Coastal Ecoregion*. New York: Springer-Verlag, 502–28.

Kautz, M., Meddens, A. J., Hall, R. J., and Arneth, A. (2017). "Biotic Disturbances in Northern Hemisphere Forests: A Synthesis of Recent Data, Uncertainties and Implications for Forest Monitoring and Modelling," *Global Ecology and Biogeography*, 26/5: 533–52.

Keddy, P. A., and Drummond, C. G. (1996). "Ecological Properties for the Evaluation, Management, and Restoration of Temperate Deciduous Forest Ecosystems," *Ecological Applications*, 6/3: 748–62.

Kerans, B. L., and Karr, J. R. (1994). "A Benthic Index of Biotic Integrity (B-IBI) for Rivers of the Tennessee Valley," *Ecological Applications*, 4/4: 768–85.

Klenk, N. L., and Wyatt, S. (2015). "The Design and Management of Multi-Stakeholder Research Networks to Maximize Knowledge Mobilization and Innovation Opportunities in the Forest Sector," *Forest Policy and Economics*, 61: 77–86.

Landres, P. B., Morgan, P., and Swanson, F. J. (1999). "Overview of the Use of Natural Variability Concepts in Managing Ecological Systems," *Ecological Applications*, 9/4: 1179–88.

Lier, M., Köhl, M., Korhonen, K. T., Linser, S., and Prins, K. (2021). "Forest Relevant Targets in EU Policy Instruments: Can Progress Be Measured by the Pan-European Criteria and Indicators for Sustainable Forest Management?" *Forest Policy and Economics*, 128: 102481.

Lindenmayer, D. B., and Likens, G. E. (2010). "The Science and Application of Ecological Monitoring," *Biological Conservation*, 143/6: 1317–28.

Lindenmayer, D. B., and Likens, G. E. (2018). *Effective Ecological Monitoring*. 2nd edn. Clayton South, Victoria, Australia: CSIRO Publishing. 209 pp.

Lindenmayer, D. B., and Luck, G. (2005). "Synthesis: Thresholds in Conservation and Management," *Biological Conservation*, 124/3: 351–4.

Lindenmayer, D. B., Margules, C. R., and Botkin, D. B. (2000). "Indicators of Biodiversity for Ecologically Sustainable Forest Management," *Conservation Biology*, 14/4: 941–50.

Lindenmayer, D. B., Woinarski, J., Legge, S., et al. (2022). "Eight Things You Should Never Do in a Monitoring Program: An Australian Perspective," *Environmental Monitoring and Assessment*, 194/10: 701.

Löfstrand, R., Folving, S., Kennedy, P., et al. (2003). "Habitat Characterization and Mapping for Umbrella Species: An Integrated Approach Using Satellite and Field Data," in P. Corona, M. Köhl, and M. Marchetti (eds), *Advances in Forest Inventory for Sustainable Forest Management and Biodiversity Monitoring*. Dordrecht: Kluwer Academic Publishers, 191–204.

Lovett, G. M., Burns, D. A., Driscoll, C. T., et al. (2007). "Who Needs Environmental Monitoring?" *Frontiers in Ecology and the Environment*, 5/5: 253–60.

McDermott, C. L., Elbakidze, M., Teitelbaum, S., and Tysiachniouk, M. (2023). "Forest Certification in Boreal Forests: Current Developments and Future Directions," in M. M. Girona, H. Morin, S. Gauthier, and Y. Bergeron (eds), *Boreal Forests in the Face of Climate Change: Sustainable Management*. Cham, Switzerland: Springer, 533–53.

McKay, A. J., and Johnson, C. J. (2017). "Identifying Effective and Sustainable Measures for Community-Based Environmental Monitoring," *Environmental Management*, 60: 484–95.

MacKenzie, W. H., and Mahony, C. R. (2021). "An Ecological Approach to Climate Change-Informed Tree Species Selection for Reforestation," *Forest Ecology and Management*, 481: 118705.

McKinley, D. C., Miller-Rushing, A. J., Ballard, H. L., et al. (2017). "Citizen Science Can Improve Conservation Science, Natural Resource Management, and Environmental Protection," *Biological Conservation*, 208: 15–28.

Matthews, S. N., and Iverson, L.R. (2017). "Managing for Delicious Ecosystem Service under Climate Change: Can United States Sugar Maple (*Acer saccharum*) Syrup Production Be Maintained in a Warming Climate?" *International Journal of Biodiversity Science, Ecosystem Services & Management*, 13/2: 40–52.

Mette, T., Brandl, S., and Kölling, C. (2021). "Climate Analogues for Temperate European Forests to Raise Silvicultural Evidence Using Twin Regions," *Sustainability*, 13/12: 6522.

Michaelian, M., Hogg, E. H., Hall, R. J., and Arsenault, E. (2011). "Massive Mortality of Aspen Following Severe Drought along the Southern Edge of the Canadian Boreal Forest," *Global Change Biology*, 17/6: 2084–94.

Millar, C. I., and Stephenson, N. L. (2015). "Temperate Forest Health in an Era of Emerging Megadisturbance," *Science*, 349/6250: 823–6.

Miller, K. M., and McGill, B. J. (2019). "Compounding Human Stressors Cause Major Regeneration Debt in over Half of Eastern US Forests," *Journal of Applied Ecology*, 56/6: 1355–66.

Moore, T., and Tink, G. (2008). "Technical Considerations in the Design of Core Habitat Patches in Forest Management: A Case Study using the Patchworks Spatial Model," *The Forestry Chronicle*, 84/5: 731–40.

Morgan, P., Aplet, G. H., Haufler, J. B., Humphries, H. C., Moore, M. M., and Wilson, W. D. (1994). "Historical Range of Variability: A Useful Tool for Evaluating Ecosystem Change," *Journal of Sustainable Forestry*, 2/1–2: 87–111.

Mulder, B. S., Noon, B. R., Spies, T. A., et al. (1999). *The Strategy and Design of the Effectiveness Monitoring Program for the Northwest Forest Plan*. General Technical Report PNW-GTR-437. Portland, OR: USDA Forest Service. 138 pp.

Nerfa, L., Rhemtulla, J. M., and Zerriffi, H. (2020). "Forest Dependence is More than Forest Income: Development of a New Index of Forest Product Collection and Livelihood Resources," *World Development*, 125: 104689.

Newsom, D., Bahn, V., and Cashore, B. (2006). "Does Forest Certification Matter? An Analysis of Operation-Level Changes Required during the Smartwood Certification Process in the United States," *Forest Policy and Economics*, 9/3: 197–208.

Niemi, G. J., and McDonald, M. E. (2004). "Application of Ecological Indicators," *Annual Review of Ecology, Evolution, and Systematics*, 35: 89–111.

Noss, R. F. (1999). "Assessing and Monitoring Forest Biodiversity: A Suggested Framework and Indicators," *Forest Ecology and Management*, 115/2–3: 135–46.

OMNDMNRF (2021). *State of Ontario's Natural Resources: Forests 2021*. Sault Ste Marie, ON: Ontario Ministry of Northern Development, Mines, Natural Resources and Forestry, 75 pp. Available online at https://www. ontario.ca/page/state-ontarios-natural-resources-forest-2021 (accessed February 16, 2024).

Pachkowski, M., Côté, S. D., and Festa-Bianchet, M. (2013). "Spring-Loaded Reproduction: Effects of Body Condition and Population Size on Fertility in Migratory Caribou (*Rangifer tarandus*)," *Canadian Journal of Zoology*, 91/7: 473–9.

Parsons, R., Morgan, P., and Landres, P. (2000). "Applying the Natural Variability Concept: Towards Desired Future Conditions," in R. G. D'Eon, J. F. Johnson, and E. A. Ferguson (eds), *Ecosystem Management of Forested Landscapes: Directions and Implementation*. Vancouver: UBC Press, 222–37.

PEFC Czech Republic (2016). *Criteria and Indicators of Sustainable Forest Management*. CFCS Technical Document 1003:2016. Praha: PEFC Czech Republic. 38 pp. Available online at https://cdn.pefc.org/pefc.org/media/ 2019-04/8d51cd48-ae7b-441f-8d86-3aa2f0b31e5c/ 369262dc-3438-5e22-a8dd-5065d027bb80.pdf (accessed February 16, 2024).

Poland, T. M., Patel-Weynand, T., Finch, D. M., Miniat, C. F., Hayes, D. C., and Lopez, V. M. (eds). (2021). *Invasive Species in Forests and Rangelands of the United States: A Comprehensive Science Synthesis for the United States Forest Sector*. Cham, Switzerland: Springer Nature. 455 pp.

Primack, R. B. (2014). *Essentials of Conservation Biology*. 6th edn. Sunderland, MA: Sinauer Associates. 603 pp.

Radford, J. Q., Bennett, A. F., and Cheers, G. J. (2005). "Landscape-Level Thresholds of Habitat Cover for Woodland-Dependent Birds," *Biological Conservation*, 124/3: 317–37.

Ravera, S., Benesperi, R., Bianchi, E., et al. (2023). "*Lobaria pulmonaria* (L.) Hoffm.: The Multifaceted Suitability of the Lung Lichen to Monitor Forest Ecosystems," *Forests*, 14/10: 2113.

Rodríguez, J. P., Beard, T. D., Bennett, E. M., et al. (2006). "Trade-offs across Space, Time, and Ecosystem Services," *Ecology and Society*, 11/1: 28.

Ruppert, K. M., Kline, R. J., and Rahman, M. S. (2019). "Past, Present, and Future Perspectives of Environmental DNA (eDNA) Metabarcoding: A Systematic Review in Methods, Monitoring, and Applications of Global eDNA," *Global Ecology and Conservation*, 17: e00547.

Rust, A. J., Hogue, T. S., Saxe, S., and McCray, J. (2018). "Post-Fire Water-Quality Response in the Western United States," *International Journal of Wildland Fire*, 27/3: 203–16.

Seidl, R., Schelhaas, M. J., Lindner, M., and Lexer, M. J. (2009). "Modelling Bark Beetle Disturbances in a Large Scale Forest Scenario Model to Assess Climate Change Impacts and Evaluate Adaptive Management Strategies," *Regional Environmental Change*, 9/2: 101–19.

Sheppard, S. R., and Meitner, M. (2005). "Using Multicriteria Analysis and Visualisation for Sustainable Forest Management Planning with Stakeholder Groups," *Forest Ecology and Management*, 207/1–2: 171–87.

Sesser, A. L., Rickhill, A. P., Magness, D. R., et al. (eds). (2019). *Drivers of Landscape Change in the Northwest Boreal Region*. Fairbanks: University of Alaska Press. 326 pp.

Simberloff, D. (1998). "Flagships, Umbrellas, and Keystones: Is Single-Species Management Passé in the Landscape Era?" *Biological Conservation*, 83/3: 247–57.

Siry, J. P., Cubbage, F. W., Potter, K. M., and McGinley, K. (2018). "Current Perspectives on Sustainable Forest Management: North America," *Current Forestry Reports*, 4: 138–49.

Spies, T. A., Long, J. W., Charnley, S., et al. (2019). "Twenty-Five Years of the Northwest Forest Plan: What Have We Learned?" *Frontiers in Ecology and the Environment*, 17/9: 511–20.

Stone, C., and Coops, N. C. (2004). "Assessment and Monitoring of Damage from Insects in Australian Eucalypt Forests and Commercial Plantations," *Australian Journal of Entomology*, 43/3: 283–92.

Terrah, S. M., Sabi, F. Z., Mosbah, O., et al. (2020). "Nonexistence of Critical Fuel Moisture Content for Flammability," *Fire Safety Journal*, 111: 102928.

Tierney, G. L., Faber-Langendoen, D., Mitchell, B. R., Shriver, W. G., and Gibbs, J. P. (2009). "Monitoring and Evaluating the Ecological Integrity of Forest Ecosystems," *Frontiers in Ecology and the Environment*, 7/6: 308–16.

Thorn, S., Bässler, C., Brandl, R., et al. (2018). "Impacts of Salvage Logging on Biodiversity: A Meta-Analysis," *Journal of Applied Ecology*, 55/1: 279–89.

Toms, J. D., and Villard, M.-A. (2015). "Threshold Detection: Matching Statistical Methodology to Ecological

Questions and Conservation Planning Objectives," *Avian Conservation & Ecology*, 10/1: 2.

Vogt, K., Gordon, J., Wargo, J., et al. (1997). *Ecosystems: Balancing Science with Management*. New York: Springer. 470 pp.

Walters, C. (1986). *Adaptive Management of Renewable Resources*. New York: MacMillan Publishing Company. 374 pp.

Wang, S., and Fu, B. (2013). "Trade-offs between Forest Ecosystem Services," *Forest Policy and Economics*, 26: 145–6.

Welden, C. W., and Slauson, W. L. (1986). "The Intensity of Competition versus its Importance: An Overlooked Distinction and Some Implications," *Quarterly Review of Biology*, 61/1: 23–44.

Westgate, M. J., Likens, G. E., and Lindenmayer, D. B. (2013). "Adaptive Management of Biological Systems: A Review," *Biological Conservation*, 158: 128–39.

Wilson, J. P. (2021). "Making Information Measurement Meaningful: The United Nations' Sustainable Development Goals and the Social and Human Capital Protocol," *Information*, 12/8: 338.

White, P. S., and Walker, J. L. (1997). "Approximating Nature's Variation: Selecting and Using Reference Information in Restoration Ecology," *Restoration Ecology*, 5/4: 338–49.

Wolff, S., and Schweinle, J. (2022). "Effectiveness and Economic Viability of Forest Certification: A Systematic Review," *Forests*, 13/5: 798.

Wotton, B. M. (2009). "Interpreting and Using Outputs from the Canadian Forest Fire Danger Rating System in Research Applications," *Environmental and Ecological Statistics*, 16/2: 107–31.

CHAPTER 11

What Resilient Forest Management Looks Like

Show me.

slogan of the US state of Missouri

11.1 Introduction

As indicated in previous chapters, resilient forest management can employ many different forward-looking strategies to address different threats and vulnerabilities overlaid on different management objectives. Combined with the unique geographical, biological, and social context of each forest management unit, this makes managing a forest for resilience a decidedly place-based process. No single approach or portfolio of tactics is easily transferrable from one situation to another. Nonetheless, a few examples and case studies may help convince practitioners that resilient forest management is feasible, and often involves relatively minor (but important!) adjustments to established forest policies and practices. Many approaches must also be considered aspirational and unproven for decades to come, because resilience-focused management is a relatively new paradigm.

With those perspectives and caveats in mind, this chapter provides a few real-world case studies as well as some topic-specific recommendations, organized according to different land-use situations and priorities. Following the triad model (see Box 2.3), the first few sections highlight north-temperate examples from protected forests (see Section 11.2), a multipurpose forest (see Section 11.3), and a production forest (see Section 11.4). Additional case studies include examples of Indigenous land management

in Australia (see Section 11.5), and progressive efforts to manage native tropical forests in Gabon (see Section 11.6). By coincidence, three case studies include pines (*Pinus* spp.) as an important component tree species. Although pine is probably the most widely managed and harvested timber genus around the world, applications of resilient forestry can also be found in several other forest types. Some recurrent themes that emerge in the selected examples include the reintroduction of fire as a management tool, diversification of tree species and habitat conditions, and efforts to restore or mitigate the effects of past exploitation, land-use change, or management practices.

In support of the UN Sustainable Development Goals (see Box 1.2) and the principles of forest landscape restoration (see Section 9.4), the management of trees and woodlands in urban and agricultural settings is also discussed, along with the importance of non-treed lands within forest management units (see Section 11.7). Additional examples are provided in figures and text boxes in other chapters (e.g. Box 9.1, Box 12.2). The management concepts and actions employed in these examples should not sound new, as they are reiterations of themes and options that have emerged repeatedly in this book. Many other supportive resources exist, illustrating the feasibility of resilient forest management and providing templates for consideration. Krumm et al. (2020) provide more than thirty European case studies of forests managed with a

Resilient Forest Management. Philip J. Burton, Oxford University Press. © Philip J. Burton (2025). DOI: 10.1093/oso/9780198832997.003.0011

strong emphasis on biodiversity conservation under various imperatives of timber production, community protection, and multifunctional forest ecosystem services provision. Palik and D'Amato (2024) provide examples and recommendations for the implementation of silvicultural systems to sustain a broad suite of ecosystem services in fifteen North American, two southern hemisphere, and two European forest types. McAfee et al. (2010) provide several examples that emphasize landscape approaches and collaborative visioning, planning, and managing for future options.

Perhaps surprisingly, resilient forest stewardship follows different degrees of emphasis, but not different approaches, in protected areas, in multifunctional forests, and on timber-production lands. The same principles apply: protecting some natural ecosystems and their legacies in every landscape, consideration and emulation of natural disturbance regimes, maintenance of wildlife/biodiversity and their habitat needs, promotion of compositional and structural diversity, accommodation of multiple community needs and values, risk reduction in response to anticipated disruptions, and a commitment to long-term forest vitality in the face of a changing climate and an uncertain future. Application of those principles typically requires some degree of compromise, forgoing the optimization of any one ecosystem service in order to foster overall system resilience (Williams and Johnson 2015).

11.2 Resilient stewardship of a protected forest landscape

Like most national parks in countries around the world, Jasper National Park in the Rocky Mountains of Alberta, Canada, is dedicated to the conservation of natural and cultural resources for future generations while also facilitating their recreational use and educational appreciation by the public. Even though most of the human-use footprint is non-consumptive, the managers of protected areas still wrestle with the universal sustainability question of finding the right balance between use and protection. It can be quite difficult to maintain a largely natural environment dominated by resilient natural ecosystem processes while

accommodating the needs and expectations of park residents, transportation and utility infrastructure, and growing numbers of outdoor recreationists and tourists.

Established in 1907, Jasper National Park (JNP) is the largest and most northerly of the Canadian Rocky Mountain national parks. The park spans 11,228 km^2 of broad montane valley bottoms, rugged mountains, glaciers, vast subalpine forests, alpine meadows, and wild rivers along the eastern side of the continental divide that runs through the Rocky Mountains. Iconic wildlife such as grizzly bears (*Ursus arctos horribilis*), caribou (*Rangifer tarandus caribou*), wolverines (*Gulo gulo luscus*), and mountain goats (*Oreamnos americanus*) are integral parts of this landscape (Parks Canada 2022). The park also includes about 4,200 permanent residents and 12,000 seasonal workers, supports important interprovincial highways, railroads, and pipelines, and accommodates an average of 2.4 million visitors per year. It is adjacent to Banff National Park to the south, designated wilderness areas and provincial parks to the east, and provincial parks to the west in British Columbia (see Figure 11.1). With its neighboring parks, JNP is part of the Canadian Rocky Mountain Parks World Heritage Site designated by UNESCO (United Nations Educational, Scientific, and Cultural Organization) in 1990. Jasper National Park is an important link in the "Yellowstone to Yukon" (Y2Y) continent-wide protected area network; that conservation initiative intends to assure spatial continuity of natural habitats along the spine of the Western Cordillera (Chester 2015; see upper left inset to Figure 11.1).

Central to meeting its mandate and management goals is a system of land-use zoning. Expanded from recognition of the usual "front-country" (road accessible) and "back-country" (road inaccessible) areas, the JNP Management Plan recognizes five categories (see Figure 11.1). Framed in a triad perspective, zones for special preservation and wilderness prioritize protection. Outdoor recreation and park services zones (incorporating transportation and utility corridors, a townsite, alpine ski facilities, and visitor services) represent intensive land uses. Natural environment is an intermediate category providing multiple ecosystem services while experiencing light and dispersed recreational use.

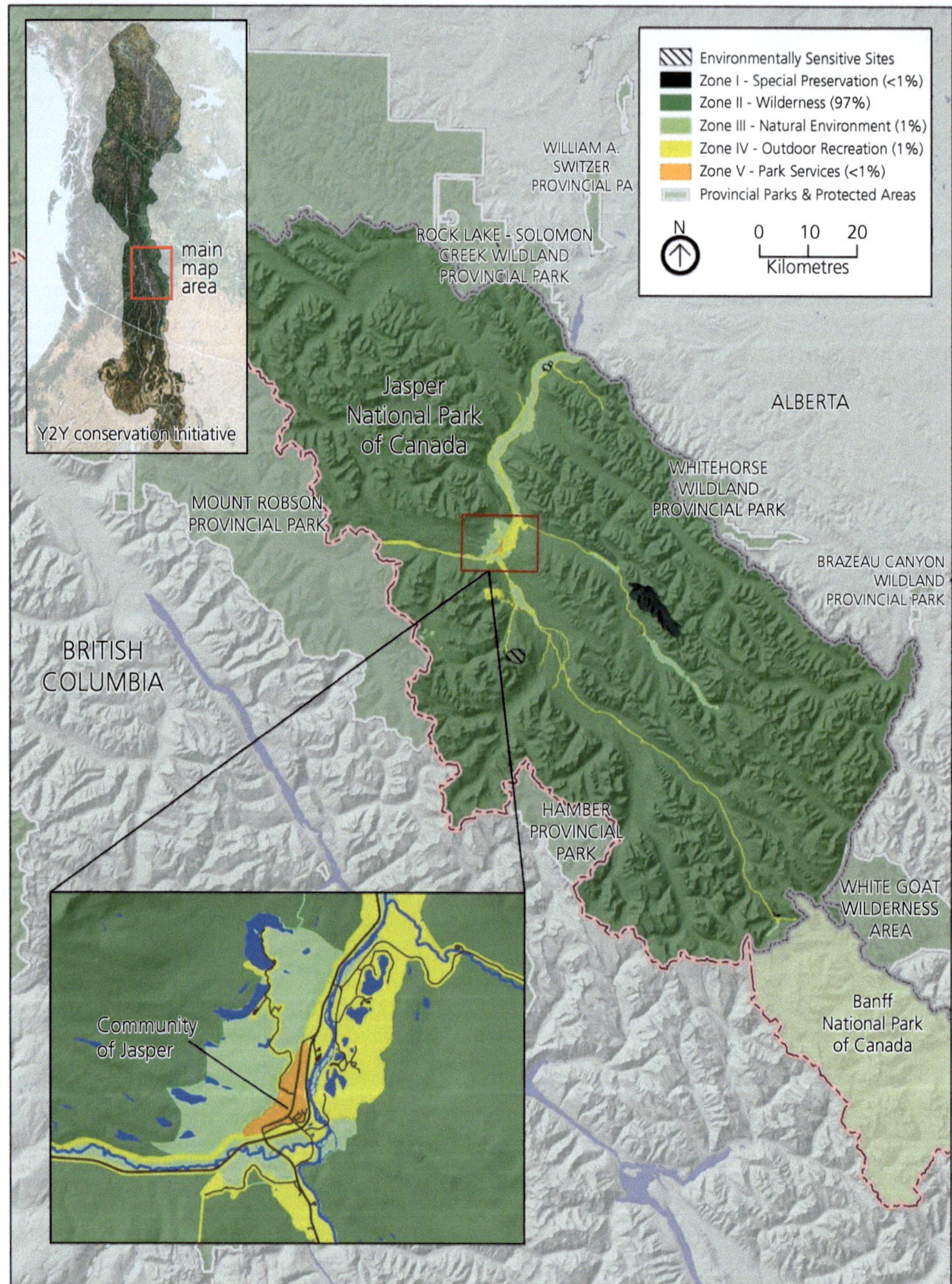

Figure 11.1 Geographic context and management zones of Jasper National Park, Canada (main map © Parks Canada (2022); Y2Y inset map © Chris Brackley/Canadian Geographic; both reproduced with permission).

Noteworthy is that JNP is not a hands-off protected area but is manipulated (that is, managed) with various levels of intensity according to the land-use zones, with fire management and fuel manipulations being the most important tools. The prohibition of wildlife hunting and trapping is also a management decision inherent to the policies of many parks and protected areas. Following a policy now endorsed by many other large national parks in North America, wildfires may be permitted to burn if they are unlikely to threaten public safety, built infrastructure, or neighboring forest lands. That sort of extensive or "modified response" wildfire management prevails in much of the wilderness back country. Fire suppression still prevails to protect infrastructure and public safety, primarily in the "natural environment" and "outdoor recreation" zones. Many valleys, however, are zoned to implement a more nuanced and situation-specific level of full or modified response to wildfire, letting wildfires burn or employing prescribed burns under controlled conditions in order to encourage more open habitats and more natural forest age-class structures.

Historical records and photographs have revealed widespread forest expansion into lower elevations that had previously been dominated by grassland and open woodlands (Rhemtulla et al. 2002). A key policy of Parks Canada is to maintain ecological integrity in its national parks, but this often requires ecological restoration (see Chapter 9) in addition to protection from human disturbance. Therefore, programs of prescribed burning have been undertaken in Jasper and Banff National Parks to restore open habitats, and to correct the growing accumulation of continuous older forest. After decades of wildfire suppression, forest age-class distributions at the landscape level were more heavily weighted to old stands and to old conifer trees (susceptible to bark-beetle attack) than expected under a natural fire regime.

Perhaps more unusual has been a program of targeted stand-level logging west and south of the Jasper townsite (see lower left inset in Figure 11.1), despite the expectation that timber harvesting normally would be prohibited in an IUCN Class II protected area. That program, conducted by commercial logging contractors and with the sale of suitable logs to a sawmill outside the park paying

for the program, was undertaken with the primary objective of removing continuous conifer fuels that might threaten the town of Jasper from prevailing wind directions. With heavy levels of stand thinning taking place throughout the treatment area, small trees, branches, and treetops were piled and burned on site, or were left dispersed to support prescribed broadcast burning treatments. The resulting open woodland has more of a savanna or parkland structure (see Figure 11.2), which has promoted meadow and grassland species, improving forage for ungulates such as elk (also known as wapiti, *Cervus canadensis*). All such fuel reduction treatments are about improving the chances of fire resistance: they are not guarantees. As this book was going to press, a large wildfire to the south of Jasper townsite unexpectedly blew up under conditions of unprecedented drought and strong winds on the late afternoon of July 24, 2024. Flying embers ignited several spot fires in town and rapidly destroyed many homes and businesses. Fire did not spread in the treated area west of town, and most of the town was saved by municipal firefighters.

Canadian national parks, particularly those in the Rocky Mountains, were early developers and practitioners of protocols for monitoring ecological integrity. Jasper National Park is now part of a nationwide program of monitoring that supports annual "state-of-the-park" reporting and the prioritization of management options (Parks Canada 2011; Wurtzebach and Schultz 2016). Data based on ground sampling and remote sensing are collected for selected indicators of ecological integrity and are synthesized for the different ecosystem types in the park. Status and trends are reported out to the public with color-coded symbology (analogous to the format described in Box 10.4).

The management of JNP also recognizes its role as part of a larger regional socio-ecological system. Its prescribed burning and stand-opening treatments at low elevations targeted lodgepole pine (*Pinus contorta* var. *latifolia*) to reduce the available habitat for mountain pine beetle (*Dendroctonus ponderosae*), which had reached epidemic proportions in nearby British Columbia (Burton 2010), with high insect populations at risk of spreading into the commercial forests of Alberta. The national park takes part in a number of regional initiatives to track and protect threatened populations of woodland

Figure 11.2 Heavily thinned forest near Jasper townsite, designed to reduce wildfire risk from the west and to provide more open, low-elevation habitat for biodiversity

caribou, grizzly bear, and whitebark pine (*Pinus albicaulis*), among other species. Its in-house program of ecological integrity monitoring parallels initiatives of the Alberta Biodiversity Monitoring Institute (Sólymos et al. 2015) being conducted on adjacent provincial lands, and contributes to annual nationwide reports under the auspices of Parks Canada (Wurtzebach and Schultz 2016). While road and rail traffic in JNP is not currently at levels experienced in the adjacent Banff National Park, the JNP transportation infrastructure may yet require isolation from natural habitats using tall fences to prevent wildlife mortalities. Fencing in Banff National Park is coupled with highway and railway underpasses and overpasses to facilitate animal population continuity and migration in the Y2Y corridor (see Denneboom et al. 2021).

Collectively, these policies are expected to improve resistance to the growing threats of elevated wildfire and insect-outbreak occurrence, while promoting forest renewal and fully complex food webs and trophic structure, all in the context of growing public demands for outdoor recreation. With an emphasis on maintaining wide expanses of wild and healthy ecosystems, it is hoped that organisms can migrate or undergo natural selection gradually to adapt to any changes in climate over the coming decades, much as

they have for millions of years. The large size and predominantly natural habitats making up JNP and its neighboring Rocky Mountain parks and wilderness areas should collectively facilitate an ongoing expression of natural resilience and adaptability by a wide range of organisms and ecosystems.

11.3 Resilient forest management in a multiple-use public forest

Many of the federally managed National Forests in the eastern US could be selected to provide examples of ecological forest management on multipurpose lands in the twenty-first century. Most have a long history of prioritizing recreation and biodiversity while also providing some sustainable timber-harvesting opportunities in support of local economies. Such policies reflect their histories and geographic context, often created from diverse landownerships after a history of land clearing or exploitative logging, and serving large urban populations with high demands for outdoor recreation and nature protection (Johnson and Govatski 2013). Apalachicola National Forest in northwestern Florida is exemplary in part because of its widespread use of prescribed fire in a manner that mimics the natural disturbance regime, and also because of the relatively modest area on which timber production and harvesting is focused.

The Apalachicola National Forest (ANF), located near the city of Tallahassee, was established in 1936 and now encompasses about 233,000 ha in a contiguous block that is managed along with two other National Forests in Florida (USFS 1999; James 2006). The ANF is a mosaic of subtropical plant communities ranging from diverse herbaceous wetlands and swamps supporting bald-cypress (*Taxodium distichum*) and mixed hardwoods to sandhill scrub oak (*Quercus incana* and *Q. laevis*) associations. Large areas were once naturally dominated by open-grown longleaf pine (*Pinus palustris*) with wiregrass (*Aristida stricta*) understories. The desirable longleaf pine trees have been mostly logged off, and much of that land was converted to slash pine (*Pinus elliotii*) and some loblolly pine (*P. taeda*) and sand pine (*P. clausa*) plantations. Largely as a result of public objections to clearcutting, Forest Service policy shifted in 1992 from promoting timber production to instead prioritizing ecosystem health and ecological sustainability (James 2006). Many rare, endangered, and threatened species are found in the ANF, particularly those associated with the longleaf pine savannas that now cover only 2–4 percent of their former range across the southern US (Jose et al. 2006; Noss et al. 2015). Examples of rare animal species include the gopher tortoise (*Gopherus polyphemus*) and the red-cockaded woodpecker (*Leuconotopicus borealis*), while Harper's beauty (*Harperocallis flava*) is a federally endangered plant endemic to the ANF. The rare Elliot's gentian (*Gentiana catesbaei*) is a pyrophyte that largely depends on recurrent fires and open conditions for persistence (O'Connor 2023). In recent years, it has been the goal of promoting population recovery of the red-cockaded woodpecker that largely shapes local ecosystem management practices (Weiss et al. 2019).

As in most National Forests in the US, the land and resource management plan for ANF reflects a highly structured, multi-year process, resulting in eleven management areas, each with specific goals, desired future conditions, standards, and guidelines (USFS 1999). Despite 90 percent of its land base being biophysically suitable for timber production, only 47 percent of the ANF is available for harvesting, mostly owing to set-asides around active or potential red-cockaded woodpecker nesting areas. Most interesting from a far-sighted stewardship perspective are ongoing efforts to convert pine plantations to fully functioning and diverse longleaf pine ecosystems,

Figure 11.3 A longleaf pine ecosystem undergoing a prescribed burn in southern Georgia, US, showing saplings and mature overstory trees of longleaf pine (*Pinus palustris*) and an understory dominated by wiregrass (*Aristida stricta*). Young seedlings look like bunchgrass for several years while they establish deep root systems, after which they grow rapidly in height. Many seedlings, saplings, and adult trees can survive light surface fires that topkill aggressive shrub and hardwood species and maintain herbaceous diversity (photo by Kevin Robertson).

with nursery-grown longleaf pine and wiregrass seedlings planted and then maintained with prescribed fire every two to four years (Jose et al. 2006; Kirkman and Jack 2017). Longleaf pine seedlings are able to persist in a unique grass-like stage under repeated burning, then undergo rapid growth to elevate foliage above the next grass-fueled fire (see Figure 11.3). Widespread use of prescribed burning—about 33,800 ha in 2021 across the ANF, for example—maintains the tall open longleaf pine canopies desired by woodpeckers, and promotes a diversity of vascular plants, insects, and vertebrates in the understory.

Principles of ecological restoration (see Chapter 9) and ecological silviculture (see Section 12.3 and Box 12.1) are also employed in the management of natural and degraded longleaf pine savannas and woodlands. Practices are inspired by the natural disturbance regime of frequent low-severity fires that restrict the development of dense shrub and hardwood layers in the understory, and occasional hurricanes that damage patches of mature trees. Large forest openings caused by logging or hurricanes can lead to dense populations of non-native invasive plant species such as cogongrass (*Imperata cylindrica*) and torpedograss (*Panicum repens*). Ongoing management activities in ANF thus include selective timber harvesting, removing exotic species (Earthbalance n.d.), and reseeding understory plants in addition to prescribed burning. The uneven-aged management specific to longleaf pine forests (frequently referred to as, or modified from, the "Stoddard–Neel approach") integrates gap-based and variable retention approaches to maintain variation in horizontal structure with clusters of regenerating and maturing pine trees, in a manner that promotes maximum biodiversity in an aesthetic park-like setting (Mitchell et al. 2006). Flexibility in cohort structure, spatial patterns, and the season of burning is encouraged (Palik et al. 2002). Although not formally adopted in the ANF, ecological silviculture for longleaf pine forests can be described as following a framework of (1) maintaining and enhancing the multi-aged structure of mature longleaf pine stands, (2) converting mature stands of other pine species, and (3) restoration plantings of longleaf pine in large disturbances (Jack et al. 2024). Frequent use of

prescribed fire, selective harvesting, and removal of undesired species are the tools applied throughout the framework, with thinning from below and variable density thinning being important in accelerating the development of a complex structure in planted stands (Fiedler et al. 1996; Jack et al. 2024).

The forests and woodlands of the ANF that are available for timber harvesting are expected to produce 1.0 million m^3/yr over their 108,800 ha, or 9.2 $m^{3 \cdot} ha^{-1 \cdot} yr^{-1}$, at least in the near term during plantation conversion to longleaf pine. Clearcut logging is practiced in sand pine stands and in slash pine or loblolly pine plantations being restored to longleaf pine. But most timber harvest is conducted using group selection and irregular shelterwood harvests that respect a visual quality objective of "canopy retention" for 70 percent of the ANF. Climate-change adaptation is not articulated as a goal in the current forest plan or its amendments, but active thinning programs in longleaf pine and slash pine stands should improve their resistance to increasing drought episodes, and uneven-aged structure in restored longleaf pine stands should facilitate rapid recovery from hurricanes (Jack et al. 2024).

Other environmentally minded policies in the ANF include restricting herbicide use to spot or strip applications, and retaining at least fifteen snags per hectare when undertaking timber-salvage operations after natural disturbances such as windstorms. There are disturbance-regime based targets for old-growth forest (or the recovery of old-growth characteristics) in each management area, and there are active programs of invasive species control and road decommissioning. Forest management must abide by specific guidelines for protecting thirteen vertebrate species as well as coarse-filter guidelines for the maintenance of habitat quality and rare and representative plant communities (USFS 1999). An integrated system of remote sensing and ground sampling has generated maps that facilitate comparisons of historical natural vegetation and current ecological conditions (Trager et al. 2018). Compliance and effectiveness of the forest plan are monitored through assessment of dozens of indicators every one to five years to allow for adjustments, and to inform the next version of the forest and resource management plan. Comprehensive reports on the status of and trends in those indicators are released to the public every two years (e.g. USFS 2022).

11.4 Resilient forest management in a private production-oriented forest

The family-owned Christinehof-Högestad estate is one of the largest in Sweden, for which commercial forestry provides the greatest source of income (Ekstrand 2020). Committed to profitable operations with a diverse line of wood-based products, the estate also incorporates several large protected areas, is undertaking restoration to native tree species, protects many rare species and encourages biodiversity, and markets forest-based experiences.

The 13,000 ha Christinehof-Högestad estate, of which 7,000 ha are productive forest land, has been owned by the same family since 1725. Approximately half of the forest area consists of native temperate broadleaf forest near the northern edge of its range, dominated by oak (*Quercus petrae*a and *Q. robur*) and beech (*Fagus sylvatica*) and other hardwoods such as lime (*Tilia cordata* and *T. platyphyllos*). The other half of the forest land base consists of spruce (*Picea abies*) and pine (*Pinus sylvestris*) plantations established on former heathlands, and short-rotation plantations of exotic hybrid poplars (*Populus balsamifera* x *trichocarpa*), hybrid larch (*Larix decidua* x *kaempferi*), and native birch (*Betula pendula* and *B. pubescens*) on former agricultural fields

(Ekstrand 2020). Large areas of native forest are maintained as nature reserves, one of which supports oaks more than three hundred years old. These formal and informal nature reserves include a bog as well as productive forest, and sum to approximately 3,250 ha in area or 25 percent of the total estate. Most of the non-forested area of the estate supports agriculture, especially cattle grazing (Paulsson 2008).

Overall stand volumes are modest (200 m^3/ha), although growth is an impressive 8.5 m^3·ha^{-1}·yr^{-1}. Even-aged management of the forest estate predominates, principally by clearcutting with reserves, in which at least 6 percent of the standing timber volume is retained for biodiversity. Large oak timbers provide the most value per tree, and cattle are sometimes grazed in harvested areas to expose mineral soil and curtail ground vegetation in order to promote acorn germination and seedling establishment. Sawlogs harvested from spruce plantations are the most important overall source of income; clearcuts are then replanted to spruce, although some openings are gradually being regenerated to broadleaf species. Pine and beech stands are naturally regenerated using the seed tree system. Except for spruce plantations, regenerating stands require fencing to protect seedlings from browsing by high populations of deer (*Dama dama* and *Capreolus capreolus*). Overall forest area is expanding, as former agricultural lands continue to be afforested. Short-rotation crops of aspen

Figure 11.4 Crop trees produced on the Christinehof–Högestad estate: (a) A representative thirty-five-year old log of hybrid larch (*Larix decidua* x *kaempferi*), which are harvested when basal diameters are 35–40 cm; (b) naturally regenerated understory of Norway spruce (*Picea abies*) and European beech (*Fagus sylvatica*) established under a planted stand of fast-growing birch (*Betula pubescens*) (photos by Anders Ekstrand).

(*Populus tremula*), exotic poplars, native birch, and hybrid larch serve as income-generating nurse crops (see Figure 11.4(a)) while establishing the more shade-tolerant, long-lived native tree species (see Figure 11.4(b)). Spruce timbers go to local sawmills, beech is exported for processing elsewhere in Europe, while thinnings, smaller timbers, and short-rotation hardwoods are directed to pulp and bioenergy markets. Mean annual increment exceeds annual harvests of 45,000–55,000 m^3, which generate some €2.5M in revenue, of which *c*. €1.0M is profit. Forest management provides employment for five to seven full-time, well-educated staff members, and for local logging contractors. Forest management and its wood products are certified under both the FSC and PEFC systems (see Box 10.1).

Biodiversity conservation and long-term planning for ecological and economic diversification are important considerations in the management of the Christinehof-Högestad estate, even while income generation is the priority. All naturally occurring snags are retained, and large woody material in the form of the tops of large trees is typically left on site after harvest. The estate supports a high diversity of bird life (150 species, including 13 raptors), as well as many species of insects and lichens that are red-listed in Sweden (Ekstrand 2020). Previously drained wetlands are being restored as aquatic habitat after harvest of the spruce plantations established there, adding to the diversity of biotopes on the landscape. The estate is open for hiking, bird watching, and mushroom and berry gathering by the public. Estate income is already diversified among forest products, livestock husbandry, and hunting permits, with hunting and tourist income for visits to the central "ecopark" around the manor house expected to be increasingly important in the future.

Although a formal system of adaptive management is not employed, growth and yield monitoring assures that wood increment exceeds wood harvests. The estate willingly hosts various scientific studies, ranging from inventories of insect diversity to carbon budget calculations (Paulsson 2008). Those investigations help managers identify stewardship options and priorities. Similarly, there are no formal plans for coping with a changing climate, although the promotion of temperate-zone broadleaf tree species (here near the northern end of their range) is a logical strategy for climate-change adaptation. Promoting two-layered forest stands (see Figure 11.4(b)) supports the option of harvesting a second timber crop a few decades after the fast-growing deciduous overstory is logged, or of retaining the naturally established native tree species indefinitely for purposes of habitat restoration. An ecologically conscious management approach that embraces biodiversity protection and forest restoration, coupled with a diverse revenue base, positions the Christinehof-Högestad estate to support a healthy, resilient forest and a vigorous, resilient enterprise for the foreseeable future. Managers credit continuity in ownership as the key to adaptation and the maintenance of multiple ecosystem services (Ekstrand 2020).

Family-owned woodlots and forest estates are generally managed with a greater emphasis on nature protection and recreational opportunities than forests run by the forest products industry or investment trusts (Bengston et al. 2011; Blanco et al 2015). Despite the challenges of the corporate governance model (see Box 2.4), there are, however, many opportunities for enhancing the resilience of industrial tree plantations. While the Christinehof-Högestad estate provides some examples (plantations of native spruce, pine, and birch species,

---

Box 11.1 Resilient forest management for industrial plantations

Although management of the Christinehof-Högestad Estate (see Section 11.4) is described as a production-oriented case study in Krumm et al. (2020), it does not represent a broad swath of industrial forestry. Therefore, opportunities

for resilient forest management in a more industrial or corporate context are explored here. The application of resilience principles in the stewardship of timber-production lands may not seem immediately obvious (or even compatible)

Box 11.1 *Continued*

where management prioritizes wood growth, harvesting efficiency, and forest estate profitability. Yet the threats of global change, risk, and uncertainty can have immediate and devastating impacts on the viability and persistence of commercial forest enterprises too. Consequently, it behooves timberland managers to prepare for such eventualities, which can be brought into focus through a broadly scoped vulnerability assessment, addressing not just climate change (see Section 6.4), but risks, threats, and uncertainties more generally (see Section 6.5). Also see Section 12.5 for a brief discussion of risk management.

The Forest Stewardship Council (FSC) wrestled with criteria and indicators by which a general agenda of environmental protection and social justice could be incorporated into the industrial model of managing fast-growing exotic tree species. Those deliberations resulted in the development of FSC's Principle #10 specifically for plantations (FSC 1996), although the latest version of the *FSC Principles and Criteria for Forest Stewardship* (FSC 2023) does not treat plantations separately. Several of the criteria under the original FSC Principle #10 are supportive of resilient forest management, embracing the precautionary principle and encouraging native species and natural ecological processes to cope with future stressors and uncertainties. Those expectations include:

- Criterion 10.2. . . . promote the protection, restoration and conservation of natural forests . . . Wildlife corridors, streamside zones and a mosaic of stands of different ages and rotation periods shall be used . . .
- Criterion 10.3 Diversity in the composition of plantations is preferred, so as to enhance economic, ecological and social stability. . .
- Criterion 10.5 A proportion . . . shall be managed so as to restore the site to a natural forest cover.
- Criterion 10.6 . . . maintain or improve soil structure, fertility, and biological activity. [plantations] . . . shall not result in long term soil degradation or adverse impacts on water quality, quantity, or substantial deviation from stream course drainage patterns.
- Criterion 10.7 . . . prevent and minimize outbreaks of pests, diseases, fire and invasive plant introductions.
- Criterion 10.8 . . . regular assessment of potential on-site and off-site ecological and social impacts. . .

The Food and Agriculture Organization of the United Nations came up with a similar set of twelve principles for responsible plantation management (FAO 2006), further emphasizing governance and stakeholder involvement. Those FAO (2006) guidelines also include explicit consideration of broader landscape and community (sociocultural) contexts, with suggestions for the operational implementation of those principles. Recognition of diverse benefits and the broader sociocultural context of planted forests also confers resilience in the form of "social license" for plantation forestry. Collectively, those international FSC and FAO guidelines are a good place to start when trying to manage plantations for resilience.

Several chapters in Bauhus et al. (2010) make the point that planted forests—including industrial plantations—can and do provide multiple ecosystem services to varying degrees, as also pointed out in Chapter 2. Foremost among these is watershed protection, with the high leaf area index and soil-holding role of big roots being superior to herbaceous or scrub cover, although (owing to periodic exposure during clearcut harvesting) timber plantations are not usually superior to natural forest in this regard (Keenan and Van Dijk 2010). Protecting soils from erosion may be a primary or secondary objective in establishing planted forests. As illustrated in Figure 11.5, however, many other ecosystem services may be inadvertently produced—even in plantations managed intensively for fiber production—without detectably reducing timber growth and yield. The emergence of those ecosystem services and opportunities can subsequently be enjoyed by forest owners or stakeholders. Collectively, this diversification of values confers greater resilience in the form of broader public support for forest plantations, and multiple pathway options for their future development.

There currently may or may not be a forest plantation program that exemplifies a broad suite of resilience-enhancing policies and practices; if such cases exist, they are not widely documented or publicized. Nonetheless, it is believed that many companies, private landholders, and government agencies implement one or more of the practices outlined here:

- as a minimum requirement, practice the same conservation measures expected of agricultural land (see Box 11.2), including the protection of soil, water, and as much native biodiversity as possible;

Box 11.1 *Continued*

Figure 11.5 Multiple ecosystem services and uses emerging in forest plantations managed for fiber production: (a) Dense yellow sweetclover (*Melilotus officinalis*) understory, highly desired by beekeepers for honey production, in a hybrid poplar (*Populus* spp.) plantation in central Saskatchewan, Canada. (b) Regeneration of Norway spruce (*Picea abies*) or silver fir (*Abies alba*) after clearfelling often requires fencing to prevent browsing by deer (*Capreolus capreolus* and *Cervus elaphus*), but also provides supportive opportunities for recreational hunting in the Plzeň Region, Czechia (Czech Republic). (c) Mature Norway spruce plantations also support recreational walking and cycling trails, and their mossy understories invite contemplative relaxation and wild mushroom foraging in Bavaria, Germany. (d) An organized off-road four-wheel drive club is permitted to explore and maintain tracks in an industrial *Pinus radiata* plantation in central North Island, New Zealand, to which access is otherwise restricted.

- avoid conversion of native vegetation (especially primary forests) to plantations of exotic tree species; rather, prioritize tree establishment on abandoned or marginal farmland, and on former industrial sites such as minespoils;
- maintain a variety of staggered stand ages across a catchment to provide more diverse habitat for biodiversity, and to avoid broad expanses of open habitat upon harvesting;
- do not log more than approximately one-third (depending on slopes, soil type, and soil depth) of a catchment basin at a time;
- maintain treed riparian buffers around streams, lakes, and ponds during harvesting operations;
- allow some patches of trees to attain greater age and size than planned for the overall stand, with such retention patches maintained indefinitely or at least until the next rotation is harvested;
- retain patches of native vegetation, or even of planted trees, centered on rare or unusual biotopes; prioritize such locations for ecological restoration;
- leave some standing snags, and some large woody material on the ground (even if of exotic tree species);

Box 11.1 *Continued*

- leave or build scattered small piles of logging slash (tree-tops, branches, and broken or decayed logs) to serve as habitat structures for a variety of terrestrial vertebrates;
- disperse most logging slash to promote the development of soil organic matter and nutrient cycling;
- follow slope contours for any site preparation furrows or trenching, in order to avoid soil erosion and contain downhill sediment movement;
- plant a diversity of tree species or clones: ideally within stands, or alternatively (and often more practically in terms of uniform time to harvestable size) in a portfolio of multiple stands/blocks in a catchment;
- minimize permanent roads and landings; employ best practices for road siting (for example, prefer ridgetops over side-slope roads), with careful drainage control in the form of appropriate bridges, culverts, cross-drains, and waterbars, especially on roads that are not being maintained;
- prepare for wildfire with fuel breaks between stands, fuel-reduction treatments, and water reservoirs on site or nearby; during extreme fire danger or when wildfire is nearby, avoid activities (industrial and public), undertake patrols to prevent ignitions and to extinguish spot fires, have fire-suppression crews and gear on standby;
- prefer tree species with growth forms (open crowns, high stem taper, deep roots) less prone to damage by high winds;

- prefer coppicing and short-rotation tree species, in order to minimize regeneration failure and to facilitate more rapid adjustment to a changing climate, respectively;
- plan for tree species, clones, and provenances to match climate projections for the full range of conditions expected throughout the rotation; and
- allow and facilitate local/community use for compatible activities: recreational hiking/tramping and cycling, dog-walking, bird-watching, hunting, berry and mushroom foraging, fuelwood gathering, and so on.

The benefits of those practices accrue in addition to the broader benefit of enhanced timber productivity that reduces the pressure to log natural forests (see Chapter 2). Some measures may require innovative engineering (for example, when logging steep sites, working around riparian buffers and other retention patches) or forgone revenues. Some practices (for example, preferring short rotations and retaining some bigger, older trees on longer rotations) may seem at odds with each other, but also constitute a diversification of product lines that may generate revenue premiums. In the long run, such practices will position a forest estate well for dealing with a changing climate and associated disturbance risks, will maintain productivity, reduce environmental impacts, and secure public support.

---

short-rotation exotic species serving as nurse crops), a number of other options exist for risk reduction and the promotion of forest vitality and persistence in a changing world (see Box 11.1), often with little or no impact on revenue-generating potential. Corporate and state forestry agencies may undertake some subset of these practices in various countries, but cases that could be described as fully embracing resilient forest management on industrial production lands are limited or non-existent at the time of writing.

11.5 Resilient stewardship of traditional Indigenous lands

After centuries of displacement and cultural prohibitions by colonizers, many Indigenous peoples around the world are now reviving traditional practices of forest and resource management. One such example can be found in the Djandak region of central Victoria, Australia, the traditional Country of the Djaara (Dja Dja Warrung People). In this context, "Country" denotes the ancestral territory of the Aboriginal First Peoples and encompasses both animate and inanimate components of the region. This region was previously a complex mosaic of floodplain forest, open woodlands, scrub, heaths, and grasslands. Over the last two centuries, most of that Country was cleared of trees and replaced with degraded gold-mining lands, cultivated croplands, and fenced sheep and cattle farms. Only a small fraction of the total area of 17,369 km^2 is nominally "public" land in the form of scattered state forests, parks, and other conservation reserves (Atkinson and Montiel-Molina 2023). It is on that diminished and degraded land base that the Djaara are now implementing a "forest gardening strategy" (DDWCAC 2022) that is part of an overall "(Healing Country) Country Plan." The strategy also calls for the Djaara to work with private landowners, municipal councils, and non-government organizations to support nature and improve the connectivity of natural habitats throughout Djandak (DDWCAC 2022).

Forest gardening, or Galk-galk Dhelkunya, means to care for and to heal trees or forests. It reflects an understanding of pre-colonial Djandak as consisting of cultural landscapes. The concept emphasizes the need for Djaara and their traditional practices to return to Djandak in order to make those landscapes (even in so-called protected areas) healthy and resilient (DDWCAC 2022). In practice, forest gardening is a landscape-scale interaction of people and nature, utilizing a toolkit of cultural practices such as selective harvesting, thinning, burning, and revegetation (DDWCAD 2022). One such scenario, outlining a multi-step process that is often required, is illustrated in Figure 11.6. The strategy is further integrated with strategies for game and water management, renewable energy, and climate change adaptation, and with multiple agreements for joint land, resource, and bushfire management with the State of Victoria (as described in multiple documents available at https://djadjawurrung.com.au/resources/). Galk-galk Dhelkunya requires working at the interface of Aboriginal and Western worldviews, with shared goals of achieving a balance of cultural, ecological, community, and economic benefits in the shared Djandak landscape (DDWCAD 2022).

Central to the Djaara forest gardening strategy is the revitalization of cultural burning practices. Aboriginal "firestick farming" is recognized as an effective means of generating fine-scale habitat mosaics that enhance the availability of traditional foods (Bliege Bird et al. 2008). Traditional food, or "bush tucker," includes edible seeds and fruits, plant roots and bulbs, and animal foods ranging from insect grubs and lizards to birds and mammals, all with particular habitat preferences. Access to and knowledge about bush tucker have nutritional and food-security benefits, especially in remote Aboriginal communities, in addition to serving an important role in cultural revitalization (Bussey 2013).

The holistic, multipurpose, and place-based aspects of cultural burning have been more widely appreciated only recently. Tangible benefits stemming from small patch burns of slow, cool, low-intensity, self-extinguishing fires include enhanced native flora and fauna, improved hunting, recovery of indigenous food and medicinal plants, the control of invasive pants, and bushfire risk mitigation. From the perspective of the Djaara, however, cultural burning is an ancestral and living practice of land management with a spiritual holistic nature requiring place-based approaches to clean and heal the land. Fire is part of Country and a sacred tool that has many Dreaming stories (deep cultural lore) attached to it (Atkinson and Montiel-Molina 2023). Victoria's Joint Fuel Management Program (FFMV 2023) now acknowledges and encourages the use of cultural fire in meeting its overall bushfire risk-reduction goal, partly in response to the catastrophic 2019–20 bushfire season. Cultural burns need to be led by Aboriginal fire knowledge holders and the Traditional Owners of Country, but are being adapted to the contemporary challenges of drastically altered land use and a changing climate (Atkinson and Montiel-Molina 2023).

Planned burns led by Traditional Owners to date largely had the ecological goal of modifying and diversifying vegetation—especially to promote desired non-tree species—and to reduce bushfire risk around communities and infrastructure. Individual burns have objectives specific to each piece of land, with ignition dates and places planned carefully on the basis of both traditional ecological knowledge (TEK) and Forest Fire Management Victoria (FFMV) agency guidelines. For example, repeated burns in Woolshed Swamp (dominated by black box, *Eucalyptus largiflorens*, and red gum, *E. camaldulensis*) are designed to assist the regeneration of rare and culturally important native grassland species such as chocolate lily (*Arthropodium strictum*), bulbine lily (*Bulbine bulbosa*), wattle mat-rush (*Lomandra filiformis*), and blue flax-lily (*Dianella longifolia*). In contrast, cultural burns at Mount Buckrabanyule, each averaging 6 ha in area, have targeted control of the invasive wheel cactus (*Opuntia robusta*) as their primary objective. The burned patches were then seeded with kangaroo grass (*Themeda triandra*), the seeds of which were traditionally used for bread making (Atkinson and

Figure 11.6 Example of woodland restoration as first steps in implementing the Djaara Galk-galk Dhelkunya forest gardening strategy in the Djandak region of Victoria, Australia. The box–ironbark forests of northern Victoria are typically dominated by grey box (*Eucalyptus microcarpa*), red ironbark (*E. tricarpa*), mugga ironbark (*E. sideroxylon*), yellow gum (*E. leucoxylon*), and red stringybark (*E. macrorrhyncha*), and are now found on a fraction of their former range (Bennett et al. 2012). The term "galka" refers to trees in general (reproduced from DDWCAD 2022 with permission of the Dja Dja Wurrung Clans Aboriginal Corporation).

Montiel-Molina 2023). Healthy stands of kangaroo grass can also restore soil structure, carbon storage, and water retention, and can slow desertification (DDWAC 2022). Although still requiring burn plans that conform to government fire agency standards, thousands of hectares of cultural burns (Djandak Wi, or Country fire) have been successfully conducted in recent years.

The Djaara forest gardening strategy is not just a backward-looking landscape restoration strategy;

it is aimed at carefully observing, hearing, and understanding what is required to heal Djandak in a post-colonial era now coping with climate change (DDWAC 2022; Atkinson and Montiel-Molina 2023). The emphasis on promoting a more diverse landscape mosaic reflects a restoration paradigm, but the fuel management implications clearly embrace forward-looking risk-reduction priorities too. "Listening to Country" is central to the TEK employed by Traditional Owners, being observant of seasonal and other ecological indicators to inform management needs and adjustments in the spirit of what Western resource managers would recognize as iterative adaptive management. The overall shift from state-mandated fire suppression and hands-off protected area management to acceptance of Indigenous landscape management using fire and other tools suggests that we are witnessing the "reorganization" phase of the adaptive cycle (see Figure 4.9(a)) in the broader socio-ecological system. Evidence in support of a forest gardening approach is provided by worldwide comparisons documenting greater biodiversity and a wider array of ecosystem services on Indigenously managed lands compared to nearby state or private lands (Fa et al. 2020; Sze et al. 2022), provided that responsive governance systems are in place (Bulkan 2017). Indigenous fire stewardship is a key driver of biodiversity in many parts of the world (Hoffman et al. 2021).

11.6 Resilient management of a tropical forest

Many arguments have been advanced that the world's remaining tropical forests should be protected from logging for reasons ranging from biodiversity conservation and safeguarding carbon stores to respect for Indigenous rights (Stone and d'Andrea 2001; Edwards et al 2019; Mackey et al. 2020; Peng et al. 2023; Roebroek et al. 2023).[1] Unfortunately, the reality is that global demand for tropical woods remains high, and impoverished rural communities and governments depend on revenue from timber harvesting from natural forests. So

timber extraction from primary and near-natural tropical forests is ongoing at the time of writing and is likely to continue in some form in the future.

In the mixed-species forests of the lowland tropics, uneven-aged management through selective logging of the very few individuals of even fewer commercial tree species can be difficult to distinguish from high-grading or timber mining but there are differences. Steps away from forest exploitation and toward sustainable management include the use of reduced-impact logging (RIL) techniques (de Blas and Pérez 2008; Putz et al. 2008; Putz and Romero 2015), harvesting at intensities and frequencies that can be sustained, and professionalization of forest workers. All of those resilience-enhancing practices are being implemented in three of the hundred-plus logging concessions in Gabon (Umunay et al. 2019). Across several sampled Gabon concessions, 60–90 percent of the pre-harvesting native forest remains intact (Putz et al. 2019), equivalent to a very high level of retention in even-aged management systems. Resilient forest management is further expressed in how the managers of these commercial operations also embrace multiple forest values and utilize practices that reduce risks and enhance climate-change mitigation and adaptation. Descriptions of these forest management enterprises, which choose to remain unnamed, are based on field observations of harvesting operations and inspection of management plans and more detailed harvest plans.

Gabon is a sparsely populated, high forest cover (>85 percent), low deforestation rate (<0.1%/yr) country in west-central Africa that is renowned for its national parks and wildlife, among which gorillas (*Gorilla gorilla*), chimpanzees (*Pan troglodytes*), and forest elephants (*Loxodonta cyclotis*) figure prominently. The lowland moist forest experiences a dry season, but it is cool with minimal moisture stress owing to persistent cloud cover. Most logging operations occur in forests that have experienced light selective logging at least once or twice before. With diminishing supplies of oil, long the economy's mainstay, the government is adopting a "green growth" development agenda ("Gabon émergent") that looks to its forest resources for a sustainable source of revenue. Policies such as the 2010 log export ban (Karsenty and Ferron 2017)

---

[1] This section is contributed by F. E. Putz and C. Romero, University of the Sunshine Coast and University of Florida.

and a more recent requirement that all concessions obtain third-party certification (Pirard et al. 2023) are intended to promote sustainable use of these resources and thereby avoid the fate of deforestation and forest degradation common to many other tropical countries. In addition to supporting sustainable forest management, the three featured concessions are managed in manners that should buffer them from a range of ecological, economic, social, and political shocks.

Recent evidence suggests that intact forests in Gabon are naturally resilient to climate change. When subjected to record temperatures and a severe drought during the 2015–16 El Niño–Southern Oscillation event, the monitored forests showed neither reductions in carbon gains nor increases in tree mortality (Bennett et al. 2021). In strong contrast to what happened elsewhere in the tropics, this lack of response has been interpreted as reflecting a history of climatic oscillations experienced by Gabon's forests over recent glacial–interglacial cycles. Coupled with the fact that Gabonese forests

are drier than elsewhere in the equatorial tropics, this means that drought-sensitive tree species are not common. Given the low logging intensities (<2 trees/ha) and employment of RIL practices, the resistance of intact forests to climate anomalies seems to carry over to harvested stands.

A component of RIL that contributes to the resilience of managed forests is the avoidance of soil compaction by careful planning and supervised use of skid trails (see Figure 11.7(a)). Soil compaction reduces plant growth by impeding root growth and therefore forest recovery after disturbances, an effect that often persists for decades. To minimize those impacts, detailed harvest plans map optimal skid-trail locations, which are then marked on the ground. Directional felling minimizes damage to other trees, with special regard to avoiding damage to future crop trees, which are paint-marked. Yarding and hauling operations are halted when the soil is wet. Through worker training, close supervision, and financial rewards for good practices, the three focal concessions all

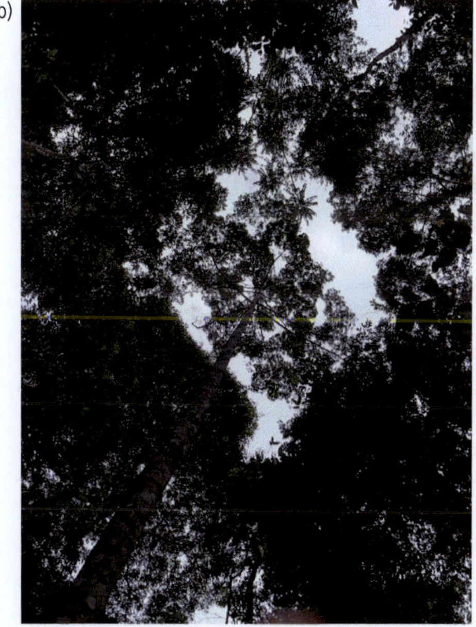

Figure 11.7   Reduced-impact logging (RIL) effectiveness in Gabonese forests: (a) A narrow skid trail in a forest concession where harvesting operations have been planned and implemented by trained personnel following RIL guidelines. Narrow paths help to minimize soil compaction and damage to neighboring vegetation including future crop trees, reducing carbon emissions, and maintaining forest attributes for forestry to persist as a sustainable land use. (b) Logging gap in a forest concession twelve years after logging in a concession managed through the adoption of RIL practices. The small central tree left standing may become a future crop tree in the next harvesting cycle, demonstrating the ability of the forest to recover from adequately planned and implemented selective management (photos by Claudia Romero).

keep soil compaction and damage to unharvested trees and vegetation to a minimum.

Over 50+ years of operations in Gabon, the three concessions displayed considerable resilience to political and economic shocks and stresses. This ability apparently derives from a number of interrelated characteristics. For example, as early adopters of forest management certification, all three avoided the stress suffered by other, less forward-thinking concessions when certification became mandatory (de Blas and Pérez 2008). Similarly, the log-export ban had only minor impacts on the two vertically integrated concessions that were already processing most of their timber in-country. It is less clear whether ownership of these three concessions by European firms confers resilience or subjects them to additional scrutiny and stress. By being large and well capitalized, these concessions are well buffered against market downturns and other economic shocks. Large operating areas support healthy animal populations having few wildlife–human conflicts, which usually involve elephants and small farms elsewhere. Instead, the wildlife provides essential seed-dispersal services for many commercial and non-commercial species across the large spatial scale at which these companies operate.

To help assure the sustainability of yields of the dozen harvested tree species, minimum cutting diameters are determined on the basis of the presence of sufficient smaller individuals (that is, future crop trees; see, for example Figure 11.7(b)) that are expected to attain harvestable size in twenty-five years when the next harvest is planned for that location (Picard et al. 2009). The models used are informed by data on growth and survival rates from permanent sample plots that help estimate reconstitution rates for each commercial species in each place. In some cases, the minimum cutting diameter is as great as 80 cm DBH; in other cases, a species is not harvested at all owing to a lack of future crop trees. By managing their forests for a variety of timber species, the concessions are buffered should harvesting a species be fully prohibited by the government owing to its importance for wildlife or to local people.

Although forest fires are not yet a major threat in Gabon, fire-maintained savannas often interdigitate with closed-canopy forests. The boundaries between the two very distinct ecosystems are dynamic: savannas expand after hot or frequent fires, and forests expand when fires are suppressed, patterns that vary locally as a function of changes in rainfall patterns. One of the tree species that colonizes savannas is *Aucoumea klaineana* (okoumé), currently the most important exported timber, exclusively as veneer (Guidosse et al. 2022). Many of the okoumé-rich stands currently being harvested were initiated after the French colonial authorities required that remote villages relocate to roadsides in the 1930s. The abandoned agricultural fields and suspended savanna-burning practices allowed okoumé to regenerate abundantly. Consequently, the challenge for today's forest managers is to promote the natural regeneration of this light-demanding species in selectively logged forests in which canopy gaps are intentionally kept small.

By treating their workers well, the companies holding these concessions enhance their resilience to labor shortages and disputes. Furthermore, managers of all three concessions argue that the substantial costs of worker training, good salaries, financial incentives for good outcomes during harvest operations, and the provision of personal protective equipment are rapidly recouped by more efficient operations. For example, trained and rewarded workers leave far less wood waste in the forest in the form of high stumps or split and unrecoverable logs. The benefits of increased worker remuneration trickle to their families, contributing to increased social well-being and community support in the socio-ecological forest system.

Overall, the insights gained after visiting these three well-managed operations prove that resilient tropical forestry is possible. The challenge now is to make these favorable practices visible to other concessions in Gabon and beyond, so that sustainable and resilient forest management can be recognized as a suitable conservation option for the tropics.

11.7 An expansive view of tree, woodland, and forest management

There are important climate mitigation and habitat provisioning roles for woody plants (and thus for foresters, arborists, and planners) in landscapes or portions of landscapes that are primarily managed for non-forest goods and services. In addition, most forests do not consist of continuous tree cover, and trees—and the ecosystem services they provide—are not restricted to lands mapped as forests. The complementary roles of trees in non-forested landscapes and of un-treed areas in forests provide an added level of diversity and resilience to the socio-ecological systems found there. This understanding is a tenet of forest and landscape restoration (Stanturf et al. 2012; see Section 9.4).

11.7.1 The great potential of our urban forests

Most towns and cities have large inventories of diverse trees in public parks and boulevards, and gracing individual homes and other private properties. Cities around the world support an average of more than 26 percent tree cover (Nowak and Greenfield 2020), with trees providing habitat for birdlife, intercepting particulate and gaseous pollutants, emitting oxygen, and moderating the effects of sun, wind, and rain (see Figure 11.8(a)). Urban and suburban trees provide immeasurable benefits in terms of aesthetics and psychological well-being, as well as carbon sequestration services that have yet to be completely tallied (Fares et al. 2017). In the US, it is estimated that treed areas store an average of 25.1 Mg C/ha within city limits, with an average of 0.46 Mg C/ha in net annual sequestration across seven representative cities (Nowak and Crane 2002). Published estimates vary, however, in the degree

to which they include the contributions of non-tree vegetation, belowground biomass, and soil carbon, and whether values are reported relative to gross urban area or for treed areas only.

Some urban forests, particularly those in gullies, ravines, and on steep terrain, constitute patches of the original natural vegetation. They are refugia of nature in a human-dominated landscape, protecting slope stability, understory plants, animals, and microbial life in regions where all land that could be cultivated or built on has now been transformed (Ramsay-Brown 2015). Much of the urban forest is a vast arboretum, with collections of trees and shrubs from all over the world, planted according to the visions of landscape architects and the whims of thousands of individual homeowners. Recognition of the importance of urban forests and their management is now reflected in a growing body of scientific research, scholarly journals dedicated to the topic, and educational programs in urban forestry at many universities (Miller et al. 2015).

As with remnant and planted trees in rural settings (see Section 11.7.2), the trees found in towns and cities provide important habitat for many woodland birds, mammals, and insects, while directly reducing the heat stress on humans and our homes through their interception of solar radiation (see Figure 11.8(a)). Afternoon air temperatures in city parks and under shade trees are frequently 3 °C lower than in open urban settings, according to a set of worldwide studies assembled by Bowler et al.

Figure 11.8 The urban forest and its many benefits: (a) A shady tree-lined street in Neepawa, Manitoba, Canada, dominated by American elms, *Ulmus americana*. (b) Native Garry oak (*Quercus garryana*) logs harvested from an urban property in Victoria, British Columbia, Canada.

(2010). Research also confirms that real-estate values and overall human quality of life in many American cities is directly correlated with tree density or canopy cover in a neighborhood (Turner-Skoff and Cavender 2019). With more frequent heatwaves occurring around the world (Perkins-Kirkpatrick and Lewis 2020), the shade benefits alone should prompt more planting of urban trees. With windstorms on the increase as well, however, a judicious selection of tree species (in terms of mature stature, rooting form, and wind resistance) and tree placement must balance their shading and other benefits.

In many ways, the urban forest is well adapted for resilience in the Anthropocene. The trees, shrubs, and associated species persisting in urban environments today have survived decades and centuries of exposure to pollution, trampling, and living in heat islands; that is, there has been a passive "urban selection" process. Planted trees in parks, roadsides, yards, and gardens typically consist of a large range of species, varieties, and genetic sources from around the world. Each city could thus be considered a largely undocumented and poorly monitored experiment in horticulture and arboriculture. The survival, growth, and vigor of those many hopeful plantings often represent tests of climate and climate-change adaptation well in advance of formalized experiments. Because urban trees and parks are an integral part of human neighborhoods, they also benefit from a dedicated constituency of citizens who regularly advocate for their health and retention. Natural forest remnants in an urban environment often face challenges from invasive weeds, marauding rats, cats, and adventurous children, however, so may require increasingly active restoration activities in order to sustain their natural values (Johnson et al. 2021).

The potential contribution of urban trees and green space to forest and community resilience remains to be fully harnessed, despite recognition of their many biophysical and health values. The leaves that regularly fall from deciduous trees are still considered a nuisance, with too much of that resource ending up in landfills rather than being used to mulch private gardens or turned into compost in support of municipal plantings and restoration activities. City planners and landscape architects today often recommend the planting of small-statured flowering trees and fruit trees rather than massive conifers, and are increasingly sensitive to the risk of storm-damaged trees falling on people, cars, and homes. Edible fruit produced by urban trees ("the urban orchard") is a key component of food security in developing countries, but much goes to waste in more prosperous countries and neighborhoods. Consequently, volunteer "gleaner" movements have sprung up in some urban centers to pick fruit from trees that would be otherwise ignored, distributing the produce to food banks, homeless shelters, and other charities (Moncrieff 2018).

Despite a high level of care and attention, trees in cities and suburbs occasionally die, are damaged, or get in the way of development plans. Urban trees also face threats from introduced pests, such as Dutch elm disease (*Ophiostoma ulmi*) and the emerald ash borer (*Agrilus planipennis*) that are decimating the urban streetscapes of eastern North America. Planning for a resilient urban forest entails provisions for a diversity of tree species and sizes or ages that cope well with biotic and abiotic disturbances. With appropriate regard for their stability and public safety, there is value in retaining some dead and dying trees in parks, especially in pocket nature reserves. Other dead trees could be felled and trimmed to provide organic benches and climbing structures in public spaces. Arborists do a brisk trade of pruning and removing dead, dying, and oversized trees (see Figure 11.8(b)); branches are regularly chipped, and the chips are often repurposed for use on trails and playgrounds. But the main boles of removed trees are rarely directed to their optimal use and value. Often consisting of a wide range of exotic hardwoods and fruitwoods, these fine materials tend to end up in landfills and fireplaces rather than in the hands of furniture-makers and other craftspeople.

Another poorly utilized resource is the great mass of timber from building demolitions and renovations that also goes to landfills, rather than being reused. Much of this material is fine-grained wood from big old trees that we will rarely see again and is much sought after for building restoration work and other fine crafting. Unfortunately, barriers regarding uncertain soundness and building codes may prevent such materials from being used again as structural timbers. Some entrepreneurial and cooperative efforts have been made at providing urban and salvaged timber brokerage services, and further opportunities exist in this area. All such activities remove harvesting pressure from natural forests, reduce transportation costs for raw materials that are used or processed in the cities, and help keep people better connected to nature and their organic environment.

For more information on urban forests and their management, see Miller et al. (2015), Salbitano et al. (2016), McBride (2017), and Pearlmutter et al. (2017). For urban forest management guidelines specifically addressing resilience to climate change and its associated stressors, see Brandt et al. (2016), plus Ordóñez Barona and Trammell (2022).

11.7.2 Promoting trees and woodlands in agricultural landscapes

An amazing 37 percent of the world's landmass is managed for food production in the form of cultivated crops or pasture and rangeland for livestock (FAO 2022). Woodland management on agricultural lands supports soil conservation, biodiversity protection, hydrological regulation, and the quality of life for rural residents, thereby contributing to resilient agricultural production with benefits that span shared landscapes. This management consists of a combination of protecting natural ecological elements and the establishment of and care for trees, shrubs, and other native vegetation. It may or may not include the purposeful production of timber, fiber, or woody biomass for harvest in a planned system of agroforestry. For example, Zhaohua et al. (1991) describe several examples of the successful intercropping of various farm products with trees in China. Patches and strips of native vegetation can be important refuges or stepping stones for many native animal species and other forms of biodiversity, especially in highly modified landscapes.

There are many reasons for farmers to integrate elements of natural vegetation, especially trees and woodlands, into their managed fields and pastures (Norton and Reid 2013; Lindenmayer et al. 2022). The partially shaded forage provided by native woodlands can be an important component of livestock grazing programs, with minimal effects on

ecosystem diversity and productivity if carefully managed (Holechek et al. 2011). Groves of trees and shrubs often host reservoir populations of insect pollinators, birds, and other predators (for example, spiders) of insect pests. Similarly, tall trees (living or dead) serve as perches and nest platforms for raptors that prey on crop-damaging rodents. Riparian buffers of perennial vegetation—preferably with diverse canopy and root structure—are essential to prevent sediment runoff and bank instability in creeks and streams, especially adjacent to croplands that are repeatedly cultivated. Fencing to exclude livestock from accessing streams and wetlands is the first step in restoring native riparian vegetation on grazing lands. Forested riparian buffers are especially effective at protecting stream ecosystems and providing multiple benefits (MacFarland et al. 2017). Efforts at soil conservation and integrated pest management can readily incorporate trees and woodlands into their programs. There can also be cost savings when marginal (rocky, steep, water-saturated, saline, or flood-prone) areas are kept in natural vegetation rather than attempting to cultivate such ground or manage livestock on problem sites.

Remnant forest groves, wetlands, hedgerows, and shelterbelts can provide important carbon sequestration services. Hedgerows have long served as field, paddock, and property boundaries in Europe, while shelterbelts situated perpendicular to prevailing winds were widely planted around the world on the open plains that are typically preferred for agriculture. For example, government programs promoted the planting of shelterbelts after the prolonged drought and wind erosion that was so widespread across central North America in the 1930s. Unfortunately, many farmers are now making the short-sighted decision to rip out those windbreaks in order to maximize production and optimize the operation of large agricultural machinery (Smith et al. 2021). To reduce the force of surface winds and provide a multitude of other ecosystem services, it is recommended that a combination of densely crowned conifers, open-branching broadleaf trees, and large-statured shrubs be planted together (Wight et al. 1991). A ramped pattern with shrubs and shorter trees on the side from which prevailing winds originate can

(a) 

(b) 

Figure 11.9 Important roles of trees and woodlands in landscapes primarily managed for agricultural production: (a) A planted shelterbelt of spruce (*Picea* sp.), willow (*Salix* sp.), poplar (*Populus* sp.), and lilac (*Syringa* sp.) in west-central Saskatchewan, Canada. (b) Cattle seeking relief from the heat in a native grove of gum trees (*Eucalyptus* sp.) in southern Victoria, Australia.

help protect tall trees (particularly conifers) from being uprooted by severe winds (see Figure 11.9(a)). Those linear woodlands and homestead groves can also serve as habitat for many forest animals, and can function as important connecting corridors or stepping stones for forest species and genes, even though they do not provide interior forest conditions and may consist of exotic tree and shrub species.

With windstorms and droughts expected to become more prevalent in future climates, one may also see increased deployment of such linear woodlands for the purposes of capturing snow and providing shade to livestock and crops. The benefits of shade trees to reduce heat stress in domestic livestock (see Figure 11.9(b)) are widely appreciated (Armstrong 1994). Groves of trees provide more benefits (and are likely to be more wind resistant) than single trees. Similar microclimate benefits have prompted the widespread use of shelterbelts and shade trees around farmyards and rural residences in otherwise open countrysides. Where livestock or dense populations of other herbivores (for example, deer, rabbits) are at large, establishing tree seedlings usually requires protective fencing for several years.

In short, maintaining woodlots, shelterbelts, riparian buffers, and other patches of woody and native vegetation contributes to sustainable, restorative, and resilient agriculture, with benefits extending to biodiversity conservation, soil protection, and hydrological regulation at the landscape scale (USDA n.d.; Smith et al. 2021). In most cases,

managing the agricultural woodland requires a shift in thinking and management practices to embrace a more holistic view of land management than taught in many production-oriented agricultural colleges. Some of the steps involve conscious protection of existing natural features, but other steps consist of plantings, fencing, and other restorative actions (see Box 11.2). The integration of trees with agricultural activities has a long history in traditional land management systems around the world (especially in the tropics), and is now receiving renewed interest and research under the broad discipline of agroforestry (Nair et al. 2021).

Box 11.2 A checklist for "natural asset farming"

The guidelines provided by Lindenmayer et al. (2022) have widespread applicability to woodland restoration and management in agricultural landscapes, even though they are designed with family-owned Australian farms in mind, and without a focus on woody vegetation alone. The following actions are modified from the table of contents of that book, with brief highlights drawn from each chapter:

1. Plan changes that make a difference

   - consider existing natural assets and identify steps and resources concurrently to improve biodiversity, productivity, and quality of life during an extinction crisis in a carbon-constrained world.

2. Enhance farm dams

   • with water an increasingly dear resource, look for opportunities to save water for use by wildlife, for irrigation, and for wildfire protection; a reservoir fringed with shade trees and other native vegetation can also serve as a nexus for biodiversity and recreational use.

3. Establish shelterbelts and other plantings

   • reduce wind and evaporation effects on soils, crops, and livestock, reintroducing native plants where possible, providing habitat for birds and other indigenous wildlife; some dead trees can be harvested for fuel, but consider leaving standing and fallen deadwood on site too.

4. Protect remnant woodland

   • retain and restore legacies of the pre-settlement vegetation, with as much compositional and structural diversity as possible to sequester and store carbon and support biodiversity; prioritize deadwood retention in these groves.

5. Protect creeks, wetlands, and riparian zones

   • restrict livestock access, stabilize creek banks, cool surface waters with tree shade, and provide flood-absorbent vegetation that is also a hotspot for biodiversity and provides connectivity and refugia of woody or other native cover across the farm.

6. Protect paddock (pasture) trees and grow new ones

   • provide shade and wind protection for livestock, where forage can sometimes remain productive longer during drought than in the open; provide perches and nest sites for birds, sequester and store carbon.

7. Maintain native perennial grasses

   • native perennial grasses typically have high root-to-shoot ratios, with greater drought tolerance than most cultivated species; root turnover contributes to deep and long-lasting carbon storage in the soil, while supporting many grassland-specific native species of insects and microbes.

8. Protect rocky outcrops

   • an important aspect of landscape complexity, rock outcrops and large boulders generate distinctive microclimates that support specialist insect and reptile species; because such locations cannot be cultivated, they can also support scattered trees and shrubs, further adding to habitat diversity.

9. Manage feral animals and overabundant native species

   • uninvited plants and animals often constrain crop and livestock productivity, and can also conflict with the rarer plant and animal species targeted for restoration; control can be achieved through a combination of habitat manipulations and carefully targeted use of manual, mechanical, and chemical methods.

10. Plan work to improve the natural assets on a farm

   • map and evaluate the multiple benefits of a portfolio of natural assets; consider how they can be better connected and mutually supportive, both within farm boundaries and as part of the larger regional landscape.

11.7.3 Importance and use of non-wooded areas in the forest landscape

A forest landscape often includes several vegetation types and land covers other than those dominated by trees. In addition to human land uses such as croplands, grazing lands, and built infrastructure, such open land can include a wide variety of wetlands, natural grasslands, rock outcrops, and scrub vegetation. Meadows and heaths support distinctive wildlife, invertebrate species, and largely unknown microbial populations that differ from those in forests. Rock outcrops and sites dominated by shallow, saline, or serpentine soils also support unique biota, and the few harvestable trees they may support are difficult to log without causing site damage and are difficult to regenerate. All such cover types with few or no trees may serve to limit the spread of wildfires or provide locations where fire control is more feasible.

Open-water lakes, graminoid-dominated marshes and fens, and peat-building bogs and mires (see Figure 11.10(a)) contribute to the hydrological regulation of water tables in the forest. Creeks and rivers are the "veins and arteries" of the forest, providing hydrological and habitat connectivity as well as distinctive habitat for prized fish and the invertebrates on which they feed. Collectively, surface water bodies denote near-surface water tables that support greater forest productivity and help avoid drought impacts. Open ponds and lakes often support recreational use in the form of fishing, water sports, camping, and general relaxation, while also providing habitat for a variety of aquatic species.

Bodies of standing water and associated landscape depressions also serve important roles in carbon sequestration and biodiversity protection. Organic matter accumulated in wetland sediments and peats is an important carbon reservoir that warrants protection (Yu et al. 2011; Nahlik and Fennessy 2016; Harenda et al. 2018; see Box 7.1). Protecting peatlands from wildfire, drainage, and land-use change is just as important as protecting forests from unintended conflagration, because these organic deposits can easily release rather than store carbon (as CO_2 and CH_4) when disturbed (Lin et al. 2021).

Just as water bodies serve as attractants to wildlife in the upland forest see (Figure 11.10(b)), so too do other treeless landscape features provide important habitat for a variety of species and ecological processes. Meadows dominated by grasses and forbs—whether natural or maintained by traditional practices of prescribed burning, livestock grazing, or haying—are often the preferred grazing grounds of ungulates and support distinctive plant and

insect species. Similarly, dense thickets of shrubs may provide preferred nesting sites for a number of birds and can persist as the dominant vegetation. Those meadows and shrublands may expand or contract during shifts in climate, so maintaining healthy reservoirs of native herb and shrub populations may be important in excluding exotic invaders. Rock outcrops, cliffs, caves, sinkholes, and springs all host distinctive flora and fauna. They and their immediate vicinities are generally unsuitable for commercial forestry, and their value to biodiversity suggests that it is best to buffer them with zones of natural vegetation free of human disturbance.

Many forest landscapes are dissected by non-treed linear corridors used for roads and railroads, electricity transmission (see Figure 11.11(a)), seismic exploration, or pipelines. Often kept free of trees to maintain access and facilitate servicing, and tens of meters in width, such corridors can be very disruptive to many forest species, values, and processes. As openings that introduce more sunlight and wind to the nearby forest, their footprint (edge effect) extends beyond their mapped area, as evidenced by windthrow, invasive species, and altered forest structure and composition in the adjacent forest (Eldegard et al. 2015; Dabros et al. 2018). The open habitat can constitute a barrier to the movement and dispersal of some forest-loving species, while it can serve as a conduit for the more rapid travel

Figure 11.10 Wetlands serving as important nodes of biodiversity and regulating hydrological and carbon cycles within the forest landscape: (a) A boreal mire with black spruce (*Picea mariana*) scattered amid cottongrass (*Eriophorum* sp.) and *Sphagnum* hummocks in northern Ontario, Canada. (b) Moose (*Alces alces*) foraging in a pond created by beaver (*Castor canadensis*) dams and now fringed with bluejoint reedgrass (*Calamagrostis canadensis*) and small snags in northwestern British Columbia, Canada.

(a)

(b)

Figure 11.11 Electrical transmission corridors in forested landscapes. Kept free of trees, such corridors provide opportunities to bundle various disruptive or non-timber uses. (a) A typical medium-scale corridor for high-voltage electricity transmission, also used for off-road motor sports, by hunters, and by berry pickers in western British Columbia, Canada. (b) Dense cover of staghorn sumac (*Rhus typhina*), a native shrub species that provides habitat for birds and small mammals while effectively reducing tree establishment and growth under an electrical transmission line in southern Ontario, Canada.

of human hunters, other predators, and invasive species (Hansen and Clevenger 2005).

The limited options for corridor management, given the priority assigned to non-forest land uses and human activities, present considerable challenges to forest land managers. It is typically assumed that road and utility rights-of-way should be dominated by grasses or other low vegetation. In planning for resilience to disturbance, such open areas can serve as fuel breaks as part of larger landscape plans to limit the spread of wildfire (Brzuszek et al. 2010; see Box 8.2). There are opportunities to promote diverse native meadow vegetation and wildflowers on roadsides (Landis et al. 2005), or shrubby vegetation under electrical transmission lines that may successfully inhibit the establishment and growth of trees that repeatedly need removal (Meilleur et al. 1994; see Figure 11.11(b)). There are advantages to combining road and utility corridors, thereby providing improved access for utility maintenance and decreasing overall forest fragmentation. Encouraging off-road motorized recreation (four-wheel drive jeeps and trucks, all-terrain vehicles or quads, snowmobiles) in utility corridors can serve to reduce environmental damage elsewhere in the forest. Locating some forest uses in power line and pipeline corridors also provides an opportunity to enhance non-timber values, goods, and services: livestock grazing, hay making, Christmas

trees, orchards, and small-fruit (for example, blueberry, *Vaccinium* sp.) production are all supportive of keeping those corridors free of tall trees. As contrasts with the adjacent forest matrix are still high, the width of utility and transportation corridors should be kept to a minimum. Directing non-timber uses to infrastructure corridors can sometimes support multiple-use objectives and may promote biodiversity and carbon sequestration at levels greater than the current default ground cover of domesticated grasses.

Box 11.3 Key points

- The case studies and guidelines provided in this chapter illustrate how resilient forest management follows similar principles in protected areas, in multipurpose forests, and on commodity-production lands in many countries, although the approach taken varies with land-use emphasis, local conditions, and priorities.
- Common to most examples of progressive, future-minded, resilience-enhancing forest and land stewardship are efforts to:

  o diversify stand and landscape composition and structure, and the portfolio of ecosystem services generated (some of which can be monetized);
  o protect natural ecosystems;

Box 11.3 *continued*

o encourage native wildlife and biodiversity and the habitats they require;

o consider (and sometimes emulate) natural disturbance regimes or Indigenous management regimes;

o reduce identifiable risks to anticipated disruptions such as droughts, wildfires, or storms;

o accommodate multiple community needs and values; and

o commit to long-term forest vitality in the face of a changing climate and an uncertain future.

• Prescribed burning is being used in many parts of the world where fire had been part of the natural or historical disturbance regime, where it supports forest regeneration and biodiversity while reducing the likelihood of destructive wildfires.

• Resilient forest management often requires ecological restoration in order to undo or mitigate decades to centuries of degrading land management policies and practices.

• Efforts at sustainable forest management are a good start, but resilient forest management requires consideration of environmental and social trends and appropriate preparations for dealing with their consequences.

• Resilience planning often requires management investments and forgoing short-term profits and efficiencies in order better to assure persistence and the attainment of long-term goals.

The reality is that the only way change comes is when you lead by example.

Anne Wojcicki (2016)

References cited

Armstrong, D. (1994). "Heat Stress Interaction with Shade and Cooling," *Journal of Dairy Science*, 77/7), 2044–50.

Atkinson, A., and Montiel-Molina, C. (2023). "Reconnecting Fire Culture of Aboriginal Communities with Contemporary Wildfire Risk Management," *Fire*, 6/8: 296.

Bauhus, J., Pokorny, B., van der Meer, P. J., Kanowski, P., and Kanninen, M. (2010). "Ecosystem Goods and Services: The Key for Sustainable Plantations," in J. Bauhus, P. van der Meer, and M. Kanninen (eds), *Ecosystem Goods and Services from Plantation Forests*. London: Earthscan, 205–27.

Bengston, D. N., Asah, S. T., and Butler, B. J. (2011). "The Diverse Values and Motivations of Family Forest Owners in the United States: An Analysis of an Open-Ended Question in the National Woodland Owner Survey," *Small-Scale Forestry*, 10: 339–55.

Bennett, A. C., Dargie, G. C., Cuni-Sanchez, A., et al. (2021). "Resistance of African Tropical Forests to an Extreme Climate Anomaly," *Proceedings of the National Academy of Sciences*, 118/21: e2003169118.

Bennett, A. F., Holland, G. J., Flanagan, A., Kelly, S., and Clarke, M. F. (2012). "Fire and its Interaction with Ecological Processes in Box–Ironbark Forests," *Proceedings of the Royal Society of Victoria*, 124/1: 72–8.

Blanco, V., Brown, C., and Rounsevell, M. (2015). "Characterising Forest Owners through their Objectives, Attributes and Management Strategies," *European Journal of Forest Research*, 134: 1027–41.

Bliege Bird, R., Bird, D. W., Codding, B. F., Parker, C. H., and Jones, J. H. (2008). "The 'Fire Stick Farming' Hypothesis: Australian Aboriginal Foraging Strategies, Biodiversity, and Anthropogenic Fire Mosaics," *Proceedings of the National Academy of Sciences*, 105/39: 14796–801.

Bowler, D. E., Buyung-Ali, L., Knight, T. M., and Pullin, A. S. (2010). "Urban Greening to Cool Towns and Cities: A Systematic Review of the Empirical Evidence," *Landscape and Urban Planning*, 97/3: 147–55.

Brandt, L., Derby-Lewis, A., Fahey, R. T., Scott, L., Darling, L., and Swanston, C. (2016). "A Framework for Adapting Urban Forests to Climate Change," *Environmental Science & Policy*, 66: 393–402.

Brzuszek, R., Walker, J., Schauwecker, T., Campany, C., Foster, M., and Grado, S. (2010). "Planning Strategies for Community Wildfire Defense Design in Florida," *Journal of Forestry*, 108/5: 250–57.

Bulkan, J. (2017). "Indigenous Forest Management," *CABI Reviews*. 16 pp. Available online at https://doi.org/10.1079/PAVSNNR201712004 (accessed February 17, 2024).

Burton, P. J. (2010). "Striving for Sustainability and Resilience in the Face of Unprecedented Change: The Case of the Mountain Pine Beetle Outbreak in British Columbia," *Sustainability*, 2/8: 2403–23.

Bussey, C. (2013). "Food Security and Traditional Foods in Remote Aboriginal Communities: A Review of the Literature," *Australian Indigenous Health Bulletin*, 13/2. 10 pp. Available online at https://healthbulletin.org.au/articles/food-security-and-traditional-foods-in-remote-aboriginal-communities-a-review-of-the-literature/ (accessed February 17, 2024).

Chester, C. C. (2015). "Yellowstone to Yukon: Transborder Conservation across a Vast International Landscape," *Environmental Science & Policy*, 49: 75–84.

Dabros, A., Pyper, M., and Castilla, G. (2018). "Seismic Lines in the Boreal and Arctic Ecosystems of North America: Environmental Impacts, Challenges, and Opportunities," *Environmental Reviews*, 26/2: 214–29.

DDWCAC (2022). *Galk–Galk Dhelkunya—Forest Gardening Strategy 2022–2034*. Bendigo, VIC, Australia: Dja Dja Wurrung Clans Aboriginal Corporation. Available online at https://djadjawurrung.com.au/resources/ (accessed February 17, 2024).

de Blas, D. E., and Pérez, M. R. (2008). "Prospects for Reduced Impact Logging in Central African Logging Concessions," *Forest Ecology and Management*, 256/7: 1509–16.

Denneboom, D., Bar-Massada, A., and Shwartz, A. (2021). "Factors Affecting Usage of Crossing Structures by Wildlife: A Systematic Review and Meta-Analysis," *Science of the Total Environment*, 777: 146061.

Earthbalance (n.d.). *Apalachicola National Forest Vegetation Management*. North Port, FL: Earthbalance Corporation. Available online at https://www.earthbalance.com/case-study/apalachicola-national-forest-vegetation-management (accessed February 17, 2024).

Edwards, D. P., Socolar, J. B., Mills, S. C., Burivalova, Z., Koh, L. P., and Wilcove, D. S. (2019). "Conservation of Tropical Forests in the Anthropocene," *Current Biology*, 29/19: R1008–R1020.

Ekstrand, A. (2020). "Christine-Högestad Estate: A Case from Scania in Southern Sweden," in F. Krumm, A. Schuck, and A. Rigling (eds), *How to Balance Forestry and Biodiversity Conservation: A View across Europe*. Birmensdorf, Switzerland: European Forest Institute (EFI) and Swiss Federal Institute for Forest, Snow and Landscape Research (WSL), 322–9.

Eldegard, K., Totland, Ø., and Moe, S. R. (2015). "Edge Effects on Plant Communities along Power Line Clearings," *Journal of Applied Ecology*, 52/4: 871–80.

Fa, J. E., Watson, J. E., Leiper, I., et al. (2020). "Importance of Indigenous Peoples' Lands for the Conservation of Intact Forest Landscapes," *Frontiers in Ecology and the Environment*, 18/3: 135–40.

FAO (2006). *Responsible Management of Planted Forests: Voluntary Guidelines*. Planted Forests and Trees Working Paper 37/E. Rome: Food and Agriculture Organization of the United Nations. 73 pp. Available online at https://www.fao.org/forestry/plantedforestsguide/38075/en/ (accessed February 17, 2024).

FAO (2022). *FAO Statistical Database (FAOstat)*. Rome: Food and Agriculture Organization of the United Nations. Available online at https://www.fao.org/faostat/en/#data (accessed February 17, 2024).

Fares, S., Paoletti, E., Calfapietra, C., Mikkelsen, T. N., Samson, R., and Le Thiec, D. (2017). "Carbon Sequestration by Urban Trees," in D. Pearlmutter, C. Calfapietra, R. Samson, et al. (eds), *The Urban Forest: Cultivating Green Infrastructure for People and the Environment*. Cham, Switzerland: Springer, 31–9.

FFMV (2023). *Joint Fuel Management Program*. Melbourne: Forest Fire Management Victoria. Available online at https://www.ffm.vic.gov.au/bushfire-fuel-and-risk-management/joint-fuel-management-program (accessed February 17, 2024).

Fiedler, C. E., Arno, S. F., and Harrington, M. G. (1996). "Flexible Silvicultural and Prescribed Burning Approaches for Improving Health of Ponderosa Pine Forests," in W. Covington and P. K. Wagner (technical coordinators), *Conference on Adaptive Ecosystem Restoration and Management: Restoration of Cordilleran Conifer Landscapes of North America*. General Technical Report RM-GTR-278. Fort Collins, CO: USDA Forest Service, 69–74.

FSC (1996). *FSC Principles and Criteria for Forest Stewardship*. FSC-STD-01-001 (version 4-0) EN. Bonn, Germany: Forest Stewardship Council International. 13 pp. Available online at https://open.fsc.org/handle/resource/392.1 (accessed February 17, 2024).

FSC (2023). *FSC Principles and Criteria for Forest Stewardship*. FSC-STD-01-001 V5-3 EN. Bonn, Germany: Forest Stewardship Council International. 33 pp. Available online at https://open.fsc.org/handle/resource/392 (accessed February 17, 2024).

Guidosse, Q., Du Jardin, P., White, L., Lassois, L., and Doucet, J. L. (2022). "Gabon's Green Gold: A Bibliographical Review of Thirty Years of Research on Okoumé (*Aucoumea klaineana* Pierre)," *Biotechnologie, Agronomie, Société et Environnement*, 26/1: 30–42.

Hansen, M. J., and Clevenger, A. P. (2005). "The Influence of Disturbance and Habitat on the Presence of Non-Native Plant Species along Transport Corridors," *Biological Conservation*, 125/2: 249–59.

Harenda, K. M., Lamentowicz, M., Samson, M., and Chojnicki, B. H. (2018). "The Role of Peatlands and their Carbon Storage Function in the Context of Climate Change," in T. Zielinski, I. Sagan, and W. Surosz (eds), *Interdisciplinary Approaches for Sustainable Development Goals: Economic Growth, Social Inclusion and Environmental Protection*. Cham, Switzerland: Springer, 169–87.

Hoffman, K. M., Davis, E. L., Wickham, S. B., et al. (2021). "Conservation of Earth's Biodiversity is Embedded in Indigenous Fire Stewardship," *Proceedings of the National Academy of Sciences*, 118/32: e2105073118.

Holechek, J. L., Pieper, R. D., and Herbel, C. H. (2011). *Range Management: Principles and Practices.* 6th edn. Upper Saddle River, NJ: Prentice Hall. 607 pp.

Jack, S. B., Knapp, B. O., and McIntyre, R. K. (2024). "Ecological Silviculture for Longleaf Pine Woodlands in the Southeastern US," in B. J. Palik and A. W. D'Amato (eds), *Ecological Silviculture Systems: Exemplary Models for Sustainable Forest Management.* Hoboken, NJ: John Wiley & Sons, 53–66.

James, F C. (2006). *An Introduction to the History, Ecology, and Management of the Apalachicola National Forest.* Fact Sheet, 1(1). Tallahassee, FL: Friends of the Apalachicola National Forest. 6 pp. Available online at https://www.bio.fsu.edu/FANF/Fact%20Sheet%201%20-%20ANF%20basics.pdf (accessed February 17, 2024).

Johnson, C., and Govatski, D. (2013). *Forests for the People: The Story of America's Eastern National Forests.* Washington: Island Press. 394 pp.

Johnson, L. R., Johnson, M. L., Aronson, M. F., et al. (2021). "Conceptualizing Social–Ecological Drivers of Change in Urban Forest Patches," *Urban Ecosystems,* 24: 633–48.

Jose, S., Jokela, E., and Miller, D. (eds). (2006). *The Longleaf Pine Ecosystem: Ecology, Restoration, and Silviculture.* New York: Springer. 450 pp.

Karsenty, A., and Ferron, C. (2017). "Recent Evolutions of Forest Concessions Status and Dynamics in Central Africa," *International Forestry Review,* 19/4: 10–26.

Keenan, R. J., and Van Dijk, A. I. J. (2010). "Planted Forests and Water," in J. Bauhus, P. van der Meer, and M. Kanninen (eds), *Ecosystem Goods and Services from Plantation Forests.* London: Earthscan, 77–95.

Kirkman, L. K., and Jack, S. B. (eds). (2017). *Ecological Restoration and Management of Longleaf Pine Forests.* Boca Raton, FL: CRC Press. 720 pp.

Krumm, F., Schuck, A., and Rigling, A. (eds). (2020). *How to Balance Forestry and Biodiversity Conservation: A View across Europe.* Birmensdorf, Switzerland: European Forest Institute (EFI) and Swiss Federal Institute for Forest, Snow, and Landscape Research (WSL). 640 pp. Available online at https://forbiodiv.wsl.ch/de/the-book.html (accessed February 14, 2024).

Landis, T. D., Wilkinson, K. M., Steinfeld, D. E., Riley, S. A., and Fekaris, G. N. (2005). "Roadside Revegetation of Forest Highways: New Applications for Native Plants," *Native Plants Journal,* 6/3: 297–305.

Lin, S., Liu, Y., and Huang, X. (2021). "Climate-Induced Arctic-Boreal Peatland Fire and Carbon Loss in the 21st Century," *Science of The Total Environment,* 796: 148924.

Lindenmayer, D. B., Macbeth, S. M., Smith, D. G., and Young, M. L. (2022). *Natural Asset Farming: Creating Productive and Biodiverse Farms.* Melbourne: CSIRO Publishing. 184 pp.

McAfee, B. J., de Camino, R., Burton, P. J., et al. (2010). "Managing Forested Landscapes for Socio-Ecological Resilience," in G. Mery, P. Katila, G. Galloway, et al. (eds), *Forests and Society: Responding to Global Drivers of Change.* Vienna: International Union of Forest Research Organizations (IUFRO), 401–39.

McBride, J. R. (2017). *The World's Urban Forests: History, Composition, Design, Function and Management.* Cham, Switzerland: Springer. 266 pp.

MacFarland, K., Straight, R., and Dosskey, M. (2017). *Riparian Forest Buffers: An Agroforestry Practice.* Agroforestry Notes 49, Lincoln, NE: United States Department of Agriculture. 8 pp. Available online at https://www.fs.usda.gov/nac/practices/riparian-forest-buffers.php (accessed February 17, 2024).

Mackey, B., Kormos, C. F., Keith, H., et al. (2020). "Understanding the Importance of Primary Tropical Forest Protection as a Mitigation Strategy," *Mitigation and Adaptation Strategies for Global Change,* 25/5: 763–87.

Meilleur, A., Véronneau, H., and Bouchard, A. (1994). "Shrub Communities as Inhibitors of Plant Succession in Southern Quebec," *Environmental Management,* 18: 907–21.

Miller, R. W., Hauer, R. J., and Werner, L. P. (2015). *Urban Forestry: Planning and Managing Urban Greenspaces.* 3rd edn. Long Grove, IL: Waveland Press. 560 pp.

Mitchell, R. J., Hiers, J. K., O'Brien, J. J., Jack, S. B., and Engstrom, R. T. (2006). "Silviculture that Sustains: The Nexus between Silviculture, Frequent Prescribed Fire, and Conservation of Biodiversity in Longleaf Pine Forests of the Southeastern United States," *Canadian Journal of Forest Research,* 36/11: 2724–36.

Moncrieff, H. (2018). *The Fruitful City: The Enduring Power of the Urban Food Forest.* Toronto: ECW Press. 240 pp.

Nair, P. K. R., Kumar, B., and Nair, V. D. (2021). *An Introduction to Agroforestry: Four Decades of Scientific Developments.* Cham, Switzerland: Springer Nature. 666 pp.

Nahlik, A. M., and Fennessy, M. S. (2016). "Carbon Storage in US Wetlands," *Nature Communications,* 7/1: 1–9.

Norton, D., and Reid, N. (2013). *Nature and Farming: Sustaining Native Biodiversity in Agricultural Landscapes.* Melbourne: CSIRO Publishing. 294 pp.

Noss, R. F., Platt, W. J., Sorrie, B. A., et al. (2015). "How Global Biodiversity Hotspots May Go Unrecognized: Lessons from the North American Coastal Plain," *Diversity and Distributions,* 21/2: 236–44.

Nowak, D. J., and Crane, D. E. (2002). "Carbon Storage and Sequestration by Urban Trees in the USA," *Environmental Pollution,* 116/3: 381–9.

Nowak, D. J., and Greenfield, E. J. (2020). "The Increase of Impervious Cover and Decrease of Tree Cover within Urban Areas Globally (2012–2017)," *Urban Forestry & Urban Greening*, 49: 126638.

O'Connor, M. R. (2023). *Ignition: Lighting Fires in a Burning World*. New York: Hachette Book Group. 384 pp.

Ordóñez Barona, C., and Trammell, T. L. (2022). "Urban Trees in a Changing Climate: Science and Practice to Enhance Resilience," *Frontiers in Ecology and Evolution*, 10: 882510.

Palik, B. J., and D'Amato, A. W. (eds). (2024). *Ecological Silvicultural Systems: Exemplary Models for Sustainable Forest Management*. Hoboken, NJ: John Wiley & Sons. 342 pp.

Palik, B. J., Mitchell, R. J., and Hiers, J. K. (2002). "Modelling Silviculture after Natural Disturbance to Sustain Biodiversity in the Longleaf Pine (*Pinus palustris*) Ecosystem: Balancing Complexity and Implementation," *Forest Ecology and Management*, 155/1–3: 347–56.

Parks Canada (2011). *Consolidated Guidelines for Ecological Integrity Monitoring in Canada's National Parks*. Gatineau, Quebec: Protected Areas Establishment and Conservation Branch, Parks Canada. 115 pp.

Parks Canada (2022). *Jasper National Park of Canada Management Plan*. Jasper, AB: Parks Canada. 36 pp. Available online at https://parks.canada.ca/pn-np/ab/jasper/gestion-management/plan/involved/plandirecteur-mgntplan (accessed February 17, 2024).

Paulsson, R. (2008). "Analysing Climate Effect of Agriculture and Forestry in Southern Sweden at Högestad & Christinehof Estate." Degree project in Physical Geography, Department of Physical Geography and Ecosystem Analysis, Lund University, Lund, Sweden. 28 pp. Available online at https://lup.lub.lu.se/student-papers/search/publication/1895520 (accessed February 17, 2024).

Pearlmutter, D., Calfapietra, C., Samson, R., et al. (eds). (2017). *The Urban Forest: Cultivating Green Infrastructure for People and the Environment*. Cham, Switzerland: Springer. 351 pp.

Peng, L., Searchinger, T. D., Zionts, J., and Waite, R. (2023). "The Carbon Costs of Global Wood Harvests," *Nature*, 620: 110–15.

Perkins-Kirkpatrick, S. E., and Lewis, S. C. (2020). "Increasing Trends in Regional Heatwaves," *Nature Communications*, 11/1: 3357.

Picard, N., Banak, L. N., Namkosserena, S., and Yalibanda, Y. (2009). "The Stock Recovery Rate in a Central African Rain Forest: An Index of Sustainability Based on Projection Matrix Models," *Canadian Journal of Forest Research*, 39/11: 2138–52.

Pirard, R., Pacheco, P., and Romero, C. (2023). "The Role of Hybrid Governance in Supporting Deforestation-Free Trade," *Ecological Economics*, 210: 107867.

Putz, F. E., Baker, T., Griscom, B. W., et al. (2019). "Intact Forest in Selective Logging Landscapes in the Tropics," *Frontiers in Forests and Global Change*, 2: 30.

Putz, F. E., and Romero, C. (2015). *Futures of Tropical Production Forests*. CIFOR Occasional Paper 143. Bogor, Indonesia: Center for International Forestry Research. 39 pp. https://doi.org/10.17528/cifor/005766.

Putz, F. E., Sist, P., Fredericksen, T., and Dykstra, D. (2008). "Reduced-Impact Logging: Challenges and Opportunities," *Forest Ecology and Management*, 256/7: 1427–33.

Ramsay-Brown, J. (2015). *Toronto's Ravines and Urban Forests: Their Natural Heritage and Local History*. Rev. edn. Toronto: James Lorimer & Company. 191 pp.

Rhemtulla, J. M., Hall, R. J., Higgs, E. S., and Macdonald, S. E. (2002). "Eighty Years of Change: Vegetation in the Montane Ecoregion of Jasper National Park, Alberta, Canada," *Canadian Journal of Forest Research*, 32/11: 2010–21.

Roebroek, C. T., Duveiller, G., Seneviratne, S. I., Davin, E. L., and Cescatti, A. (2023). "Releasing Global Forests from Human Management: How Much More Carbon Could Be Stored?" *Science*, 380: 749–53.

Salbitano, F., Borelli, S., Conigliaro, M., and Chen, Y. (2016). *Guidelines on Urban and Peri-urban Forestry*. FAO Forestry Paper, No. 178. Rome: Food and Agriculture Organization of the United Nations. 158 pp.

Smith, M. M., Bentrup, G., Kellerman, T., MacFarland, K., Straight, R., and Ameyaw, L. (2021). "Windbreaks in the United States: A Systematic Review of Producer-Reported Benefits, Challenges, Management Activities and Drivers of Adoption," *Agricultural Systems*, 187: 103032.

Sólymos, P., Morrison, S. F., Kariyeva, J., et al. (2015). "Data and Information Management for the Monitoring of Biodiversity in Alberta," *Wildlife Society Bulletin*, 39/3: 472–9.

Stanturf, J., Lamb, D., and Madsen, P. (eds). (2012). *Forest Landscape Restoration: Integrating Natural and Social Sciences*. Dordrecht: Springer, 330 pp.

Stone, R. D., and d'Andrea, C. (2001). *Tropical Forests and the Human Spirit: Journeys to the Brink of Hope*. Berkeley and Los Angeles: University of California Press. 315 pp.

Sze, J. S., Carrasco, L. R., Childs, D., and Edwards, D. P. (2022). "Reduced Deforestation and Degradation in Indigenous Lands Pan-Tropically," *Nature Sustainability*, 5/2: 123–30.

Trager, M. D., Drake, J. B., Jenkins, A. M., and Petrick, C. J. (2018). "Mapping and Modelling Ecological Conditions of Longleaf Pine Habitats in the Apalachicola National Forest," *Journal of Forestry*, 116/3: 304–11.

Turner-Skoff, J. B., and Cavender, N. (2019). "The Benefits of Trees for Livable and Sustainable Communities," *Plants, People, Planet*, 1/4: 323–35.

Umunay, P. M., Gregoire, T. G., Gopalakrishna, T., Ellis, P. W., and Putz, F. E. (2019). "Selective Logging Emissions and Potential Emission Reductions from Reduced-Impact Logging in the Congo Basin," *Forest Ecology and Management*, 437: 360–71.

USDA. (n.d.). *Windbreaks for Conservation*. Agriculture Information Bulleting 339. Washington: Natural Resources Conservation Service, United States Department of Agriculture. 25 pp. Available online at https://www.fs.usda.gov/nac/assets/documents/morepublications/windbreaksforconservation.pdf (accessed February 26, 2024).

USFS (1999). *National Forests in Florida: Land and Resource Management Plan*. Management Bulletin R8-MB-83A. Tallahassee, FL: USDA Forest Service. Available online at https://www.fs.usda.gov/detail/florida/landmanagement/planning/?cid=stelprdb5269793 (accessed February 17, 2024).

USFS (2022). *Biennial Monitoring Evaluation Report for the National Forests in Florida: Fiscal Years 2020–2021*. Tallahassee, FL: USDA Forest Service. 50 pp. Available online at https://www.fs.usda.gov/Internet/FSE_DOCUMENTS/fseprd1077276.pdf (accessed February 17, 2024).

Weiss, S. A., Toman, E. L., and Corace, R. G. (2019). "Aligning Endangered Species Management with Fire-Dependent Ecosystem Restoration: Manager Perspectives on Red-Cockaded Woodpecker and Longleaf Pine Management Actions," *Fire Ecology*, 15/1: 1–14.

Wight, B., Boes, T. K., and Brandle, J. R. (1991). *Windbreaks for Rural Living*. EC 91-1767-X. Lincoln, NE: University of Nebraska Extension. 6 pp. Available online at https://www.fs.usda.gov/nac/practices/windbreaks.php (accessed February 17, 2024).

Williams, B. K., and Johnson, F. A. (2015). "Optimization and Resilience in Natural Resources Management," in C. R. Allen and A. S. Garmestani (eds), *Adaptive Management of Social-Ecological Systems*. Dordrecht: Springer, 217–33.

Wurtzebach, Z., and Schultz, C. (2016). "Measuring Ecological Integrity: History, Practical Applications, and Research Opportunities," *BioScience*, 66/6: 446–57.

Yu, Z., Beilman, D. W., Frolking, S., et al. (2011). "Peatlands and their Role in the Global Carbon Cycle," *Eos*, 92: 97–108.

Zhaohua, Z., Mantang, C., Shiji, W., and Youxu, J. (eds). (1991). *Agroforestry Systems in China*. Singapore: Chinese Academy of Forestry and Canadian International Development Research Centre. 216 pp.

CHAPTER 12

Forest Stewardship Based on Socio-Ecological Principles

It's the end of the world as we know it, and I feel fine.

Rock band R.E.M. (1987).

12.1 Common threads

Contents of the previous chapters will have been familiar to any ecologist or any student of complex adaptive systems. In many ways, resilient forest management expands upon concepts of ecological forestry that have been advocated for several decades (Franklin 1989; Maser 1990; Hammond 1991; Seymour and Hunter 1999; Larocque 2016; Palik and D'Amato 2017; Franklin et al. 2018). These concepts are increasingly being adapted to address uncertainty and inevitable change (Folke et al. 2009; Messier et al. 2013), explicitly recognizing the need for flexibility and adaptability in order for ecosystem services to be sustainable over the long term. This chapter attempts to link some of the concepts of resilience-based management and ecological forestry with foundational principles in ecology—principles responsible for the development, adaptation, and persistence of biological lineages and communities around the planet and throughout its history.

The recurrent themes of diversity, flexibility, and adaptive management to promote resilience may resonate with advocates of social justice as well, for they are common to democratic and responsive governance. This is a good reminder that resource management is ultimately about managing people, directing and constraining human interactions with the environment, for which effective policies and practices must successfully meet socioeconomic needs too. The adoption of resilience strategies for forest management requires a simultaneous adoption of the reliance philosophy in the institutions (governments, communities, corporations, families) that own, govern, and benefit from the world's forests. Consequently, resilient resource management is about the broad socio-ecological system, not just forest ecosystems. This is a principle long embraced by the Resilience Alliance (https://www.resalliance.org/), as expressed in its various workshops and publications. Botkin (1990) posits that the only "balance of nature" we can maintain is in humankind's balancing its impacts within the bounds of natural variability and adaptability.

Terms such as sustainable forest management, ecological forestry, forest ecosystem management, and ecosystem-based management enjoy considerable overlap and flexibility in their interpretation (see Box 1.1). The implicit assumptions behind such terms, particularly about system stability and the degree to which we can engineer our world, lead to a reconsideration of the appropriate terminology to describe resilient and adaptive forestry. Recognizing limits to our ability fully to understand and effectively control natural systems (Holling and Meffe 1996) prompts redirection of all terminology concerning "management"—which implies direction and a command-and-control mentality—to "stewardship"—which instead denotes a humbler level of guidance and care as well as responsibility. That humility, a recognition that the inner workings

Table 12.1 Resilience-based ecosystem stewardship contrasted with steady-state ecosystem management and steady-state production forestry

Characteristic	Ecosystem stewardship	Steady-state ecosystem management	Steady-state production forestry
Reference point	Trajectories of ecological change	Historical ecosystem condition	Physiological potential for timber yields
Central goal	Socio-ecological sustainability benefits	Ecological integrity	Financial returns
Predominant approach	Managing, stabilizing, and amplifying feedbacks to nurture resilience	Managing resource stocks and condition for sustainability	Managing harvesting and silviculture to maximize sustainable timber yield
Role of uncertainty	Actions maximize flexibility to adapt to an uncertain future	Research reduces uncertainty before taking action	Intensive interventions maintain desired conditions
Role of resource manager	Facilitator who engages stakeholders to respond to and shape socio-ecological change	Decision maker who sets course for sustainable management	Deployment, scheduling, and enforcement of harvesting and silvicultural treatments
Response to disturbance	Adapt to changes and sustain options	Minimize disturbance probability and impacts	Prevention, suppression, and post-disturbance salvage
Resources of primary concern	Biodiversity, socio-ecological well-being, and adaptive capacity	Species composition and ecosystem structure	Commercially desirable tree species

Based on Chapin et al. (2009b), Puettmann et al. (2009), and Franklin et al. (2018).

and interactions of ecosystems and our impacts on them are imperfectly known, is a central tenet of ecological forestry (Franklin et al. 2018). Sexton et al. (1999) equate ecological stewardship and ecosystem management, but Chapin et al. (2009b) explicitly consider resource stewardship as a necessary evolution of ecosystem management, applying ecological principles for resilience in a changing world (see Table 12.1). Ecological forestry differs from timber-focused forestry in its economic and social implications, as well as it ecological implications (Franklin et al. 2018; Palik et al. 2021). Stewardship also implies an ethical responsibility not to damage the world, its resources, and its ecosystem services—something akin to Leopold's (1949) "land ethic".

12.2 Some key ecological principles

Might it be feasible to devise a template for resilience-promoting ecological forestry based on some set of universal laws of ecology? Unlike the "hard sciences"' (mathematics, physics, and chemistry), the genetic variability and innumerable species interactions possible in ecological systems tend to defy strict adherence to underlying rules; that may be the key to their resilience. Nonetheless, recognition of a number of widely applicable ecological principles can provide useful guidance for forest stewardship.

Barry Commoner (1971) proposed the following four Laws of Ecology: (1) Everything is connected to everything else; (2) everything must go somewhere; (3) nature knows best; and (4) there is no such thing as a free lunch. Though largely conceived to address issues of persistent organic pollutants and waste management at the time, these principles are relevant to forest stewardship as well. We continue to learn how extensively species, ecosystems, landscapes, and even continents are functionally connected to each other, in support of Law 1 (e.g. Snyder 2010; Simard et al. 2013; Scherer-Lorenzen et al. 2022; see also Section 7.3.4). Those connections are reflected in the trade-offs and knock-on effects implied in Laws 2 and 4, as efficiencies gained in support of one ecosystem service inevitably result in losses to others, and unintended consequences abound (Aryal et al. 2022; Pearson et al. 2022). "Nature" can be incredibly variable and unpredictable (Botkin 1990), so whether nature or naturalistic approaches are always best, as per Law 3, depends on human values and priorities (see

Chapter 2). Nonetheless, a billion years of evolution and co-evolution, plus millennia of ecosystem development, must be respected if introducing novel chemicals, species, genes, and composition or structural reconfigurations into forest systems. This is the basis for evaluating ecological integrity against the standard of natural ecosystems (Tierney et al. 2009; Karr et al. 2022). Recent research (e.g. Scherrer et al. 2023) is increasingly confirming the greater resilience of natural and near-natural forests compared to those that have been heavily modified by human interventions.

Although perhaps not qualifying as natural laws, there are several bio-ecological constraints under which forest ecosystems have developed and persisted. Those same constraints now limit, and can serve to guide, the actions of forest managers who are trying to safeguard resource sustainability and promote forest resilience. Outlined below are a few of those general principles, bundled into a set of broad categories, with some implications to resilience-based forest stewardship.

12.2.1 Biodiversity is foundational to adaptive evolution

Biodiversity in all its forms—the composition, structure, and function of genes, species, communities, and ecosystems—helps confer resilience through ecological niche differentiation (including different vulnerabilities to environmental and human stressors) and risk dispersion. All species exhibit a range of tolerable environmental or resource conditions (in response to temperature or nitrogen availability, for example), with optimal levels somewhere between those extremes (see Figure 10.2(c)); this is sometimes referred to as the "Goldilocks principle." Tolerances and optima can vary within a species for establishment, growth, and reproduction, and among populations that have evolved in somewhat different environments. That full response spectrum, when characterized for multiple resources and multiple environmental factors, and coupled with a species' role in the ecosystem, describes the ecological niche of a species (see, e.g., Figure 6.2 as a simplified example). No two species can occupy precisely the same ecological niche, a fact described as the "competitive exclusion principle." This constrains the assembly of biological communities and promotes evolutionary radiation and adaptive specialization.

Organic evolution can increase the frequency of adaptive traits and novel organisms suited to new or changing environments or to biotic challenges, by means of multiple generations of natural selection that filter variation generated by genetic recombination and mutations. Collectively, biodiversity, niche ecology, and their potential are the foundation of all ecosystems we see today, and for their resilience into the future. Unless that diversity is maintained and encouraged with the potential to change and evolve, forests and their associated ecosystem services have a very poor chance of persisting in a changing world.

12.2.2 Energy flows and matter cycles

Energy flows from one trophic level to the next, with some energy lost through respiration at each level, such that the biomass supported at higher trophic levels must always be less than that found at lower levels. Economies dependent on higher trophic levels (herbivores and especially carnivores) are more inefficient, and often more precarious, than those based on primary producers (plants). Although not usually a concern for timber management, this principle often seems to be forgotten in fisheries management (Pauly et al. 2000), and needs to be considered in maintaining habitat and resources in support of keystone species that may be higher up the food chain in forests too.

In contrast to energy, matter cycles within ecosystems. Organic matter decomposes, complex organic compounds are broken down into simpler and more mobile molecules, nutrients are released in soils and waters, and even final breakdown products of CO_2 and H_2O can be used again by other organisms in the biosphere. At greater scales, organic carbon can be incorporated into geological strata (for example, as fossil fuels such as coal and petroleum, and as coral reefs that become limestone), where it can remain static for long periods of time, or until released by humans burning fossil fuels or manufacturing cement. Maintaining processes of decomposition, nutrient cycling, and soil health is critical for the sustainability of trees, other plants, and the animals on which they depend. Forests are being increasingly valued for their carbon uptake and storage potential, and for their importance in absorbing and gradually releasing water—ecosystem services that can be more important than timber production in many parts of the world, especially as the climate heats up.

12.2.3 There are multiple limits to growth

Liebig's "law of the minimum" proposes that plant growth is constrained by the most limiting nutrient or other growth resource (for example, light, water, phosphorus), making supplements of non-limiting resources inefficient or pointless. Exactly which resource is limiting to a plant or animal can vary widely with climate, soil, and species. Understanding such limitations and constraints is central to place-based vulnerability assessments, renewable resource planning, and setting realistic expectations in prioritizing ecosystem services.

The saturation kinetics of photosynthesis mean that only so much CO_2 can be fixed, often limited by the concentration of the ribulose bisphosphate carboxylase-oxygenase (RuBisCO) enzyme in plant leaves, even if more light or CO_2 is available. Nitrogen is often the most limiting nutrient in terrestrial ecosystems (LeBauer and Treseder 2008) because RuBisco and other enzymes are proteins built from amino acids. This makes the contribution of N-fixing bacteria, cyanolichens, and those herbs, shrubs, and trees symbiotic with *Rhizobium* and *Frankia* N-fixers so important to retain in managed ecosystems.

The "−3/2 thinning law," simply put, means that you can either have a few large trees or many small ones supported by any given site. This principle is based on observations that the relationship of mean biomass or volume per tree (or plant in general) in fully stocked stands has a slope of approximately −1.5 when plotted against stand density on logarithmic scales. It has provided the basis for silvicultural thinning guidelines to utilize growing space most efficiently or to redirect potential resources from suppressed trees (doomed to die soon) to more efficient growth by crop trees (Newton 2021). Stand thinning can accelerate the growth of individual trees to the larger sizes desired for commercial harvesting (although wood density and quality may be lower), for wildlife habitat, or for fire resistance, but thinning cannot increase total biomass production or carbon sequestration in trees.

The reproductive potential of organisms is such that exponential population growth is possible, until that population experiences environmental constraints, at which point mortalities balance the birth of new organisms. This "carrying capacity" principle is usually applied in the context of animal populations and wildlife management, but it also underlies the −3/2 thinning law. The inherent carrying capacity of land can be increased by external subsidies, such as the transfer of water and nutrients from elsewhere (even upslope), or the application of fertilizer. One might also say that modern human society and economies are subsidized by the burning of fossil fuels, and we do not know what the realistic operating capacity of a forest enterprise, forest community, or nation state would be without that subsidy.

12.2.4 Geographic context and spatiotemporal connectivity matter

The characteristic "species versus area curve" (of species richness plotted against the area sampled) rises rapidly then gradually flattens out. Larger areas of land (plots sampled, contiguous area, or habitat islands) typically support more species, usually because of the incorporation of more habitats but also because of a greater chance of encountering individuals of rare species. Isolated areas are likely to support fewer species than those well connected to larger habitat areas and populations. Habitat connectivity facilitates dispersal and genetic interchange, and the ability to recolonize habitats after local extirpations caused by disturbance or disease. Forest fragmentation and associated edge effects can greatly limit the area of effective interior forest conditions (Voller and Harrison 1998; Harper et al. 2005). Collectively, geographic context, spatial topologies, and spatial connectivity can influence phenomena as diverse as disturbance vulnerability and species viability. Connectivity over time is important too, both in the form of biological legacies from pre-disturbance forests to boost-start ecosystem recovery (Franklin et al. 2000), and in planning for the feasible persistence of conservation elements, forests, and forest plans in the future climate space (Rose and Burton 2009).

Conservation planning, forest planning, land-use planning, and the promotion of sustainable and resilient human communities are all expressed in terms of maps at one or more stages in their respective processes. All planning for the development and sustainability of natural resources involves landscape ecology (see Box 5.3) and spatial ecology (Rai 2013). The spatial portrayal and allocation of values and priorities is an important step in resilient forest stewardship. Similarly, if they are to make informed decisions, forest managers and stakeholders need to evaluate clearly presented differences and trends from past forest conditions (see Figure 10.3, for example) and the best-available projections of future conditions (see Figure 10.6, for example) based on current understanding and management options.

12.3 Applying ecological principles in forest planning

As with any fundamental law of nature, it is impossible for forest managers to circumvent the constraints and implications of the above ecological principles. Puettmann et al. (2009: 73) further list more than twenty important ecological concepts and their application in forestry. The management of all renewable natural resources clearly depends on ecological drivers and constraints, but that still allows considerable latitude in approaches for meeting different human objectives (McPherson and DeStefano 2003). Some ecological principles inevitably have been incorporated into silvicultural practices over past centuries, although typically with a command-and-control agronomic (cropping system) perspective applied at the stand level (e.g. Toumey 1928; Smith 1962). An updated perspective

on stand manipulations based on ecological principles is provided in Section 12.4.

As wild forests became scarcer, as biodiversity loss became apparent, and with a growing appreciation of the role of natural disturbance regimes in sculpting the evolutionary adaptations and habitat requirements of forest organisms, ecological principles have been further articulated for forest ecosystem management, with a greater emphasis on the landscape level (Franklin 1989; Kohm and Franklin 1997; Hunter 1999; Larocque 2016; Franklin et al. 2018). Similar ecological principles have been articulated for land-use planning in general (Dale et al. 2000) and are reflected in guidance for management of the US National Forests (Aber et al. 2000). With a strong foundation in landscape ecology, land-use planning guidelines are also applicable to the mapping out of resource emphasis zones (for example, triad zoning) in forest planning too (see Box 12.1). Unfortunately, there has been surprisingly little application of these various ecological principles beyond token efforts in most of the world's managed forests.

Traditional forestry with its agricultural steady-state perspective focuses on simplification and homogenization in order to maximize fiber productivity and harvesting efficiency. In contrast, forest stewardship from an ecological perspective focuses on complexity and diversity to promote forest ecosystem resilience (Puettmann et al. 2009; Filotas et al. 2014; Messier et al. 2015). The following features that characterize forest ecosystems as complex adaptive systems (as introduced in Section 4.4) confer many of the traits supportive of resilient forests and resilient forest management (Messier et al. 2013: 338–9):

Box 12.1 Ecological principles and guidelines for managing land use

The following principles and guidelines are borrowed from Dale et al. (2000), for which they were developed in the context of broad land-use planning. However, they are also useful in mapping out resource emphasis zones within forest landscapes. Such principles should be applied along with those of landscape analysis and design (Diaz and Apostol 1992; Diaz and Bell 1997; Bell and Apostol 2007), especially where flows of people, wildlife, and water are prevalent, or where barriers to the spread of disturbances are important. See also Chapter 2 for further discussion of forest values and ecosystem services that may conflict, and how compatible ecosystem services can be bundled in land-use zoning.

Box 12.1 *Continued*

Ecological principles and implications

1. *The Time Principle*: "Ecological processes function at many time scales, some long, some short; and ecosystems change through time." (Dale et al. 2000: 649).

   - The current composition, structure, and function of an ecological system are, in part, a consequence of historical events or conditions, with the imprint of land use potentially persisting for a long time.
   - Long-term effects of land use or management may be difficult to predict owing to variation and change in ecosystem structure and processes, with the full ecological effects of human activities often not seen for many years.

2. *The Species Principle*: "Particular species and networks of interacting species can have key, broad-scale ecosystem-level effects" (Dale et al. 2000: 650).

   - Different focal species affect or reflect ecological systems in different ways: keystone species, ecological engineers, link species, umbrella species, or indicator species.

3. *The Place Principle*: "Local climatic, hydrologic, edaphic, and geomorphologic factors as well as biotic interactions strongly affect ecological processes and the abundance and distribution of species at any one place" (Dale et al. 2000: 651).

   - Environmental constraints, productive capacity, flora, and fauna all vary geographically.
   - Consequently, applying the same management approach everywhere cannot be expected to elicit the same results.

4. The Disturbance Principle: "The type, intensity, and duration of disturbance shape the characteristics of populations, communities, and ecosystems" (Dale et al. 2000: 653).

   - The frequency and severity of both natural and anthropogenic disturbances are agents of mortality and displacement, but also provide opportunities for renewal, reorganization, and improved adaptation.

5. *The Landscape Principle*: "The size, shape, and spatial relationships of land-cover types influence the dynamics of populations, communities, and ecosystems" (Dale et al. 2000: 654).

   - The spatial arrangement, size, isolation, or connectivity of habitats affect species differently, as does the balance of edge and interior conditions.

Ecological guidelines for land use and resource emphasis planning

1. "Examine the impacts of local decisions in a regional context" (Dale et al. 2000: 656).

   - Consider effects of activities intended for a parcel of land on the ecosystem services and values elsewhere in the landscape, and how that parcel may be affected by activities elsewhere.

2. "Plan for long-term change and unexpected events" (Dale et al. 2000: 659).

   - Natural disturbances, climate change, delayed effects of human activities, and cumulative effects of natural processes and management decisions can all influence the future viability of species and the sustainability of ecosystem services in unforeseen ways.
   - Linear extrapolation of historic trends is often unreliable.

3. "Preserve rare landscape elements and associated species" (Dale et al. 2000: 659).

   - Protected areas should be part of every forest landscape. Examples of those landscape features and ecological elements infrequent in the managed forest will be most effective in protecting vulnerable biodiversity.
   - Rare landscape elements may be defined by site type or vegetation type, but also by stand development stage. For example, in some landscapes, burned forest may be rarer than old-growth forest.

4. "Avoid land uses that deplete natural resources over a broad area" (Dale et al. 2000: 659).

   - For example, large or contiguous clearcuts can compromise watershed hydrology if occupying too much area in a catchment or if created over too short a time.

5. "Retain large contiguous or connected areas that contain critical habitats" (Dale et al. 2000: 660).

   - Large areas that are free of edge effects and human influence are important for some species.
   - The effectiveness of and need for corridors to facilitate migration and gene flow between critical habitats vary by species and the degree to which the landscape matrix is inhospitable to them.

6. "Minimize the introduction and spread of nonnative species" (Dale et al. 2000: 660).

Box 12.1 *Continued*

- Exotic insects, fungal pathogens, weeds, and omnivorous vertebrates can often threaten indigenous biodiversity, and sometimes threaten forest productivity and sustainability of valued ecosystem services.

7. "Avoid or compensate for effects of development on ecological processes" (Dale et al. 2000: 661).

   - Minimize disruption caused by road building, drainage modification, and timber harvesting.
   - Remediate past damage and unavoidable impacts, restoring productivity and diversity as feasible.

8. "Implement land-use and land-management practices that are compatible with the natural potential of the area" (Dale et al. 2000: 661).

   - Utilize appropriate inventory and ecosystem classification and mapping methods to apply the Place Principle in recognition of the productivity, habitat values, and environmental constraints of different sites.
   - Recognize that some sites may have great value for multiple ecosystem services (e.g. timber production and as habitat for a rare species), so trade-offs, compatible land-use bundling, and negotiated prioritization will still have to take place.

Openness: forest systems and forest stewardship activities are affected by, and have influences on, many spheres of the biophysical and social world outside the management unit under consideration; both internal and external connectedness is important.

Heterogeneity and diversity: biological and spatial diversity support ecosystem production and habitat functions but are especially important in promoting recovery and adaptation in response to disturbances and changing conditions.

Hierarchy: forests consist of individual organisms, populations, communities, ecosystems, and landscapes that interact across a range of scales in space and time. Forest stewardship takes place through myriad interventions at local levels and over short periods of time but has impacts and implications at broader scales and over long periods of time.

Memory: residual structure and biological legacies provide habitat, populations, and propagules that provide ecological continuity over time, and contribute to the recovery and adaptability of the forest ecosystem after disturbance.

Self-organization: environmental filtering and niche complementarity of species traits from the pool of available flora, fauna, and genomes mean that communities can reassemble after disruption in a manner well adapted to prevailing conditions, so long as those pools are diverse and healthy.

Uncertainty: non-linearities, unknown thresholds, and unforeseen disturbances and changes in social values collectively mean that forest stewardship requires flexible management options and openness to multiple possible futures for the forest.

Adaptation: through the above inherent features of forests, forest ecosystems and the sociocultural systems in which they are embedded are changing constantly (though often imperceptibly) as they adjust in manners that confer resilience and, ultimately, persistence.

These and other properties of complex adaptive systems are invoked in support of general approaches for promoting forest resilience in Chapter 5. It is only by recognizing and supporting these attributes of complexity and adaptive capacity that forest resilience can be encouraged and the delivery of important ecosystem services sustained into an uncertain future.

By their mere persistence to the present day, we know that the species, populations, and communities that make up natural forests represent successful solutions to the problems of survival. Millions of years of evolution and millennia of repeated colonization, stand development, natural disturbance, recovery, and succession have resulted in today's natural forests—which are persistent (if not "optimal") examples of sustainable ecosystems. In the absence of better information, natural forests therefore set the bar for forest integrity and vitality (Wurtzebach and Schultz 2016). We must explicitly recognize that any human manipulations compromise ecological integrity, weighing whether that trade-off is justified to satisfy society's needs and wants (Seymour and Hunter 1999). Fewer and less severe interventions than practiced under extractive or agricultural models of forestry are possible when we embrace a wider range of forest outcomes (Puettmann et al. 2009).

Proceeding from a biocentric worldview (in which we assume that all species have the right to persist) or simply a utilitarian desire to save potentially useful species, it follows that a central axiom of ecological forestry is to conserve biodiversity. Ecosystem stewardship should first and foremost provide for the needs of biodiversity—not just endangered species, but the full sweep of genes, species, communities, and their interacting structures and functions—with only resources surplus to those needs harvested for human use (Grumbine 1994). In order for indigenous biodiversity to persist and for forests to remain productive and resilient, any manipulation of forests would have to retain the full suite of habitat conditions found prior to extensive human alteration of the landscapes. This "coarse filter" approach to biodiversity conservation (Hunter et al. 1988) represents a precautionary approach in the absence of detailed information regarding the needs and tolerances of the hundreds of species and processes to be found within any given forest. Although the concept has been widely accepted and cited for decades, it fails to be rigorously applied, while research has raised as many questions as answers regarding the ecological needs of individual species.

Resilient forest management emphasizes silvicultural activities that are expected to reduce risks to important forest values and to increase future societal options in the management and use of the forest. Resilience-minded stewardship abandons the command-and-control approach (Holling and Meffe 1996) and accepts variability in space and time as an inherent attribute that allows forests to adapt to new biotic and abiotic conditions (Puettmann et al. 2009: 145). A broadly defined diversity of diversities is the key to any shock-resistant portfolio of forest holdings (Knoke et al. 2005; Maier 2012). Risk reduction, rather than growth or profit maximization, and promoting a resilient forest providing a diversity of

long-term future options are at the heart of managing forests for general resilience (see Chapter 5). In the short term, this may mean undertaking forest management activities that include primary or co-objectives of reducing wildfire risk, enhancing drought and storm tolerance, surviving outbreaks of insects and invasive pests, and coping with other changes induced by a changing climate (see Chapters 7 and 8).

12.4 Principles and applications of ecological silviculture

Applications of the above ecological principles ultimately find their expression in how we manipulate forests on the ground—namely, in activities such as inventory, road building, timber harvesting, and silviculture. The following guidelines are compiled primarily from Franklin et al. (2018) and Palik et al. (2021), with an emphasis on stand-level forest practices. The ecological foundations of silviculture, whether following classical or ecological approaches, are those of natural disturbances and biological legacies, and of forest development patterns and processes. Their application incorporates management for complexity, diversity, and heterogeneity at stand and landscape scales, with recognition that different tree species have evolved and forest types have developed under different natural disturbance regimes (Palik et al. 2021). This foundational understanding can be blended with lessons from complex adaptive systems to provide some general guidance in implementing silvicultural systems that not only are more ecological than the classical systems developed in Europe, but also set the stage for a more resilient forest management.

The first priority is to maintain temporal, spatial, and process continuity in forest structure, function, and biota between pre- and post-harvest ecosystems (Franklin et al. 2018: 93–6). This guideline is primarily concerned with constraining and more carefully and creatively implementing timber-harvesting activities, and is usually expressed in variable levels and patterns of canopy retention (Franklin et al. 1997; Gustafsson et al. 2012; see Figure 12.1). Forest managers are advised to consider as wide a variety of ecosystem components (that is, more than trees) and functions as

Figure 12.1 Various configurations of variable retention practiced in western Canada: (a) mixed group and single-tree retention in lowland coastal forest (photo by Bill Beese); (b) group and linear retention in machine-harvested boreal forest (photo provided by Ellen Macdonald, University of Alberta, courtesy of the EMEND project); (c) retention of scattered non-commercial canopy trees, regeneration clusters, and stubbed aspen (*Populus tremuloides*) stems in sub-boreal forest; (d) uniform seed tree and shelterwood silvicultural systems are usually employed to promote natural regeneration, but long-lived overstory trees such as these interior Douglas-firs (*Pseudotsuga menziesii* var. *glauca*) can also be retained indefinitely after regeneration is successful. Unharvested trees making up the stand matrix following single-tree or group selection (gap-based) extraction can also be interpreted as a high level of retention.

possible, recognizing that many non-tree components and functions play critical supportive roles in maintaining forest health, productivity, and resilience (Puettmann et al. 2009: 145). Many green trees, understory plants, small mammals, birds, amphibians, invertebrates, microbes, and fungi that otherwise would not survive in a clearcut environment can survive in association with large residual trees. Uncut patches of living trees subsequently serve as nodes ("lifeboats") from which those organisms can expand into the surrounding disturbed area, thereby contributing resilience to the entire forest stand (Franklin et al. 2000). How much retention is sufficient, and in what spatial configuration, constitutes an active area of research. Appropriate solutions are likely to be very ecosystem- and

site-specific and will often represent a compromise between ecological values and forgone harvesting revenues or future crop tree growth and yield.

The next consideration is to create and maintain structural complexity and biological richness, including spatial heterogeneity at multiple spatial scales (Franklin et al. 2018: 97–9). In other words, active measures are undertaken to maintain, develop, or restore the within- and among-stand heterogeneity in ecosystem composition, structure, and function throughout the life of the stand (Puettmann et al. 2009: 145). Although partially implemented through variable retention in harvesting (above), the principle is applicable to silvicultural activities such as site preparation, brushing, and thinning too. This guideline also needs to be implemented when restoring the vitality and resilience of homogenous second-growth stands and intensively managed plantations (see Chapter 9). A much greater range of retention levels and patterns—not just of overstory trees, but of the whole primary or mature forest community—is needed in order to maintain and re-create the structures and habitats found after natural disturbances (Lindenmayer and Franklin 2002; O'Hara 2014). Retaining and creating standing and leaning snags, fallen logs, and decadent features such as dead branches and hollows in living trees are critical contributors to forest complexity and ecosystem resilience (Hodge and Peterken 1998; Gibbons and Lindenmayer 2002; Vítková et al. 2018).

The timing and frequency of silvicultural activities, not just their spatial patterns or the severity of their disturbance, need to reflect ecological processes (Franklin et al. 2018: 99–100). Rotation lengths (regeneration harvests) should match not only the average but the variation in natural disturbance return intervals (Seymour and Hunter 1999). This guideline supports a key principle of forests as complex adaptive systems, in which stands are encouraged to develop within the full envelope of possible conditions in order to foster resilience (Puettmann et al. 2009: 145). Intermediate disturbances such as prescribed understory burns or crown thinning ("thinning from above") may be needed to maintain or restore the health, regeneration, or diversity of some forest ecosystems (Fiedler et al. 1996; Bauhus et al. 2017). Multiple vegetation control or thinning treatments, applied following "variable density thinning" guidelines (Carey 2003), can be used to generate "skips and patches" of tree cover and understory vegetation (Comeau and Fraser 2018).

As outlined in Section 12.2.4 and in the Place Principle presented in Box 12.1, the spatial context for decisions in forests and other socio-ecological systems is important. Silvicultural activities are the tools for implementing a broader vision of the future forest, so they need to be planned and undertaken in the context of plans developed at larger (landscape) spatial scales. Allowances for wide variation in stand-level attributes and multiple development trajectories confer resilience, with success at achieving fundamental management objectives being most appropriately set and evaluated at the landscape or forest estate level (per Puettmann et al. 2009: 145; see also Price et al. 2009; Franklin et al. 2018). As in traditionally practiced production forestry, trees or forest stands of all ages are required in order to provide a relatively even supply of timber and habitat within any forest estate or sustainability unit (see Chapter 2). The timing and extent of regeneration harvests, commercial thinnings, and growth-enhancing silvicultural treatments can all contribute to that forest-level goal. Similarly, a wide complement of tree and stand ages, compositions, and structures is needed to conserve, maintain, or restore healthy populations of all plants and animals in a forest landscape. As at the stand level, forest zoning and harvesting need to retain the conservation and resilience value of primary or mature forest cover. Reserve design should consider fragmentation, connectivity, edge effects, and the protection of interior old-forest conditions.

Many forest researchers have arrived at a consensus regarding the refinement, application, and extension of the principles outlined above in a manner that would promote resilient forest stewardship. These tenets of ecological forestry and complex adaptive systems provide guidance in thinking carefully about the assumptions and context for any human interventions. Nonetheless, each principle needs to be customized in its application in different ecosystems, with consideration to local place-based history and geographic context, and in recognition of the range of (often conflicting) values and expectations held by forest owners, managers, and stakeholders (McAfee et al. 2010; Rutt et al. 2015). As introduced in Chapter 2, forest zoning to emphasize different sets of compatible forest values in different areas can help resolve some incongruous objectives.

Forests are hierarchies of interacting organisms, biological communities, ecosystems, landscapes, and bioregions, all with distinctive traits, constraints, and dependencies on the Earth systems of climate, terrain, and soils. Within that hierarchy, some elements and processes operate on fast cycles (for example, insect populations) while others (for example, forest succession) take place much more slowly, and typically act as constraints to other components (Gunderson and Holling 2002). The modular nature of lower levels (for example, trees, stands), combined with their agency or adaptability, helps provide resilience through flexibility and adaptability to the whole system. This modularity and hierarchical structure further support the flexibility in achieving "means objectives" at the stand or local watershed level while supporting successful attainment of more immutable "fundamental objectives" or goals at the forest estate or landscape level (Gregory et al. 2012; Franklin et al. 2018: 494).

A forest is thus a system composed of many moving parts, for which context is everything. The unique "ecology of place" determines not only the relative role of different site limitations or ecological processes, but also the pool of species available, the legacies of past disturbances or land uses, and the influence of (and effect on) neighboring elements. Spatial topology is an important aspect of that context, with disturbed or otherwise modified forest resulting in edge effects and altered edge-to-interior forest habitat area, and different degrees of connectivity, isolation, and barriers to movement provided by managed stands, transportation or utility infrastructure, and alternative land uses. Landscape context can determine a stand's exposure to desirable (for example, native propagules) and undesirable (for example, invasive species) agents of change, while cumulative effects at a broader scale can be expressed in hydrology and habitat suitability for wide-ranging animals.

The coarse filter approach is further developed through the emulation of natural disturbance in forest stand and landscape management (Perera et al. 2004; Kneeshaw and Bergeron 2016). Forest managers cannot precisely mimic the attributes of natural disturbances; no natural disturbance transports logs to a mill, after all! Nonetheless, the natural disturbance regime can provide guidance in terms of the disturbance return interval (rotation length or cutting cycle), the retention of biological legacies, the cohort structure of trees within stands, the age structure (time since disturbance) of stands in the landscape, and the spatial patterning of removals and retention at multiple scales (Seymour and Hunter 1999; Kimmins 2004; Kneeshaw and Bergeron 2016). What is retained on site following a disturbance (biological legacies; Franklin et al. 2000) is more important to ecological resilience than what is removed. Biological

legacies ("ecological memory") of well-developed soils, intact seed banks and microbial populations, patches of living trees and other plants, coupled with standing dead trees, fallen logs, and forest litter collectively provide continuity over time, facilitating the recovery and potential continuity of a forest ecosystem after disturbance. This is why dead trees are just as important to retain as living ones (Seymour and Hunter 1999; Fenger et al. 2006), providing habitat for many species, hydrological and erosion protection, and making important contributions to carbon pools and soil development (Lindenmayer et al. 2008).

Although natural disturbances, disturbance regimes, and their legacies are infinitely variable, Palik et al. (2021) recognize four "natural disturbance archetypes" that provide guidance in the management of different forest types. In forests historically subject to infrequent severe disturbances (for example, the western hemlock (*Tsuga heterophylla*) rainforests of western North America), legacy retention is paramount, natural regeneration by all tree species is desirable, thinning can encourage complexity, and regeneration harvests should be infrequent (Palik et al. 2021: 163–82). Where forests were historically characterized by frequent low-severity fire disturbances (for example, the inland ponderosa pine (*Pinus ponderosa*) and dry mixed-conifer forests of inland western North America), open canopy conditions and densities should be maintained or restored. Furthermore, surface and ladder fuels need to be reduced using prescribed fire or mechanical methods to prevent crown fires, and a fine-scale mosaic of old trees and regeneration clusters needs to be maintained or restored through variable density thinning and multiage management (Palik et al. 2021: 183–205). Where forests historically experienced frequent gap-scale disturbances (for example, forests containing sugar maple (*Acer saccharum*) in northeastern North America, or beech (*Fagus sylvatica*) in western Europe), gap-scale regeneration harvests should span a range of sizes, with large tree and large woody material retention. Management should also employ techniques that expose mineral soil to encourage early successional species and that release crop trees to promote a multiage structure (Palik et al. 2021: 207–27). In managing forests with a history of mixed-severity disturbances (for example, the red pine (*Pinus resinosa*) forests around North America's Great Lakes, or the Scots pine (*P. sylvestris*) forests of Fennoscandia), legacy retention is important, site preparation and thinning should include large untreated areas, regeneration harvests should be infrequent, and deadwood retention or creation may be required (Palik et al. 2021: 229–49). The goal in all ecologically managed forests is to promote native biodiversity, heterogeneity within stands and across the landscape, and a multitude of options for an uncertain future. Palik and D'Amato (2024) have expanded upon the four templates outlined above to collate guidelines for ecological silvicultural systems customized to fifteen forest types in North America, two forest types in Europe, and two forest types in the southern hemisphere. That contribution is a welcome update to the Burns (1983) handbook (which has much more of a timber focus and splits its descriptive prescriptions among forty-eight forest cover types found in the continental US) and the Matthews (1991) textbook description of the classical silvicultural systems as applied around the world.

The predominance of mixed-severity natural disturbances in many of the world's forests means that there are biota and forest ecosystem processes dependent on a wide array of residual stand structures and patterns, and a range of tree age or cohort structures within and among stands (Seymour and Hunter 1999). There are many variations to harvesting and silvicultural systems other than the extremes of clearcut even-aged versus single-tree selection uneven-aged management options, including two-cohort and multi-cohort stand structures in various uniform or clustered arrays (O'Hara 2001). Variability at all stages of forest management is preferable to uniformity (Puettmann et al. 2009; Palik et al. 2021). This premise can be applied in varying opening (patch) sizes (e.g. Coates 2000), variable retention harvesting (Martínez Pastur et al. 2020), and variable density thinning (Carey 2003). By accepting some disturbances and within-stand mortality, space and resources are freed up for organisms (which may be species or genotypes of trees or other plants and animals) better suited to conditions now prevailing, thereby conferring the ability for the forest to adapt (Puettmann et al. 2009). There are some species (not only trees, but associated flora and fauna) adapted to severe "stand-replacing disturbances" from which trees recover by inseeding, seed banks, or vegetative sprouting. There are also species adapted to "releasing disturbances" that leave much of the forest floor and understory vegetation intact; the tree layer then recovers from a bank of pre-existing seedlings and saplings. Most forests can be managed under both disturbance regimes and their intermediates, and probably have species and processes adapted to different disturbance attributes, resulting in different stand compositions, structures, and productivity (levels and allocations among species) according to the disturbance regime or management approach undertaken.

12.5 Integrating economic, social, and cultural perspectives

Franklin et al. (2018) organize their ecological forest management textbook around five major themes that can help manage forests for resilience. Paraphrased and expanded upon, these themes can be expressed as follows:

1. The philosophy and principles of ecological forestry are based on natural systems, as contrasted with those of modern production forestry, which are based on managed systems keyed to rate of return on investment.
2. There is an ever-expanding understanding of forests as complex ecosystems that can be drawn upon to assure provision of multiple services and goods, not only timber.
3. All forest management occurs within a public policy framework that may or may not favor ecological forestry over forest exploitation or production forestry.
4. Because of the multiple values and expectations of forests, effective negotiation is important in forest management, especially (but not exclusively) on public forest lands.
5. Strategies for coping with change and uncertainty are an essential ingredient in forest management planning in the twenty-first century.

The above concepts are universal in their relevance to promoting forest resilience, even though Franklin et al. (2018) are primarily addressing an American audience and draw most examples of applications from US National Forests. These themes represent the logical extension of ecological facts and principles if one's goal is the persistence of healthy, productive forests capable of providing multiple ecosystem services (Palik et al. 2021; Palik and D'Amato 2024). Despite the focus on ecological principles, it is noteworthy, however, that these themes also have human values at their core: expectations for economic viability or profit; viewing the natural world in terms of ecosystem services; a need for negotiations; all within the sociocultural constraints of public policy and institutions.

A similar set of socio-ecological principles for ecosystem-based planning is espoused by Hammond (2009: 23):

1. focus on what to protect, then on what to use;
2. recognize the hierarchical relationship between ecosystems, cultures, and economics (e.g. as in Figure 3.4(b));
3. apply the precautionary principle to all plans and activities;
4. protect, maintain, and (where necessary) restore ecological connectivity and the full range of composition, structure, and function of enduring features, natural plant communities, and animal habitats and ranges;
5. facilitate the protection and/or restoration of Indigenous land use;
6. ensure that the planning process is inclusive of the range of values and interests;
7. provide for diverse, ecologically sustainable, community-based economics; and
8. practice adaptive management (see Chapter 10).

Hammond (2009) then outlines a framework to operationalize those principles in an iterative manner across a hierarchy of geographic scales and with recurrent consideration of cultural, social, and economic implications. Simply put, the process of assessing, designing, integrating, and implementing ecologically responsible human activities is followed up by more assessing, designing, and so on in classic adaptive management fashion (see Figure 10.1).

The application of ecological forestry and resilience-based stewardship is not without its trade-offs and critics. A shift from a timber-optimizing mindset to one based on promoting a wider range of values, precautionary risk reduction, and a more naturalistic approach to forest management is an uphill battle where profit maximization is the primary objective. Despite good intentions, the forgone revenue of unharvested trees could make commercial forestry economically unfeasible in some situations. It has also been suggested that continuous-cover forestry practices, characterized by very small openings and frequent stand entries as widely adopted in Europe (Gresh and Courter 2021), may result in too narrow a range of forest attributes for resilience in a changing climate (Bauhus et al. 2013; O'Hara 2015). Klenk et al. (2009) warn that the emulation of natural disturbances and "following nature" will result in "reproducing a snapshot of the past" that is not responsive to global change and shifts in societal values.

Another criticism is that ecological forestry is normative in its objectives, but without articulating a clear set of underlying ethical principles (Batavia and Nelson 2016). This criticism is easily addressed by reviewing the rationale for coarse-filter biodiversity management in general. That is, we do not understand the specific habitat and life-cycle requirements of all organisms, but (in the absence of better information) can assume they will be met if we can successfully maintain the environment in which they evolved (Hunter et al. 1988; Hunter 1993). Indeed, Palik and D'Amato (2017: 51) responded to Batavia and Nelson's (2016) challenge with the statement that "the goal of ecological forestry is to sustain healthy productive forests … replete with native species diversity and a full array of ecosystem services." This perspective can be bolstered by adoption of the precautionary principle—namely, that it is preferable to err on the side of conservation when faced with uncertainty (and the potential for irreversible impacts such as the extinction of species or distinctive populations) regarding the impacts of proposed development. One might say that coarse-filter and precautionary approaches are the foundation of resilience-based forest stewardship, proceeding from the normative assumption that humankind wishes to avoid biological extinction events or the loss of forest ecosystem services. As outlined in Chapter 2, the precise balance of resource use and conservation values to be promoted in a given landscape can still vary according to landowner priorities and community values, including differences in risk tolerance. That flexibility is a strength rather than a weakness of the approach (Palik and D'Amato 2017), allowing for greater experimentation, site-specific solutions, and a more responsive way forward when implemented on a local basis.

People are diverse, with many different values, needs, and expectations of forests. In the spirit of democracy and universal human rights, it can be argued that all such expectations are legitimate, if they can be sustained and do no harm to other people (Kimmins 1992). In particular, there is often tension—and sometimes conflict—between forest policies imposed by centralized governments or corporate tenure holders and the Indigenous or local community members who wish to continue practicing traditional resource use and management rights (e.g. Owubah et al. 2001; Teitelbaum et al. 2023). The challenge in forest stewardship is to meet as many of those expectations as possible, while safeguarding the sustainability, persistence, and resilience of valued ecosystem services. As in ecosystems, there is value to diversity among the stakeholders, employees, managers, and elected representatives in those institutions governing forest stewardship. Further expanding the diversity concept to respect universal equality and human rights in a democratic context, it is demonstrable that equity, diversity, and inclusion (EDI) of multiple ethnicities, ages, handicaps, and gender identities contribute to broader perspectives, innovation, and robust decision-making (Østergaard et al. 2011; Bennett and Satterfield 2018).

There is extensive recent literature on risk management in forestry, much of it focused on climate-change adaptation (Yousefpour et al. 2012) or dealing with market uncertainty for forest products (e.g. Alonso-Ayuso et al. 2018). Other studies identify preferred strategies for dealing with specific threats such as wildfire or storm damage; vulnerability analysis (see Section 6.4) can help focus such efforts. Strategies for general resilience (see Chapter 5) and the growing sophistication of risk management in business, civic, cyber networks, and defense contexts (Ettouney and Alampalli 2016; Sadgrove 2020) have collectively embraced a common set of strategies to address risk and uncertainty. Based on the balance of general uncertainty and specific risks identified, managers can develop policies and practices that avoid, accept, control, or share risks (Sadgrove 2020). Maintaining a diversified portfolio of assets, inputs, and outputs (products and markets) is a classic risk-management strategy that can be applied to forestry too. Individuals, communities, and businesses vary in their risk tolerance, often balanced against the expectation of potentially high or low rewards. That reality is one of the sociological considerations specific to every forest management area and decision.

The principles of ecological forestry and the precautionary principle as applied to biodiversity conservation would suggest that we avoid, reduce, and mitigate risks to forest integrity as much as possible. One strategy is to spread or balance high-risk situations or decisions with low-risk ones, the classic "balanced portfolio" approach. For example, the Coast Information Team (2004) developed a hierarchical approach to maintaining overall low risk to hydrological function and old-growth dependent wildlife across the Great Bear Rainforest region of coastal British Columbia. Within that framework, uncertainty in response functions and the fuzziness of critical thresholds (see Figure 12.2) is acknowledged. Low overall risk is maintained if a few moderate risk landscapes in the region can be balanced against more low-risk and protected landscapes. Likewise, the legacies of a few high-risk watersheds can be accepted along with some new high-risk stand treatments, so long as they are offset by a greater abundance of protected, low- and moderate-risk watersheds and stands (Coast Information Team 2004). Planning for the Great Bear Rainforest then proceeded with the articulation of collaboratively derived management targets and high-risk limits in an adaptive management framework (Price et al. 2009), with commitments for the monitoring of selected indicators (Coast Information Team 2004). Uncertainties have subsequently been reduced as a result of monitoring, research, and insights gained throughout the region.

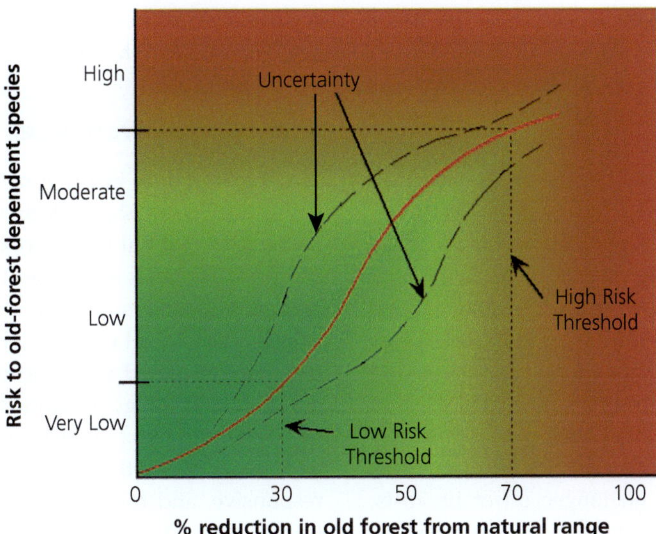

Figure 12.2 Conceptual portrayal of uncertainty in risk thresholds, illustrating three aspects of uncertainty: (1) imperfect understanding of what constitutes "high risk" (often denoting losses that are irreversible in a management context) in the value of interest on the y-axis; (2) a fuzzy determination of the level of a driving variable (x-axis) that would induce a high risk; and (3) uncertainty in the shape of the response function, especially with respect to moderate risk. Shown here with old forest loss as the driver, similar uncertainties prevail in predicting risks associated with climate change, habitat fragmentation, and other drivers (reproduced with permission from Coast Information Team (2004)).

Although it is tempting to decry the exploitative motives of people more closely affiliated with the utilitarian side of Leopold's (1949) A–B schism (see Figure 2.4), we all need to meet material needs for survival. Forest stewardship in a socio-ecological context is most successful when it can satisfy legitimate short-term needs (typically for resource harvesting or employment) while also protecting species, ecological processes, and ecosystem services over the long term. Identifying and achieving this balance has been the great challenge of conservation movements and the sustainability philosophy (see Chapter 2). Following centuries of forest conversion and unsustainable utilization, greater conservation (protection) of forest resources and ecosystems is now urgently needed, especially in times of rapid change and uncertainty. Though necessary, any constraints to past norms of forest use are bound to prompt resentment, protest, and pushback, for which forest stewards must prepare.

Another conceptual divide in worldviews and economies is that of "farmers" compared to "hunter-gatherers" (Brody 2001), which has relevance to the consideration and stewardship of forests. Both approaches to the production and harvesting of renewable resources can be sustainable, and both have their efficiencies. Farming—whether of swine, fish, wheat, or trees—depends on a much greater degree of environmental manipulation and resource inputs, which generate superior yields but compromise natural biodiversity and ecosystem processes. Hunting of animals and the gathering of plant products consist of harvesting resources (which can also be somewhat husbanded or nurtured) from largely natural ecosystems, but typically require a large land base and may involve greater uncertainty. Along with the underappreciation of subsistence and non-market economies (Emerton 1997), this difference in approaches contributes to the ongoing conflict between Indigenous peoples and European imperialism. The colonial model for "proper forest management" is based on concepts of regulated forests and farmed plantations, while Indigenous approaches tend to be less intensive, working with natural variability and processes rather than trying to control them. Households in many rural areas of the world practice a mixture of foraging, growing, and wage earning, for which a healthy forest landscape can be important. Furthermore, such a mixed strategy—sometimes adopted out of necessity, sometimes as a cultural preference—provides the diversity and flexibility to be a very resilient approach to survival (Elbakidze and Angelstam 2007; Bauer et al. 2022).

There is also a widespread human tendency—perhaps psychological, perhaps cultural—that resists change, or, more precisely, that reflects a reluctance to admit mistakes. Best expressed as "the Concorde Fallacy" (e.g. Dawkins and Brockman 1980), policymakers, bureaucrats, and managers often feel compelled to continue a course of action in which much has been invested, even after its benefits no longer offset its costs. Whether reflecting a cultural rigidity, institutional inertia (Poffenberger 1990), or personal ego, this tendency is anathema to the adaptive management needed for resilience (Folke et al. 2009).

Finally, another aspect of human nature that needs to be recognized and incorporated in forest policy and management is that many people derive spiritual inspiration from forests and trees, as fellow beings that are rooted in the ground and reach to the sky, joining heaven and earth (Nadkarni 2009). Forests and forest resources are central to the identity of many Indigenous cultures, in which people treat their relationship to plants, animals, and nature in general as a personal one of respect and reciprocity (Berkes 1999). To some people, even those living in wooden houses, the felling of a tree will always be considered a crime against nature or at least a sacrifice, an enterprise not to be undertaken lightly. Such perspectives may be driven by the aesthetics of cathedral-like grandeur in old forests, or simply a respect for long-lived organisms (as we extend to other humans, and also to elephants and whales, for example). Such motivations may drive campaigns for the protection of old-growth forests as much as material arguments to conserve carbon stores and biodiversity. That perspective is found not only in some Indigenous cultures but increasingly among nature-deprived urbanites. An inclusive approach to forest stewardship must then treat the harvesting of trees with respect and recognition that public values may call

for more forest preservation or restrictive management practices in the future.

12.6 Forest governance

Forest governance institutions, structures, and protocols (see Figure 12.3) collectively are an expression of the sociocultural framework within which decisions about forests are made. The decision-making power, constraints, and mechanisms by which public input, social values, and traditional resource use are considered in forest management are key aspects of forest governance (see Section 2.7). In general, there must be institutions, structures, and processes in place that promote effective, equitable, responsive, and robust pathways to sustainability (see Figure 12.3; Bennett and Satterfield 2018). Forests and the ecosystem services they provide are so important locally (and more widely) that there is always a public interest in sustainable forest management and a resilient future for forests. Freehold land ownership, exclusive harvesting rights, or comprehensive management responsibilities would seem to grant clear and unobstructed rights to making forest management decisions. But broader impacts and responsibilities (for example, to watershed hydrology and the protection of endangered species) empower a public interest in all forests. Any entity—corporate, government, or community—risks outrage, protests, boycotts, and a general loss of the social license to operate if it undertakes exploitative, degrading, or short-sighted decision-making with regard to those public interests.

There are many models of forest ownership and jurisdiction around the world, including small family holdings, corporate ownership, public land with volume-based tenures or area-based tenures or logging concessions, community forests, church forests, and Indigenous management areas. The breadth of forest ownership, oversight, and governance is beyond the scope of this book, for which the reader is referred to other sources (e.g. Arts and Visseren-Hamdkers 2012; Kishor and Rosenbaum 2012; Katila et al. 2020). Nonetheless, the policy direction provided through forest governance probably provides some of the strongest threats to — or, hopefully, the strongest motivations for —

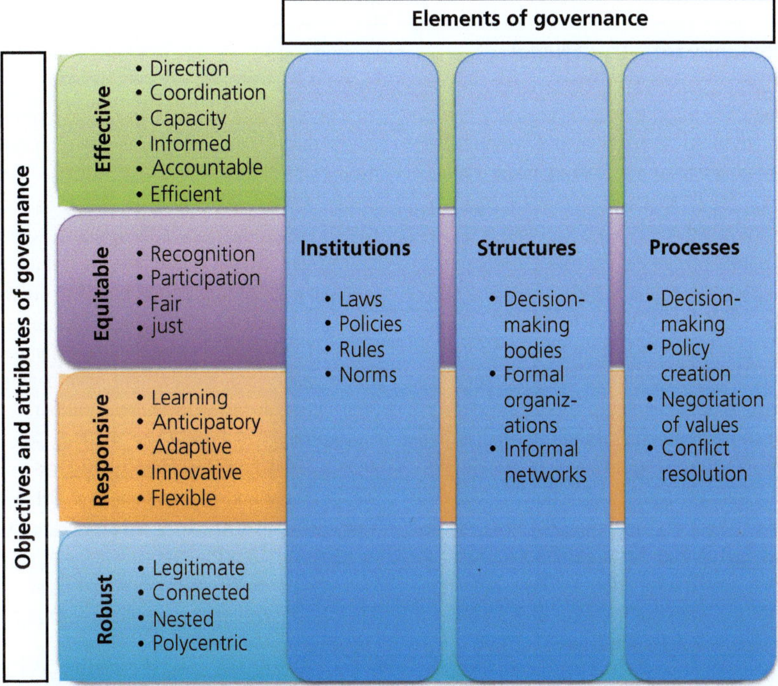

Figure 12.3 A practical framework for describing the objectives, attributes, and elements of environmental governance, applicable to managing forests for sustainability and resilience (Bennett and Satterfield (2018), reproduced under Creative Commons CC BY 4.0 DEED).

resilience-based stewardship. Of particular concern is the widespread drive for short-term economic gain from wood extraction, whether in a purely exploitative (non-renewable, non-sustainable) manner, or to optimize financial returns from sustainably managed rotational forests. This profit motive is best expressed in forest management organized under a corporate structure, which might be represented by industry-owned land in Scandinavia, an investment trust in the US, or long-term corporate tenures over public forest lands in Canada. It is tempting to fear that corporate ownership or control of forest management conflicts with the sustainability of environmental values (Klein 2015; see Box 2.4). Yet the prioritization of short-term profits by a corporation (or royalties and employment benefits on the part of government) at the expense of long-term persistence and resilience is not unique to corporate governance or influence.

Localized place-based ownership or governance systems such as family-held forest holdings (see, for example, Section 11.4) and community forests (see, for example, Box 12.2) may be more superior in retaining economic benefits and promoting locally relevant approaches to sustainability. Such local forms of forest tenure promote the "adapting mosaic" scenario of the Millennium Ecosystem Assessment (see Figure 1.5). Of course, many family- and community-owned enterprises also find themselves in the situation where immediate needs—or their perception through the lens of political expediency—take priority, with nature protection and precautionary approaches representing a real or perceived cost (e.g. Owubah et al. 2001). It takes a disciplined administration and patient, educated, and well-empowered stakeholder engagement to arrive at management decisions that pass the test of making no threat to socio-ecological resilience or the options for future generations.

An alternative perspective on forest governance and stewardship is provided by Indigenous peoples around the world. A common feature of many traditional, non-industrial cultures is a worldview

---

Box 12.2 Community forestry in Nepal

The Nepalese community forest process is arguably one of the world's most advanced and progressive examples of participatory forest governance and management (Chhetri 2010; Paudel et al. 2022). Nepal has seen a progressive devolvement of centralized management of national forest lands to thousands of community forest user groups (CFUGs) a process that was initiated in the 1970s and institutionalized by the 1993 Forest Act (Acharya 2002; Kanel and Kandel 2004). Its original goals were to halt deforestation and provide sustainable supplies of forest products for local subsistence use. Although many CFUGs have not been exploiting the commercial potential of their forests (Thoms 2008), recent years have seen a growing emphasis on providing economic benefits and poverty reduction through the sale of timber and fuelwood (Upadhyay 2012; KC et al. 2014). More than two million hectares of forest (almost one-third of Nepal's total forest area) are now managed by some 22,380 CFUGs, with participation of 3.34 million households (Paudel et al. 2022; Cadman et al. 2023).

With Nepal supporting tropical and temperate forests over a wide range of elevations, forests and associated management priorities and practices vary considerably among the thousands of CFUGs. Forest ownership remains with the Government of Nepal, and forest governance is structured by the 2019 version of the Forest Act and must comply with land-use restrictions and the operational plan of the Division Forest Office (Cadman et al. 2023). Households pay membership fees or provide labor to be part of the CFUG, and usually elect an executive committee (Joshi 2003; Kanel and Kandel 2004; Poudel et al. 2015; Cadman et al. 2023). Recognized as autonomous corporate entities, CFUGs have the right to direct the harvesting and sale of forest products, although those products are subject to provincial and national taxes. There are now also opportunities to capture international revenue for carbon storage and sequestration through the REDD+ program (Devkota and Mustalahti 2018), and potentially other payments for ecosystem services such as hydrological regulation (Poudyal et al. 2021). At least 25 percent of revenues must be reinvested in forest protection, renewal (for example, tree nursery production and plantation establishment), and other management activities (Thoms 2008; Upadhyay 2012), with the remainder used in support of community benefits such as water and electricity infrastructure, and to help build schools, community halls, and temples (Joshi 2003).

Community forests in Nepal have been successful in meeting many of their original objectives, but still face a number of challenges now and going forward. First and foremost, they have succeeded in reversing deforestation, and Nepal's total forest area and condition has improved over recent decades (Paudel et al. 2022; Cadman et al. 2023). With forests now more actively managed and with villagers using them to meet material needs, community forestry does not seem to be improving conditions for wildlife and biodiversity, however (Shrestha and McManus 2008; Cadman et al. 2023). Economic benefits to community projects and in providing local employment have been steadily increasing, although forest revenues contribute only 7.4 percent of household incomes (Chhetri 2010). As such, community forests have had a modest success as a vehicle for rural development and the attainment of UN Sustainable Development Goals such as poverty reduction (Upadhyay 2012; see Box 1.2). But their revenue-generating potential largely depends on the presence of valuable commercial tree species such as *Shorea robusta* or *Pinus roxburghii* (Chhetri et al. 2012b).

Ongoing access to timber, fuelwood, animal fodder, and non-timber forest products for subsistence use has been mixed. Maintaining a forest ecosystem in close proximity to most villages has been beneficial, but forest use has sometimes been restricted by CFUGs, ostensibly for purposes of conservation. It is primarily poor households that depend on subsistence use of forest products, and it is primarily poor people who are penalized for modest infractions of harvesting restrictions (Chhetri et al. 2012a). Poor households, women, and ethnic minorities are also underrepresented in CFUGs and in leadership positions (Devkota and Mustalahti 2018). The most powerful (wealthy, educated) members of each community have maintained their positions of domination, in which community forests are treated as an extension of municipal governance. Some of these weaknesses have been addressed by REDD+ requirements for democratically elected representation, but leadership and management skills still need to be developed (Devkota and Mustalahti 2018). Most communities do not have the expertise or the capacity to meet the legal and technical requirements for forest management plans, so plans tend to follow generic templates regardless of local objectives and priorities (Rutt et al. 2015; Baral et al. 2020). There is still great dependence on government foresters and administrators. Adaptive resilience (to climate change and socioeconomic pressures) varies greatly among communities, and is correlated with both community size and the size and diversity of its forest holdings (Karki et al. 2022).

Box 12.2 *Continued*

With the Nepalese having a history of community-governed forest use for over a millennium (Paudel et al. 2022), it could be said that the current approach to community forest management is a blend of Indigenous and modern forestry (Rutt et al. 2015). Community forests and participatory governance are not panaceas in resolving the challenges of sustainable forest management (Arts and Visseren-Hamdkers 2012) nor in promoting forest resilience. As in most public forests, Nepal's experience shows that there continue to be ongoing differences of opinion concerning the preferred balance between forest protection, access for subsistence use, and economic development. Nonetheless, community forestry seems to be successful, as illustrated in its long history in Nepal, and its reversal of forest exploitation and the draining of forest-derived revenues to corporate shareholders or central government coffers. Community forests need to wrestle with issues of social equity, conflicting forest values, and uncertain environmental change. Their resilient management, within the constraints of any given forest land base, will depend on inclusion, diversity, education, and leadership.

---

of regarding the forest as "home," as "neighborhood," or as "place," not primarily as a resource (Beckley 2003). While the forest may be cleared for small areas of farming and gardening and is certainly used for building materials and fuelwood as well as a source of food and medicines, the concept of wholesale forest liquidation or conversion to plantations is foreign to the traditions of responsible resource management in such societies (e.g. Dauvergne 1994; Savyasaachi 2008). Forested blocks of land may be owned or under the exclusive jurisdiction of individual households, larger family groups, or clans, or can be held in common at the village level. Forest stewardship and the regulation of forest product harvesting may be the responsibility of individual hereditary or elected leaders or may require broader clan and community consultation and consensus. Decision-making around land and resource issues is often part of larger cultural and ceremonial activities, sometimes explicitly referencing rights and responsibilities conferred by ancestors and owed to descendants. Those responsibilities usually embrace the need for multigenerational forest sustainability over a defined area, as briefly discussed in Sections 1.2 and 3.3.

Many indigenous, rural, and remote communities today can be described as suffering from rural poverty, characterized by low cash incomes and often with poor nutrition and limited lifestyle options (De La O Campos et al. 2018). On the other hand, low material demands with ongoing subsistence use of the forest and its lands and waters can support healthy households and culturally vibrant communities. Resilience is provided by utilizing a broad diversity of natural resources and by being open to a variety of wage-based opportunities that may present themselves (Bauer et al. 2022). Traditional family- or community-based forest stewardship can also incorporate inherent consideration of uncertainty through a long history of place-based knowledge of environmental (for example, drought or storm) and social (for example, conflict-based) risks. Even under a changing climate and in a world of many accelerating demands and opportunities, that deep knowledge of place, underpinned by cultural values and traditions, typically promotes some combination of forest protection and sustainable use adapted to local conditions (Elbakidze and Angelstam 2007; Joa et al. 2018; Wyllie de Echeverria et al. 2019; Bauer et al. 2022). The combination of flexible household economies and culturally empowered traditional ecological knowledge makes for both resilient communities and resilient forest stewardship. There is a widespread danger, however, of Indigenous leadership being swayed (often out of economic necessity) by corporate and government agendas to exploit timber values for short-term benefits.

It can be difficult to apply the tenets of ecological forestry, principles of managing forests as complex adaptive systems, and guidelines for promoting resilience in forests, forest enterprises, and forest communities as per the concepts outlined above and in previous chapters. The greatest challenges are often socioeconomic: finding the resources for treatments that do not generate revenues, and garnering acceptance from landowners, forest managers, and community stakeholders for activities that run counter to well-ensconced assumptions about "proper" forest management. For now, most forest revenues are derived from the sale of wood and fiber, but payment systems for carbon storage and sequestration are well developed (Michaelowa et al. 2019). Other income sources can include the harvest of non-timber forest products, payment for recreational use, watershed protection, and biodiversity protection (Mercer et al. 2011; Poudyal et al. 2021). But we are also seeing the growing recognition that the world's forests serve a public good, so a growing segment of the public is willing to have their taxes support forest protection and the stewardship of ecosystem services (Jo et al. 2021; Hafer and Ran 2022).

> • Forest management (or even a gentler, kinder, more modest perception of forest interventions in the form of forest stewardship) is ultimately about people management. Forests are incredibly resilient and will persist, in one form or another, quite well without us. Humankind, on the other hand, needs forests and their ecosystem services, so resilient forests are needed if we wish to have a resilient forest sector and resilient forest communities.

The difficulty lies, not so much in the new ideas, but in escaping from the old ones.

John Maynard Keynes (1936)

References cited

Aber, J., Christensen, N., Fernandez, I., et al. (2000). *Applying Ecological Principles to Management of US National Forests.* Issues in Ecology, No. 6. Washington: Ecological Society of America. 20 pp. Available online at https://www.esa.org/wp-content/uploads/2013/03/issue61.pdf (accessed February 27, 2024).

Acharya, K. P. (2002). "Twenty-Four Years of Community Forestry in Nepal," *International forestry Review,* 4/2: 149–56.

Alonso-Ayuso, A., Escudero, L. F., Guignard, M., and Weintraub, A. (2018). "Risk Management for Forestry Planning under Uncertainty in Demand and Prices," *European Journal of Operational Research,* 267/3: 1051–74.

Arts, B., and Visseren-Hamdkers, I. (2012). "Forest Governance: A State of the Art Review," in B. Arts, S. Bommel, M. Ros-Tonen, and G. Verschoor(eds), *Forest–People Interfaces: Understanding Community Forestry and Biocultural Diversity.* Wageningen, Netherlands: Wageningen Academic Publishers, 241–57.

Aryal, K., Maraseni, T., and Apan, A. (2022). "How Much Do We Know about Trade-offs in Ecosystem Services? A Systematic Review of Empirical Research Observations," *Science of the Total Environment,* 806: 151229.

Baral, S., Hansen, C. P., and Chhetri, B. B. K. (2020). "Forest Management Plans in Nepal's Community Forests: Does One Size Fit All?" *Small-Scale Forestry,* 19: 483–504.

Batavia, C., and Nelson, M. P. (2016). "Conceptual Ambiguities and Practical Challenges of Ecological Forestry: A Critical Review," *Journal of Forestry,* 114/5: 572–81.

Bauer, T., de Jong, W., Ingram, V., Arts, B., and Pacheco, P. (2022). "Thriving in Turbulent Times: Livelihood Resilience and Vulnerability Assessment of Bolivian

---

Box 12.3 Key points

• Managing forests for resilience has much in common with previous incarnations of forest ecosystem management (Kohm and Franklin 1997), holistic forestry (Hammond 1991), restorative and regenerative forestry (Stanturf 2016), managing forests as complex adaptive systems (Messier et al. 2015; Filotas et al. 2014), ecosystem stewardship (Chapin et al. 2009a), and ecological forest management in general (Larocque 2016; Franklin et al. 2018; Palik and D'Amato 2017; Palik et al. 2021).

• A combination of ecological and social drivers, constraints, and principles are fundamental to any forest management decision. When the laws of ecology are not respected, or the scientific foundations of ecosystem processes not supported, then environmental degradation is inevitable, with social consequences to compromised forest values and the loss of ecosystem services.

• The capacity and potential for applying principles of resilient forest stewardship vary with location. Its implementation must be place-based in order to navigate change and uncertainty successfully in a given biophysical and socioeconomic setting.

Indigenous forest Households," *Land Use Policy*, 119: 106146.

Bauhus, J., Forrester, D. I., Pretzsch, H., Felton, A., Pyttel, P., and Benneter, A. (2017). "Silvicultural Options for Mixed-Species Stands," in H. Pretzsch, D. I. Forrester, and J. Bauhus (eds), *Mixed-Species Forests: Ecology and Management*, New York: Springer, 433–501.

Bauhus, J., Puettmann, K. J., and Kühne, C. (2013). "Close-to-Nature Forest Management in Europe: Does It Support Complexity and Adaptability of Forest Ecosystems?" in C. Messier, K. J. Puettmann, and K. D. Coates (eds), *Managing Forests as Complex Adaptive Systems: Building Resilience to the Challenge of Global Change*. New York: Routledge, 187–213.

Beckley, T. M. (2003). "The Relative Importance of Sociocultural and Ecological Factors in Attachment to Place," in L. E. Kruger (ed.), *Understanding Community-Forest Relations*. General Technical Report PNW-GTR-566. Portland, OR: USDA Forest Service, 105–26.

Bell, S., and Apostol, D. (2007). *Designing Sustainable Forest Landscapes*. London: Taylor & Francis. 368 pp.

Bennett, N. J., and Satterfield, T. (2018). "Environmental Governance: A Practical Framework to Guide Design, Evaluation, and Analysis," *Conservation Letters*, 11/6: e12600.

Berkes, F. (1999). *Sacred Ecology: Traditional Ecological Knowledge and Resource Management*. Philadelphia: Taylor & Francis. 209 pp.

Botkin, D. B. (1990). *Discordant Harmonies: A New Ecology for the Twenty-First Century*. New York: Oxford University Press. 241 pp.

Brody, H. (2001). *The Other Side of Eden: Hunter-Gatherers, Farmers, and the Shaping of the World*. London: Faber & Faber. 374 pp.

Burns, R. M. (1983) (technical compiler). *Silvicultural Systems for the Major Forest Types of the United States*. Agriculture Handbook 445. Washington: USDA Forest Service. 191 pp.

Cadman, T., Maraseni, T., Koju, U. A., Shrestha, A., and Karki, S. (2023). "Forest Governance in Nepal Concerning Sustainable Community Forest Management and Red Panda Conservation," *Land*, 12: 493.

Carey, A. B. (2003). "Biocomplexity and Restoration of Biodiversity in Temperate Coniferous Forest: Inducing Spatial Heterogeneity with Variable-Density Thinning," *Forestry*, 76/2: 127–36.

Chapin, F. S., Kofinas, G. P., and Folke, C. (eds). (2009a). *Principles of Ecosystem Stewardship: Resilience-Based Natural Resource Management in a Changing World*. New York: Springer. 401 pp.

Chapin, F. S., Kofinas, G. P., Folke, C., et al. (2009b). "Resilience-Based Stewardship: Strategies for Navigating Sustainable Pathways in a Changing World," in F.

S. Chapin, G. P. Kofinas, and C. Folke (eds), *Principles of Ecosystem Stewardship: Resilience-Based Natural Resource Management in a Changing World*. New York: Springer, 319–37.

Chhetri, B. B. K. (2010). *Community Forestry in Nepal*. Saarbrücken, Germany: Lambert Academic Publishing. 108 pp.

Chhetri, B. B. K., Larsen, H. O., and Smith-Hall, C. (2012a). "Law Enforcement in Community Forestry: Consequences for the Poor," *Small-Scale Forestry*, 11: 435–52.

Chhetri, B. B. K., Lund, J. F., and Nielsen, Ø. J. (2012b). "The Public Finance Potential of Community Forestry in Nepal," *Ecological Economics*, 73: 113–21.

Coast Information Team (2004). *Ecosystem-Based Management Planning Handbook*. Victoria, BC: Cortex Consultants. 80 pp. Available online at https://coastfunds.ca/resources/ecosystem-based-management/ (accessed March 1, 2024).

Coates, K. D. (2000). "Conifer Seedling Response to Northern Temperate Forest Gaps," *Forest Ecology and Management*, 127/1–3: 249–69.

Comeau, P. G., and Fraser, E. C. (2018). "Plant Community Diversity and Tree Growth Following Single and Repeated Glyphosate Herbicide Applications to a White Spruce Plantation," *Forests*, 9/3: 107.

Commoner, B. (1971). *The Closing Circle: Nature, Man, and Technology*. New York: Knopf Press. 326 pp.

Dale, V. H., Brown, S., Haeuber, R. A., et al. (2000). "Ecological Principles and Guidelines for Managing the Use of Land," *Ecological Applications*, 10/3: 639–70.

Dauvergne, P. (1994). "The Politics of Deforestation in Indonesia," *Pacific Affairs*, 66/4: 497–518.

Dawkins, R., and Brockmann, H. J. (1980). "Do digger Wasps Commit the Concorde Fallacy?" *Animal Behaviour*, 28/3: 892–6.

De La O Campos, A. P., Villani, C., Davis, B., and Takagi, M. (2018). *Ending Extreme Poverty in Rural Areas: Sustaining Livelihoods to Leave No One Behind*. Rome: Food and Agriculture Organization of the United Nations. 84 pp. Available online at https://www.fao.org/3/ca1908en/CA1908EN.pdf (accessed February 29, 2024).

Devkota, B. P., and Mustalahti, I. (2018). "Complexities in Accessing REDD+ Benefits in Community Forestry: Evidence from Nepal's Terai Region," *International Forestry Review*, 20/3: 332–45.

Diaz, N., and Apostol, D. (1992). *Forest Landscape Analysis and Design: A Process for Developing and Implementing Land Management Objectives for Landscape Patterns*. R6 ECO-TP-043-92. Portland, OR: USDA Forest Service. 118 pp. Available online at https://www.fs.usda.gov/research/treesearch/6268 (accessed January 30, 2024).

Diaz, N. M., and Bell, S. (1997). "Landscape Analysis and Design," in K. A. Kohm and J. F. Franklin. (eds), *Creating a Forestry for the 21st Century: The Science of Ecosystem Management*. Washington: Island Press, 255–69.

Elbakidze, M., and Angelstam, P. (2007). "Implementing Sustainable Forest Management in Ukraine's Carpathian Mountains: The Role of Traditional Village Systems," *Forest Ecology and Management*, 249/1–2: 28–38.

Emerton, L. (1997). "Valuing Household Use of Non-Timber Forest Products," in S. A. Crafter, J. Awimbo, and J. Broekhoven (eds), *Non-Timber Forest Products: Value, Use and Management Issues in Africa, Including Examples from Latin America*. Gland, Switzerland: International Union for the Conservation of Nature, 17–22.

Ettouney, M. M., and Alampalli, S. (2016). *Risk Management in Civil Infrastructure*. Boca Raton, FL: CRC Press. 528 pp.

Fenger, M., Manning, T., Cooper, J., Guy, S., and Bradford, P. (2006). *Wildlife and Trees in British Columbia*. Edmonton: Lone Pine Publishing. 336 pp.

Fiedler, C. E., Arno, S. F., and Harrington, M. G. (1996). "Flexible Silvicultural and Prescribed Burning Approaches for Improving Health of Ponderosa Pine Forests," in W. Covington and P. K. Wagner (technical coordinators), *Conference on Adaptive Ecosystem Restoration and Management: Restoration of Cordilleran Conifer Landscapes of North America*. General Technical Report RM-GTR-278. Fort Collins, CO: USDA Forest Service, 69–74.

Filotas, E., Parrott, L., Burton, P. J., et al. (2014). "Viewing Forests through the Lens of Complex Systems Science," *Ecosphere*, 5/1: 1

Folke, C., Chapin, F. S., and Olsson, P. (2009). "Transformations in Ecosystem Stewardship," in F. S. Chapin, G. P. Kofinas, and C. Folke (eds), *Principles of Ecosystem Stewardship: Resilience-Based Natural Resource Management in a Changing World*. New York: Springer, 103–25.

Franklin, J. F. (1989). "Towards a New Forestry," *American Forests*, 95/11–12: 37–44.

Franklin, J. F., Berg, D. R., Thornburgh, D. A., and Tappeiner, J. C. (1997). "Alternative Silvicultural Approaches to Timber Harvesting: Variable Retention Harvest Systems," in K. A. Kohm and J. G. Franklin (eds), *Creating a Forestry for the 21st Century: The Science of Ecosystem Management*. Washington: Island Press, 111–39.

Franklin, J. F., Johnson, K. N., and Johnson, D. L. (2018). *Ecological Forest Management*. Long Grove, IL: Waveland Press. 646 pp.

Franklin, J. F., Lindenmayer, D., MacMahon, J. A., et al. (2000). "Threads of Continuity," *Conservation Biology in Practice*, 1/1: 8–16.

Gibbons, P., and Lindenmayer, D. (2002). *Tree Hollows and Wildlife Conservation in Australia*. Melbourne: CSIRO Publishing. 217 pp.

Gregory, R., Failing, L., Harstone, M., Long, G., McDaniels, T., and Ohlson, D. (2012). *Structured Decision Making: A Practical Guide to Environmental Management Choices*. Chichester, UK: Wiley-Blackwell. 314 pp.

Gresh, J. M., and Courter, J. R. (2021). "In Pursuit of Ecological Forestry: Historical Barriers and Ecosystem Implications," *Frontiers in Forests and Global Change*, 4: 571438.

Grumbine, R. E. (1994). "What Is Ecosystem Management?" *Conservation Biology*, 8/1: 27–38.

Gunderson, L. H., and Holling, C. S. (eds). (2002). *Panarchy: Understanding Transformations in Human and Natural Systems*. Washington: Island Press. 507 pp.

Gustafsson, L., Baker, S. C., Bauhus, J., et al. (2012). "Retention Forestry to Maintain Multifunctional Forests: A World Perspective," *BioScience*, 62/7: 633–45.

Hafer, J. A., and Ran, B. (2022). "Exploring the Influence of Attitudes and Experience on Valuation of State Forest Lands via Contingent Valuation," *Public Performance & Management Review*, 45/6: 1461–86.

Hammond, H. (1991). *Seeing the Forest among the Trees: The Case for Wholistic Forest Use*. Vancouver: Polestar Book Publishers. 309 pp.

Hammond, H. (2009). *Maintaining Whole Systems on Earth's Crown: Ecosystem-Based Conservation Planning for the Boreal Forest*. Slocan Park, BC: Silva Forest Foundation. 389 pp.

Harper, K. A., Macdonald, S. E., Burton, P. J., et al. (2005). "Edge Influence on Forest Structure and Composition in Fragmented Landscapes," *Conservation Biology*, 19/3: 768–82.

Hodge, S. J., and Peterken, G. F. (1998). "Deadwood in British Forests: Priorities and a Strategy," *Forestry*, 71/2: 99–112.

Holling, C. S., and Meffe, G. K. (1996). "Command and Control and the Pathology of Natural Resource Management," *Conservation Biology*, 10: 328–37.

Hunter, M. L. (1993). "Natural Fire Regimes as Spatial Models for Managing Boreal Forests," *Biological Conservation*, 65: 115–20.

Hunter, M. L. (ed.). (1999). *Maintaining Biodiversity in Forest Ecosystems*. Cambridge: Cambridge University Press. 698 pp.

Hunter M. L., Jacobson, G. L., and Webb, T. (1988). "Paleoecology and the Coarse-Filter Approach to Maintaining Biological Diversity," *Conservation Biology*, 2/4: 375–85.

Jo, J. H., Lee, C. B., Cho, H. J., and Lee, J. (2021). "Estimation of Citizens' Willingness to Pay for the Implementation of Payment for Local Forest Ecosystem Services: The Case of Taxes and Donations," *Sustainability*, 13/11: 6186.

Joa, B., Winkel, G., and Primmer, E. (2018). "The Unknown Known: A Review of Local Ecological Knowledge in

Relation to Forest Biodiversity Conservation," *Land Use Policy*, 79: 520–30.

Joshi, M. J. (2003). "Community Forestry Programs in Nepal and their Effects on Poorer Households." Paper presented at the XII World Forestry Congress, Quebec City. Available online at https://www.fao.org/3/xii/0036-a1.htm (accessed February 29, 2024).

Kanel, K. R., and Kandel, B. R. (2004). "Community Forestry in Nepal: Achievements and Challenges," *Journal of Forest and Livelihood*, 4/1: 55–63.

Karki, G., Kunwar, R., Bhatta, B., and Devkota, N. R. (2022). "Climate Change Effects, Adaptation and Community-Based Forest Management in the Mid-Hills of Tanahu and Kaski Districts, Nepal," *International Forestry Review*, 24/4: 573–93.

Karr, J. R., Larson, E. R., and Chu, E. W. (2022). "Ecological Integrity is Both Real and Valuable," *Conservation Science and Practice*, 4/2: e583.

Katila, P., McDermott, C., Larson, A., Aggarwal, S., and Giessen, L. (2020). "Forest Tenure and the Sustainable Development Goals: A Critical View," *Forest Policy and Economics*, 120: 102294.

KC, B., Stainback, G. A., and Chhetri, B. B. K. (2014). "Community Users' and Experts' Perspective on Community Forestry in Nepal: A SWOT–AHP Analysis," *Forests, Trees and Livelihoods*, 23/4: 217–31.

Keynes, J. M. (1936). *The General Theory of Employment, Interest and Money*. London: Macmillan & Co. 403 pp.

Kimmins, J. P. (1992). *Balancing Act: Environmental Issues in Forestry*. Vancouver: UBC Press. 244 pp.

Kimmins, J. P. (2004). "Emulating Natural Forest Disturbance: What Does This Mean?" in A. H. Perera, L. J. Buse, and M. G. Weber (eds), *Emulating Natural Forest Landscape Disturbances: Concepts and Applications*. New York: Columbia University Press, 8–28.

Kishor, H., and Rosenbaum, K. (2012). *Assessing and Monitoring Forest Governance: A User's Guide to a Diagnostic Tool*. Program on Forests (PROFOR). Washington: World Bank. 115 pp. Available online at https://www.profor.info/sites/profor.info/files/AssessingMonitoringForestGovernance-guide.pdf (accessed February 18, 2024).

Klein, N. (2015). *This Changes Everything: Capitalism vs the Climate*. Toronto: Penguin Random House. 564 pp.

Klenk, N. L., Bull, G. Q., and MacLellan, J. I. (2009). "The 'Emulation of Natural Disturbance' (END) Management Approach in Canadian Forestry: A Critical Evaluation," *Forestry Chronicle*, 85/3: 440–5.

Kneeshaw, D., and Bergeron, Y. (2016). "Applying Knowledge of Natural Disturbance Regimes to Develop an Ecosystem Management Approach in Forestry," in G. R.

Larocque (ed.), *Ecological Forest Management Handbook*. Boca Raton, FL: CRC Press, 3–31.

Knoke, T., Stimm, B., Ammer, C., and Moog, M. (2005). "Mixed Forests Reconsidered: A Forest Economics Contribution on an Ecological Concept," *Forest Ecology and Management*, 213/1–3: 102–16.

Kohm, K. A., and Franklin, J. F. (eds). (1997). *Creating a Forestry for the 21st Century: The Science of Ecosystem Management*. Washington: Island Press. 475 pp.

Larocque, G. R. (ed.). (2016). *Ecological Forest Management Handbook*. Boca Raton, FL: CRC Press. 604 pp.

LeBauer, D. S., and Treseder, K. K. (2008). "Nitrogen Limitation of Net Primary Productivity in Terrestrial Ecosystems Is Globally Distributed," *Ecology*, 89/2: 371–9.

Leopold, A. (1949). *A Sand County Almanac, and Sketches Here and There*. New York: Oxford University Press. 240 pp.

Lindenmayer, D. B., and Franklin, J. F. (2002). "Transitions to Ecological Sustainability in Forests—a Synthesis," in Lindenmayer, D. B., and Franklin, J. F. (eds), *Towards Forest Sustainability*. Melbourne: CSIRO Publishing, 205–13. 240 pp.

Lindenmayer, D. B., Franklin, J. F., and Burton, P. J. (2008). *Salvage Logging and its Ecological Consequences*. Washington: Island Press. 227 pp.

McAfee, B. J., de Camino, R., Burton, P. J., et al. (2010). "Managing Forested Landscapes for Socio-Ecological Resilience," in G. Mery, P. Katila, G. Galloway, et al. (eds), *Forests and Society: Responding to Global Drivers of Change*. Vienna: International Union of Forest Research Organizations (IUFRO), 401–39.

McPherson, G. R., and DeStefano, S. (2003). *Applied Ecology and Natural Resource Management*. Cambridge: Cambridge University Press. 165 pp.

Maier, D. S. (2012). *What's So Good About Biodiversity?* Dordrecht: Springer. 568 pp.

Martínez Pastur, G. J., Vanha-Majamaa, I., and Franklin, J. F. (2020). "Ecological Perspectives on Variable Retention Forestry," *Ecological Processes*, 9: 12.

Maser, C. (1990). *The Redesigned Forest*. 2nd edn. Don Mills, ON: Stoddart Publishing. 224 pp.

Matthews, J. D. (1991). *Silvicultural Systems*. Oxford: Clarendon Press. 284 pp.

Mercer, D. E., Cooley, D., and Hamilton, K. (2011). *Taking Stock: Payments for Forest Ecosystem Services in the United States*. Asheville, NC: USDA Forest Service, Southern Research Station. 49 pp. Available online at https://www.srs.fs.usda.gov/pubs/ja/2011/ja_2011_mercer_001.pdf (accessed February 18, 2024).

Messier, C., Puettmann, K., Chazdon, R., et al. (2015) (2015). "From Management to Stewardship: Viewing

Forests as Complex Adaptive Systems in an Uncertain World," *Conservation Letters*, 8/5: 368–77.

Messier, C., Puettmann, K. J., and Coates, K. D. (2013). "The Complex Adaptive System: A New Integrative Framework for Understanding and Managing the World Forest," in C. Messier, K. J. Puettmann, and K. D. Coates (eds), *Managing Forests as Complex Adaptive Systems*. New York: Routledge, 327–41.

Michaelowa, A., Shishlov, I., and Brescia, D. (2019). "Evolution of International Carbon Markets: Lessons for the Paris Agreement," *Wiley Interdisciplinary Reviews: Climate Change*, 10/6: e613.

Nadkarni, N. (2009). *Between Earth and Sky: Our Intimate Connections to Trees*. Berkeley and Los Angeles: University of California Press. 336 pp.

Newton, P. F. (2021). "Stand Density Management Diagrams: Modelling Approaches, Variants, and Exemplification of their Potential Utility in Crop Planning," *Canadian Journal of Forest Research*, 51/2: 236–56.

O'Hara, K. L. (2001). "The Silviculture of Transformation: A Commentary," *Forest Ecology and Management*, 151: 81–6.

O'Hara, K. L. (2014). *Multiaged Silviculture: Managing for Complex Forest Stand Structures*. Oxford: Oxford University Press. 213 pp.

O'Hara, K. L. (2015). "What is Close-to-Nature Silviculture in a Changing World?" *Forestry*, 89: 1–6.

Østergaard, C. R., Timmermans, B., and Kristinsson, K. (2011). "Does a Different View Create Something New? The Effect of Employee Diversity on Innovation," *Research Policy*, 40/3: 500–9.

Owubah, C. E., Le Master, D. C., Bowker, J. M., and Lee, J. G. (2001). "Forest Tenure Systems and Sustainable Forest Management: The Case of Ghana," *Forest Ecology and Management*, 149/1–3: 253–64.

Palik, B. J., and D'Amato, A. W. (2017). "Ecological Forestry: Much More than Retention Harvesting," *Journal of Forestry*, 115/1: 51–3.

Palik, B. J., and D'Amato, A. W. (2024). "The Context of Ecological Silviculture," in B. J. Palik and A. W. D'Amato (eds), *Ecological Silvicultural Systems: Exemplary Models for Sustainable Forest Management*. Hoboken, NJ: John Wiley & Sons, 1–10.

Palik, B. J., D'Amato, A. W., Franklin, J. F., and Johnson, K. N. (2021). *Ecological Silviculture: Foundations and Applications*. Long Grove, IL: Waveland Press. 343 pp.

Paudel, G., Carr, J., and Munro, P. G. (2022). "Community Forestry in Nepal: A Critical Review," *International Forestry Review*, 24/1: 43–58.

Pauly, D., Froese, R., and Palomares, M. L. (2000). "Fishing down Aquatic Food Webs: Industrial Fishing over the Past Half-Century Has Noticeably Depleted the Topmost Links In Aquatic Food Chains," *American Scientist*, 88/1: 46–51.

Pearson, D. E., Clark, T. J., and Hahn, P. G. (2022). "Evaluating Unintended Consequences of Intentional Species Introductions and Eradications for Improved Conservation Management," *Conservation Biology*, 36/1: e13734.

Perera, A. H., Buse, L. J., and Weber, M. G. (eds). (2004). *Emulating Natural Forest Landscape Disturbances: Concepts and Applications*. New York: Columbia University Press. 315 pp.

Poffenberger, M. (1990). "Facilitating Change in Forest Bureaucracies," in M. Poffenberger (ed.), *Keepers of the Forest: Land Management Alternatives in Southeast Asia*. West Hartford, CT: Kumarian Press, 101–18.

Poudel, N. R., Fuwa, N., and Otsuka, K. (2015). "The Impacts of a Community Forestry Program on Forest Conditions, Management Intensity and Revenue Generation in the Dang District of Nepal," *Environment and Development Economics*, 20/2: 259–81.

Poudyal, B., Upadhaya, S., Acharya, S., and Khanal Chhetri, B. B. (2021). "Assessing Socio-Economic Factors Affecting the Implementation of Payment for Ecosystem Services (PES) Mechanism," *World*, 2/1: 81–91.

Price, K., Roburn, A., and MacKinnon, A. (2009). "Ecosystem-Based Management in the Great Bear Rainforest," *Forest Ecology and Management*, 258/4: 495–503.

Puettmann, K. J., Coates, K. D., and Messier, C. C. (2009). *A Critique of Silviculture: Managing for Complexity*. Washington: Island Press. 189 pp.

Rai, V. (2013). *Spatial Ecology: Patterns and Processes*. Oak Park, IL: Bentham Science Publishers. 148 pp.

Rose, N.-A., and Burton, P. J. (2009). "Using Bioclimatic Envelopes to Identify Temporal Corridors in Support of Conservation Planning in a Changing Climate," *Forest Ecology and Management*, 258: S64–S74.

Rutt, R. L., Chhetri, B. B. K., Pokharel, R., Rayamajhi, S., Tiwari, K., and Treue, T. (2015). "The Scientific Framing Of Forestry Decentralization in Nepal," *Forest Policy and Economics*, 60: 50–61.

Sadgrove, K. (2020). *The Complete Guide to Business Risk Management*. 2nd edn. New York: Routledge. 578 pp.

Savyasaachi (2008). "Deforestation, Nature and Conservation: An Indigenous Discourse," *Contemporary Perspectives*, 2/2: 279–313.

Scherer-Lorenzen, M., Gessner, M. O., Beisner, B. E., et al. (2022). "Pathways for Cross-Boundary Effects of Biodiversity on Ecosystem Functioning," *Trends in Ecology & Evolution*, 37/5: 454–67.

Scherrer, D., Baltensweiler, A., Bürgi, M., Fischer, C., Stadelmann, G., and Wohlgemuth, T. (2023). "Low Naturalness of Swiss Broadleaf Forests Increases their Susceptibility to Disturbances," *Forest Ecology and Management*, 532: 120827.

Shrestha, K. K., and McManus, P. (2008). "The Politics of Community Participation in Natural Resource Management: Lessons from Community Forestry in Nepal," *Australian Forestry*, 71/2: 135–46.

Sexton, W. T., Szaro, R. C., and Johnson, N. C. (1999). "The Ecological Stewardship Project: A Vision for Sustainable Resource Management," in N. C. Johnson, A. J. Malk, R. C. Szaro, and W. T. Sexton (eds), *Ecological Stewardship: A Common Reference for Ecosystem Management*, i. *Key Findings*. Kidlington, UK: Elsevier Science, 1–8.

Seymour, R., and Hunter, M. (1999). "Principles of Ecological Forestry," in M. L. Hunter (ed.), *Maintaining Biodiversity in Forest Ecosystems*. Cambridge: Cambridge University Press, 22–61.

Simard, S., Martin, K., Vyse, A., and Larson, B. (2013). "Meta-Networks of Fungi, Fauna and Flora as Agents of Complex Adaptive Systems," in C. Messier, K. J. Puettmann, and K. D. Coates (eds), *Managing Forests as Complex Adaptive Systems: Building Resilience to the Challenge of Global Change*. New York: Routledge, 133–64.

Smith, D. M. (1962). *The Practice of Silviculture*. 7th edn. New York: John Wiley & Sons. 578 pp.

Snyder, P. K. (2010). "The Influence of Tropical Deforestation on the Northern Hemisphere Climate by Atmospheric Teleconnections," *Earth Interactions*, 14/4: 1–34.

Stanturf, J. A. (ed.). (2016). *Restoration of Boreal and Temperate Forests*. 2nd edn. Boca Raton, FL: CRC Press. 547 pp.

Teitelbaum, S., Asselin, H., Bissonnette, J. F., and Blouin, D. (2023). "Governance in the Boreal Forest: What Role for Local and Indigenous Communities?" in M. M. Girona, H. Morin, S. Gauthier, and Y. Bergeron (eds). *Boreal Forests in the Face of Climate Change: Sustainable Management*. Cham, Switzerland: Springer, 513–32.

Thoms, C. A. (2008). "Community Control of Resources and the Challenge of Improving Local Livelihoods: A Critical Examination of Community Forestry in Nepal," *Geoforum*, 39/3: 1452–65.

Tierney, G. L., Faber-Langendoen, D., Mitchell, B. R., Shriver, W. G., and Gibbs, J. P. (2009). "Monitoring and Evaluating the Ecological Integrity of Forest Ecosystems," *Frontiers in Ecology and the Environment*, 7/6: 308–16.

Toumey, J. W. (1928). *Foundations of Silviculture upon an Ecological Basis*. New York: John Wiley and Sons. 438 pp.

Upadhyay, S. (2012). "Community Based Forest and Livelihood Management in Nepal," in D. Bollier and S. Helfreich (eds), *The Wealth of the Commons: A World beyond Market & State*. Amherst, MA: Levellers Press. Available online at https://wealthofthecommons.org/contents (accessed February 29, 2024).

Vítková, L., Bače, R., Kjučukov, P., and Svoboda, M. (2018). "Deadwood Management In Central European Forests: Key Considerations for Practical Implementation," *Forest Ecology and Management*, 429: 394–405.

Voller, J., and Harrison, S. (eds). (1998). *Conservation Biology Principles for Forested Landscapes*. Vancouver: UBC Press. 243 pp.

Wurtzebach, Z., and Schultz, C. (2016). "Measuring Ecological Integrity: History, Practical Applications, and Research Opportunities," *BioScience*, 66/6: 446–57.

Wyllie de Echeverria, V. R., and Thornton, T. F. (2019). "Using Traditional Ecological Knowledge to Understand and Adapt to Climate and Biodiversity Change On The Pacific Coast of North America," *Ambio*, 48/12: 1447–69.

Yousefpour, R., Jacobsen, J. B., Thorsen, B. J., Meilby, H., Hanewinkel, M., and Oehler, K. (2012). "A Review of Decision-Making Approaches to Handle Uncertainty and Risk in Adaptive Forest Management under Climate Change," *Annals of Forest Science*, 69: 1–15.

Index